BIOLOGY
OF
DROSOPHILA

CONTRIBUTING AUTHORS

Dietrich Bodenstein   Albert Miller

Kenneth W. Cooper   D. F. Poulson

G. F. Ferris   B. P. Sonnenblick

Warren P. Spencer

BIOLOGY

OF

DROSOPHILA

EDITED BY

M. Demerec

Department of Genetics
Carnegie Institution of Washington
Cold Spring Harbor, New York

HAFNER PUBLISHING COMPANY
New York and London

1965

Printed and Published by
Hafner Publishing Company, Inc.
31 East 10th Street
New York, N.Y. 10003

Library of Congress Catalog Card Number: 64-66008

Printed in U.S.A. by
NOBLE OFFSET PRINTERS, INC.
NEW YORK 3, N. Y.

Preface

Although *Drosophila melanogaster* is one of the most important organisms used in biological research during the past three decades, its complete development and general biology have not been fully described. The organs of the adult have not been traced in detail to their primordia, and in most instances even fragmentary knowledge of their plan of development has not been available. Studies of other Diptera have been used to obtain comparative accounts of Drosophila ontogeny; but often this is not satisfactory. In some cases research workers have had to abandon the study of certain problems because of lack of information about the normal ontogeny of the organs in which they were interested, or they have interrupted their experiments in order to devote much time to a study of the necessary phase of ontogeny.

Consequently there has been an outstanding need for a descriptive volume covering the morphology and development of Drosophila, both to bring together the scattered information available in print and to fill in the remaining gaps in our knowledge. Plans for the preparation of such a volume were made about ten years ago, in consultation with several Drosophila investigators; and a research program was initiated to obtain material for a complete account of the development of this organism. The aims were to supply a comprehensive and detailed treatment of the anatomy, histology, and development of a single species of insect that had become an important laboratory animal; to present its basic norm as a standard for the analysis of experimentally induced genetic variations and for other physiological studies; to provide detailed descriptions that should be of considerable value in comparative morphological and embryological studies of other invertebrates; and to give a brief summary of the methods used for collecting material in the field and culturing it in the laboratory. In carrying out these aims, emphasis has been placed on the volume's usefulness as a

reference manual. The sequence of events in organic development has been depicted by means of a series of photographs of
cytological and histological preparations, as well as diagrams.
Timetables are given for the appearance of various stages in
the development of important organs under specified laboratory
conditions, and the external and internal anatomy of the fly is
illustrated by an extensive series of drawings.

The original plans for this work were made in cooperation
with Professor Alfred F. Huettner, who with the collaboration
of his students intended to prepare the section dealing with embryology and development. Unfortunately, a serious illness and
other circumstances forced him to drop the project shortly after
it was started; and he passed his share of the responsibility on
to his former student B. P. Sonnenblick. Ultimately the subject matter of this section was treated in four chapters, written
by Dietrich Bodenstein, Kenneth W. Cooper, D. F. Poulson,
and B. P. Sonnenblick. Dr. Huettner generously placed at their
disposal a considerable body of material accumulated during
his many years of work on the spermatogenesis and embryology
of Drosophila. It is with pleasure that I acknowledge here this
broadminded action. An additional delay in the progress of
the work was caused by the fact that the author of one chapter
was engaged for a number of years in war research.

Various sections of the manuscript have been critically examined by Th. Dobzhansky, Jack Schultz, and Curt Stern, for
whose suggestions the authors and I wish to express apprecia-
tion. I wish to acknowledge also the help of Miss Agnes C.
Fisher, who as my secretary spent a great deal of effort in
checking over the manuscripts.

M. DEMEREC

Cold Spring Harbor
May, 1950

Contents

1

Normal Spermatogenesis in Drosophila

KENNETH W. COOPER *

INTRODUCTION

Spermatogenesis in Drosophila has been of special cytogenetic interest ever since Morgan (1912, 1914) made his surprising discovery that genetic crossing over is rare or absent in male *Drosophila melanogaster* yet regularly present in the female. Much is now known regarding the significance of the post-diakinetic stages of meiosis in male Drosophila, owing chiefly to the theoretical analysis made by Darlington (1934). Early consideration of the difference in genetic crossing over between the sexes, however, had brought with it the hope of a cytological revelation of both the physical mechanism of crossing over and something of the nature of the differences leading to alternative meiotic pathways in male and female Drosophila. The hope was a natural one, for surely a microscopic examination of the maturation process in the two sexes would seem to offer a ready means for discovering the essential chromosomal behavior causing genetic crossing over. Nevertheless but small success has been had in this undertaking. Although much is known of meiosis in the male of a number of species of Drosophila, little is known of the corresponding stages in the females. Worst of all, the very stages of greatest importance, namely the early meiotic prophases, have so far proved beyond detailed and convincing analysis in both sexes.

The principal cytological difficulties presented by Drosophila species are the very small size of the nuclei and chromosomes,

* Department of Biology, Princeton University.

their poor fixability and stainability with most of the reagents so far employed, and lastly the formidable task of objectively seriating the meiotic prophase stages in the male. Most of these difficulties may prove surmountable by phase-contrast microscopy of living material or by improved cytological methods, but at the present time these drawbacks account for the wholly inadequate state of our knowledge of oögenesis, of the early spermatocyte prophase, and of the details of spermiogenesis. Even so, the main features of most other stages of spermatogenesis in Drosophila are now clear, although a complete and accurate description for any one species is still lacking.

For the present account *D. melanogaster* has been chosen as a type. This has been done for a number of reasons. For one thing, it is the principal form used in cytogenetic studies the world over, and it is likely to remain so. For another, many more authors have studied spermatogenesis in this species than in any other. It is perhaps needless to say that great value is to be placed upon independently derived, corroborative evidence when a cytological problem is technically so fraught with difficulty as this one. Unfortunately most other species are known only through single studies, often of fragmentary nature or limited scope.

In the account which follows, information likely to be of service to one studying spermatogenesis of Drosophila firsthand is given for *D. melanogaster*. Included are notes on the recognition of sex in the larva, the microscopic structure of the testis, and so on. For the most part these sections are applicable in at least a general way to other species. Hence, except for the general discussions following the detailed account of spermatogenesis in *D. melanogaster,* other species are discussed only where they especially illuminate the problems or are known to differ so strikingly from our type as to merit special comment.

The figures accompanying this account have been collected to form an atlas of the more important structures and stages of interest to the cytologist and geneticist. Most of these are not original but have been redrawn from the works of others in a single style for this article. Each figure is accredited to its originator in the legend, but in certain cases only a portion or a single element of the original illustration has been copied. The fact that the original figure has been only partially reproduced is indicated by the Latin preposition *ex*. Thus Fig. 72, "After Guyénot and Naville (1929), *ex* Fig. 112," represents only a portion of the original Fig. 112. On the other hand, Fig. 71, "After Huettner (1930c), Fig. 48," is a complete copy in so far as is possible. Needless to say, the reader should refer to the original figures in critical cases of paramount interest to him. Figures not bearing an acknowledgment are published here for the first time.

THE RECOGNITION OF SEX IN THE LARVA AND PUPA OF DROSOPHILA MELANOGASTER

A fact most fortunate for geneticist and cytologist alike is that male and female larvae of *D. melanogaster* may be told apart shortly after hatching. The paired gonads lie laterally in the posterior third of the body cavity of the larva, between the third and fourth (from the posterior end) main lateral branches of the longitudinal tracheal trunks, adherent to or embedded in the adjacent fat body. The ovary is much smaller than the testis at every instar (Table 1). It is more spherical in shape and

TABLE 1

Gonad Length (L), Smallest Diameter (l), and Proportion L/l in Larvae Raised at $25° \pm 1°C$

Measurements are in microns, and each value represents the mean of not less than 70 nor more than 118 gonads (Data from Gloor, 1943)

Hours after Laying	Stage Attained	Ovaries			Testes		
		L	l	L/l	L	l	L/l
25	First-instar larva	21.5	18.7	1.15	40.3	34.4	1.17
50	Second-instar larva	39.3	32.6	1.21	90.0	60.7	1.48
75	Second- and third-instar larvae	53.5	44.7	1.19	145.9	107.7	1.36
100	Third-instar larva ready for pupation	93.8	84.7	1.11	231.9	181.3	1.27

lies embedded in the tissues of the fat body. The testis, on the other hand, is very much larger than the ovary (cf. Figs. 1 and 2), generally more noticeably ovoid, more transparent, and usually more loosely attached to the fat body (Bridges in Sturtevant, 1921; Geigy, 1931; Kerkis, 1931, 1933; Gloor, 1943; Aboim,

1945). Because the differences between the larval testis and ovary are so striking, intact and living larvae may easily be sexed by merely examining their gonads through the transparent integument with an ordinary binocular dissecting microscope.

During pupal life most of the larval organs either undergo histolysis and destruction or are considerably modified as the organs of the adult fly are formed. The gonads, however, continue their growth and differentiation without interruption (Robertson, 1936). At all stages of pupal life the testes are larger than the ovaries (Kerkis, 1931). By the eleventh hour of pupation (at 25°C) the testes start to elongate posteriorly, attaching to the gonoducts from the thirtieth to the thirty-sixth hour (Stern, 1940, 1941). Thereafter their considerable linear growth involves coiling of the testis, only the uncoiled cephalad tip lying close to or at its original position. Although the testes are fully formed and contain mature spermatozoa by the time of eclosion, the ovaries are still in a juvenile state and grow enormously during early adult life.

HISTOLOGY OF THE TESTIS OF DROSOPHILA MELANOGASTER

The somatic structure of the testis may be viewed in both larva and imago as a sac enclosing within its lumen the closely packed germ cells, certain somatic cellular elements, and cellular debris and fluid depending upon the age of the individual. In the adult *D. melanogaster* the sac is coiled, tubular, and open to the seminal vesicle at its originally caudad extremity, whereas in the larva it is merely a slightly elongated and closed ellipsoid. In a small number of Drosophila species the testes do not become coiled but remain elliptical vesicles in the adult. This is the case, for example, in *D. pseudoobscura* and *D. miranda*.

Aside from the germ cells, not less than six discrete histological elements have been described in the testis of *D. melanogaster*. Assuredly not all these are unrelated elements; nevertheless the detailed interrelations of the somatic, mesentodermal elements of the testis still appear to be somewhat uncertain. That differentiation of some of the testicular somatic tissues, such as the testis coat, the apical, interstitial, and terminal cells, will take place in the absence of germ cells appears to be clearly shown by the observations of Geigy (1931), Geigy and Aboim (1944), and

Aboim (1945) on ultraviolet-castrated embryos. The following account, while tentative, will therefore serve to draw attention to the principal histological elements of the testis and to some of the problems associated with them.

The Adult and Pupal Testis

The adult testis is ensheathed by a two-layered epithelial coat or capsule (Fig. 7), the following description of which is drawn chiefly from the work of Stern (1941). The thickness of the two-layered coat is approximately 1.5–2 μ (Miller, 1941), but the inner of the two membranes comprises but a fraction of a micron in thickness. The external layer of the testicular coat is composed of large squamous cells with bulging nuclei (Fig. 7, sc). The cytoplasm of these outer epithelial cells contains the pigment which colors the testis wall. The "inner membrane" forms a continuous sheath showing little visible structure yet apparently containing within it small scattered nuclei or their relics —perhaps nucleoli (Fig. 7, im). This internal coat confers upon the adult testis its coiled shape, and Geigy (1931) has suggested that it may be largely chitinous in the fully formed testis. In any event it is perhaps best regarded tentatively as a basal membrane derived from a much flattened and possibly degenerated inner epithelial layer.

At the posterior extremity of the adult testis, in the region just before the testicular duct or true *vas efferens* (Miller, 1941), there occurs a single-layered cuboidal or low columnar epithelium of large cells with large nuclei (Shen, 1932). This "terminal epithelium" (Fig. 7, te) lies internally to the inner membrane of the testis (Stern, 1941; Aboim, 1945). According to Shen (1932), it is in the vicinity of this epithelium that maturing spermatozoa first gain their capacity for full motility.

Apically, at the blind cephalad end, the adult testis (Fig. 5) has a small area, perhaps 50–100 μ long, which is occupied by predefinitive spermatogonia, by spermatogonia, and by apical cells (see below). Behind the tip there occur, generally along the inner membrane, islands of spermatocytes and bundles of transforming spermatids (Fig. 5). Spermatocytes are rare beyond the uncoiled cephalad part of the adult testis, and the lumen of most of this part, as well as all the coiled regions thereafter, contains

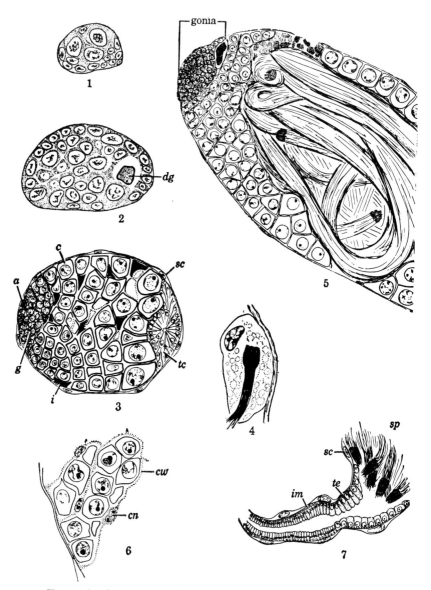

FIGS. 1–7. Histology of the testis of *Drosophila melanogaster*.

Fɪɢ. 1. Ovary of 24-hr larva. Longitudinal section; cephalad pole to left. ca. 780✕. After Gloor (1943), Fig. 1a.

Fɪɢ. 2. Testis of 24-hr larva. Longitudinal section; cephalad pole to left; dg, degenerating spermatogonia. ca. 780✕. After Gloor (1943), Fig. 1b.

Fɪɢ. 3. Diagrammatic representation of testis of 4-day larva. Longitudinal section; cephalad pole to left. Apical cells, a, are represented by small dark circles at apical pole; spermatogonia, g, form the zone of small cells of the apical fifth; middle three fifths made up of primary spermatocytes, c, and scattered interstitial cells, i, solid black polygons; caudad fifth comprised of terminal cells, tc. The testis is encircled by squamous cells, sc, and presumably an inner membrane as well. ca. 210✕. Diagram based upon Gloor's (1943) Fig. 2 and Geigy's and Aboim's (1944) Fig. 2a.

Fɪɢ. 4. Bundle of spermatozoa represented by Guyénot and Naville (1929) as within a nutritive cell. ca. 455✕. After Guyénot and Naville (1929), Fig. 115.

Fɪɢ. 5. Semidiagrammatic representation of the apex of a late pupal testis, showing the apical zone of spermatogonia and the cysts of spermatocytes tending to become appressed against the wall of the testis as they push caudally. In the anterior end of the testis the cysts of spermatocytes surround the more central aggregates of bundles of maturing sperm. The latter course through the testis in lazy, irregular spirals. ca. 260✕.

Fɪɢ. 6. Section through two peripheral cysts of early primary spermatocytes in a late larval testis. The outlines of the cyst walls, cw, and the nuclei, cn, of three cyst cells (interstitial cells?) are shown. ca. 450✕.

Fɪɢ. 7. Terminal epithelium of late pupal testis at entrance to testicular duct. Note orientation of the sperm bundles, sp, the sheath cells, sc, terminal epithelium, te, and inner membrane, im, showing "nuclei." ca. 450✕.

bundles of maturing and matured spermatozoa, as well as cellular debris, passing backwards to the seminal vesicles.

In the anterior portion, especially in pupal testes where the early stages of spermatogenesis are so abundant, septa or laminae enclose and separate groups of late generations of spermatogonia and clusters of spermatocytes. The related germ cells are thus enclosed in cysts (Fig. 6), just as in the testes of many other insects. The fine septae are probably of protoplasmic nature because they contain small, flat nuclei (Fig. 6, *cn*) which, however, are few in number (Stern, 1941; Gloor, 1943; Aboim, 1945). Such septae seem to be lost in the hinder regions of the testis where only later stages of spermiogenesis and mature spermatozoa are encountered (Stern, 1941).

One last category of cells, the so-called "nutritive cells," has been briefly discussed by Guyénot and Naville (1929), Geigy (1931), and Aboim (1945). According to these authors, quite advanced spermatids (Fig. 74) form compact bundles and become implanted in groups of fifty or more within giant nutritive cells [1] having a vacuolated cytoplasm (Fig. 4). The presumptive nutritive cells are said to occur in the second half of the testis, scattered along its wall, and to be recognizable even in young pupal testes by their large size, often triangular clear nucleus, and voluminous chromatic nucleolus. These cells, analogous to mammalian Sertoli cells, degenerate or disappear as the sperm mature within them. They are said to be represented finally by only their nuclei which may remain for a time adherent to the packets of matured spermatozoa (Geigy, 1931). Stern and Hadorn (1938), investigating the cause for spermatozoan immotility (and hence infertility) in *D. melanogaster* males lacking one or both of the fertility gene complexes in the differential arms of the Y chromosome, suggest that a function of the nutritive cell may be the determination of sperm motility.

The Larval Testis

The juvenile testis (Fig. 3) differs from the adult and pupal testes just described not only in the fact that spermatogenesis

---

[1] It is a curious fact that other authors have apparently either overlooked these cells or failed to find them.

is largely in abeyance, but also in the greater prominence of two collections of cells, the apical cells and the terminal cells (Gloor, 1943; Aboim, 1945), as well as by the occurrence of clearly defined but scattered interstitial cells (Geigy and Aboim, 1944; Aboim, 1945).

In the newly hatched larva there is no regular arrangement of the ensheathed central cells of the testis (Fig. 2). By the end of the first day, however, the arrangement of the cells within the testicular capsule becomes a polarized one (Aboim, 1945). Cephalically there is a small cluster of apical cells (Fig. 3, a), of very small size when compared with the spermatogonia (Fig. 3, g), which have only a scant cytoplasm. Behind these the germ cells successively increase in size as the caudal end of the testis is approached (Fig. 3), and among them occur scattered interstitial cells which give rise to the septae (Fig. 3, i) (Geigy and Aboim, 1944; Aboim, 1945). At the hindmost extremity there is another cap of cells (Fig. 3, tc), the terminal cells (Geigy, 1931; Gloor, 1943; Geigy and Aboim, 1944; Aboim, 1945).[2] Both the apical cells and the terminal cells lie internally to the sheath of the testis.

The terminal cells, at least in part, transform to the terminal epithelium of the adult testis (Fig. 7, te). From the work of Stern and Hadorn (1939) and Stern (1941) it is evident that the terminal cells do not form gonoducts, as formerly maintained by Geigy (1931).

The functions and fates of the apical cells are more obscure. Gloor (1943) and Geigy and Aboim (1944) have suggested that the interstitial cells (parietal and septal) are derived from the apical cells, and Aboim (1945) gives strong support to this view. Geigy (1931) had pointed out that the greater size and apparent detachment of some cells from the terminal mass suggested an origin of the nutritive cells from the terminal cells. Geigy and Aboim (1944) and Aboim (1945), however, now hold that the nutritive cells are more probably transformed interstitial cells. Thus the apical cell, the interstitial cell (hence septal cell), and the nutritive cell may represent three steps in the evolution of the same cell line, all perhaps being devoted in a trophic fashion to the welfare of the developing germ cells.

---

[2] The terminal cells are referred to by Geigy (1931) and Aboim (1945) as "cellules des canaux"; by Gloor (1943) as "Calottenzellen."

THE CHROMOSOME SET OF DROSOPHILA MELANOGASTER

The first satisfactory account of the general morphology of the chromosomes of male *Drosophila melanogaster* is that of Bridges (1916), who showed that there are two pairs of large V-shaped autosomes which are nearly equal in size, a pair of dotlike chromosomes, and an unequal pair of sex chromosomes in the spermatogonial complement (Fig. 9). Since that time many observations have been made on the morphology of the chromosomes of *D. melanogaster*, and each chromosome has been more or less satisfactorily identified with a corresponding genetic linkage group (Figs. 8, 10, 11). Thus the unpaired rod is known to be the X, the J-shaped chromosome is the Y, and the dotlike-chromosomes correspond to the fourth linkage group. The genetic identities of the two large V-shaped chromosomes are by no means certain (cf. Heitz 1933); but such evidence as exists points to the conclusion that the very slightly smaller of the two contains the second group of linked genes, while the larger corresponds with the third linkage group (Dobzhansky, 1930; Kaufmann, 1934; Bridges, 1935; etc.).[3] Hereafter each chromosome may from time to time be indicated by its contained group of linked genes. The haploid complements derivable from a male diploid set by normal spermatogenesis therefore are: X(I), II, III, IV, and Y, II, III, IV respectively.

For the purpose of this account the principal morphological peculiarities of the different chromosomes are alone of interest. Information concerning chromosomal peculiarities, such as the constrictions, has, as often as not, been obtained from the study of mitoses in female flies, and for the most part is further derived from ganglion cell preparations where the chromosomes are large, attenuated, and most favorable for detailed analysis. In

---

[3] Nevertheless the matter cannot be considered closed, nor the slight size difference between the two a reliable means for judgment. Hinton (1942), for example, has given measurements which suggest that the second chromosome is larger than the third. Hinton (in correspondence) is inclined to attribute little significance to the comparative value of his measurements of the two chromosomes (only 4 third chromosomes were measured, as compared with nearly 200 second chromosomes), and the fact remains that the size difference, if it exists, is too small for consistent differentiation between the two chromosomes.

these, four quite different sorts of constrictions may be noted: (1) primary or kinetochore constrictions, (2) constrictions marking sites of nucleolus formation, (3) constrictions of unknown function which represent intercalary points of prophasic and anaphasic stretch or extension in chromosomes, and lastly (4) constrictions which apparently merely mark heterochromatic-euchromatic boundaries. Although the constrictions dealt with below are in general agreed upon by all recent authors, it should be borne in mind that quite a number of other constrictions in all chromosomes but IV have been described by some authors and denied by others. Those interested will find the essential evidence and arguments in the papers of Bridges (1927), Dobzhansky (1930a, b, 1931, 1932a, b), Muller and Painter (1932), Prokofyeva (1935), and especially Kaufmann (1933, 1934) and Heitz (1933).

The X Chromosome

(Figs. 8–11, 13–15, 19)

In ordinary preparations the condensed X chromosome appears as a rodlike element ranging in length from 1.56 μ to perhaps 3.5 μ (Bridges, 1916, 1928; Gowen and Gay, 1933; Hinton, 1942; etc.), with Bridges' estimate of 1.8 μ perhaps representing its average length. The X is about 0.3 μ in girth, which is approximately the same thickness as that of the other chromosomes with the possible exception of IV (Bridges, 1928).[4]

Although for years the X chromosome was considered to be a simple rod in shape, Kaufmann (1933, 1934) and Prokofyeva (1935) have conclusively shown the kinetochore (spindle attachment) to be interstitial (Figs. 11, 14), setting off a small limb to its genetic right (i.e., the kinetochore lies at a constriction between the genetically mapped "left" region of the X and the newly discovered right arm). It is of historical interest that Stern, in his now classic studies of the Y chromosome, clearly figured the right arm of the X, although he did not specifically

[4] This estimate may well be too generous because it probably involves a measurement across *two* somewhat divergent chromatids. How well any of the published estimates agrees with the true dimensions of the living chromosomes is not known. The measures cited in this chapter must be construed as nothing more than orders of magnitude.

comment upon his discovery (e.g., see Figs. 1, 2, 3, 14, etc., in Stern, 1927).

In prophase, and rarely in metaphase, approximately the proximal half of the X chromosome is heterochromatic (Figs. 11, 13) (Heitz, 1933; Kaufmann, 1934; Prokofyeva, 1935; etc.) and corresponds with a relatively geneless, possibly uncoiled length of chromonema (Muller and Painter, 1932; Muller and Prokofyeva, 1935; etc.). This heterochromatic length is frequently referred to as the "inert region" of the X, but as Muller, Raffel, et al. (1937) point out, it is better to refer to this portion of the X as its "chromocentral region," because it is not genetically inert and its most characteristic property is the formation of a chromocenter in polytene nuclei.

Cutting across the chromocentral region of the X chromosome, at a point about a third the full length of the X from the proximal end, is a prominent constriction (Figs. 8, 10, 11, 13–15, 19) often easily visible at metaphase (Bridges, 1927; Dobzhansky, 1932a; etc.). Kaufmann (1933, 1934) and Heitz (1933) have shown this constriction to mark the site of a nucleolus organizer (Fig. 15).

The Y Chromosome

(Figs. 8–11, 15–17, 19)

The Y chromosome totals from 1.85 μ to about 2 μ in length and has a girth of approximately 0.3 μ (Bridges, 1928; Gowen and Gay, 1933). The two arms are quite unequal in length (Figs. 8–10, etc.). The short arm (Y^S) is about 0.8 μ long, whereas the long arm (Y^L) is approximately 1.2 μ long (Bridges, 1928). The kinetochore is located at a constriction at the juncture of Y^S and Y^L, and it is said to be less marked than the kinetic constrictions of the large autosomes. So far as can be told, the Y chromosome is almost totally heterochromatic (Heitz, 1933), and proximal regions in each of its arms are at least partially homologous with portions of the chromocentral region of the X. The heterochromaticity of the Y is not to be construed as an indication of genetic inertness, for not only do some genes homologous with those of X occur in the proximal regions of both arms, but also the distal extremities of each arm of Y possess gene complexes essential for fertility in the male (Stern, 1929; Neuhaus, 1939; etc.).

In addition to the kinetic constriction, the Y chromosome has a pronounced, strong constriction located approximately in the middle third of the short arm (Figs. 8, 10, 11, 15–17, 19) (Bridges, 1927). Kaufmann (1933, 1934) and Heitz (1933) have demonstrated this constriction to mark the location of a nucleolus organizer (Fig. 15).

The Second and Third Chromosomes

(Figs. 8–11, 18–20)

Each of the two V-shaped autosomes is strikingly similar in morphology, but Heitz (1933), Kaufmann (1934), and Prokofyeva (1935) have demonstrated that in one arm of one of them there is a very pronounced constriction about one-fifth of the arm length from the kinetochore (Figs. 8, 10, 11, 18, 19). This is the most prominent of all the constrictions which occur in any of the chromosomes and serves for immediate identification of the chromosome and arm in which it lies. It seems to be a region at which the chromosome is easily stretched at prophase and anaphase (Figs. 18, 19). Although similar to the prominent constrictions of X and Y, this secondary constriction is apparently not associated with a nucleolus organizer. According to Kaufmann (1934) and Hinton (1942), it marks the left arm of the second chromosome (i.e., II L). In the other arm of this chromosome and in both arms of the other V, there occur at approximately one-fifth of the respective arm lengths from the kinetochores less marked constrictions which show up characteristically in late prophase, as well as occasionally in metaphase (Figs. 8, 11) (Bridges, 1927; Dobzhansky, 1932a, b; works cited above; etc.). These constrictions mark the junctions of the heterochromatic regions lying proximal to the kinetochore with the distal euchromatic sections.

The two V-shaped autosomes are closely similar in size. The smaller of the two measures from 2.2 μ to 5.8 μ in length, with a diameter about 0.3 μ (Bridges, 1916, 1928; Gowen and Gay, 1933; Hinton, 1942). Bridges has estimated the average length to be 2.6 μ. The pronounced constriction referred to above is said to lie in this smaller chromosome which, as noted, has been identified as chromosome II. The larger chromosome ranges from 2.8 μ to

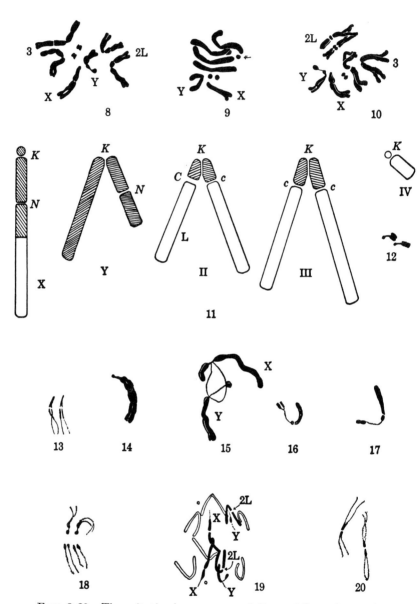

FIGS. 8–20. The mitotic chromosomes of *Drosophila melanogaster*.

Fig. 8. Metaphase in neuroblast cell of male larval brain. Acetocarmine squash. ca. 3700X. After Kaufmann (1934), Fig. 24.

Fig. 9. Spermatogonial metaphase plate. The body indicated by an arrow is probably of no significance, according to Bridges. Technique? ca. 4095X. After Bridges (1916), plate 1, Fig. 4.

Fig. 10. Early metaphase in neuroblast cell of male larval brain. Note the somatic pairing of chromatids in 2R, resulting in a chiasma-like configuration. Acetocarmine squash. ca. 3700X. After Kaufmann (1934), Fig. 18.

Fig. 11. Diagrammatic representation of the locations of the primary constrictions or kinetochores, *K*, heterochromatin (cross-lined), nucleolar constrictions, *N*, and secondary constrictions, *c*. The prominent secondary constriction, *C*, said to mark 2L, is tentatively represented in the shorter arm. Figures such as 18 suggest that heterochromatin occurs to the left of *C* as well. The diagrams are scaled to Bridges' size estimates, but are here represented as though the chromosomes were composed of but one chromatid each.

Fig. 12. Chromosomes IV showing their short arms. Prophase of a neuroblast cell in female larval brain. Acetocarmine squash. ca. 2585X. After Kaufmann (1934), *ex* Fig. 4.

Fig. 13. Somatically paired X chromosomes at prophase in a neuroblast cell in female larval brain. Note the division of the X chromosomes into heterochromatic (*dark*) and euchromatic (*light*) halves. The prominent constriction through the heterochromatic sections is the nucleolar constriction. Acetocarmine squash. Magnif.? After Heitz (1933), plate 1, *ex* Fig. 3.

Fig. 14. Metaphasic X chromosome from a neuroblast cell. Note the small right arm. Acetocarmine squash. ca. 4620X. After Kaufmann (1934), Fig. 22.

Fig. 15. X and Y chromosomes associated with the nucleolus during prophase in a neuroblast cell. Acetocarmine squash. ca. 4620X. After Kaufmann (1934), Fig. 45.

Figs. 16 and 17. Y chromosomes at prophase in neuroblast cells. Acetocarmine squashes. Magnif.? After Heitz (1933), plate 1, Figs. 13 and 15 respectively.

Fig. 18. Prophasic autosomes (II?) from a neuroblast cell of a female larva, showing somatic pairing, division of the chromosomes into proximal heterochromatic and distal euchromatic regions, the pronounced secondary constriction (*below*), and the smaller kinetochore constriction (*above*). Compare with Fig. 20. Acetocarmine squash. Magnif.? After Heitz (1933), plate 1, Fig. 4.

Fig. 19. Anaphase in a neuroblast cell of a male larva. Chromosomes which are identifiable by their constrictions are figured in solid black and appropriately labeled. Acetocarmine squash. Magnif.? After Heitz (1933), Fig. 4a.

Fig. 20. Somatically paired large autosomes (III?) from a male neuroblast cell. Note proximal heterochromatic regions and the pronounced kinetochore constriction. Acetocarmine squash. Magnif.? After Heitz (1933), *ex* Fig. 11g.

5.2 μ in length (see footnote on p. 10), and is approximately 0.3 μ in girth (Bridges, 1916, 1928; Gowen and Gay, 1933; Hinton, 1942). Bridges' estimate of 3.2 μ may prove to be an average length for this chromosome, tentatively identified as III. In each of the autosomes one arm is said to be slightly smaller than the other [cf. Hinton's (1942) numerical data for the two arms of II].

Chromosome IV

(Figs. 8–12)

The dotlike chromosomes are from 0.2 to 0.28 μ in length and approximately equally wide at full metaphase (Bridges, 1916, 1928; Gowen and Gay, 1933). Kaufmann (1934) was first to demonstrate cytologically the existence of two unequal arms in IV (Figs. 11, 12). Prokofyeva (1935) has made a similar claim, and Panshin and Khvostova (1938) and Griffen and Stone (1940) have provided conclusive genetic and cytological evidence to this effect. So far as can be made out from mitotic preparations, IV is not heterochromatic (Heitz, 1933), and this is confirmed by Griffen and Stone's (1940) study of the polytene IV.

Concluding Remarks

It was earlier believed that the metaphase chromosomes of *Drosophila melanogaster* tapered in such a way that distal segments generally have a wider diameter than proximal ones. [See, for example, Bridges (1927).] Dobzhansky (1930b) pointed out that, inasmuch as there is little difference in thickness of the parts of a chromosome at prophase, quite likely the apparent distal thickening is due to a premature separation of the chromatids at the ends of the chromosomes. There seems little doubt that this is the principal cause of the tapered appearance. Kaufmann (1934) has shown that the linear halves of a chromosome are essentially isodiametric throughout at prophase (with the possible exception of the frequently swollen end of Y^s) and that it is their progressive distal to proximal separation that gives rise to the appearance of taper. Kaufmann has likewise shown that the chromosomes are clearly bipartite in anaphase and early prophase, at least in ganglion cells. It is his contention, furthermore, that the process of doubling probably takes place in prophase, giving rise to a quadripartite metaphase chromosome.

When the mitotic chromosomes of *D. melanogaster* are prepared by ordinary methods such as sectioning or smearing (in contrast to acetic-orcein or acetocarmine squash methods), the constrictions may be quite obscure. Even so, most of the constrictions described above as landmarks for the identification of chromosomes may at times be located, even though not clearly evident as constrictions, since the chromosomes frequently tend to bend sharply or angularly at a constriction, whereas they bend smoothly elsewhere (Bridges, 1927, 1928).

When squash methods are used, the chromosomes are subject to considerable forces of shear, pressure, and stretch. It is accordingly not surprising that constrictions are best shown when squashes are studied because then the achromatic region of a constriction may be greatly stretched. But it is most important that sectioned material be employed as a check to determine whether such profound separation of chromosome segments as has been figured by Kaufmann (1934) and Hinton (1946) truly occurs in the living nucleus. Until this is done, Hinton's (1946) evidence for "long-range specific attractive forces" as the underlying mechanism of somatic pairing must be considered inconclusive (cf. Cooper, 1948).

A last few points to be noted in the study of these chromosomes is that their size may vary considerably in different tissues, and with this variation may go a corresponding increase or decrease in the details that can be worked out. Thus constrictions are more easily seen in large ganglion cells than in small ganglion cells or in gonial cells (Dobzhansky, 1930b, 1932b; Heitz, 1933; Kaufmann, 1934; etc.). Furthermore, if Prokofyeva (1935) is correct, all chromosomes are not equally favorably prepared by the same fixative; fixatives which gave her the best preparations of the sex chromosomes gave poorer preservation of autosomal detail and vice versa. The conclusion to be drawn from all this seems to be that, while much is now known of the structure of mitotic chromosomes of *D. melanogaster*, there is still a crying need for a detailed and comparative statistical analysis of the sizes, segments, and constrictions of these chromosomes in the various tissues, employing, where possible, living, smeared, sectioned, and squashed material. When this is done, the many disputed points, largely omitted from this discussion of the chromosome set, should at last be resolved. Needless to say, the mitotic chromosomes of other species of Drosophila are less well known than those of *D. melanogaster*.

SPERMATOGENESIS IN DROSOPHILA MELANOGASTER

The Course of Spermatogenesis

Spermatogenesis includes all the successive divisions and transformations undergone by germ cells in the formation of spermatozoa. In the testis of the newly emerged larva of *Drosophila melanogaster* only spermatogonia are found (Kerkis, 1933; Sonnenblick, 1941; Gloor, 1943), and these show no noticeable order of arrangement (see Fig. 2). Approximately 28 hours after hatching (at 25°C) both spermatogonia and primary spermatocytes occur, and this is the condition of the larval testis until shortly before pupation (Kerkis, 1933; Gloor, 1943; Geigy and Aboim, 1944). Related cells now lie in transverse, scalelike cysts one or two cells in thickness. The sizes of cells in successive cysts increase in an apical to caudal direction (Fig. 3). The more caudally located primary spermatocytes are ready to undergo the first maturation division—an event not infrequently occurring in larvae commencing to pupate (Huettner, 1930c). Indeed Gleichauf (1936) has figured the testis of a fully grown larva as containing stages up to and including sperm bundles. In any event the testis of the early pupa has all stages of spermatogenesis, and sperm are already present 24–30 hours after the onset of pupation (Huettner, 1930c; Kerkis, 1933; Gleichauf, 1936; Gloor, 1943; and others). At eclosion the testis contains principally transforming spermatids and maturing spermatozoa, but there still remain spermatogonia and fairly numerous cysts of spermatocytes of various stages in the apical portion of the testis. These cells continue their normal course of successive divisions and differentiation to form sperm, and, as pointed out by Duncan (1930), males may still be undergoing spermatogenesis even after they have reached a state of senile infertility (i.e., more than 32 days after eclosion) brought about apparently by failure of the ejaculatory mechanism. For most cytological purposes, however, imaginal testes—even of recently emerged males—are often quite unsatisfactory, although they may be used for certain cytogenetic purposes, as Philip (1942a) has pointed out. Early pupal testes in which the successive stages are still in fairly orderly sequence, in which all stages of meiosis are frequent, and in which the great bulk of cells is not one of spermatids and spermatozoa are the testes of choice for cytological work. On the other hand, adult gonads from species having ellipsoidal testes in the imago are almost as satisfactory for cytological purposes as are pupal testes.

The Spermatogonial Divisions

Mitosis of the spermatogonial cells adds to an accumulating bulk of growing primary spermatocytes during the greater part of larval life. For the most part the meiotic mitoses of these spermatocytes are held in abeyance until the larva is ready for metamorphosis. Thereafter, during pupal and adult life, spermatogonia give rise to an unending succession of spermatocytes which complete their growth and undergo meiosis apparently without noticeable delay. The postmeiotic descendants of these cells transform to mature spermatozoa, the steps in this transformation being spoken of as "spermiogenesis," or "spermateleosis." The absolute numbers of cells undergoing meiosis seem to reach a maximum during midpupal life in *Drosophila melanogaster;* thereafter testes contain smaller numbers of active spermatocytes. During adult life a relatively constant but small aggregate of spermatogonial cells caps the apex of each testis.

As with many insects, two sorts of spermatogonia occur in the apex of the testis of *D. melanogaster.* One type of spermatogonial cell is characterized by not occurring in groups and by undergoing asynchronous mitoses. These are the "primary" spermatogonia or "predefinitive" spermatogonia of Tihen (1946). In number Tihen estimates them to comprise about eleven cells per testis,[5] or about as many cells as originally present in the embryonic testis.

[5] Nine adult testes among the author's own slides gave: 16, 15, 14, 13, 10, 10, 8, 7, 6 *apical cells* (of 5–6 µ in diameter) per testis, with an average number of 11—the same as for Tihen's *"predefinitive gonia."* But the apical cells are somatic and not germinal elements, as Geigy (1931) and Aboim (1945) have shown. Did Tihen count apical cells?

Contrasting with the above are the "secondary" or "definitive" spermatogonia. In life these range in size from about 7 to 10 μ in greatest diameter. They occur (excepting the initial one) in fairly well-defined cysts, all members of each cyst being derived from the same original secondary spermatogonium. In *D. melanogaster* four such secondary gonial divisions occur, giving rise to cysts containing sixteen primary spermatocytes entering upon meiotic prophase (Pontecorvo, 1944; Tihen, 1946).[6] The nearly absolute mitotic synchrony of all members in a given cyst of secondary spermatogonia offers a reliable means for identifying these cells and hence a means for determining their proximity to the onset of meiosis.

The functions of the two sorts of gonial cells are quite different. One, the primary or predefinitive spermatogonium, has a maintenance function. It preserves a gonial reserve in the active gonad along with a continuous production of secondary spermatogonia. On the other hand, the latter cells seem to have their fates determined and serve solely as propagators of nests of primary spermatocytes. The relation between the two types of gonial cell has been interestingly discussed by Tihen (1946), who makes the attractive suggestion that the primary or predefinitive spermatogonia behave in an analogous fashion to the cambial cells of higher plants. That is, the mitotic division of a primary spermatogonium would have as its two products one primary spermatogonium plus one secondary spermatogonium. Such hypothetical cambial divisions would have as their direct consequence the maintenance of a nearly constant reserve of primary spermatogonia along with the continued production of new secondary spermatogonia replacing those transformed to primary spermatocytes.

Little has been written of the cytology of the spermatogonia in *D. melanogaster*, and the best account of male gonial cytology in a species of Drosophila is that by Dobzhansky (1934) for *D. pseudoobscura*. In *D. melanogaster* the secondary gonial cells of the living, intact testis show themselves to be closely packed

---

[6] In *Drosophila pseudoobscura* and *D. miranda,* on the other hand, there are five divisions of the secondary spermatogonia, giving rise to cysts of thirty-two primary spermatocytes (Dobzhansky, 1934, 1935; Sturtevant and Dobzhansky, 1936). This seems to be the case in *D. subobscura* also (Philip, 1944).

in their cysts. The cell bodies are somewhat polygonal in shape, and the longest diameters average from 7 to 10 μ, with the nearly spherical nuclei only slightly smaller, being from 5.6 to 7.1 μ in diameter.[7] The interphase or resting stage of the living spermatogonial cell has a nucleolus (more rarely two smaller nucleoli) nearly a third the diameter of the nucleus in size. The rest of the nucleus seems an empty void under the ordinary light-field microscope. In the cytoplasm, generally to one side and appressed against the nucleus, there occur numerous granules which are probably of mitochondrial nature, as Dobzhansky has remarked. A smaller number of such granules are scattered through the cytoplasm.

When such resting-stage cells are fixed, the nucleus shows a reticulum of varying coarseness and staining density dependent upon the fixative and dye employed. With onset of prophase the reticulum is replaced by threadlike chromosomes which, even in these early stages of condensation, seem to be twisted together in pairs of homologues. The paired condition persists as the chromosomes continue their condensation but becomes looser with disappearance of the nuclear membrane and onset of metaphase. Such pairing of homologues, which seemingly occurs in virtually all the diploid mitotic nuclei of *D. melanogaster* and most other flies (see, for example, Figs. 8–10), is known as "somatic pairing."[8] It was first discovered and remarked upon by Stevens in 1907 (1908) in her study of *D. melanogaster*.

A fairly frequent and striking feature of the somatically paired autosomes (II and III) in spermatogonial prophases, and X chromosome somatic pairs in oögonia, is that a change in association of chromatids may occur between homologues (Cooper,

[7] Measurements from fixed preparations depart widely from these figures and from one another. Guyénot and Naville (1929) give spermatogonial diameters of 4–5 μ, with the nucleus perhaps 3 μ in diameter. Woskressensky and Scheremetjewa (1930) record spermatogonial diameters as 16–21 μ, with nuclei up to 12 μ in diameter.

[8] There are several cases in which somatic pairing is not uniformly expressed in all tissues. For example, somatic pairing is suppressed in the spermatogonia and first spermatocytes of Sciara (Metz, Moses, and Hoppe, 1926) and is absent in the early oögonia and variable in the early spermatogonia of Tipula (Bauer, 1931). Other peculiarities in the expression of somatic pairing will be found in the article by Cooper (1944).

1948, 1949). These changes in chromatid association give rise to chiasmata just as Kaufmann (1934) first discovered in the somatic mitoses of larval neuroblast cells (e.g., note *right* limbs of II in Fig. 10). Apparently X and Y pairs, however, cannot give rise to such chiasmata, because they do not occur between the X and Y chromosomes in males or XXY females. In any event these chiasmata are for the very most part unrelated to a prior act of genetic crossing over (Kaufmann, 1934; Cooper, 1949). They are accordingly of exceptional theoretical importance for, contrary to the beliefs of most modern geneticists, they demonstrate that not all chiasmata arise from genetic crossing over, nor are all chiasmata to be equated to genetic crossing over. Such changes in association generally lapse before full metaphase, sister chromatids coming to lie side by side or nearly so.

The nucleolus gradually disappears from view, no vestige of it normally remaining by the end of prophase. Dobzhansky (1934) believes that in *D. pseudoobscura* the sex chromosomes condense in association with the waning nucleolus. In view of Kaufmann's (1933, 1934) and Heitz's (1933) demonstrations of a similar association of the nucleolus with the sex chromosomes in ganglion cell mitoses of *D. melanogaster* (Fig. 15), most likely the sex chromosomes here also condense from a similar association with the nucleolus.

With dissolution of the nuclear membrane the chromosomes come to lie on a flat equatorial plate at full metaphase (Fig. 9). At this stage somatic pairing is still evident, but the chromosomes are least closely associated in pairs. Of all the chromosomes the sex chromosomes show the greatest tendency to be unassociated at metaphase (Metz, 1926). During anaphase the homologues are said again to become intimately associated in pairs, and presumably they enter the resting-stage nucleus in such paired association (Metz, 1926). One consequence of such persistent somatic pairing is that related gonial cells in *D. pseudoobscura* tend to have similar patterns of chromosomes at metaphase (Dobzhansky, 1936). This is possibly the case in *D. melanogaster* also, but such a study has not been made for this species. Another consequence of great significance is that the homologous chromosomes of male Drosophila possibly enter the first spermatocyte nucleus in a paired state (Stevens, 1908; Metz, 1926; etc.). Confirmation of this suggestion is needed.

The Meiotic Prophase of the First Spermatocyte

The only aspects of the first meiotic mitosis of Drosophila melanogaster upon which most authors seem agreed are the late diakinetic and the anaphasic stages. The early and growing spermatocytes have proved almost hopelessly refractory to most of the older cytological methods. Add to this unfortunate fact the extraordinary difficulty of accurately and objectively seriating these early stages, and it becomes a matter of little surprise that there have been expressed almost as many views of the early prophase phenomena in the first spermatocytes of D. melanogaster as there are scientists who have investigated these stages. The account given here is largely a synthetic and tentative one, drawing from earlier works descriptions of particular aspects about which there is general agreement (as, for example, the growth and subsequent dissolution of the nucleolus) or reinterpreting structures to which one or more authors had ascribed another significance.

It may properly be asked at the very outset on what grounds the account which follows is preferred, and how a choice was made among published descriptions or possible interpretations. The answer is as follows.

After a close study of the published reports for D. melanogaster and other species it became only too evident that no average or simple sum of the accounts could be made which would lead to a consistent interpretation of the growth stages in Drosophila. Accordingly some forty smear preparations of pupal testes were made and studied. These were fixed in B-15 or San Felice and stained by the Feulgen technique or with gentian violet or Heidenhain's iron haematoxylin.[9] In addition to the above slides a number of studies were made of living spermatocytes through the testis wall; these proved of inestimable value when

---

[9] The method of preparing such true "smears" is described elsewhere (Cooper, 1946). B-15 diluted with five volumes of distilled water proved better than full-strength B-15 or San Felice, but it is possible that some other fixatives would do as well or better.

judging the significance of certain evidence from fixed preparations.

The permanent slides to which greatest significance is attached are those stained by the Feulgen method. The necessary control slides prepared without preliminary hydrolysis were made (see Lison, 1936), and comparison of these with hydrolyzed material gave strong assurance that in the latter at least a partial cytologic history of the desoxyribose nucleic acids could be made out. With a fairly high probability of validity, it is further assumed that by following the Feulgen-positive structures in these slides the gross prophasic history of the chromosomes may at least in part be deciphered. By a fortunate preservation of cytoplasmic details and mitochondria in advanced growth stages, and by the utilization with these of cell, nuclear, and nucleolar sizes, most of the stages have been quite objectively seriated. Studies of the Feulgen preparations were followed by the examination of gentian violet and iron haematoxylin preparations. Finally, a small collection of very beautifully prepared sections of the testes of XY and XYY male *D. melanogaster* was lent by Professor Curt Stern.[10] These were of enormous value in the study of structures earlier described by Metz, Guyénot and Naville, Huettner, and others.

Before closing these introductory remarks it must again be emphasized that *the account which follows should be regarded as wholly tentative.* Many checks must be carefully made before the prophasic behavior is finally established and well understood. Most important of these are a restudy of the alleged somatic pairing of the last gonial anaphase and telophase (Metz, 1926) [11] and the unraveling of the conditions described by Huettner (1930c) as leptotene. As a matter of record, Stevens (1908) and Woskressensky and Scheremetjewa (1930) have given descriptions of these stages which are, of all accounts, in closest agreement with that which follows. The latter workers used the Feulgen technique also, and Stevens' study was made very largely by the acetocarmine squash technique.

Early or late in the growth period, the living spermatocytic nucleus is nearly spherical in shape in cells that are not greatly

[10] Belar's (1928) excellent photographs (Plate 2, Figs. 5a, 5b, 5c, and 6) are from these slides.

[11] Guyénot and Naville (1929) maintain that such pairing occurs in anaphase but is lost during telophase in spermatogonial divisions.

elongated. The descriptions of markedly lobed or indented nuclei, based upon fixed preparations, are therefore incorrect. The cytoplasm of the very young spermatocyte shows an excentric idiozomal accumulation more or less appressed to one side of the nucleus. Fixed preparations may, in favorable cases, show this to include an accumulation of granular or rodlike mitochondria and perhaps also platelike Golgi bodies or dictyosomes (Fig. 22) (see Huettner, 1930c). Within the nucleus little more than a nucleolus and perhaps one or two smaller inclusions may be seen in the living, young spermatocyte.

The nucleolus in living cells appears as though short segments of cords lie embedded in its surface, irregularly protruding from it. In Flemming-fixed material the nucleolus may have a very lifelike appearance, with the "cords" staining darkly with haematoxylin and the remainder of the nucleolus being pale gray (Fig. 22). When the Feulgen stain is employed, however, the nucleolus remains colorless, thus proving to be a plasmosome, but its surface may be tinged or have appressed to it several darkly staining, Feulgen-positive clods (Figs. 21, 22, 33, 34). Inferring from Heitz's (1933) and Kaufmann's (1933, 1934) demonstration of nucleolus organizers in the sex chromosomes, from Metz's (1926) and Woskressensky's and Scheremetjewa's (1930) observations on growth of the spermatocyte, and from manifestations to be discussed subsequently, it seems quite probable that these clumps appressed to the nucleolus are heteropycnotic elements of the sex chromosomes.

Fixed cells show a nuclear reticulum which is coarse or fine, apparently depending upon the nature of fixation. The reticulum may be resistant to stain (Figs. 22, 24), or it may show threadlike segments through its substance when it is heavily overstained or when Feulgen's method is used (Figs. 21, 23, 33, 34). If cells are studied which are intermediate between the smallest spermatocytes (Figs. 21, 24) and those perhaps twice their linear size (Fig. 27), then some are found in which the reticulum is clearly cut into three principal accumulations (Figs. 34–36). Thereafter the growing spermatocyte at each size may generally be shown to possess three such clumps (Figs. 28, 29, 35–38). One of the clumps, at all but the latest stages (Figs. 32, 38), is associated with or close to the nucleolus (Figs. 28, 29, 35–37). It is almost certainly made up of parts or the whole of the sex chromosomes.

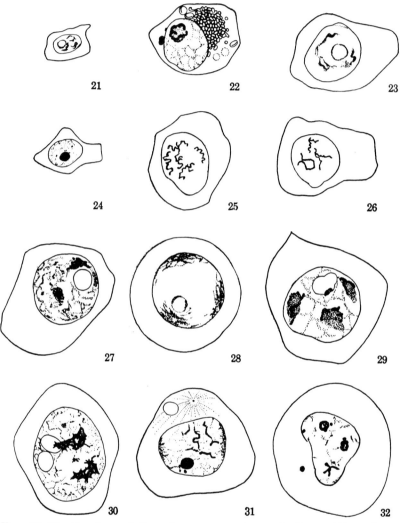

FIGS. 21–32. Prophase of the primary spermatocyte in *Drosophila melano-gaster*. (All figures from sectioned material.)

Fig. 21. Early primary spermatocyte at onset of growth. Fixative(?) and Feulgen. ca. 0.8× original. After Huettner (1930c), ex Fig. 3.

Fig. 22. Early growth stage represented semidiagrammatically. Note the idiozome to upper right, the pale reticulum within the nucleus, and the "corded" plasmosome. Flemming, and Heidenhain's iron haematoxylin. ca. 1650×.

Fig. 23. Early growth stage. Fixative(?) and Feulgen. ca. 0.8× original. After Huettner (1930c), ex Fig. 3.

Fig. 24. Early growth stage. Carothers, and Heidenhain's iron haematoxylin. ca. 1390×. After Guyénot and Naville (1929), ex Fig. 39.

Fig. 25. Reputed formation of leptotene threads (cf. Fig. 31). (Technique?) ca. 0.8× original. After Huettner (1930c), Fig. 13.

Fig. 26. Reputed synaptotene stage. (Technique?) ca. 0.8× original. After Huettner (1930c), ex Fig. 15.

Fig. 27. Advanced growth; note the two more heavily staining clumps in the nucleus, which are probably paired autosomes. Fixative(?) and Biondi triple stain. ca. 0.8× original. After Huettner (1930c), ex Fig. 11.

Fig. 28. Advanced growth, showing the three large Feulgen-positive clumps, probably representing the two large autosomes and the sex chromosome bivalent (below nucleolus). Sublimate acetic and Feulgen. ca. 1× original. After Woskressensky and Scheremetjewa (1930), Fig. 12.

Fig. 29. Late growth, showing the appearance of twisted strands within the three clumps, representing the large autosomes and sex chromosome bivalent (partially attached to and to the right of nucleolus). San Felice, and Heidenhain's iron haematoxylin. ca. 1650×.

Fig. 30. Late growth, showing a double plasmosome and two clumps, probably representing the large autosomal bivalents. Flemming, and Heidenhain's iron haematoxylin. ca. 1455×. After Guyénot and Naville (1929), Fig. 45.

Fig. 31. Reputed preleptotene stage. Note that, in addition to the alleged chromosomal threads and the nucleolus, there are diffuse clumps at each side of the nucleus, which possibly represent the large autosomal bivalents. A chromatoid body, aster, and centriole lie above the nucleus. Carothers, and Heidenhain's iron haematoxylin. ca. 1455×. After Guyénot and Naville (1929), ex Fig. 57.

Fig. 32. Diakinesis. The condensing, unequal sex chromosome bivalent which lies below shows only a small surface of conjunction. Carothers, and Heidenhain's iron haematoxylin. ca. 1455×. After Guyénot and Naville (1929), Fig. 68.

The other two clumps generally lie separately, appressed against or close to the nuclear membrane (Figs. 28, 29, 35–38). They may also be free in the nuclear interior (Figs. 27, 30). In some preparations fixed with San Felice and stained by Heidenhain's iron haematoxylin these bodies appear to contain a paler, diffuse matrix over or through which course apparently tangled or twisted threads (Fig. 29). In sectioned material only two such clumps may be evident (Figs. 27, 30); presumably the third is to be found in another section. There is little doubt that the two which are unassociated with the nucleolus represent the large autosomes during the growth stages. The dotlike autosomes have not been identified with certainty before diakinesis (cf. Figs. 38, 39).

This clodlike stage of the chromosomes has been observed by all who have closely studied the growth stages, but the probable origin of the clods from a typical early interphase reticulum seems to have escaped notice. This is partly owing to the fact that in sections prepared after most fixations and with most stains the nucleus is, curiously, either a hodge-podge of assorted non-characteristic artifacts, polychromatic masses, and inclusions [read especially Huettner's (1930a, b, c) account] or else virtually empty except for the nucleolus. When slides show nuclei with abundant differentiations within them [as in Guyénot's and Naville's (1929) and Huettner's (1930c) preparations], then such clumps have invariably been seen (Figs. 27, 29, 30, 31). But in these particular cases they were specifically held to be non-chromosomal, the chromosomes being believed to arise as threads somewhat later in growth. Stevens (1908), using aceto-carmine, Metz (1926), using iron haematoxylin, and Woskressensky and Scheremetjewa (1930), using Feulgen, all agree that in middle or somewhat later growth stages these diffuse bodies are the chromosomes. That Huettner, also using the Feulgen technique, should miss the early stages and indeed misconstrue the nature of the clods can easily be understood by the fact that only in slides in which the Feulgen stain is very intense (in the sperm heads, for example) do the diffuse bodies show a suffused pink, positive reaction. Unless care is taken to obtain a maximal Feulgen reaction, no evident staining of the diffuse masses may occur. Care must also be taken to no more than tinge the cells with a counterstain, as the Feulgen-positive masses are strongly oxyphilic in early stages. A heavy counterstain with light green for example, may totally mask a weak but positive Feulgen reaction.

During the growth stages the nucleolus (there are rarely two large nucleoli as in Fig. 30) may increase somewhat in size. Aside from the nucleolus, little more than an occasional small globule or nebulous patch may be seen in the nucleus of living cells at advanced growth. The diffuse clods have never been seen by the present author in such cells, but, as the fresh preparations become moribund, a close tangle or several skeins of coarse threads come to view in some, but not all, preparations. These bring to mind the so-called "leptotene" stages described by Guyénot and Naville (1929) [12] and Huettner (1930c), especially as they occur at an apparently comparable stage of the growth process. Indeed Flemming-fixed and Heidenhain-stained sections lent by Professor Stern not infrequently show figures such as they describe (i.e., like Figs. 25, 26, 31), and these *seem* closely similar to the configurations of moribund cells. But are these threads chromosomal?

If 70 per cent acetic acid containing 3 per cent orcein is pulled under the coverslip of such a moribund preparation, the threads in question vanish quickly from view. Only in cells verging on full diakinesis do three bodies stain sharply with orcein in such nuclei, and these prove to be the bivalents. They show, so far as could be observed in a small number of trials, no positional relation to the earlier existent threads. The chromosomes arise in a peripheral location, whereas the threads are generally, if not always, more centrally located. Now, if the threads were nucleoprotein, they should be fixed and stained by the aceto-orcein treatment. They are not. But if they are albumin or globulin precipitates, they would be expected to dissolve in excess acetic acid (Baker, 1945), which they do. Consequently these threadlike configurations within moribund cells, and probably those in fixed cells at this stage, are not chromosomal in nature but are perhaps pure artifacts. In agreement with this conclusion, (1)

[12] Guyénot and Naville (1933) have retracted their statements concerning the existence of leptotene in male *Drosophila melanogaster,* but they still subscribe to the regular occurrence of a stage in which the chromosomes are threadlike.

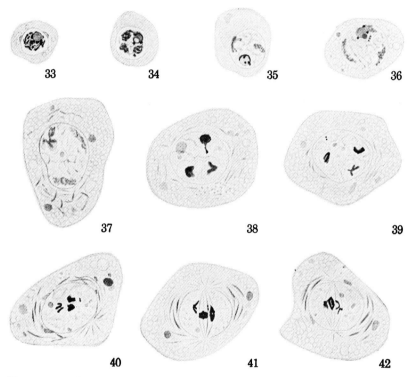

FIGS. 33–42. Prophase of the primary spermatocyte in *Drosophila melano-gaster*. (All figures from smears fixed in dilute B-15 and stained with Feulgen and light green. ca. 1430✕.)

FIG. 33. Early primary spermatocyte. Note the tangle of chromatic threads about the plasmosome.

FIG. 34. Early growth, showing the loosening of the tangle of chromatic threads into three principal regions. The upper clump may be associated with the nucleolus, which is partially rimmed by densely heterochromatic material.

FIG. 35. Early growth, showing two large but weakly staining elements, probably representing the large autosomes near nuclear margin. Nucleolus with Feulgen-positive associated elements.

FIG. 36. Midgrowth, showing three large, weakly stained clumps. The one associated with the nucleolus and the three dark spots against the nucleolus are probably elements of the sex chromosomes. The two elongate blobs against the nuclear periphery probably represent the two large autosomal bivalents.

FIG. 37. Late growth. The bodies probably representing the autosomal bivalents are at two and six o'clock. The sex chromosome bivalent is associated with the fragmenting nucleolus. Mitochondria and other bodies are visible in the cytoplasm.

FIG. 38. Early diakinesis. The sex chromosome bivalent above is no longer associated with the nucleolus, which lies at nine o'clock. The two large autosomal bivalents lie below.

FIG. 39. Diakinesis. Dotlike bivalent (at eleven o'clock), large autosomal bivalents (at nine and three o'clock), and sex chromosome bivalent (below); nucleolus fragmented. The Y chromosome is lowermost in the sex bivalent.

FIG. 40. Diakinesis. Note orientation of bivalents within nucleus, the mitochondria enveloping the nucleus, and the aster at five o'clock. The sex chromosome bivalent is to the left, with X below and connected with Y as though by a thread.

FIG. 41. Prometaphase. Note the difference in configuration of the large autosomal bivalents. The one to the right has "terminalized" its formerly lengthwise association (cf. Figs. 39, 40), while the other is in the initial stage of doing so. The sex chromosome bivalent is to the left, with X uppermost. Asters, spindle, mitochondria, nucleolar fragments, nuclear outline, and chromatoids (?) are visible.

FIG. 42. Metaphase or nearly so. The sex chromosome bivalent is to the right, with X below. Note persistence of nuclear outline and the central location of the spindle in the old nuclear space.

similar but stained threads were not observed in Feulgen preparations,[13] yet at all stages of growth other, non-threadlike, weakly
Feulgen-positive bodies could be found; and (2) in non-Feulgen
preparations the diffuse clodlike bodies may occur simultaneously
with the problematical threads. For example, Fig. 31, copied
from Guyénot and Naville (1929), shows the threads centrally
located; but against the nuclear membrane at each side of the
figure occur two pale clumps which probably represent the two
largest pairs of autosomes.

By the time full growth has been reached, threadlike mitochondria and other inclusions are quite conspicuous in the cytoplasm. They may be seen in living or slightly moribund cells,
as well as in cells fixed with diluted B-15 (Figs. 37–42). In the
vicinity of an aster the mitochondria become oriented as though
parallel to the individual astral rays. This gives a very pretty
configuration in the fresh preparation and is of aid in fixing the
stage of a cell as late prophase or diakinesis. The remaining
mitochondria seem to course over and about the nucleus, fore-
shadowing their characteristic location at prometaphase (cf. Figs.
37–42) when they form a capsule about the nucleus. Most of
the mitochondria at prometaphase and thereafter lie essentially
parallel to the future long axis of the spindle, clustered about the
latter when fully formed, and generally removed from the astral
spheres.

Within the nucleus the chromosomal clumps become more compact and, as a rule, stain more deeply (Figs. 37, 38). The nucleolus becomes dissociated from the condensing sex chromosome bivalent (Figs. 32, 38–40), generally fragmenting into a variable

---

[13] Huettner's account suggests that he has stained such threads by the
Feulgen technique but does not explicitly state this. If Huettner is correct that the "leptotene" he describes is very rare (about 1 in 10,000 first
spermatocytes may be expected to be in this stage, according to Huettner),
it is quite possible that the stage has been missed in this tentative study.
But it is not likely that it sets in so late in growth as he describes, nor does
it seem probable (from the size of the cells he figures) that it occurs in
the very small spermatocyte. If the description given above of the
chromosomes as clumps from early to late growth should prove correct,
then it seems very unlikely that the course of events becomes suddenly
interrupted with the paired chromosomes resolving themselves into unpaired threads, only to resume their state quickly as a haploid number
of clumps.

number of elements.[14]   The chromosomes themselves give the impression of rapidly condensing (Figs. 32, 38–40), and diakinetic nuclei are quite frequently suitable for a study of the structure of all three autosomal bivalents and the sex chromosomes (Figs. 39, 40).   In such nuclei the dotlike chromosomes are paired but show a gap between them somewhat less in width than the diameter of one of the chromosomes (Figs. 39, 40).   The two large V-shaped autosomal pairs may be flexed near their midpoints, or kinetochores, but appear to be associated side by side along their whole lengths (Figs. 38–40).   In contrast to these, the sex chromosomes are conjoined to form a bivalent by only a short interstitial segment in each chromosome (Figs. 32, 39, 40).   The region of conjunction may be pulled out into a threadlike connection (Fig. 40).   Nevertheless, there is no evidence for kinetochore repulsions of homologous chromosomes at this time, among either the autosomes or the sex chromosomes.

The First Meiotic Division

With completion or near completion of their condensation, the bivalents move from a peripheral to a more central position in the nucleus (Figs. 38–40).   The asters lie at opposite poles of the nucleus, and a spindle forms between them (Figs. 41, 42).   The mitochondria which closely encapsulate the diakinetic nucleus remain through prometaphase as though encircling an intact nucleus, even after the spindle has definitely formed and received the chromosomes (Figs. 40–42).   At such a stage a very fine line demarcates the innermost rim of the mitochondrial sheath (hence the outermost margin of the nuclear space) in the author's slides. It has not been possible to determine with certainty whether this is artifactual or whether it is the nuclear membrane which persists through this stage.[15]

[14] For detailed discussions of various alleged emissions from the nucleus to the cytoplasm, and of the history of non-chromosomal nuclear inclusions, consult especially the papers by Guyénot and Naville (1929) and Huettner (1930c).

[15] Stevens (1912), Metz (1926), and Zujtin (1929) maintain that the nuclear membrane persists through the entire division, Metz and Zujtin holding this also to be the case for the second meiotic mitosis.   On the other hand, Guyénot and Naville (1929), Huettner (1930c), and Woskressensky and Scheremetjewa (1930) all agree that the nuclear membrane

As the spindle forms and the chromosomes congress and orient upon it, as well as during chromosomal segregation, the nucleolus, its fragmentation products, and other inclusions scatter over and through the substance of the spindle. Indications of these bodies occur in many of the illustrations included with this account (Figs. 40–43, 45–48, 52). It is important to understand that these inclusions may greatly confuse prometaphase through anaphase figures. Unless special staining procedures are employed (such as Feulgen's, gentian violet, Kasanzew's, or Auerbach's, for example) which differentiate the chromosomes from the inclusions, or unless great care is given to the differentiation of the haematoxylin stains, these bodies may completely obscure the mitotic events of the first division and give the impression of utter cytologic chaos (Fig. 50).[16] In the subsequent account of the mitosis the inclusions will be ignored.

As remarked above, the diakinetic chromosomes show no recognizable indication of repulsions between homologous kinetochore regions (Figs. 38–40). As soon as the chromosomes go onto the spindle, however, homologous kinetic regions appear as though mutually repellent (Figs. 41, 42). The result is a coörientation of homologues on the spindle, during which a lengthwise conjunction of the homologous arms in the large autosomal bivalents is converted to a terminal association (Figs. 41–49, 51, 52). Successively, (1) a large autosomal bivalent appears as though made up of two widely open (or flexed) V's whose arms lie in parallel apposition during diakinesis; (2) the arms form an open cross during congression and coörientation of prometaphase; lastly, (3) each V has its arms folded together and stuck to those of its partner by corresponding homologous, free ends during metaphase and earliest anaphase.

The sex chromosomal bivalent gives no sign of such a shifting of the locus of conjunction during coörientation (Figs. 41, 42, 51,

disappears before metaphase. They are not agreed, however, at what stage it vanishes.

[16] And so it has been taken to be by Jeffrey (1925, 1931) and Jeffrey and Hicks (1925), who have contributed a large number of papers based upon simultaneous misconceptions of genetics and these inclusions. Metz (1926), Belar (1928), Guyénot and Naville (1929), League (1930), Huettner (1930a, b), and Woskressensky and Scheremetjewa (1930) have dealt very adequately with Jeffrey's notions.

53), but the chromosomes may part from one another totally in so far as visible connections are concerned (Figs. 43, 46, 47) or remain connected by only a thread (Figs. 49, 53). The dotlike autosomes become more widely separated during their movement to the equatorial plate.

Although most authors seem agreed that *Drosophila melanogaster* does not form a well-defined equatorial plate at first metaphase, this is not an easy matter to establish. For one thing, the flat-plate stage may be a fleeting one in the course of the mitosis. For another, figures are not uncommon which *do* represent a reasonably regular metaphase plate stage (Figs. 42, 44, 47, 49). These occur even in technically poor preparations (e.g., see Zujtin, 1929). Since the best fixed smear preparations consistently show at least some metaphase plates, and since these also occur in more poorly fixed, sectioned material (where the spindle may be badly distorted), it seems reasonable to conclude that the first spermatocytic metaphase in *D. melanogaster* is characterized by the usual equatorial configuration of bivalents (see also Cooper, 1949).

Anaphase may start irregularly. The sex chromosomes and dotlike fourth chromosomes may precede the large autosomes in anaphasic movement (Fig. 54), or the sex chromosomes may lag in their disjunction (Fig. 51). What the average or usual condition may be requires a statistical study which is yet to be made, but the impression is gained that the most frequent single condition involves the disjunction of all the large chromosomes at approximately the same time (Figs. 45, 47, 49) (Guyénot and Naville, 1929). In anaphase the arms of the V's may open considerably from their folded metaphase configuration (compare Figs. 46, 48, 51, 52, and 47, 49, 55, 56), all chromosomes becoming progressively more slender and elongated as the poles are attained. By late anaphase (Fig. 56) the large chromosomes may show a divergence of their chromatids, although this is not always evident in sectioned material (Fig. 55). To judge from Stevens' (1908) account and figures (her Figs. 65–67, 69–70, etc.; our Figs. 49, 51, 53, 54, 56), at least one of the sex chromosomes (X?) may have its chromatids divergent at one extremity as early as late diakinesis, also showing this peculiarity at metaphase and anaphase. In any event, by telophase the large autosomes often appear crosslike, owing to a divergence of chromatids at all points

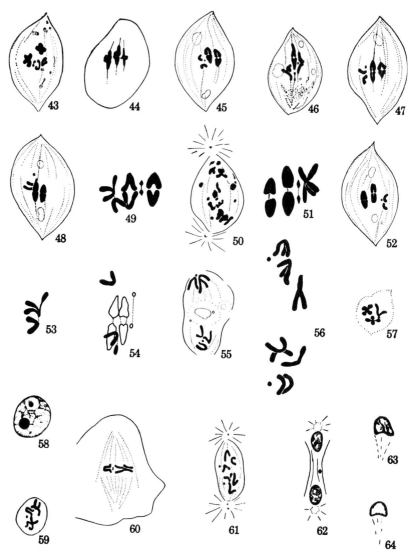

FIGS. 43–64. First and second spermatocyte divisions in *Drosophila mel-anogaster.* (Figures 49, 51, 53, 54, and 56 are from squashes; the others are from sectioned material.)

FIG. 43. Prometaphase —1. X-Y bivalent to right, X above. Carothers, and Heidenhain's iron haematoxylin. ca. 1585×. After Guyénot and Naville (1929), *ex* Fig. 71.

FIG. 44. Metaphase —1. X-Y bivalent to right. Flemming, and Heidenhain's iron haematoxylin. ca. 840×. After Metz (1926), Fig. 28.

FIG. 45. Transition to anaphase —1. X (*left, below*) disjoined from Y. Carothers, and Heidenhain's iron haematoxylin. ca. 1585×. After Guyénot and Naville (1929), *ex* Fig. 78.

Fig. 46. Transition to anaphase —1. Sex chromosome bivalent to left, X probably below. Flemming, and Heidenhain's iron haematoxylin. ca. 1585×. After Guyénot and Naville (1929), ex Fig. 80.

Fig. 47. Transition to anaphase —1. Sex chromosome bivalent to left, X probably above. Carothers, and Heidenhain's iron haematoxylin. ca. 1585×. After Guyénot and Naville (1929), ex Fig. 81.

Fig. 48. Metaphase —1. Carothers, and Heidenhain's iron haematoxylin. ca. 1585×. After Guyénot and Naville (1929), ex Fig. 79.

Fig. 49. Transition to anaphase —1. Sex chromosome bivalent to left, X probably above. Acetocarmine. ca. 1260×. After Stevens (1908), Fig. 75.

Fig. 50. First meiotic division with the nucleolar and other inclusions heavily stained, giving rise to haphazard configurations. Jeffrey and Hicks mistook the fragments for chromosomes. Carnoy, and Heidenhain's iron haematoxylin. ca. 1585×. After Guyénot and Naville (1929), ex Fig. 128.

Fig. 51. Transition to anaphase —1. The sex chromosome bivalent is to the right, and the two arms projecting above probably represent the Y chromosome. Acetocarmine. ca. 1260×. After Stevens (1908), Fig. 69.

Fig. 52. Metaphase —1. Sex chromosomes have apparently disjoined. The two are to the right, with the X above. Carothers, and Heidenhain's iron haematoxylin. ca. 1585×. After Guyénot and Naville (1929), ex Fig. 130.

Fig. 53. Early anaphase —1 disjunction of the sex chromosome; X above. Acetocarmine. ca. 1260×. After Stevens (1908), ex Fig. 70.

Fig. 54. Anaphase —1. Sex chromosomes black, with X below. Acetocarmine. ca. 1260×. After Stevens (1908), Fig. 72.

Fig. 55. Late anaphase —1, X at upper pole. Flemming, and Heidenhain's iron haematoxylin. ca. 1585×. After Guyénot and Naville (1929), ex Fig. 83.

Fig. 56. Late anaphase —1, X and Y rightmost chromosomes with X above. Acetocarmine. ca. 1260×. After Stevens (1908), Fig. 76.

Fig. 57. Telophase —1, in polar view, showing the open-cross configuration of the large autosomes, owing to a separation of their chromatids except at the kinetochores. Y chromosome (?) to right. Duboscq-Brazil, and Heidenhain's iron haematoxylin. ca. 1585×. After Guyénot and Naville (1929), ex Fig. 86.

Fig. 58. Interphase nucleus. Flemming, and Heidenhain's iron haematoxylin. ca. 1585×. After Guyénot and Naville (1929), ex Fig. 90, lower pole.

Fig. 59. Prophase —2, showing persistent divergence of homologous chromatids in the large autosomes and the X (?) between them. Carothers, and Heidenhain's iron haematoxylin. ca. 1585×. After Guyénot and Naville (1929), ex Fig. 94. These authors regard the sex chromosome figured as the Y.

Fig. 60. Metaphase —2, showing the nearly parallel orientation of homologous chromatids before anaphase. Fixative (?) and Feulgen. ca. 0.8× original. After Huettner (1930c), Fig. 23. Huettner wrongly describes this as "metaphase of first meiosis."

Fig. 61. Anaphase —2 of Y-bearing secondary spermatocyte. Carothers, and Heidenhain's iron haematoxylin. ca. 1585×. After Guyénot and Naville (1929), ex Fig. 102.

Fig. 62. Telophase —2, showing the formation of the spermatid nuclei. Carothers, and Heidenhain's iron haematoxylin. ca. 1585×. After Guyénot and Naville (1929), ex Fig. 106.

Figs. 63 and 64. Telophase —2 nuclei before their growth and rounding up. Sublimate acetic and Feulgen. ca. 0.8× original. After Woskressensky and Scheremetjewa (1930), ex Figs. 37 and 38 respectively.

excepting the kinetochores (Fig. 57). According to Guyénot and Naville (1929) the chromatid separation is suppressed in the sex chromosomes until the second meiotic prophase, but it is doubtful that this is regularly the case.

During anaphase the threadlike mitochondria lie over the surface and parallel to the long axis of the spindle. With division of the cell at telophase to form two secondary spermatocytes, the mitochondria are passively separated into two essentially equivalent groups. Thereafter the mitochondria come to encapsulate the developing telophasic nuclei. The behavior of the Golgi bodies or dictyosomes at this division is unknown.[17] It may be assumed, on analogy with other forms, that they are likewise passively parceled to the two secondary spermatocytes in roughly equivalent groups.

Stevens (1908) believed that the chromosomes promptly entered the second meiotic mitosis at the end of the first. However, Guyénot and Naville (1929), Zujtin (1929), Huettner (1930c), and Woskressensky and Scheremetjewa (1930) have shown that an interphase nucleus, possessing one or two nucleoli (or heteropycnotic segments?) and a diffuse, Feulgen-positive reticulum, intervenes between the two meiotic divisions (Fig. 58).

The Second Meiotic Division

The course of division of the secondary spermatocyte presents few notable attributes. The chromosomes appear in the prophase nucleus with their chromatids divergent as in the earlier telophasic group (Fig. 59). When the univalents are fully condensed, they form a regular metaphase plate at the equator of the spindle (Fig. 60). A matter of interest is the fact that sister chromatids are no longer divergent but lie in nearly parallel orientation to one another.

The mitochondria, which enveloped the interkinetic nucleus, now encapsulate the second division spindle. Just as in the first meiotic division, the mitochondria lie over the spindle surface with their lengths directed poleward.

Anaphase is brought about by the nearly synchronous disjunc-

---

[17] Brief mention of the Golgi is given by Zujtin (1929).

tion and separation of the kinetochores of the univalents. During midanaphase the chromosomes are sufficiently sharp and well spread in some cells, despite their very small size, to allow individual identification of the chromosomes (Fig. 61). Study of such groups confirms the conclusion that the sex chromosomes disjoin reductionally at the first division, as each anaphase possesses but one or the other type of sex chromosome. At telophase the chromosomes at each pole form a small, darkly staining nucleus which has its chromatin distributed peripherally (Figs. 62–64).

The mitochondria are passively divided into two nearly equal groups by constriction of the cell across its equator into two daughter cells. These new cells are the *spermatids* whose remaining action is self-conversion into motile spermatozoa. Since there are four divisions of the secondary spermatogonia, one meiotic division of the first spermatocyte, and one meiotic division of the secondary spermatocyte, the spermatids which are the end-products of this chain of multiplication occur in cysts of sixty-four cells each. Without further division each spermatid is converted into a spermatozoön, and this transformation takes place synchronously for all members of a cyst of spermatids. Spermatozoa thus come to maturity in bundles of sixty-four gametes each.

Spermiogenesis

It will be recalled that the early and growing primary spermatocyte has a well-formed idiozomal complex placed against and to one side of the nucleus (Fig. 22). This complex includes the granular or rodlike mitochondria and without doubt, on analogy with other forms, the Golgi bodies and centrosomal apparatus. Before first prometaphase the idiozomal complex breaks up, and the mitochondria scatter about and over the nucleus. The centrioles are located at the centers of pronounced asters, and the spindle is formed between them (Figs. 40–42). As earlier described, the mitochondria lie over the spindle, on the average their threads paralleling the spindle fibers. The state of the Golgi apparatus at this time can only be guessed. Possibly it is in the form of dictyosomes or dictyoconts scattered over and among the mitochondria or free in the more peripheral cytoplasm about the spindle. In any event they are approximately equally divided among the four spermatids resulting from the two meiotic divisions. Like that of the mitochondria described above, the division of the Golgi probably is a passive one, equality resulting from the random scattering of the bodies over or about the division apparatus.

In addition to these cytoplasmic elements at least two others have been remarked upon, but their natures are not clear. Huettner (1930c) has noted the occurrence of "certain persistent vacuoles which divide in the first meiosis." Zujtin (1929) has described chromatoid bodies, and these are possibly what Huettner (1930c) referred to as "condensed cytoplasmic bodies." Chromatoid bodies behave passively at each division, being single or few in number in the first spermatocyte, so that varying percentages of the spermatids possess at least one of them. No function is yet known for chromatoids, but where they have been closely studied they prove ultimately to be eliminated when cytoplasmic detritus is cast from the maturing sperm (see Wilson, 1934).

In the early spermatid, as it is divided from its sister cell, there occur scattered Golgi bodies, a cone or partial cone of filamentous mitochondria surrounding the spindle rest, an apical centriole, and, between centriole and mitochondria, the enlarging nucleus. In addition other elements, such as a chromatoid body, for example, may be included in the cytoplasm. Figure 66 depicts such 'a condition as has been just described, and represents *in diagram* what may be inferred as a stage antecedent to Fig. 65 from Huettner (1930c). Now the known steps are few in the transformation of such an early spermatid into the mature spermatozoön of *Drosophila melanogaster,* and only the following sketchy account can be given. It is drawn from the descriptions of Guyénot and Naville (1929), Huettner (1930c), Woskressensky and Scheremetjewa (1930), and Shen (1932) for *D. melanogaster,* and of Dobzhansky (1934) for *D. pseudoobscura.* As with the growth stages of the first spermatocyte, many details—especially those relating to telokinetic movements—remain to be worked out before a wholly satisfactory account of spermiogenesis in Drosophila can be given.

As cleavage of the two cells from one another is completed, the spermatid nucleus enlarges rapidly in volume to perhaps three times its telophasic diameter (Figs. 62–66). When stained by the Feulgen technique, the chromatin forms chiefly a thin peripheral coat on the inner surface of the "nuclear membrane," only a pale reticulum lacing its structure through the nuclear interior.

During nuclear growth a series of changes confer visible polarity upon the spermatid. The mass of mitochondria becomes rounded and more compact, forming a spheroidal body or nebenkern about the same size as the nucleus (Fig. 65). In the meantime the centriole has come to lie between the nucleus and the nebenkern. For the most part the centers of the nucleus, centriole, and nebenkern all lie roughly on one axis (Figs. 65–68). The cytosome, or cell body of the spermatid, becomes drawn out along the length of this axis. In general the apices or nuclear ends of the spermatids are directed against the outer cyst wall, and the long axis of the sister spermatids within a cyst come to be roughly parallel during their subsequent elongation.

Before appreciable linear growth of the spermatid has occurred, a plasmosome appears in the nucleus (Figs. 66, 67). As the spermatid undergoes its initial lengthening, the chromatin of the nucleus becomes largely concentrated in the basal part just above the centriole (Figs. 68, 69). In the meantime the nebenkern commences to elongate (Figs. 67, 69). In the living spermatid, as well as in favorable permanent preparations, the nebenkern gives the impression of being composed of complex spiral and concentric laminae (Fig. 67; consult Wilson, 1934). At this time the centriole can be shown to be doubled. The one remaining close to the nuclear wall is termed the "proximal centriole," its sister product being the "distal centriole." From the latter grows an axial filament which passes over or between the substance of the nebenkern (Figs. 67, 69–71). So far as has been determined, the proximal and distal centrioles remain close together and in their original positions with respect to the nucleus as the spermatid matures. The nebenkern itself becomes divided into two progressively elongating halves (Figs. 67, 69, 70). As the spermatid and axial filament grow in length, the paired nebenkern derivatives pull out into axially twisted, ribbonlike structures (Fig. 71). Thereafter they become difficult or perhaps impossible to follow (Figs. 72, 73). Presumably, as in other spermiogeneses, they form an element of the tail sheath or filament.

The Golgi material performs equally spectacular maneuvers and changes in form during spermatid elongation. Initially it is represented by perhaps both dictyosomes and dictyoconts, the

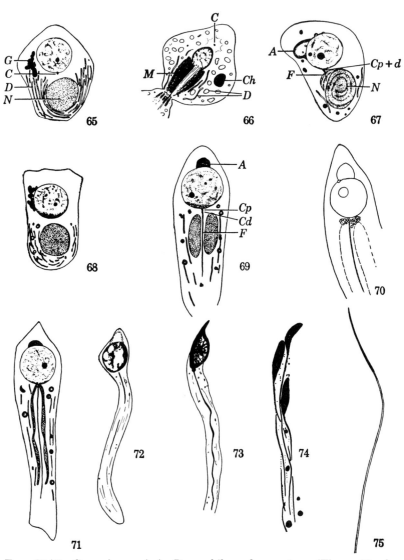

FIGS. 65–75. Spermiogenesis in *Drosophila melanogaster*. (Figure 66 schematic, Fig. 70 from living, others from sectioned, material.)

FIG. 65. Early spermatid, showing the mitochondria mass or nebenkern, *N*, encircled by Golgi material (dictyosomes, *D*). Golgi bodies, *GA*, forming acroblast to side of nucleus, and the centriole, *C* (undivided?), already between nucleus and nebenkern. Technique and magnification? Reproduced ca. 0.8× original. After Huettner (1930c), Fig. 42.

FIG. 66. One pole of a hypothetical telophase showing centriole, *C*, in apical position; mitochondrial envelope, *M*, partially encircling spindle rest; a chromatoid body, *Ch*; and Golgi bodies, *D*, scattered randomly in the cytoplasm. The Golgi bodies are portrayed both as threadlike and platelike forms.

FIG. 67. Initial elongation stage of spermatid. The acroblast, *A*, is still lateral; the centriole is clearly doubled, *Cp + d*, into proximal and distal elements; the nebenkern, *N*, seems divided and to be internally spiral in structure; the axial filament, *F*, is projecting from the distal centriole between the nebenkern's halves. Flemming, and Heidenhain's iron haematoxylin. ca. 1350×.

FIG. 68. Elongating spermatid. Some of the threadlike Golgi bodies appear to be replaced by ringlike forms. Technique and magnification? Reproduced ca. 0.8× original. After Huettner (1930c), Fig. 43.

FIG. 69. Elongating spermatid, having the acroblast, *A*, fully formed and in apical position; proximal, *Cp*, and distal, *Cd*, centrioles; axial filament, *F*; and divided nebenkern. Technique and magnification? Reproduced ca. 0.8× original. After Huettner (1930c), Fig. 47.

FIG. 70. Apical extremity of much elongated, living spermatid. Note the very characteristic shape of the apical acroblast, the elongate halves of the nebenkern enclosing the axial filament which arises from the distal centriole, and the cluster of granular material encircling region of the centrioles. *In vitro*. ca. 1350×.

FIG. 71. Ribbon-shaped and twisted elements of the nebenkern, elongating with growth of the axial filament. Technique and magnification? Reproduced ca. 0.8× original. After Huettner (1930c), Fig. 48.

FIG. 72. Condensation of chromatin along one (*right*) surface of nucleus. Carothers, and Heidenhain's iron haematoxylin (?). ca. 1550×. After Guyénot and Naville (1929), *ex* Fig. 112.

FIG. 73. Further condensation of the nucleus. Flemming, and Heidenhain's iron haematoxylin (?). ca. 1550×. After Guyénot and Naville (1929), Fig. 113.

FIG. 74. Three of a group of very advanced spermatids at a stage just before their alleged fixation to a nutritive cell. Carothers, and Heidenhain's iron haematoxylin (?). ca. 1550×. After Guyénot and Naville (1929), *ex* Fig. 114.

FIG. 75. Head and first portion of tail of mature sperm from the vas of an adult male. Full representation of missing portion of tail would require an additional length of more than thirteen feet. B-15, and Harris' haematoxylin. ca. 2000×.

latter being threadlike Golgi bodies (Fig. 65). The dictyoconts are replaced by ringlike bodies, according to Huettner (1930c), which enter the developing tail, wherein they persist as such to very late stages (Figs. 65, 68, 69, 71). On the other hand, the more apical Golgi material forms one or two large masses which adhere closely to the nucleus (Figs. 65, 67–72). These form an acroblast as they slide over the nuclear surface to a point opposite the centrioles (Figs. 67–72). During elongation of the spermatid the apical cytoplasm becomes greatly reduced, and the nucleus, capped by the acroblast, now forms the head of the developing sperm (Figs. 72–75). In living spermatids of this stage a cluster of granules may occur near the acroblast or in a position close to the centrioles (Fig. 70). The ultimate fate of these granules and the actual formation of the acrosome by the acroblast remain to be worked out.

As the elongation of the spermatid becomes extreme, the chromatin within the nucleus becomes located against one side of the nuclear membrane (Fig. 72). Here it is very condensed, forming what appears to be a short, curved bar bearing the rest of the nuclear vesicle on its concave face (Fig. 73). Progressively the nucleus lengthens and becomes more uniformly and darkly staining (Fig. 74). The cytoplasm forms but the slenderest envelope over the nucleus, and the cytoplasmic tail which threads away from the nucleus becomes much too slender to allow any determination of structure within it.

The Mature Spermatozoön

The mature spermatozoön of *Drosophila melanogaster* proves to be a most impressive gamete. When fixed in B-15 its head is a mere 7.4–8.8 μ long, and perhaps 0.28–0.37 μ in width (Fig. 75).[18] Head size is, however, partially a function of the mode of preparation; aceto-orcein preparations of sperm from the testicular duct give a length of 13.0 μ by a width of 0.5 μ. In the B-15 preparations the tail is more than 200 times the length of the head. Five mature sperm (four from the vas of a male, one from the ventral receptacle of a female) gave a mean length of

[18] Gowen and Gay (1933) give 7.36×0.368 μ.

1.76 mm with a standard error of ±0.04 mm. The girth of the tail apparently lies between 0.1 and 0.2 μ.[19] It is perhaps noteworthy that the mature egg is but 0.5 mm long (Morgan, Bridges, and Sturtevant, 1925).

Just what the selective value of such a tremendously elongated spermatozoön may be is a matter of pure conjecture. Perhaps the great length serves to increase the area of contact of the spermatozoön with the genital-tract epithelium in the female, in effect giving the spermatozoön the potentiality of exerting a rather powerful *thrust*. Capacity for thrust may well be an important factor if a narrow micropyle is to be penetrated, and a small spermatozoön incapable of thrusting from a purchase point might be expected to be unable to penetrate a micropylar egg by swimming movements alone.

THE COMMON ATTRIBUTES OF MEIOSIS IN THE MALE OF
DIFFERENT SPECIES OF DROSOPHILA

The description just given of spermatogenesis in *Drosophila melanogaster* represents a fairly complete rendering of facts and literature to date.[20] In so far as spermatogenesis in other species of Drosophila is concerned, only a small number have been investigated, and these, for the most part, in a very limited manner. Aside from *D. melanogaster* the species best worked out are *D. pseudoobscura* (Metz, 1926; Koller and Townson, 1933; Darlington, 1934; Dobzhansky, 1934; Koller, 1936), *D. persimilis* [21] (Darlington, 1934; Dobzhansky, 1934; Koller, 1936), *D. virilis, D. funebris,* and *D. willistoni* (Metz, 1926), and *D. miranda* (Dobzhansky, 1935; Koller, 1939; MacKnight and Cooper, 1944; Cooper, 1946). Fragmentary observations are also on record for

[19] Dobzhansky (1934) records a length of 0.4–0.5 mm for the mature spermatozoön of *Drosophila pseudoobscura* when extracted from the seminal receptacle of a fertilized female. The data given by Gowen and Gay (1933) in their Fig. 2 leads to an estimated 0.44 mm as the *minimum* length of a spermatozoön similarly extracted from a female of *D. melanogaster.*

[20] Omitted are the papers of Frolova (1931) and Woskressensky (1933), which the present author has not seen.

[21] *Drosophila persimilis* is the name given by Dobzhansky and Epling (1944) to race B of *D. pseudoobscura;* but see Sturtevant (1944).

D. subobscura (Philip, 1942a, b, 1944; Philip, Rendel, *et al.*, 1944) and *D. prosaltans* (Dobzhansky and Pavan, 1943).[22]

Where comparable meiotic stages have been studied, meiosis in the males of these species follows reasonably closely the pattern described above for *D. melanogaster*. For example, in *D. pseudoobscura* (Metz, 1926; Koller and Townson, 1933; Koller, 1936), *D. virilis*, *D. funebris*, and *D. willistoni* (Metz, 1926) and *D. subobscura* (Philip, 1944), leptotene, zygotene, and pachytene seem to be either absent or undemonstrable.[23] Before diakinesis the chromosomes in all these species, as well as in *D. miranda* (Dobzhansky, 1935; Cooper, 1946), appear as flocculent, spongy masses when they can be made out at all. The sex chromosomes are frequently associated with a nucleolus, the size, staining properties, and behavior of which may vary somewhat from species to species (Metz, 1926). At diakinesis the autosomes have each limb in parallel association, whereas the sex chromosomes conjoin over very restricted lengths in sharp contrast with the autosomal bivalents. Such differences as are known to exist among the species seem to represent no more than relatively minor variations upon a single theme. In so far as the accounts go, therefore, what seem to be the principal features of meiosis common to male Drosophila may be summarized as given below. In this summary each statement is followed by a list of species for which it is believed to hold on the basis of accounts or figures given by the authors listed above. The state of affairs in species not mentioned remains to be ascertained.

1. Homologous chromosomes may enter the primary spermatocyte nucleus in a paired association, owing to somatic pairing during the last spermatogonial anaphase (*D. melanogaster, D. pseudoobscura, D. persimilis, D. virilis, D. funebris, D. willistoni*). Correlated with the precocious pairing seems to be the absence of true leptotene, zygotene, and pachytene stages [above species, probably also *D. miranda* (see p. 48) and *D. subobscura*].

[22] See footnote 2, p. 182, in Cooper (1946).

[23] It must be recorded, on the other hand, that Dobzhansky (1934) has described and illustrated nuclei held by him to be suggestive of leptotene, pachytene, and diplotene in *Drosophila pseudoobscura* races A and B (race B later named *D. persimilis*). Nevertheless he is reluctant to evaluate these figures ". . . because they are rare, and are apparently produced by some unusual circumstances in fixation and staining that are not reproducible. These cells may even be pathological."

2. Until diakinesis the chromosomes are represented in the growing primary spermatocyte nucleus by a haploid number of bodies (except for the dotlike chromosomes) in a variable state of diffusion and flocculency (*D. melanogaster, D. willistoni, D. miranda, D. subobscura*). The clump corresponding with the sex chromosome bivalent is adherent to a nucleolus (*D. melanogaster, D. pseudoobscura, D. persimilis, D. virilis, D. funebris, D. willistoni*).

3. At diakinesis the bivalents condense. The nucleolus fragments at this time (*D. melanogaster, D. pseudoobscura, D. persimilis, D. virilis, D. funebris, D. willistoni*) or perhaps earlier (*D. miranda*), the sex chromosomes becoming freed from it. At this stage the large autosomes have each homologous limb in parallel association. In marked contrast to the large autosomes, the sex chromosomes are conjoined over very short, more or less proximal regions (*D. melanogaster, D. pseudoobscura, D. persimilis, D. virilis, D. funebris, D. willistoni, D. miranda, D. subobscura*).

4. During coörientation and congression the kinetochores of homologues become divergent (*D. melanogaster, D. pseudoobscura, D. persimilis, D. virilis, D. funebris, D. willistoni, D. miranda*), changing the parallel association of arms in the large autosomal bivalents to a terminal or subterminal one in *D. melanogaster* and *D. virilis*. Although the region of association in the sex chromosome bivalent remains the same, by metaphase there may or may not be a visible contact or connection between X and Y (*D. melanogaster, D. pseudoobscura, D. persimilis, D. virilis, D. funebris, D. willistoni, D. miranda, D. subobscura*).

5. A metaphase plate is formed (*D. melanogaster, D. pseudoobscura, D. persimilis, D. virilis, D. willistoni, D. miranda*).

6. The first meiotic division is reductional and regular, although the sex chromosomes may precede or lag (*D. melanogaster, D. pseudoobscura, D. persimilis, D. virilis, D. funebris, D. willistoni, D. miranda*).

7. Sister chromatids may freely disjoin during the first anaphase, except in the vicinity of the kinetochores (*D. melanogaster, D. pseudoobscura, D. persimilis, D. miranda*).

8. There may (*D. melanogaster, D. pseudoobscura, D. persimilis*) or may not (*D. miranda*) be a true interphase stage. In *D. melanogaster* interphase resembles a resting stage; in *D. pseu-*

doobscura and *D. persimilis* the chromosomes remain visible throughout interphase.

9. During the second meiotic prophase sister chromatids come to lie more or less parallel once again (*D. melanogaster*), or they may be irregularly disposed and held together only at the kinetochores (*D. pseudoobscura, D. persimilis, D. miranda*). The second metaphase is regular, and the ensuing anaphase is normal (*D. melanogaster, D. pseudoobscura, D. persimilis, D. virilis, D. funebris, D. miranda*).

It should perhaps be noted here that *D. miranda* is unique among the species of Drosophila so far studied in having an X_1YX_2 sex chromosome complement in the male (Dobzhansky, 1935). Yet this lends no especial peculiarity to the meiotic process, as a trivalent is formed just as in most other organisms having a compound X and Y (MacKnight and Cooper, 1944; Cooper, 1946). Thus X_1 and X_2 both conjoin with Y and as a rule segregate together from Y at first anaphase. However, there is one notable feature shown by the *D. miranda* X_1, namely, the possession of pronounced relic coils in both of its arms as late as diakinesis (Cooper, 1946). Since these coils are the vestige of the mitotic coils of the ultimate gonial division, they offer strong evidence that at least the X_1 chromosome has not passed through a leptotene stage in the process of forming an X_1Y bivalent. If leptotene is not essential for conjunction of X_1 and Y, then it is reasonable to suppose that it is also not a necessary step in bivalent formation for the other chromosomes. Evidence for a similar state of affairs has recently been discovered in the case of the XY bivalent of *D. melanogaster* (Cooper, 1949).

CHIASMATA, CROSSING OVER, AND MEIOSIS IN DROSOPHILA

What now can be said concerning (1) the nature of the meiotic differences which lead to genetic crossing over in female Drosophila but not in the male, (2) the relations of bivalent structure to metaphase association and the absence of crossing over in the male, and finally (3) the origin of the Drosophila-like type of meiosis? These questions have proved puzzling for many years, and it is unfortunately true that the answers given below are inconclusive.

1. With regard to the meiotic difference between male and female this much can be said. Although very little has been published concerning meiosis in female Drosophila or closely related forms, the observations of Metz and Plough (in Metz, 1926) and of Guyénot and Naville (1929, 1933) on *D. melanogaster,* of Stella (1936) on *D. immigrans,* and of Naville (1932) on Calliphora and Phormia, all seem to agree concerning a regular occurrence of leptotene-, zygotene-, and pachytene-like threads in the female. It was pointed out in the preceding section that probably no such stages occur in the male. Now Bauer (1931) and Wolf (1941) have shown that customary leptotene and zygotene stages may be omitted from meiosis in certain flies. Nevertheless a visibily normal pachytene stage occurs, followed by the formation of chiasmate bivalents. But where pachytene is absent, as in male Drosophila, the bivalents formed are of the non-crossover type. Hence the primary cytological difference between spermatogenesis and oögenesis in Drosophila, which is correlated with the restriction of crossing over to the female sex, appears to be the presence of a pachytene stage in the developing oöcyte and its absence in the first spermatocytic prophase. However, it is important to note that, whereas pachytene is a visible and probably necessary antecedent to crossing over, it does not guarantee it. This follows from Wolf's (1941) observations on *Penthetria holosericea,* a nematoceran fly, wherein a normal pachytene regularly occurs, but nevertheless meiosis is abnormal, and so far as can be told non-crossover bivalents are formed. It seems quite possible, therefore, that in Drosophila, antecedent to the elimination of early meiotic stages in spermatogenesis, a genetic restriction or loss of the *capacity* to produce crossing over occurred in the male. Given the loss of such a faculty, selection would no longer be maintained for the perpetuation of any chromosomal maneuvers the sole functions of which are the provision of configurations and states necessary for crossing over between homologous chromosomes or subsequent phenomena dependent upon prior crossing over. On the simplest interpretation of this thesis, the genetic restriction or loss of capacity for crossing over in spermatogenesis could have as its evolutionary consequence a more or less gradual elimination of now unessential prophase stages. In those cases of male Drosophila which have been cytologically investigated, leptotene through pachytene seem to

have become just such unessential stages and to have been eliminated from the meiotic sequence.

2. Coming now to the structure of the metaphase bivalent in male Drosophila, the autosomes and sex chromosomes may best be dealt with separately. In the very first study of meiosis in male Drosophila, Stevens (1908) pointed out that "bivalents" are formed by the large autosomes rather than the usual cross- and ring-shaped "tetrads," and that this departure is correlated with the absence of the customary pre-diakinetic meiotic stages. Nevertheless it was only after Darlington (1934) turned to the analysis of meiosis in male Drosophila that the probable significance of these facts was appreciated, and the usual structure of the autosomal bivalent in male Drosophila was shown to correspond with a non-crossover tetrad. Darlington's interpretation is essentially as follows.

Consider first a rod-shaped autosome such as occurs in *D. pseudoobscura*, *D. persimilis*, *D. miranda*, *D. virilis*, *D. funebris*, etc. (Fig. 76). Were a single crossover to occur in an autosome of this type and, as customary, non-sister strands to become mutually repellent but sister strands to adhere, then an open-cross bivalent of the sort shown in Fig. 76 *xa* would be formed. This is in fact the kind of configuration regularly seen in other organisms when a single exchange has occurred. However, in the Drosophila male the customary configuration is otherwise (Fig. 76 *Dros.*) and does not differ from what would be expected if (*a*) no crossover were to occur, and (*b*) all four chromatids were to remain in parallel association except in the vicinity of the divergent kinetochores.

Where V-shaped autosomes are concerned (Fig. 77), such as those of *D. melanogaster* and *D. willistoni*, the parallel association of all four chromatids in each arm leads directly to the formation of a bivalent (Fig. 77 *Dros.*) which closely resembles cytologically the ordinary open-cross configuration due to a single chiasma in a rod-shaped bivalent (cf. Fig. 76 *xa* with 77 *Dros.*). Indeed Woskressensky and Scheremetjewa (1930) mistook the large autosomal bivalents in *D. melanogaster* to be characteristic tetrads. Such they are not, because the kinetochore is nearly median in these autosomes and hence each arm of the cross must be composed of two pairs of chromatids, although in this case the individual chromatids may not be microscopically demonstrable until first anaphase. Were a single crossover to occur interstitially in one arm of such a bivalent, non-sister chromatids would be expected to repel one another to form a cross, the open

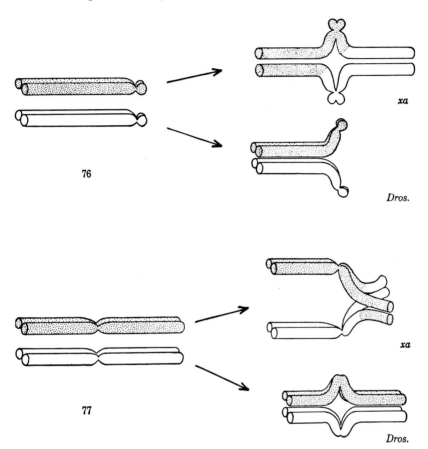

FIGS. 76–77. The structure of autosomal bivalents (see text).

FIG. 76. Rod-shaped chromosome, forming a bivalent with a single proximal chiasma, *xa*, or, alternatively, without chiasmata, *Dros.*, as in male *Drosophila* (e.g., *D. pseudoobscura*).

FIG. 77. V-shaped chromosome, forming a bivalent with a single chiasma, *xa*, or with no chiasmata, *Dros.*, as in male *Drosophila* (e.g., *D. melanogaster*). Note the superficial similarity of 77 *Dros* to 76 *xa*.

arms of which would tend to be more or less vertical to the plane of the non-crossover arms (Fig. 77 *xa*). A single exchange in each arm, on the other hand, would give a ring bivalent made up of two such open crosses joined at the kinetochores. Since the crossover configurations (Figs. 76 *xa*, 77 *xa*) are ordinarily not found in Drosophila males,[24] whereas the non-crossover types (Figs. 76 *Dros.*, 77 *Dros.*) are the rule, the cytological evidence from metaphase autosomal bivalents is in good agreement with the general absence of genetic crossing over in the male. It may also be pointed out that the dot-shaped autosomes form a bivalent the structure of which likewise corresponds with a non-crossover tetrad (Figs. 43, 46, 49, 51, etc.).

The sex chromosomal bivalent in Drosophila species has awakened considerable discussion. Although the illustration (Fig. 78) diagrams the situation in *D. melanogaster*, it serves equally well to demonstrate the argument for the other species. It is clear that X and Y are not associated by a single exchange. This follows because the pairing loci are interstitial in both X and Y, and each chromosome is heteromorphic (Fig. 78A, B). Were a single crossover to occur, the result would be the formation of a free arm of Y having an X kinetochore, and the attachment of one chromatid of the other arm of Y to X (Fig. 78B). During first anaphase such a configuration would be easily discernible cytologically. Such an event has never been recorded and must be very rare if it occurs at all.

Darlington (1931, 1934, 1939) nevertheless believes that X and Y do conjoin by chiasmata, but he conceives these to be regularly double and reciprocal (Fig. 78C). Such an arrangement of chiasmata would allow exchanges regularly to occur between X and Y without cytological or genetical disturbance of the identity or peculiarities of either chromosome, since the second chiasma reëstablishes the chromatid association destroyed by the first. Were such reciprocal chiasmata to be formed, a small loop, theoretically visible from above (arrow, Fig. 78C), might

---

[24] Nevertheless such apparently chiasmate autosomal bivalents *do occur* in some undetermined but probably small percentage of the meioses of male *Drosophila melanogaster* (Cooper, 1949). Their occurrence does not appreciably alter the validity of Darlington's analysis, inasmuch as such bivalents are not the rule, as they almost certainly represent a persistence to metaphase of a particular pattern of prophasic somatic pairing first discovered by Kaufmann (1934) in the neuroblasts of *D. melanogaster,* and since the evidence suggests that, like the gonial chiasmata (cf. pp. 21–22), they most probably *do not* represent the cytological consequences of crossing over. For an extensive evaluation of these rare meiotic chiasmata and their theoretical significance, see Cooper (1949).

provide a cytological criterion by which this configuration of chiasmata could be recognized. However, this loop has never been recorded from cytological material. All the evidence for the reciprocal chiasma hypothesis has been shown to be sus-

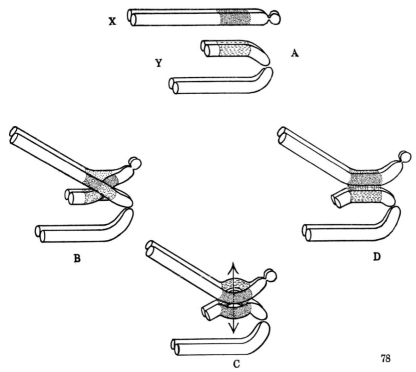

Fig. 78. Possible modes of meiotic conjunction of X and Y chromosomes in *Drosophila melanogaster*. A: Homologous pairing segments of X (stippled) and Y^S (broken lines). B: Bivalent having a single chiasma in the pairing segments. C: Bivalent with two reciprocal chiasmata in the pairing segments. The arrow passes through a loop formed by the reciprocal chiasmata. D: Bivalent formed without chiasmata. Schematic representation of a binding material between the homologous pairing segments.

ceptible to alternative interpretations, and several lines of disproof have been brought to light (Cooper, 1944, 1945, 1946). Indeed the simplest interpretation compatible with the facts known today appears to be that X and Y meiotically conjoin over their short, mutually homologous lengths by the very same mechanism that causes a non-chiasmate yet paired association of homologous autosomes. The paired regions therefore have all four chromatids in parallel association (Fig. 78D). When X pulls away from Y, a thread of connecting material frequently extends between the previously conjoined homologous regions, and this is frequently also the case for the autosomes. Whether this thread represents the remnants of a binding material or common matrix is not known, but this possibility should be borne in mind and is represented hypothetically in Fig. 78D.

From all this it may be concluded that in the male Drosophila so far investigated the cytological structure of both autosomal and sex chromosomal bivalents usually corresponds with non-crossover configurations. To this extent the cytological evidence is in complete agreement with the genetically observed absence of crossing over in the male. It should perhaps be noted that the very rare (ca. $2-8 \times 10^{-4}$) genetically detectable exchanges which do occur between the sex chromosomes or between homologous autosomes in male Drosophila are most probably the result of mitotic rather than meiotic events (see discussions and literature lists in Bauer, 1937; Cooper, 1944). In this connection Philip (1944) has briefly reported chromatin bridges in a single cyst of first spermatocytes at anaphase in a male of *D. subobscura*. These she interprets as direct evidence of crossing over. Since, from her observation, ten of the thirty-two anaphases would have to be interpreted as having come from nuclei in which crossing over occurred, another interpretation seems more likely. Such cysts are not unknown in the spermatogenesis of *D. pseudoobscura* (unpublished observations), but in these cases seem due to an abnormal "stickiness" of the chromosomes—perhaps artifactual. Because genetic exchanges are so rare in males, it is very improbable that such a burst of crossing over would occur in a single cyst.

3. The belief that meiosis in the Drosophila male is abnormal, since crossing over no longer regularly occurs in it, has long been coupled with the notion that meiosis in the Drosophila female is probably wholly normal. Were these views correct, the problem of deriving the male meiotic sequence from the usual course of meiosis would at least be well defined. Unfortunately it cannot yet be said that the meiotic processes of female Drosophila are "entirely normal," as White (1945), among others, has done. The reason is a twofold one.

First, although those who have studied meiosis in Drosophila females are agreed on the regular occurrence of leptotene-, zygotene-, and pachytene-like threads in the early prophase, none of these studies has given a wholly acceptable and satisfactory account of post-pachytene configurations where alone the behavior of chiasmata can be properly analysed. This is, of course, principally owing to the peculiar difficulties in working out the nuclear cytology of such refractory material, but the fact remains that too few details of critical stages are yet known to evaluate those leading directly to the metaphase bivalent. As a matter of fact, the "diaphase" phenomena described by Guyénot and Naville (1929, 1933) are highly aberrant if correctly portrayed.

Second, Darlington's (1934, p. 117) conclusion that oögenic meiosis is normal and conditioned by chiasmata, inasmuch as pachytene pairing occurs in the female, is almost assuredly wrong. Genetic evidence (Sturtevant and Beadle, 1936; Cooper, 1945) shows that normal segregation occurs in female Drosophila even when crossing over is suppressed. Furthermore the few figures that have been published of bivalents in females (Guyénot and Naville, 1929, Figs. IVh, i, j, k; 1933, Fig. IV; Naville 1932, Fig. 11(16), pl. 16, Fig. 75, for Calliphora) show no recognizable criteria of chiasmata whatsoever, even though at least some of the bivalents figured almost certainly contain crossover chromatids. Whether this means that chiasmata are not cytologically detectable (e.g., as cross configurations) in the bivalents of Drosophila females is not certain because too little study has been made of tetrads in the female. If it does prove to be the case, then the arguments given earlier (p. 50 et seq.) for the agreement of bivalent structure with the absence of crossing over in the male would be largely or completely deprived of their force.

In any event this much seems clear: Neither male nor female Drosophila require chiasmata for the metaphase association of their homologous chromosomes or for normal segregation, even though bivalent formation is still necessary for meiotic segregation as in most other organisms. The mechanism of metaphase association is probably the same in the two sexes and, since it requires neither the persistence nor the presence of chiasmata, has made possible the genetic loss of crossing over in the male. In other words, this type of meiotic association probably preceded the loss of crossing over in the male. It thus seems possible, as argued above (p. 49), that the origin of the meiotic and cytological differences between male and female Drosophila lies in the genetic suppression of crossing over in the male sex. The suppression of crossing over in the male may be viewed as having brought with it an evolutionary deletion of most or all of the meiotic phenomena in spermatogenesis required only for crossing over.

Such an argument in turn prompts the question of whether non-chiasmate association of homologues in Diptera is a consequence of their unique faculty for "somatic pairing," as suggested by Darlington (1934), among others. As Bauer (1939) and Wolf (1941) have pointed out, this is probably not the case, for today a fairly large number of unrelated organisms (certain bugs, mantids, lepidopterans, a scorpion, a mite, etc.—see Cooper, 1944) are known, none of which show somatic pairing, yet all of which form chiasma-free bivalents at meiosis in one or the other sex. Those interested in the phylogenetic aspects of the evolution of meiosis in the Diptera should consult Wolf's (1941) fine paper.

LITERATURE CITED

ABOIM, A. N. 1945. Développement embryonnaire et post-embryonnaire des gonades normales et agamétiques de *Drosophila melanogaster*. *Rev. suisse zool.* **52**:53–154.

BAKER, J. R. 1945. *Cytological Technique.* 2nd ed.   Methuen, London.

BAUER, H. 1931. Die Chromosomen von *Tipula paludosa* Meig. in Eibildung und Spermatogenese. *Z. Zellforsch. u. mikroskop. Anat.* **14**:138–193.

BAUER, H. 1937. Cytogenetik. *Fortschr. Zoöl.*, N.F. **2**:547–570.

BAUER, H. 1939. Chromosomenforschung (Karyologie und Cytogenetik). *Fortschr. Zoöl.*, N.F. **4**:584–597.

BELAR, K. 1928. Die Cytologischen Grundlagen der Vererbung. *Handb. Vererbungsw.* **1**(B)5. Borntraeger, Berlin.

BRIDGES, C. B. 1916. Non-disjunction as proof of the chromosome theory of heredity. *Genetics* **1**:1–52, 107–163.

BRIDGES, C. B. 1927. Constrictions in the chromosomes of *Drosophila melanogaster*. *Biol. Zentr.* **47**:600–603.

BRIDGES, C. B. 1928. The chromosomes of *Drosophila melanogaster*. *Coll. Net* **3**:10–11.

BRIDGES, C. B. 1935. Salivary chromosome maps. *J. Heredity* **26**:60–64.

COOPER, K. W. 1944. Analysis of meiotic pairing in Olfersia and consideration of the reciprocal chiasmata hypothesis of sex chromosome conjunction in male Drosophila. *Genetics* **29**:537–568.

COOPER, K. W. 1945. Normal segregation without chiasmata in female *Drosophila melanogaster*. *Genetics* **30**:472–484.

COOPER, K. W. 1946. The mechanism of non-random segregation of sex chromosomes in male *Drosophila miranda. Genetics* **31**:181–194.

COOPER, K. W. 1948. The evidence for long-range specific attractive forces during the somatic pairing of dipteran chromosomes. *J. Exptl. Zoöl.* **108**:327–336.

COOPER, K. W. 1949. The cytogenetics of meiosis in Drosophila. Mitotic and meiotic autosomal chiasmata without crossing over in the male. *J. Morphol.* **84**:81–122.

DARLINGTON, C. D. 1931. Meiosis. *Biol. Rev.* **6**:221–264.

DARLINGTON, C. D. 1934. Anomalous chromosome pairing in the male *Drosophila pseudoobscura. Genetics* **19**:95–118.

DARLINGTON, C. D. 1939. *The Evolution of Genetic Systems.* Cambridge University Press, Cambridge.

DOBZHANSKY, TH. 1930a. Translocations involving the third and the fourth chromosomes of *Drosophila melanogaster. Genetics* **15**:347–399.

DOBZHANSKY, TH. 1930b. Cytological map of the second chromosome of *Drosophila melanogaster. Biol. Zentr.* **50**:671–685.

DOBZHANSKY, TH. 1931. Translocations involving the second and fourth chromosomes of *Drosophila melanogaster. Genetics* **16**:629–658.

DOBZHANSKY, TH. 1932a. Cytological map of the X-chromosome of *Drosophila melanogaster. Biol. Zentr.* **52**:493–509.

DOBZHANSKY, TH. 1932b. Studies on chromosome conjugation. I. Translocations involving the second and the Y-chromosomes of *Drosophila melanogaster. Z. indukt. Abstamm.-u. Vererbungslehre* **60**:235–286.

DOBZHANSKY, TH. 1934. Studies on hybrid sterility. I. Spermatogenesis in pure and hybrid *Drosophila pseudoobscura. Z. Zellforsch. u. mikroskop. Anat.* **21**:169–223.

DOBZHANSKY, TH. 1935. *Drosophila miranda,* a new species. *Genetics* **20**:377–391.

DOBZHANSKY, TH. 1936. The persistence of the chromosome pattern in successive cell divisions in *Drosophila pseudoobscura. J. Exptl. Zoöl.* **74**:119–135.

DOBZHANSKY, TH., and C. EPLING. 1944. Taxonomy, geographic distribution, and ecology of *Drosophila pseudoobscura* and its relatives. *Carnegie Inst. Wash. Pub.* **554**:1–46.

DOBZHANSKY, TH., and C. PAVAN. 1943. Chromosome complements of some South Brazilian species of Drosophila. *Proc. Natl. Acad. Sci.* **29**:368–375.

DUNCAN, F. N. 1930. Some observation on the biology of the male *Drosophila melanogaster. Am. Naturalist* **64**:545–551.

FROLOVA, S. L. 1931. Spermatogenesis and oogenesis in *Drosophila melanogaster. Zhur. eksp. Biol.* **7**:368–376. [In Russian.]

GEIGY, R. 1931. Action de l'ultraviolet sur le pôle germinal dans l'œuf de *Drosophila melanogaster.* (Castration et mutabilité.) *Rev. suisse zool.* **38**:187–288.

GEIGY, R., and A. N. ABOIM. 1944. Gonadentwicklung bei Drosophila nach frühembryonalen Ausschaltung der Geschlechtszellen. *Rev. suisse zool.* **51**:410–417.

GLEICHAUF, R. 1936. Anatomie und Variabilität der Geschlechtsapparates von *Drosophila melanogaster. Z. wiss. Zoöl.* **148**:1–66.

GLOOR, H. 1943. Entwicklungsphysiologische Untersuchungen an den Gonaden einer Letalrasse (lgl) von *Drosophila melanogaster. Rev. suisse zool.* **50**:339–394.

GOWEN, J. W., and E. H. GAY. 1933. Gene number, kind, and size in Drosophila. *Genetics* **18**:1–31.

GRIFFEN, A. B., and W. S. STONE. 1940. The second arm of chromosome-4 in *Drosophila melanogaster. Univ. Texas Pub.* 4032:201–208.

GUYÉNOT, E., and A. NAVILLE. 1929. Les chromosomes et la réduction chromatique chez *Drosophila melanogaster.* (Cinèses somatiques, spermatogenèse, ovogenèse.) *Cellule* **39**:25–82.

GUYÉNOT, E., and A. NAVILLE. 1933. Les bases cytologiques de la théorie du "crossing-over." Les premières phases de l'ovogenèse de *Drosophila melanogaster. Cellule* **42**:211–230.

HEITZ, E. 1933. Die somatische Heteropyknose bei *Drosophila melanogaster und ihre genetische Bedeutung. Z. Zellforsch. u. mikroskop. Anat.* **20**:237–287.

HINTON, T. 1942. A comparative study of certain heterochromatic regions in the mitotic and salivary gland chromosomes of *Drosophila melanogaster. Genetics* 27:119–127.

HINTON, T. 1946. The physical forces involved in somatic pairing in the Diptera. *J. Exptl. Zoöl.* 102:237–252.

HUETTNER, A. F. 1930a. Meiosis in *Drosophila melanogaster. Coll. Net* **5**:112–113.

HUETTNER, A. F. 1930b. Recent criticisms concerning meiosis in *Drosophila melanogaster. Science* **71**:241–243.

HUETTNER, A. F. 1930c. The spermatogenesis of *drosophila* [*sic!*] *melanogaster. Z. Zellforsch. u. mikroskop. Anat.* **11**:615–637.

JEFFREY, E. C. 1925. Drosophila and the mutation hypothesis. *Science* **62**:3–5.

JEFFREY, E. C. 1931. Cytological evidence as to the status of *Drosophila melanogaster. Am. Naturalist* **65**:19–30.

JEFFREY, E. C., and G. C. HICKS. 1925. Evidence as to the cause of so-called mutations in Drosophila. *Genetics* **7**:273–286.

KAUFMANN, B. P. 1933. Interchange between X- and Y-chromosomes in attached-X females of *Drosophila melanogaster. Proc. Natl. Acad. Sci.* **19**:830–838.

KAUFMANN, B. P. 1934. Somatic mitoses of *Drosophila melanogaster. J. Morphol.* **56**:125–155.

KERKIS, J. 1931. The growth of the gonads in *Drosophila melanogaster. Genetics* **16**:212–224.

KERKIS, J. 1933. Development of gonads in hybrids between *Drosophila melanogaster* and *D. simulans. J. Exptl. Zoöl.* **66**:477–509.

KOLLER, P. C. 1936. Structural hybridity in *Drosophila pseudoobscura. J. Genetics* **32**:79–102.

KOLLER, P. C. 1939. A new race of *Drosophila miranda. J. Genetics* **38**:477–492.

KOLLER, P. C., and T. TOWNSON. 1933. Spermatogenesis in *Drosophila obscura*. I. The cytological basis of suppression of crossing-over. *Proc. Roy. Soc. Edinburgh* **53**:130–146.

LEAGUE, B. B. 1930. The normality of the maturation divisions in the male *Drosophila melanogaster*. *Science* **71**:99.

LISON, L. 1936. *Histochimie animale. Méthodes et problèmes.* Gauthier-Villars, Paris.

MACKNIGHT, R. H., and K. W. COOPER. 1944. The synapsis of the sex chromosomes of *Drosophila miranda* in relation to their directed segregation. *Proc. Natl. Acad. Sci.* **30**:384–387.

METZ, C. W. 1926. Observations on spermatogenesis in Drosophila. *Z. Zellforsch. u. mikroskop. Anat.* **4**:1–28.

METZ, C. W., M. S. MOSES, and E. N. HOPPE. 1926. Chromosome behavior and genetic behavior in Sciara (Diptera). I. Chromosome behavior in the spermatocyte division. *Z. indukt. Abstamm.-u. Vererbungslehre* **42**:237–270.

MILLER, A. 1941. Position of adult testes in *Drosophila melanogaster*. *Proc. Natl. Acad. Sci.* **27**:35–41.

MORGAN, T. H. 1912. Complete linkage in the second chromosome of the male of Drosophila. *Science* **36**:719–720.

MORGAN, T. H. 1914. No crossing-over in the male of Drosophila of genes in the second and third pairs of chromosomes. *Biol. Bull.* **26**:195–204.

MORGAN, T. H., C. B. BRIDGES, and A. H. STURTEVANT. 1925. The genetics of Drosophila. *Bibl. Genet.* **2**:1–262.

MULLER, H. J., and T. S. PAINTER. 1932. The differentiation of the sex chromosomes into genetically active and inert regions. *Z. indukt. Abstamm. u. Vererbungslehre* **62**:316–365.

MULLER, H. J., and A. A. PROKOFYEVA. 1935. The structure of the chromonema of the inert region of the X-chromsome of Drosophila. *Compt. rend. acad. sci. URSS.* **9**:658–660.

MULLER, H. J., D. RAFFEL, S. M. GERSHENSON, and A. A. PROKOFYEVA-BELGOVSKAYA. 1937. A further analysis of loci in the so-called "inert region" of the X chromosome of Drosophila. *Genetics* **22**:87–93.

NAVILLE, A. 1932. Les bases cytologiques de la théorie du "crossing-over"; Étude sur la spermatogenèse et l'ovogenèse des Calliphorinae. *Z. Zellforsch. u. mikroskop. Anat.* **16**:440–470.

NEUHAUS, M. 1939. A cytogenetic study of the Y-chromosome of *Drosophila melanogaster*. *J. Genetics* **37**:229–254.

PANSHIN, I. B., and V. V. KHVOSTOVA. 1938. Experimental proof of the subterminal position of the attachment point of the spindle fiber in chromosome IV of *Drosophila melanogaster*. *Biol. Zhur.* **7**:359–380.

PHILIP, U. 1942a. Meiosis in Drosophila. *Nature* **149**:527–528.

PHILIP, U. 1942b. Spermatogenesis in Drosophila. *Drosophila Information Service* 16:65.

PHILIP, U. 1944. Crossing-over in the males of *Drosophila subobscura*. *Nature* **153**:223.

Philip, U., J. M. Rendel, H. Spurway, and J. B. S. Haldane. 1944. Genetics and Karyology of *Drosophila Subobscura*. *Nature* **154**:260–262.

Pontecorvo, G. 1944. Synchronous mitoses and differentiation, sheltering the germ track. *Drosophila Information Service* 18:54–55.

Prokofyeva, A. A. 1935. Morphologische Struktur der Chromosomen von *Drosophila melanogaster*. *Z. Zellforsch. u. mikroskop. Anat.* **22**:255–262.

Robertson, C. W. 1936. The metamorphosis of *Drosophila melanogaster*, including an accurately timed account of the principal morphological changes. *J. Morphol.* **59**:351–399.

Shen, T. H. 1932. Cytologische Untersuchungen über Sterilität bei Männchen von *Drosophila melanogaster* und bei F_1-Männchen der Kreuzung zwischen *D. simulans*-Weibchen und *D. melanogaster*-Männchen. *Z. Zellforsch. u. mikroskop. Anat.* **15**:547–580.

Sonnenblick, B. P. 1941. Germ cell movements and sex differentiation of the gonads in the Drosophila embryo. *Proc. Natl. Acad. Sci.* **27**:484–489.

Stella, E. 1936. Études génétiques et cytologiques sur *Drosophila immigrans* Sturt. *Rev. suisse zool.* **43**:397–414.

Stern, C. 1927. Ein genetischer und zytologischer Beweis für Vererbung im Y-Chromosom von *Drosophila melanogaster*. *Z. indukt. Abstamm. u. Vererbungslehre* **44**:187–231.

Stern, C. 1929. Untersuchungen über Aberrationen des Y-Chromosoms von *Drosophila melanogaster*. *Z. indukt. Abstamm. u. Vererbungslehre* **51**:253–353.

Stern, C. 1940. Growth *in vitro* of the testis of Drosophila. *Growth* **4**:377–382.

Stern, C. 1941. The growth of testes in Drosophila. I, II. *J. Exptl. Zoöl.* **87**:113–158; 159–180.

Stern, C., and E. Hadorn. 1938. The determination of sterility in Drosophila males without a complete Y-chromosome. *Am. Naturalist* **72**:42–52.

Stern, C., and E. Hadorn. 1939. The relation between the color of testes and vasa efferentia in Drosophila. *Genetics* **24**:162–179.

Stevens, N. M. 1908. A study of the germ cells of certain Diptera, with reference to the heterochromosomes and the phenomena of synapsis. *J. Exptl. Zoöl.* **5**:359–374.

Stevens, N. M. 1912 (1907). The chromosomes of *Drosophila ampelophila*. *Proc. VII Intern. Congr. Zoöl.* 380–381.

Sturtevant, A. H. 1921. The North, American species of Drosophila. *Carnegie Inst. Wash. Pub.* 301.

Sturtevant, A. H. 1944. *Drosophila pseudoobscura*. *Ecology* **25**:476–477.

Sturtevant, A. H., and G. W. Beadle. 1936. The relations of inversions in the X-chromosome of *Drosophila melanogaster* to crossing-over and disjunction. *Genetics* **21**:554-604.

Sturtevant, A. H., and Th. Dobzhansky. 1936. Geographic distribution and cytology of "sex-ratio" in *Drosophila pseudoobscura* and related species. *Genetics* **21**:473–490.

TIHEN, J. A. 1946. An estimate of the number of cell generations preceding sperm formation in *Drosophila melanogaster*. *Am. Naturalist* **80**:389–392.

WHITE, M. J. D. 1945. *Animal Cytology and Evolution*. Cambridge University Press, Cambridge.

WILSON, E. B. 1934. *The Cell in Development and Heredity*. 3rd ed., with corrections. Macmillan, New York.

WOLF, E. 1941. Die Chromosomen in der Spermatogenese einiger Nematoceren. *Chromosoma* **2**:192–246.

WOSKRESSENSKY, N. M. 1933. Maturation in the gametes of *Drosophila melanogaster*. *Biol. Zhur.* **2**:97–109.

WOSKRESSENSKY, N. M., and E. A. SCHEREMETJEWA. 1930. Die Spermiogenese bei *Drosophila melanogaster* Meig. *Z. Zellforsch. u. mikroskop. Anat.* **10**:411–426.

ZUJTIN, A. I. 1929. On the peculiarities of spermatogenesis in *Drosophila melanogaster*. *Bull. Bur. Genet. Leningrad* **7**:97–107.

2

The Early Embryology of Drosophila Melanogaster*

B. P. SONNENBLICK†

MATERIALS AND TECHNIQUES

The majority of larvae of *Drosophila melanogaster* raised at 25°C hatch from the embryonic envelopes between 20 and 22 hours after fertilization. If in such a rapidly developing organism critical cytological and embryological studies are to be made, it is obviously essential that good material and carefully maintained conditions be employed. For the benefit of those who may find it necessary at some time to prepare eggs and embryos of Drosophila for microscopical study, a rather detailed review of some of the techniques used by other investigators and by the writer will be presented here.

Since the uterus of a Drosophila female contains but one egg at a time (Nonidez, 1920), and since freshly deposited eggs may be at a level of development ranging anywhere from the period before pronuclear conjugation to just before the emergence of the larva (Huettner, 1923), various egg-collection methods have had to be devised in order to obtain eggs and embryos of known age. For certain precise cytological studies and for some ex-

*This work was done for the most part at Queens College, Flushing, New York, and at Columbia University, New York, during the tenure of a Guggenheim Fellowship. A debt of gratitude is due to Professor A. F. Huettner of Queens College for his encouragement of this study and for permission to use certain of his published photomicrographs (Figs. 14 and 20–23 in this text).

† The Newark College of Arts and Sciences, Rutgers University, Newark, N. J.

62

perimental embryological work, it is essential to know the age of each egg utilized by the investigator. To do this, virgins of both sexes are segregated and kept for the ensuing 4 or 5 days in vials containing fresh food media onto which some chips of yeast are added. Some time before the collection period is to begin, the unmated males and females are released into a vial containing either old food or a moistened bit of blotting paper on which is a meager amount of food. Usually, mating quickly follows. Later, the flies are etherized, and single females are placed in each of a series of vials. In every vial is then inserted a glass slide on which is placed a slip of dark blotting paper over whose surface is spread a thin film of a fermented banana yeast mash. The vials are put in an incubator which has a glass observation top and whose temperature is controlled. With the aid of a powerful hand lens, the observer can with great exactness determine the time of deposition of every egg. As each egg is laid, its position and the time of deposition are noted on an outline drawing. Under the conditions described, eggs are often laid in quick succession. It is not unusual, for example, for a female to lay from 8 to 10 eggs within a 20-minute period, thereby introducing a very small error in the age computation of the various eggs. However, the females are occasionally un-coöperative.

For the study of older embryonic stages and for most experiments in which large numbers of eggs are desired, the above general method of meticulously determining the age of every egg is impracticable. A number of mass-collection methods have been used. Parks (1936) put as many as 20 virgin males and females into vials containing "yeast-food spoons" to which some diluted acetic acid had been added. The eggs were fixed at half-hour intervals. Poulson (1937) used metal trays filled with the regular cornmeal, molasses, agar medium plus some powdered charcoal. He collected eggs at 2- to 4-hour intervals, thereby introducing a fairly large error of 1–2 hours in the timing, if the age of the organisms is calculated from the midpoint of the laying period. Kaliss (1939) poured onto the paper caps of half-pint bottles a medium consisting of 2 per cent agar dissolved in 20 per cent molasses. Some 40 pairs of flies were put into the bottles and permitted to oviposit for periods of 2 hours.

The writer has found the following method to be practicable and useful for the securing of relatively large numbers of eggs and embryos which are, on the whole, at similar stages of development. The material used in these studies was obtained from a vigorous, fertile Oregon-R strain of wild-type *Drosophila melanogaster* which had been sustained in mass cultures for several years and had consistently given from 94 to 97 per cent egg hatchability. Preparatory to the collection period, groups of 8–12 pairs of flies are kept in vials containing media on the surface of which some bits of dried yeast are added. After 4–6 days the flies are released from the vials into bottles containing strips of banana skin. The inside surface of the skin of an over-ripe banana is an attractive substrate on which eggs can be deposited and normal development can occur. The surface is sufficiently moist to furnish the developing zygotes with the required moisture, but occasionally, to guard further against desiccation, the strips are placed in a moist chamber made by lining the inside of a Petri dish with dampened filter paper. On the banana skin, counts of egg deposition and subsequent hatchability may be made by the investigator, and furthermore, if it is so desired, the strips with the eggs may be subjected to agents such as X-rays.

The flies are first permitted to lay eggs for 1 hour, and the work is continued only if many eggs are deposited during this preliminary period, for then it may be assumed that most, if not all, of the females have deposited the eggs of uncertain ages which are present in their uteri at the beginning of the collection period. If many eggs are deposited, the strip containing them is discarded, and a fresh strip introduced into the bottles. The second laying period does not exceed 30 minutes, and the age of the zygotes is calculated from the midpoint of this period, the maximum error introduced thereby being small. After the laying period, the strips are placed in an incubator regulated to function at 25 ± .5°C. Observations on living, dechorionated eggs and on groups of sectioned eggs and embryos indicate that most of the material collected in this manner is at the same level of development.

To observe the developmental processes occurring in the living egg, the opaque chorionic membrane must be removed. The eggs are placed on dark blotting paper which is moistened with 0.65 per cent salt solution. In this fashion not only are the organisms kept moist during the process of removal of the envelopes but also the rough fibers of the blotting paper help to keep the eggs in position. Stroking of the chorion with the blunt end and the sides of an ordinary dissecting needle or with a camel's-hair brush causes the chorion to rupture. The dechorionated eggs, or embryos, are then placed in depression dishes containing salt solution or mounted on glass slides. Observations with high-power magnification can be made on zygotes so mounted and covered with thin coverslips, the latter supported by stiff bristles or by bits of clay. Because of evaporation, it is necessary, as Poulson (1937) found, to replenish at intervals the salt solution surrounding the organisms. Abnormalities of development are occasionally met with; these are probably, in the main, the result of injuries induced during the process of rupturing and removing the outer membranes. A number of methods for removing the chorion and for the preparation of wet mounts for study are described by Child and Howland (1933), Huettner and Rabinowitz (1933), Parks (1936), Poulson (1937), and Kaliss (1939). Rabinowitz (1941a) found that eggs placed in distilled water, in 33 per cent sea water, or in $0.33M$ or $0.5M$ glycerine can develop normally. Recently, Hill (1945) reports that, when eggs of *D. virilis* are placed for 2 minutes in a 3 per cent solution of sodium hypochlorite (or in commercial Clorox) the chorions are removed without apparent harm to the embryos and without any delay in hatching.

As the early developmental processes were followed in living, dechorionated eggs, sketches were made by the writer and timings taken. From observations on numerous eggs, it was possible to obtain a fairly accurate picture of the sequence of the various events and of the relationships of the diverse furrows and grooves evident in the zygotes. However, it is essential to supplement the observations on living organisms with studies on fixed and prepared material. Eggs and embryos incubated for known periods of time were, therefore, killed by puncturing with carefully sharpened steel needles while they were immersed in a fixative.

To ensure a quick, uniform penetration of the fixing fluid such as will effectively bring out the nuclear details, it is necessary to puncture the egg envelopes. Even with the chorion removed, it is still essential to puncture the vitelline membrane in order to obtain satisfactory penetration of the killing agent. The fixative consistently employed throughout the writer's work was Huettner's modification of the formol-alcohol-acetic acid mixture

(6:16:1, plus 30 parts water) used by Kahle (1908) in his studies on Miastor. It is not necessary to prepare a fresh solution of the fixative every time it is to be used, and material may be allowed to remain in it for at least several weeks.

Punctures of the early eggs have to be made with special care because of the fluidity of the egg contents. A successful puncture will permit only a delicate strand of oöplasm to escape. The fixative (F.A.A.) penetrates quite rapidly, for in eggs which are punctured midway between the poles, a whitish opaque area is seen to pass anteriorly and posteriorly within 5 or 6 minutes after the operation. When young eggs are fixed, the punctures are made in the posterior portion of the eggs in order to prevent the disturbance and distortion of the maturation or early cleavage figures which lie anteriorly. Older eggs and embryos should be punctured in diverse regions, since the embryonic organization in the region of any one puncture may be more or less disturbed. The study of embryos of the same age which have been pierced in different areas will thus enable one to construct a composite picture of a particular developmental stage.

Hot F.A.A. (70°C and 80°C) has been used in some instances to fix embryos. The heated fluid almost immediately coagulates the embryonic contents. However, when embryos so treated are sectioned, it is found that, although the contours of the individual cells stand out strikingly, making these preparations very valuable for topographical studies, the nuclei stain poorly. Fixing with hot fluids is inadvisable if the investigator is interested in observing nuclear details or the achromatic figure but is excellent for some aspects of embryogenesis.

Among the investigators who have reported their observations on Drosophila embryos, Parks and Poulson make it a practice to dechorionate eggs before fixation. The present author's experience suggests that the advantages gained by not removing the chorionic membrane may outweigh possible advantages obtained by its removal. It is not fully proved as yet that the Clorox treatment does not injure the organism. Furthermore, eggs without chorions are quite slippery and difficult to puncture; the punctures may therefore be too drastic. Finally, dechorionated eggs are troublesome to orient properly when embedding. It is much simpler to puncture eggs which possess the outer membranes, and the anterodorsal filaments of the chorion provide markers which facilitate orientation in the paraffin.

After fixation the eggs are dehydrated by passing them, at 5- to 10-minute intervals, through a graded series of alcohols and then clearing in xylol. Before being transferred to 95 per cent alcohol, they are placed in an alcoholic solution of eosin or erythrosin to make them easily visible and so further aid in orientation. Embryos were cut transversely, horizontally, and sagittally at a uniform thickness of 8 μ. The yolk spheres and the chorion were no impediments to successful sectioning. The stains tested were gentian violet, Harris's haematoxylin, Mayer's haemalum, and Heidenhain's iron haematoxylin, the latter sometimes counterstained with eosin. The finest results were obtained with Heidenhain's stain (with no counterstain), and this was consistently employed throughout. The technique was simple: mordant with 4 per cent iron alum for 1 hour, wash for several minutes, and then stain for 45–90 minutes. Destaining in 2 or 4 per cent iron alum must be done carefully, for sections which are treated as described destain rapidly. Poulson (1945) employs a modification of Bodian's method of silver impregnation which gives interesting polychromatic effects, staining nerve fibers, many cytoplasmic inclusions, nucleoli, and chitin black, while yolk spheres are pinkish to light red, nuclear material exclusive of nucleoli appears deep red, and the cytoplasm generally has a pinkish to faint purplish tinge. Details of nuclear structure are not well preserved.

Approximately one thousand eggs and embryos in stages ranging from the maturation divisions and entrance of the sperms to hatching of the young larva were fixed and sectioned for the present work. A few drawings are shown, but an extensive series of photomicrographs of organisms at the various developmental levels comprises the main pictorial material. The pictures have served as a framework on which to base the account, but at appropriate intervals critical discussions concerning certain features of embryogenesis which are of interest to the geneticist and the embryologist have been introduced. It is hoped that this method of presentation, utilizing a series of photomicrographs to depict as clearly as possible the rapid morphological changes which take place as the relatively simple egg becomes ever more complex and heterogeneous in structure, will aid the reader in following the complicated changes which occur in development.

THE UNFERTILIZED EGG, MATURATION, AND FERTILIZATION

The position of the egg of Drosophila within the uterus, described by Nonidez (1920), is found to obey the so-called law of orientation of Hallez (1886). The cephalic and caudal poles, the right and left sides, and the dorsal and ventral surfaces are orientated coincidently with those of the mother. The axes of the egg cell will, furthermore, correspond with those of the presumptive embryo; the embryonic axes are therefore determined in the egg before deposition. This generalization of Hallez, holding for Drosophila and many other insects, is not universal, and Richards and Miller (1937) and Johannsen and Butt (1941) cite a number of exceptions to it.

The egg is bilaterally symmetrical, and distinction between the dorsal and ventral surfaces is indicated by distinctions in curvature, for the dorsal side is flattened while the ventral side is somewhat convex. These differences in curvature are visible in living eggs and are readily observed in well-cut sections of eggs and embryos. The anterior and posterior poles can be distinguished by certain differentiations, the micropylar opening for the passage of sperms, for example, being always anteriorly and somewhat dorsally located. Paired delicate filaments which are extensions of the chorion and which vary in number in different species (Sturtevant, 1921) are situated on the dorsal surface not far from the anterior end of the egg. The dimensions of the egg are variable; an average length is 420 μ and width 150 μ.

The mature egg is invested by two envelopes, an inner homogeneous vitelline membrane, probably secreted by the egg itself, and an outer tough, opaque chorion. Through the transparent, structureless vitelline membrane, which is closely apposed to the protoplasmic contents, one may directly observe certain of the developmental processes. The chorion is not structureless but rather ornamented with hexagonal and pentagonal figures, although examination of wet mounts of the D. melanogaster membrane indicates that occasional quadrilateral and septagonal figures are formed. These ornamental markings have been considered to be the impressions of the ovarian follicle cells on the originally soft, plastic membrane secreted by the follicular epithelium. A third membrane is observed in prepared material. The fixed egg shrinks slightly, disclosing a surface which Huettner (1923) calls a "plasma membrane." Even at present we have virtually no knowledge of the factors regulating permeability at the surface of the egg. We do know that in order to ensure rapid and delicate fixation the vitelline membrane has to be punctured with a needle.

From the investigations of Huettner (1923, 1924, 1927) we secured the first description of the internal structure of the Drosophila egg. His observations have been confirmed by subsequent studies. Throughout the cytoplasm are scattered yolk spheres of different sizes, each surrounded by a clear area which is probably the result of shrinkage. These inclusions appear nearly black when stained with Heidenhain's iron haematoxylin. A large number of vacuoles which vary in diameter and which take no stain are also observed in the cytoplasm. Whether these occur solely from tangential cutting or are actual vacuoles is difficult to determine. At the periphery of the egg is a thin protoplasmic layer, finely granular and free from yolk inclusions, which is continuous with the large inner cytoplasmic mass in which the metaplasmic materials lie. Except at the cephalic and caudal poles, where it broadens considerably, the cortical layer forms a delicate boundary about the eggs. To this peripheral layer or periplasm, Weismann (1863) gave the name "Keimhautblastem," a term still commonly encountered in the literature, for in this region he saw large clear spots which he considered to be spontaneously derived and which he believed were to become the nuclei of the future blastodermic cells. This view he later retracted.

The posterior polar plasm in the mature oöcyte and in the young fertilized egg is an area of distinct appearance, visibly different from the other cortical regions of the egg, for this area contains an aggregate of darkly staining granules. These granules are incorporated in the germ cells in a haphazard manner, their distribution not being precise, and varying from cell to cell. In fact, Rabinowitz (1941a) has figured germ cell primordia which are completely cut off from the main cytoplasmic mass and which do not contain any polar granules at all. He notes, further, that some granules may remain behind in the posterior oöplasm and never become included in the germ cell primordia. These observations are interesting in the light of Hegner's interpretation (see his book, *The Germ-cell Cycle in Animals*, 1914) of the polar granules found in various insect orders as "germ cell determinants." The evidence in Drosophila is hardly such as to support Hegner's belief. The nature of the polar granules is as yet undetermined, although Huettner (1923) presents some evidence which leads him to conclude that they are neither mitochondria nor minute yolk particles. The granules are still of interest, however, because they are present in many insects and in some manner, which may or may not be significant, are usually associated with the formation of the primordial germ cells. The posterior cytoplasmic region, wherein the polar granules are consistently differentiated, is necessary for the formation of germ cells, but it hardly follows that the granules are organ-forming.

Fertilization of the egg occurs in the uterus. As indicated by Nonidez (1920), the egg occupies most of the uterine cavity, with the paired dorsal filaments remaining in the oviduct, the appendages thus precluding the passage of sperms into the oviduct after the release of the gametes from the seminal receptacles. A minute space is left in the forward portion of the uterus, in which area the sperms are released for fertilization. The micropylar opening at the anterior end of the egg is directed toward the opening of the ventral receptacle to receive the sperms. Little or no space remains between the outer envelope of the egg and the uterine walls, tending to prevent the escape of the gametes from the cephalic pole where penetration of the egg takes place.

Polyspermy is the general rule in Drosophila, each egg usually being penetrated by more than one spermatozoön, some receiving thirty or more sperms. The latter condition is probably an aberrant one, since about five to eight sperms are typically found in eggs. Instances such as those described and drawn by Huettner (1927) in which large numbers of spermatozoa have passed into the egg are rare and undoubtedly abnormal; such eggs are incapable of normal development. Even after the exposure of sperms to heavy doses of X-rays, the gametes are apparently unhindered in their functioning, since treated sperms penetrate the ova and take part in initiating development (Sonnenblick, 1940; Kaufmann and Demerec, 1942). As in most insects, the sperm tails enter the eggs but degenerate, as do those sperm heads other than the one which resolves itself into the vesicular pronucleus. The first meiotic division is in process at the time the sperms enter the egg. The chromosomes are oriented in a metaphase plate on a spindle located close to the dorsal surface of the egg in a clear protoplasmic island situated approximately one-third of the distance from the anterior pole. This area can be readily recognized, since it is virtually free from yolk inclusions which surround the island. Upon the entrance of the spermatozoa, the first division continues (Fig. 1).[1] It may be noted at this point that in the maturation-division figures no centrioles or asters are present in distinction to the later cleavage mitoses as well as those observed in embryonic tissues and organs, where such cytoplasmic differentiations are clearly and consistently observed. The centrioles seen in all mitoses from the first

FIG. 1. Anaphase of the first meiotic division shortly after entrance of spermatozoa.

cleavage on are, in all likelihood, derived from the division center or centers originally associated with the spermatozoön which contributed the fusion nucleus. In the female the precise and speedy meiotic chromosomal movements occur independently of centriole or aster, seemingly making no contribution to those division centers observed in the ensuing cleavages. The chromosomes separate on the elongating meiotic spindle, and when telophase is reached the sperm head which is to become the male pronucleus is in a clear area close to the region where meiosis is taking place (Fig. 2). The spindle, oriented almost perpendicularly to the long axis of the egg, usually is pointed in the direction of the sperm head close by. The chromosomal complexes which separate during the first meiotic division do not pass through any interkinetic stage but, after a few minutes, pass directly upon the second polar spindles (Fig. 3). The spindle fibers are distinct, and the arrangements of the spindles are such that they lie

[1] Because they were rarely seen and because of their size, the minute fourth chromosomes are not indicated in the drawings.

more or less in a single straight line, tandem fashion. At telophase of the second division four haploid chromosomal complements are observed, each complement forming a compact clumped mass (Fig. 4). The innermost of the four groups, that is, that complement furthest from the surface of the egg, will become the

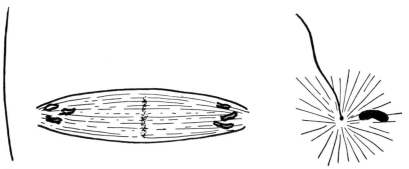

FIG. 2. Telophase of the first meiotic division. The spindle has elongated somewhat, and the midbody is noted at the center of the spindle. The midbody disappears during the second division. The detached sperm head, the centriole and aster that form about it, and a portion of the sperm tail are in an island near the spindle.

female pronucleus. The other three are the polar body nuclei; these are not extruded from the egg but remain peripherally disposed in the anterodorsal island and there eventually disintegrate. A recent detailed description of the changes in and the fate of the polar body chromosomes has been given by Rabino-

FIG. 3. Spindles of the second meiotic divisions with separating haploid complements.

witz (1941a). He has followed the history of the polar body nuclei to their fragmentation and disappearance at the time when the syncytial cleavage nuclei are dividing at the egg periphery. From his account it is evident that the chromosomes of the polar body nuclei play no role in normal development. The polar body chromosomes are absorbed and enter the cytoplasm of the egg.

A membrane is formed about the haploid complement which is to serve in fertilization as the female pronucleus.   Meanwhile,

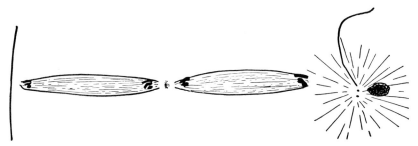

FIG. 4. Telophase of the second meiotic division with the innermost of the four haploid groups usually becoming the female pronucleus. The centriole has doubled, and the sperm head is slightly vesicular.

the compact sperm head has become more and more vacuolated and spherical, and at the time when the two pronuclei have approached and lie close to one another in the interior of the egg they appear as two faintly staining vesicles (Fig. 5). For a brief

FIG. 5. Pronuclei closely opposed with chromosomes gradually becoming apparent. Three polar-body nuclei are at the egg periphery.

period nothing of the nuclear structure can be observed, but as the pronuclei increase in volume their contained chromosomes become evident. This is the first time since spermiogenesis (when the spermatid is transformed into the mature spermatozoön) that the chromosomes of the male gamete can be observed. The chromatic material of the delicate, elongate sperm head is not discernible to the investigator, and as Kaufmann (1941) points out, attempts to determine the arrangement of the chromosomes within the Drosophila sperm head have failed.

FIG. 6. Pronuclei lengthening and spindle fibers appearing within the nuclei. Male and female chromosomal complements are separate.

The apposed pronuclei tend to become drawn out and pointed, and their membranes gradually grow faint and more difficult to observe. The spindle fibers now appear, originating within the gamete nuclei (Fig. 6). The maternal and paternal chromosomal complements which are to conjugate are readily distinguishable at this time and, as separate groups, pass onto the first cleavage spindle, which, as noticed by Huettner, is really composed of two units. Each of the spindle groups holds one of the haploid complexes (Fig. 7). The chromosomal halves disjoin, the elements introduced by each parent still separate, and it is only at the end of first cleavage that mixture of the chromosomes occurs. This instance of gonomeric grouping in early Drosophila ontogeny thus persists only for a limited time during the initial mitotic cleavage before the chromosomes become indiscriminately intermingled.

During the interval before conjugation, when the gamete nuclei are becoming vesicular and approaching one another, significant chromosomal movements may occur. It is at this period that nuclear regrouping, whether of entire chromosomes or sections of chromosomes, could and probably does take place.

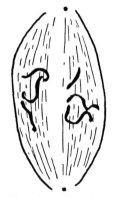

FIG. 7. The first cleavage spindle with the chromosomes still in separate groups (gonomery). Soon the chromosomes reproduce, disjoin, and pass to the poles, where they finally unite.

It would appear unlikely that any chromosomal rearrangements could be established within the compact sperm head, but as the sperm head becomes vacuolated the opportunity for mobility and possible regrouping becomes greater. An early and opportune time to study induced or spontaneous chromosomal alterations, many of which would act later as dominant lethals, would be at the period just before the gonomeric stage. Such material in sufficient numbers would, however, be difficult to obtain, for events at this time happen with great rapidity, and individual eggs would have to be carefully timed for fixation. This would involve large-scale and time-consuming work. There is also one possible hindrance toward making effective use of material so laboriously prepared. In three preparations of eggs at this level of development (gonomery), when the introduced complexes are still separate, the writer has noticed a tendency on the part of the paternal chromosomal complement toward compactness and clumping, while the group contributed by the female gamete is, at the same time, decidedly more diffuse. Should this tendency on the part of the sperm chromosomes be a consistent and regular phenomenon, it would mean that cytological analyses of alterations in these chromosomes at this early period could not be successfully carried out. Of course, salivary-gland chromosomes may be quite precisely studied for alterations in the hereditary material, although selection plays a role here, since alterations so analysed are the uneliminated remainder of those surviving the embryonic period.

Like other characters, meiosis in Drosophila appears to be genotypically controlled. Genotypically conditioned abnormalities of meiosis have been reported for many organisms, chiefly plants (see the classification and literature list in Darlington, 1937, pp. 404–405). There may be reduced chiasma formation and lack of metaphase pairing, or breakdown of the spindle and failure of the chromosomes to separate, or, if the chromosomes disjoin, the cell wall may not form between them. Gowen (1933) has reported that in *D. melanogaster* females homozygous for a recessive factor (*c*III*G*) in the third chromosome crossing over is greatly reduced in all the chromosomes, not alone in the chromosome in which the factor is located. This reduction in crossing over is, furthermore, accompanied by the formation of mature ova containing abnormal numbers of chromosomes. The effect of the gene is clearly sex-limited, since oögenesis and not spermatogenesis is influenced. [Dobzhansky (1934) has observed aberrant meioses resulting in the non-production of functional sperms during spermatogenesis of the hybrids between the races of *D. pseudoobscura*.] In triploid females crossing over is inhibited when three doses of *c*III*G* are present but is normal when at least one wild-type allele is in the complex.

Aberrant maturation phenomena have also been described by Wald (1936) in the claret mutant strain of *D. simulans*. Sturtevant (1929) had shown earlier that homozygous claret females laid the normal number of eggs, but many of these were inviable. Few of the zygotes which completed their development were normal, and exceptional males and females, gynandromorphs, and flies with mosaic areas were obtained in relatively large proportions. Wald noted one general type of abnormality in the first meiotic division of the homozygous claret females, namely, the chromosomes were separated in scattered groups at either end of an abnormally wide and distorted spindle. Failure of the second division and subsequent reorganization of the chromosome groups into many nuclei each with a few chromosomes (sometimes only one) were consistently noted. Instead of four nuclei resulting from the divisions, one of which would ordinarily serve as the female gamete nucleus, eggs with four to twelve nuclei were found. On the basis of her observations, Wald presents a hypothesis to account for the formation of gynandromorphs in the claret strain. These, according to her, may result from the addition of an X chromosome to either (*a*) a chromosome group which is XO in constitution, because of the earlier exclusion of an X chromosome from the female pronucleus, producing XX and XO types of cleavage nuclei, or (*b*) one pole of a normal XY complement, producing XXY and XY types of cleavage nuclei.

Excessive polyspermy may be a factor, as reported by Huettner (1927), in disturbing the meiotic processes. A spermatozoön or several sperms may enter the area where the divisions are taking place and set up independent spindles, disorganizing polar body formation in various ways. Meiotic aberrations are also induced by treatment of mature oöcytes with X-rays (Sonnenblick, 1940). The first spindle may not form at all, or occasionally only a part of the spindle is evident; multipolar spindle configurations rather than the normal bipolar arrangements are sometimes present. When no spindle or a half-spindle is formed, the chromosomes are scattered about in the clear protoplasmic island, and on the multipolar spindles they are distributed over the several poles. If as a result of disorganized divisions no female pronucleus is formed, the sperm head may resolve itself into its haploid complement of chromosomes, but no instance was ever observed in which a haploid sperm complement set up an independent spindle and began to cleave. Aberrations of meiosis, whether chromosomal or in the achromatic figure, are rarely ob-

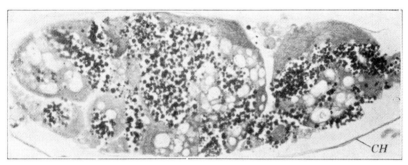

FIG. 8. Unfertilized egg 12 hours after deposition, showing characteristic necrotic appearance. 390×. *CH,* chorion, external covering of egg.

served in normal eggs but are induced in large numbers in irradiated oöcytes.

In eggs which do not receive sperms, the maturation divisions do not go to fruition, and subsequent differentiation does not occur. The egg becomes progressively disorganized, and at the time when fertilized, developing embryos have reached an advanced stage of organization the sterile egg is quite aberrant. Three hours after deposition the distribution of yolk is irregular throughout the cytoplasm, and one or more fissures will begin to traverse the organism from the surface. In the regions through which the fissures pass, particles of the egg mass begin to round up and form spherical droplets. Some hours later the sterile egg has a characteristic appearance, evident in Fig. 8, which shows an egg 12 hours after deposition by a virgin female. The egg is split into a number of parts, some of which are spherical, while the cytoplasm may be either homogeneous or fibrous in appearance and may contain deeply staining yolk granules or other lightly staining masses presumably also lipoidal in nature.

This typical necrotic appearance of the sterile egg is unmistakable once it has been observed.

EARLY CLEAVAGE

Figures 9 and 10 are longitudinal sections of two young, freshly deposited eggs. The ventral side is convex, and the dorsal side is flattened. The heavy outer chorionic envelope and the inner

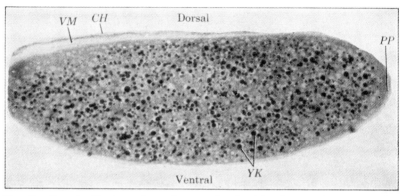

FIG. 9. Longitudinal section of an early, freshly deposited egg, showing the contours of the organism, a portion of the outer chorion and of the inner vitelline membrane, and a view of scattered yolk inclusions in the cytoplasm. 390×. *CH,* chorion; *PP,* posterior pole of egg; *VM,* vitelline membrane; *YK,* yolk.

delicate vitelline membrane can be seen in these and other preparations. Development proceeds normally after the removal of the chorion, but the organism dies when the inner vitelline membrane is damaged. Yolk inclusions in the form of spheres of different sizes are distributed throughout the general cytoplasmic mass. Some take an intense stain with iron haematoxylin and, because of shrinkage when fixed, often show in section a clear area about them. Other vacuoles, not containing any of the black spheres, are also evident in the egg. A thin layer of relatively yolk-free cytoplasm ("Keimhautblastem") is observed at the egg cortex, but at the poles (see posterior pole in Fig. 9) the peripheral layer broadens somewhat. Anterodorsally in Fig. 10 may be noticed a portion of the filaments which are extensions of the chorion. These filaments will be observed in many of the photomicrographs.

Figure 10 illustrates particularly well the position of the clear protoplasmic island, situated dorsally and approximately one-third of the distance from the fore-end of the egg, where the maturation divisions occur. In this egg, the second meiotic division is taking place, but the minute configurations are barely visible at this magnification. The male gamete nucleus is present in an island close by, which is at a level slightly anterior to the area wherein the maturation divisions are occurring. The position of the sperm head is variable with respect to that of the maturation spindles.

Figures 11 and 12 represent high-power views, in two different eggs, of the clear cytoplasmic islands, surrounded by yolk spheres, in which fertilization is taking place. The pronuclei, usually bearing haploid chromosomal complements, are in juxtaposition. In Fig. 11 the upper gamete nucleus is very likely the male pronucleus, for a faintly staining sperm tail is evident in rather close association with it. In Fig.

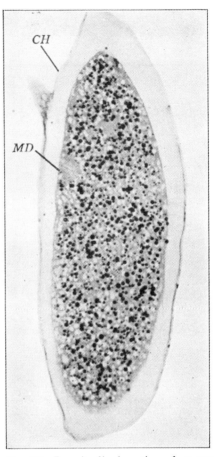

Fig. 10. Longitudinal section of an egg in which the second meiotic division is occurring in a clear anterodorsal island. The chorion and a portion of the chorionic filament are seen. 300×. *CH*, chorion; *MD*, site of maturation divisions.

12 the male gametic nucleus is at the left; part of a sperm tail may be seen just above it. Material at this stage is difficult to obtain, and in the few preparations which the writer has studied, the chromosomes in the paternal nucleus appear to be in a compact, almost clumped condition, whereas those of the maternal

gamete nucleus are decidedly more diffuse. It is very possible that because of their compact orientation in the sperm head the male chromosomes lag somewhat behind the female complement in becoming more diffuse.

The nuclear membranes become faint, and, as indicated in Fig. 12, the spindle fibers appear while the fusion nuclei progressively elongate. The first cleavage spindle is comprised of two units,

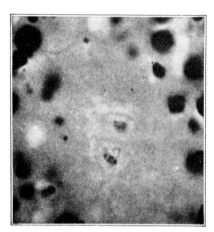 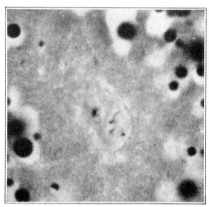

FIG. 11. Meeting of the male and female pronuclei. The sperm nucleus is uppermost, with a portion of the sperm tail close by. 1200×.

FIG. 12. Fertilization, gonomery. Spindle fibers of the first cleavage have appeared within the nuclei. Male and female chromosomal complexes are still separate. The male complex is at the left with a portion of the sperm tail evident. 1200×.

with the maternal and paternal complements remaining apart on the separate spindles (gonomery) until near the end of the first cleavage. This gonomeric relationship, reported for a number of organisms (Wilson, 1928), was initially noted in Drosophila by Huettner (1924). It is during the period when the gamete nuclei tend to become vesicular and approach one another that shifting and relocation of chromosomes and chromosome segments probably first occur. It is doubtful that such nuclear movements can take place in the attenuated sperm head.

Figure 13 is a longitudinal section through an egg showing the diploid early cleavage nuclei. During these initial cleavages, the nuclei may be noted in clear protoplasmic areas (the so-called protoplasmic islands) which are distributed in the egg interior.

At this time, the mitotic cleavage divisions are synchronous—the yolk nuclei and the pole cell nuclei, segregated posteriorly, are the first to depart from the simultaneous succession of divisions— and the nuclei, with the exceptions just noted, increase by powers of two. The velocity of these mitotic cycles, which include an interkinetic period during which the chromosomes, as recognizable entities, fade from view and give a negative Feulgen reaction is extraordinarily high (Huettner, 1933; Rabinowitz, 1941a). Approximately 10 minutes is required, at ordinary temperatures, for the completion of a mitotic cycle, that is, for the passage from one mitotic phase to the same phase in the next division. This is certainly one of the most rapid mitotic cycles, and perhaps the most rapid, reported in biological literature. A more extended discussion of this feature will be presented in a subsequent section.

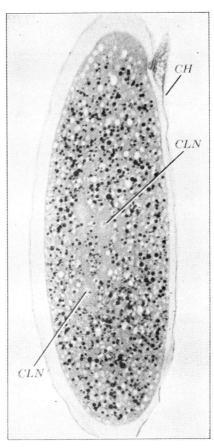

Fig. 13. Longitudinal section of an early egg (50–60 minutes old). The nuclei divide synchronously in the protoplasmic islands. Fifth cleavage is just beginning. 300×. *CH,* chorion; *CLN,* cleavage nucleus.

Throughout this early syncytial period of development, when the egg mass is yet unsegmented, the majority of nuclei, as noted above, divide in unison. By the time the 512-nuclei stage is reached, the segmentation products have arrived at the egg periphery, where they will divide, still in synchrony, at least three times more.

A portion of a section of an egg fixed while the seventh cleavage was in progress is indicated in Fig. 14. Some of the 64 ana-

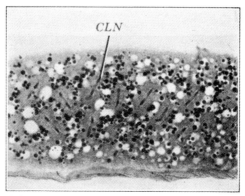

FIG. 14. Syncytial nuclei are dividing synchronously in the protoplasmic islands. Late anaphase figures. About 300×. *CLN,* cleavage nucleus. From Huettner (1933).

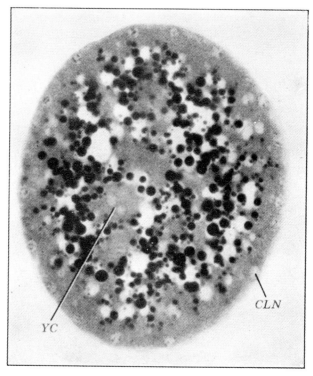

FIG. 15. Transverse section of an egg (1½ hours old), showing nuclei at the periphery during eighth cleavage. Not all nuclei reach the periphery; note several in the interior of the egg. 600×. *CLN,* cleavage nucleus; *YC,* yolk cells.

phasic figures, in separate islands, are seen; 128 daughter nuclei will result from this division. It will be observed that all the cleavage figures are at the same mitotic stage. Figures 15 and 16

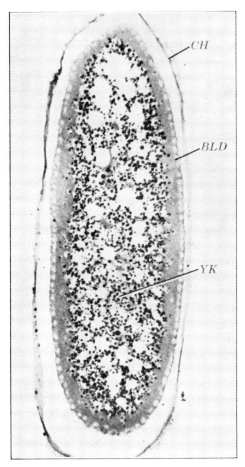

FIG. 16. Horizontal section of an embryo (eleventh and twelfth cleavages), showing the central yolk mass with a clear peripheral protoplasmic layer. The organism is still a syncytium with no cell membranes formed as yet. 300×. *CH,* chorion; *BLD,* blastoderm, nuclei and cells; *YK,* yolk.

show the nuclei, under different conditions, in the cortical layers of eggs which have been sectioned transversely and horizontally. When the nuclei reach the periphery, they do not cover the entire surface area (Fig. 15), and it is only after they have divided in the cortex that the configuration noted in Fig. 16 becomes evident.

It is possible to arrest the activity of nuclei which are in various stages of the mitotic cycle. The fixative penetrates the region punctured and gradually spreads throughout the egg. If the puncture is made at one end of the egg, the fixative will penetrate to the opposite region in about 10–12 minutes, during the course

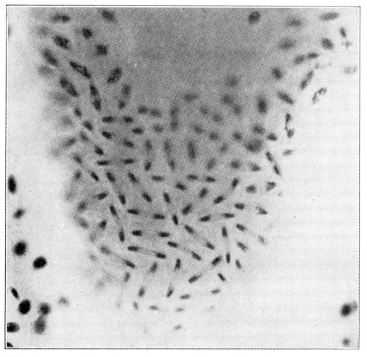

FIG. 17. A tangential section of an egg while twelfth cleavage is in progress. Most of the synchronous figures are in telophase or late anaphase, but a gradient of mitotic stages is noted as a result of diffusion of fixative through the egg. 600×.

of which the mitotic phases are differentially stopped. It is possible to note a transition or gradient extending away from the area punctured. Especially interesting and somewhat spectacular are the gradients observed in tangential sections of eggs fixed while the nuclei are in late anaphase in the region last reached by the advancing fluid. This is well shown in Fig. 17. Forward of this posterior area of late anaphases, early anaphases with the daughter halves barely separated and some metaphases can be seen. It should be remembered that this is but one section of the egg and that many other peripherally disposed cleavage figures can be noted in the remaining sections.

Reported observations concerning the number of synchronous cleavages which take place in the egg have been in close but not exact agreement, with 10, 11, and 12 successive mitoses having been noted by various investigators. Counts made by the writer on organisms which were in the true blastodermic stage, that is, no longer a syncytium but one with formed columnar cells, indicate that 12 cleavages occur. Had the nuclei increased in geometric progression for 11 divisions, one would expect to find 2048 nuclei in the blastoderm, and after 12 cleavages 4096 might be expected. Counts in favorable material show that some 3200–3500 nuclei are present after blastulation. Considering that not all the cleavage products are destined to become the nuclei of the definitive blastodermal cells, since yolk and pole cell nuclei are earlier set aside, then 12 cleavages ought to produce a blastoderm with more than 2048 but less than 4096 nuclei. That is what is found.

The incipient blastodermal stage with the nuclei in the unsegmented peripheral oöplasm (Figs. 15, 16) is the "blastema" stage, employing the terminology of Patten (1884). In this stage, a narrow nucleated layer of protoplasm surrounds the yolk mass and is characterized by the absence of cell walls. Such eggs as are illustrated in these figures can be considered as multinucleate cells, or, if each nucleus and its adjacent cytoplasm is regarded as a cell, as multicellular organisms. During the progress of the mitoses at the egg cortex, which Rabinowitz (1941a) has studied in detail, the outer protoplasmic layer broadens while the yolk becomes progressively more centralized. Compare the cortical layers in Figs. 13, 15, 16, 18. The nuclei are closely packed, occupying all the available space in the cortical layer.

YOLK NUCLEI

Before the concentration, centrally, of the yolk inclusions is completed, as evident in Fig. 18, islands of protoplasm are still noted in the yolk mass. Numbers of the cleavage products remain in these areas, never migrating to the peripheral protoplasm to become part of the blastema. Several nuclei in such interior islands may be noted in Fig. 15, at the time when the cortical nuclei are in prophase. The nuclei which are left behind are the primary yolk nuclei. Studies on various insect forms indicate that yolk nuclei can also be derived in another manner.

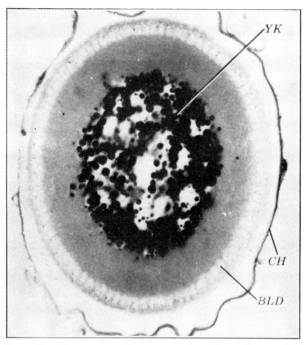

FIG. 18. Central yolk concentration is completed in embryo, which is about 2½ hours old. A wide, clear protoplasmic layer surrounds the yolk. Nuclei, originally spherical, are beginning to elongate. The wavy contour indicates that furrows have appeared which will press inward toward the yolk mass. Here the furrows are at the outer edge of the nuclei. Nucleoli are now visible. 600×. *BLD,* blastoderm, nuclei and cells; *CH,* chorion; *YK,* yolk.

All the cleavage nuclei may migrate to the egg periphery, after which a variable number may wander back to penetrate the yolk mass. The yolk nuclei are derived solely in the latter fashion in Sciara (DuBois, 1932; Butt, 1934). In this fly all the nuclei of the somatic line reach the cortical oöplasm and there undergo chromosome elimination as described by DuBois. During the incipient blastodermal stage, no nucleus is evident anywhere in the yolk mass, but later several nuclei migrate back into the yolk to play no part in subsequent development.

Other dipterous forms, including Phormia (Auten, 1934) and Drosophila (Rabinowitz, 1941b), derive their yolk nuclei in a combination of ways. Cleavage nuclei may be left behind in the yolk (primary yolk nuclei), while the remainder pass to the egg periphery; having reached the cortex, some nuclei may return from this area back into the yolk mass (secondary yolk nuclei). Rabinowitz has reported still another category, the so-called tertiary yolk nuclei, as did Lassmann (1936), working with Melophagus, which originate from pole cell nuclei segregated at the hindend of the organism. These, according to Rabinowitz, push between the posterior elongate blastoderm nuclei and return to the yolk mass. The literature on the yolk nuclei, or "vitellophags," as they are usually called, is quite extensive. For a comprehensive literature list and a summary in tabular form of the mode of vitellophag genesis in a series of forms belonging to ten insect orders, the reader is referred to the paper by Sehl (1931).

The reason why certain of the cleavage nuclei play the role of vitellophags in Drosophila and other insects is far from understood. There does not appear to be any microscopically discernible difference between the nuclei which become part of the germinal and somatic lines and those which remain in or enter the yolk mass.

Yolk nuclei have been observed to divide mitotically (Johannsen and Butt, 1941) as well as undergo amitosis before their disintegration (Auten, 1934; Rabinowitz, 1941b). To the vitellophags has been ascribed the rather indefinite and uncertain function of liquefying the yolk and thus "rendering it readily assimilable by the embryo," while some insect embryologists have derived the midgut epithelium from vitellophags (Stuart, 1935). In Drosophila we have no means at present of knowing whether the yolk nuclei do assist in the processes of yolk resorption. In fact, the role of the yolk nuclei is not clear, for, despite Rabinowitz's observation of disintegrating yolk nuclei, there exists other evidence, shown to the writer by D. F. Poulson, that the yolk nuclei persist until such time as the gut wall encloses the yolk and that some of these remaining vitellophags possibly do play a part in the formation of the midgut epithelium. This is discussed by Poulson in Chapter 3.

CLEAVAGE PATTERNS AND MOSAIC DEVELOPMENT

Unfortunately, it is not possible in Drosophila to trace directly the movements and wanderings of the descendants of the first cleavage nuclei, as in certain cell-lineage studies in other organisms, and to determine thereby the precise morphological value of the early nuclei or cells in the building of the body. For certain problems in developmental genetics, particularly for the interpretation of conditions observable in different types of mosaics, it would be especially helpful were it feasible to follow and plot the formation of the embryo cell by cell so as to establish the relationship existing between the initial segmentation products and the structures ultimately derived from them. There is no possibility, however, of predicting the role to be played by the early nuclei, for despite the rhythmicity of the mitoses (Figs. 14, 17), the cleavage patterns are quite inconstant, varying from egg to egg. Cleavage patterns have been studied comprehensively by Parks (1936), who fixed and stained a great deal of material and then constructed models to assist him in following and interpreting nuclear migrations and the directions and angles made by the spindles with respect to the egg axes. Among the conclusions reached by Parks relative to cleavage pattern types are the following:

1. Beginning with the very first cleavage, there is no constancy in the cleavage pattern. This observation supports Sturtevant (1929), who suggested that the early divisions are not oriented in a fixed pattern. The orientation of the first cleavage spindle is indeterminate, random, arranged in different eggs at various angles to the egg axes. The spindles may be situated in anteroposterior, dorsoventral, and bilateral positions; they may, furthermore, assume varied intermediate positions. The first spindles are not centrally located but occur in the anterior half of the egg at varying levels along the long axis.

2. In the following cleavages, which occur in the egg interior, the orientation of the nuclei and of the spindles indicates continuing random, indefinite positional relationships with one another and with the long axis of the egg, although it is important to notice that the location of the first cleavage spindle will deter-

mine the position of the nuclei in the ensuing cleavages. Random movement and "drifting" of nuclei bearing genetically diverse complements (elimination of an X chromosome at first cleavage in a potential female, resulting in XX and XO cleavage products) may result in the intermingling of prospective male and female cells during the early cleavages.

3. Although the cleavage nuclei enter the peripheral protoplasmic layer in random fashion, more enter the cortex at the ventral surface. It is at the ventral surface that the germ band will later form.

4. The nuclei do not at first completely cover the egg surface (see our Fig. 15) but will divide in unison several times, becoming closely packed cortically. Considering that the nuclei may be genetically diverse, as from the loss of a chromosome or chromosome segment, and that they apparently enter the peripheral cytoplasm indeterminately, the mitoses at the egg surface are a mechanism for further spreading the mixed nuclei which soon, after cell formation, are to become part of the multicellular blastoderm. With blastulation completed, the stage is set for the processes of organogenesis.

It would seem that these observations and conclusions of Parks, which still require confirmation in certain details, provide some foundation for elucidation of the mechanism leading to the varied and distinct types of distribution of male and female tissue in sex mosaics (gynandromorphs). Knowledge of the indeterminate nature of the cleavage processes in Drosophila, including the factors of chance orientation of the initial cleavage spindle and the random movements and penetration of the segmentation products into the egg cortex to form the incipient blastoderm, is of aid in understanding this mechanism. Detailed analysis of a large number of gynandromorphs, produced for the most part by irradiating flies with normal chromosomal complements and then mating to individuals carrying recessive genes scattered along the X chromosome, indicates that the probability of finding sex mosaics with identical patterns is rather small (Patterson, 1931; Patterson and Stone, 1938). However, the tendency in many, and perhaps the majority, of the mosaics is toward a bilateral arrangement in the distributional pattern of male and female parts. Since the position of the first cleavage spindle, indeterminate though it is, will directly influence the cleavage patterns which follow, analysis of sufficient material containing first cleavage spindles should indicate a preponderance of spindles arranged bilaterally, or nearly so, rather than in other possible positions.

It is generally believed that a majority of gynandromorphs can be explained on the basis of an elimination of all or part of an X chromosome from a female cell, resulting in daughter groups one of which contains two X chromosomes while the other has either one or one plus a segment of an X chromosome. The latter cells may give rise to sectors of male tissue in otherwise female animals. However, certain cases are very complex and difficult to analyze; other possibilities of gynandromorphic origin, such as double fertilization or double elimination, must be assumed in interpreting these instances (Patterson and Stone, 1938).

Another factor functioning in apparently random fashion may be added to those mentioned by Parks to relate the indeterminate early developmental processes with the variable configurations noticed in the sex mosaics. We have seen above that not all the cleavage products arrive in the cortical layer, an indefinite number remaining behind in islands located in the yolk. In the early cleavages all the nuclei—those destined to remain in the yolk mass as well as those which are to play a part in bodily development—appear structurally identical. Transmission to each daughter nucleus of a complement identical with that found in the original fusion nuclei is the rule. Should genetic diversity, by whatever means, occur in early cleavage, it would be a matter of chance whether particular daughter nuclei bearing altered and unaltered chromosomal complements become vitellophags or the nuclei of blastodermal cells. If mixture of the different nuclei (XX or XO or X plus combinations) can occur in the early cleavages, if chance determines whether any one of these nuclei will persist in the yolk or will have a role in determining the external phenotype, and if the blastoderm nuclei are already in a mixed mosaic condition, it is hardly to be wondered that a wide assortment of gynandromorphic patterns can result. The cellular movements occurring during gastrulation, to be described later, add further complications to the attempts to analyze pattern types in Drosophila sex mosaics. Several large cell clusters invaginate, carrying into the interior material which is either destroyed at metamorphosis or, if persisting, forms internal tissue. Again, chance seems to determine whether certain cells will partake in the formation of imaginal discs which will give rise to surface tissues amenable to analysis by the geneticist.

Mosaics with quite irregular distribution of parts have been noted, and, most interestingly, Patterson and Stone (1938) have found small isolated patches of female tissue within male tissue derived from one imaginal disc. This proves that a disc can be of mixed origin, comprised of male and female cells; elimination in a late division in an imaginal cell will not account for these instances. Despite the indeterminate nature of most of the early developmental processes, checkerboard patterns with complex distribution of parts are seldom found. The first cleavage spindle must have significant directive capacity in determining the position of the ensuing cleavage products. The cytoplasm may, furthermore, have sufficient rigidity to prevent indiscriminate drifting and nuclear movement; otherwise crazy-quilt distribution patterns ought to be more prevalent. All in all, the evidence indicates that the indeterminate early developmental processes are the basis for the varied gynandromorphic types reported.

Elimination of a part of or an entire X chromosome from a female cell in early ontogeny may be chiefly responsible for the origin of individuals that are mosaic for sex characters. There are other causes for mosaicism, however, and among these may be mentioned somatic mutations (Patterson, 1929), somatic crossing over (Stern, 1936) and certain unstable genes (Demerec, 1941) which frequently mutate in germinal as well as somatic cells, giving rise to individuals with tissues of different genetic constitutions.

THE MITOTIC CYCLE

General Features

The mitotic cycle in the early egg of Drosophila presents a number of interesting biological features. The cleavage nuclei divide simultaneously, in a precise rhythm, asynchrony becoming apparent first in the divisions of the segregated pole cells and in the yolk nuclei. With regard to the mass of peripherally located nuclei (Fig. 19), asynchrony sets in only after the blastodermal cells are delimited by furrows which press in from the surface to mark off the cells. Synchronization of the early mitoses is typical not only for Drosophila but also for many other insect forms, although an occasional exception has been reported. For example, cleavages in *Calandra oryzae* are from the first char-

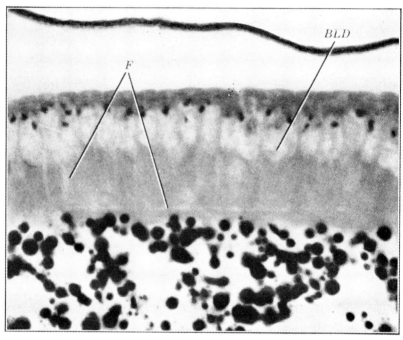

Fɪɢ. 19. Enlarged view of a portion of the blastodermic wall. Nuclei are lengthening, as furrows demarcating the individual cells have reached almost to the central yolk mass. Ends of furrows are uniting to form a continuous inner basement for the cells; this is seen as a whitish streak in the photomicrograph. Nucleoli are quite prominent at the periphery of the nuclei. 1200×. *BLD,* blastoderm, nuclei and cells; *F,* lateral and inner cell limits of blastoderm cells.

acterized by a complete absence of synchronization (Tiegs and Murray, 1938). The induction with X-rays of alterations in male or female germ cells may also lead to a loss of synchrony in Drosophila eggs developing with the treated chromosomes (Sonnenblick, 1940).

The factors governing this simultaneous type of mitosis in the syncytial egg of Drosophila and of other insects are not known. Except for the comparatively few instances where genetic diversity arises, for example, through mutation or chromosomal elimination in one of the initial cleavages, the cleavage nuclei may be considered to be structurally identical. Thus, assuming no differences to exist between the nuclei scattered throughout the egg interior, it may be inferred that synchronous mitoses result from the influence of a uniform environment upon a population of structurally similar nuclei. Is the environment homogeneous? Such an interpretation must be considered in the light of experimental embryological studies on the eggs of Diptera (especially Reith, 1925; Howland and Child, 1935; and Howland and Sonnenblick, 1936) which show that the egg is determinative. Before visible differentiation is evident, regions are set aside to produce specific structures of the embryonic-larval system. Chemodifferentiation occurs before blastoderm formation, and the various regions, if injured by cauterization or puncture, can show only a slight amount of regulative capacity. The environment in the determinative egg is therefore, from an early age, hardly uniform with respect to differentiation. However, a distinction must be made between factors influencing cleavage and those influencing differentiation. Cleavage and differentiation of eggs of multicellular animals have been found to be differently and independently affected by artificial treatment (Lillie, 1906). It is possible for cleavage to be suppressed, while differentiation can occur. It has been found that in the absence of the X chromosomal genes of Drosophila cellular proliferation can continue, but differentiation is adversely affected (Poulson, 1940). Irradiation of Drosophila gametes may also eventuate in embryos which are virtually sacs of cells, but which show no structural organization (Sonnenblick and Henshaw, 1941). Even if the Drosophila egg assumes a mosaic nature quite early, that is, with respect to organization it is non-homogeneous, this does not necessarily preclude the fact that the nuclear environment in the different regions of the egg may be uniform with respect to cleavage. So long as organic continuity exists between the nuclei, the mitoses proceed in unison, but as soon as protoplasmic continuity is broken (by the pole cells constricted off posteriorly from the main egg mass, as in Fig. 24, or by the formation of cell walls, as in Fig. 19) asynchrony follows.

Divisions in unison are generally observed in syncytial organisms but are also found to occur in tissues blocked off into cells. In testes of Orthoptera, for instance, which are organized into cysts containing a number of cells, synchronous divisions of the cyst components are noted. It is possible in such instances that the barriers between adjacent cells are not impassable, and that continuity is maintained by way of intercellular protoplasmic bridges. Intercellular strands have been described for a number of plant and animal cells (Sharp, 1934). It is also possible that simultaneity of cleavage may be a fundamental property of the nucleus. In an egg whose development is initiated by the fusion of an irradiated male complement and an untreated female chromosomal complement, normal (synchronously dividing) and aberrant figures in various mitotic phases may be noted side by side; total synchrony has been lost (Sonnenblick, 1940).

This brief discussion of cleavage synchrony is conjectural for the most part. Similar or additional data may lead others to still different interpretations. However, the origin of asynchrony from synchrony is perhaps a more fundamental question, for with it we enter upon the problems of differentiation and organization.

Rate of Mitosis; Chromosome Reduplication

The time required for the completion of one mitotic cycle in the early Drosophila egg is extremely small, only some 10 minutes being necessary for the passage from one phase of the cycle to the same phase in the next division. This rapid rate can be checked and confirmed in various ways until there is no question of error concerning the duration of the cleavage mitoses. The velocity of mitosis in the egg of the fruit fly is one of the highest ever reported for any organism, plant or animal. After but 1 hour of incubation at 25°C, an egg which was in first cleavage at the start of the incubation period will contain from 64 to 128 nuclei distributed throughout the unsegmented cytoplasm. Within the few minutes required for one mitotic cycle, the visible chromosome, and presumably the genic substance which occupies a minute portion of the bulk of the chromosome, is reduplicated. Furthermore, in this remarkably short period, a mechanism for the separation of the replicas is organized. The centrioles reproduce by division, the spindle is formed, the daughter chromosomes pass to the poles, and, after the nuclear membrane is formed, the segmentation products undergo an interkinetic stage during which the chromosomes gradually lose their affinity for stains. All these phenomena of the nuclear cycle are repeated, in synchrony, some twelve times until the blastoderm is formed. The asynchronous mitoses, for example, in the underlayer of the germ band (Fig. 30) and in the nervous tissue (Fig. 40) where the giant neuroblasts are almost always observed in mitosis, cannot be accurately timed. During organogenesis, however, cellular movements occur with such rapidity that it is quite possible that the asynchronous mitoses, although they cannot be timed, have as rapid a mitotic rate as have the cleavage cycles.

Many studies on the velocity of mitosis and the length of time occupied by the various phases of the process have been made on living material, and some investigators believe that accurate observations can be made only on material which has not been fixed. There are decided limitations to the study of nuclear cycles in the living Drosophila egg, and recourse is of necessity made to prepared material. The methods devised and utilized by a number of investigators (see the section on materials and techniques) enable one to determine with great accuracy the age of different eggs and to secure a wealth of material for microscopical study.

With such material Huettner (1933) originally fixed "the time involved in one complete cycle at approximately 10 minutes at 23°C." The writer and others have since confirmed this many times. Rabinowitz (1941a), in his detailed study, proceeded further and made estimates of the frequency and duration of the different mitotic phases. The estimates were obtained from the study of many eggs which had developed at one of three temperatures: 24°C, 29°C, and 30°C. Rabinowitz's results are indicated in Table 1.

TABLE 1

FREQUENCIES WITH WHICH THE DIFFERENT MITOTIC PHASES OCCUR IN EGGS DEVELOPING AT VARIOUS TEMPERATURES (Rabinowitz, 1941a)

Numbers of eggs used: 24°C, 341; 29°C, 307; 30°C, 179

Stage of Mitosis	Percentage of Eggs at This Stage			Duration (in minutes)		
	24°	*29°*	*30°*	*24°*	*29°*	*30°*
Interkinesis	36	28	31	3.4	2.5	2.7
Prophase	42	41	33	4.0	3.6	3.0
Metaphase	3	6	8	0.3	0.5	0.7
Anaphase	10	16	12	1.0	1.4	1.1
Telophase	9	9	16	0.9	0.8	1.4

In the table the frequencies with which the different phases of the cycle occur are given as percentages of the number of eggs observed. The percentage of eggs at each stage being known, and the average time necessary for the completion of a division at each of the three temperatures employed (at 24°C mitosis is carried through in an average of 9.5 minutes, while at 20°C and at 30°C, 8.9 and 8.8 minutes, respectively, are required) having been computed earlier, the duration of the various phases could be determined. These data are also presented in Table 1. Prophase is the longest stage, with interkinesis, anaphase, telophase, and metaphase occupying progressively lesser periods of the nuclear cycle. Such data are in accord with others derived from the study of plant and animal material (see Sharp, 1934; Schrader, 1944). Of the mitotic phases, prophase is usually longest, while anaphase and metaphase are passed through relatively quickly. The nuclei are in interkinesis for approximately one-third of the cycle. It is during this stage, as will be described shortly, that they increase in size, while the chromosomes become more and more faint. Eventually, the reticulum can hardly be discerned, and the nucleus appears homogeneous, taking no common basic stains and reacting negatively to Feulgen's nucleal stain.

Search through a good deal of literature has not uncovered any reported mitotic velocity comparable to the one discussed above. There is some evidence, however, that in certain other dipterous forms the cleavage divisions may be equally rapid, although the timing of one complete mitotic cycle has not been given. For instance, Lowne (1890–1895), working with the blowfly Calliphora, states that 2–3 hours after fertilization a blastoderm is formed, comprising a single layer of columnar cells which enclose the yolk. Auten (1934), studying the early embryology of the fly *Phormia regina*, observes that the period from fecundation to blastoderm formation occurs, in the majority of eggs, within 2 hours after deposition. Since these early developmental rates are comparable to those obtaining in Drosophila and since the developmental processes in these flies are quite similar, it is likely that the cleavage divisions in these and perhaps other dipteran forms require but a few minutes for completion.

A chromosome of the original diploid group can be reduplicated some twelve times within 120 minutes. Since the descendants of that chromosome multiply in unison, except for those set aside in the germ cells and in the yolk nuclei, increasing by powers of two, the egg soon contains several thousand replicas of the ancestral element. This refers, of course, to only one of the original complement, and we must remember that every chromosome introduced by the gametic nuclei will normally give rise to its quota of descendants. A chromosome, with its genic substance, has the property of specific self-multiplication, but little is known concerning the mechanism normally bringing about an exact mirroring of structure. A theory of chromosomal reproduction, Delbrück (1941) has indicated, must meet certain requirements. It must propose a catalytic mechanism which will be specifically autocatalytic; it must indicate the source of energy required for the necessary syntheses; it must explain why synthesis customarily ceases with the production of one duplicate. The phenomena of self-multiplication, as well as those of anaphasic separation, endomitosis, and other cellular processes, still await interpretation (see Muller, 1941).

One possibility has been suggested (Painter, 1940) in answer to the question of how the chromosomes in rapidly dividing cleavage nuclei can so quickly utilize from their surroundings the materials required in the reduplication process. Because of its method of growth, the cytoplasm of the Drosophila ovum is the recipient, before fertilization, of the contents of fifteen of the accessory nutritive nurse cells. Before their dissolution, the nurse cell nuclei have become highly polyploid, containing several hundred chromosomal complements as a result of a series of endomitotic cycles (Painter and Reindorp, 1939). Thus, when the nutritive cell contents are absorbed, the substance of thousands of chromosomes enters the egg cytoplasm. Painter cites evidence indicating that the derivatives of the absorbed nuclear material may persist in a partially polymerized form and proposes that "the rapid building up of the cleavage chromosomes is possible in the segmenting egg because the synthesis is more in the nature of a reassembling of already existing materials, such as nucleotides, etc., under the guidance of the active chromosomes, rather than an actual synthesis of the building blocks from relatively simple substances." This is, of course, speculative, and we still require an explanation of how a chromosome, and specifically the minute portion which is the gene, reacts in the cell environment to produce a "copy" of itself.

Description of a Mitotic Cycle

INTERKINESIS (FIG. 20A)

The interkinetic nucleus passes through a stage where it is vacuolar in appearance and optically homogeneous. As the nucleus grows in volume, the chromosomal material becomes ill defined and progressively disappears, until such time as the nucleus has no affinity for common stains and does not give the characteristic reaction with Feulgen's nucleal stain. The two centrioles, which appear to be in virtual contact with the nuclear membrane, are distinct and apart from each other. No aster is as yet evident about the centrioles. It is particularly interesting to note that no nucleolus or pycnotic chromosomal material is evident in the interkinetic nucleus. Later, when the period of rapid, synchronous mitoses is over and the furrows are delimiting the blastodermic cells, nucleoli are suddenly noticed in every nucleus (Figs. 18, 19, 24) in association with pycnotic chromosomal regions (Sonnenblick, 1947).

PROPHASE TO METAPHASE (FIG. 20B–G)

Though their positions are not constant, the central bodies have usually passed to the opposite poles of the nucleus with the onset of prophase. Prophase is first indicated by the loss of optical homogeneity of the nuclear contents. Deeply staining granular threads appear within the nucleus (Fig. 20B) and, as they contract, begin to be recognized as the definitive chromosomes (Fig. 20C). While contraction of the chromosomes is taking place, the nuclear volume has increased, and the centrioles move slightly out into the cytoplasm away from the nuclear wall (Fig. 20D). The diverging rays of the aster, seen in Fig. 20D, form about each division center. Contraction of the chromosomes has resulted in a shortening of each element, as well as an increase in its thickness. Dissolution of the nuclear membrane begins initially in that sector closest to the centrioles. The spindle fibers, intranuclear in origin, have also made their appearance (Fig. 20E). As dissolution of the nuclear membrane proceeds, pendu-

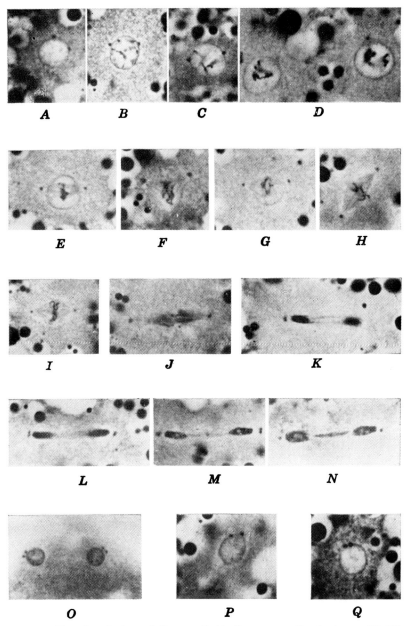

FIG. 20. For description of these mitotic figures see the text, pp. 98–101. From Huettner (1933).

lous spindle configurations, such as that represented by Fig. 20F, may be seen. Finally, the nuclear-cytoplasmic surface will entirely disappear, and, as the centrioles move slightly apart (Fig. 20G), a lengthened, well-defined spindle is evident. The nuclear cycle is now in metaphase (Fig. 20H).

METAPHASE (FIG. 20H–I)

The chromosomes have now attained their greatest degree of contraction and come to lie on the equatorial plate between the poles of the spindles (Fig. 20H–I). The chromatids or "daughter chromosomes" are noticed in occasional plates.

ANAPHASE (FIG. 20J–L)

As chromosomal disjunction begins, the centriole at each pole splits into two. The centriole pairs are seen with clarity at later anaphase (Fig. 20K–L). The asters at this time are quite prominent (Fig. 20K). An alteration in the character of the spindle from that noticed in metaphase (Fig. 20H–I) has taken place. As the corresponding chromosomal groups move to the poles, the spindle narrows transversely and increases in length. Some cytologists have maintained that the lengthwise expansion of the anaphase spindle assists in completing the polar movement of the chromosomes. It is possible in certain favorable anaphase figures to make out the individual chromosomes, but because of the minuteness of the components much care and study are necessary in order to make such an analysis.

TELOPHASE TO INTERKINESIS (FIG. 20M–Q)

A centriole pair is present at each pole, while granular differentiations of the spindle fibers in the equatorial region form the so-called midbody (Fig. 20M). Eventually, the spindle and the astral rays fade from view. The chromosomal group at each pole forms at first a compact deeply staining mass. Stages resembling prophase, but in reverse, then occur. Regions which do not take stain appear within the elongate clumped mass which gradually becomes spherical (Fig. 20N, O). The non-staining regions increase, a nuclear membrane is organized, and the nucleus becomes granular in appearance (Fig. 20O). As the nucleus increases in volume, the chromosomal contents form a reticulum which becomes progressively indefinable (Fig. 20P) until such time as it can hardly be demonstrated (Fig. 20Q). The centriole pairs are somewhat apart but close to the nuclear membrane. These bodies, reproducing by division once per cell generation, can thus be traced through each phase of the mitotic cycle (Wilson and Huettner, 1931; Huettner, 1933).

Mitotic Abnormalities, Spontaneous and Induced

The entire sequence of events just reported, involving chromosomal reduplication and disjunction, occurs normally within about 10 minutes. It is noteworthy that, on the whole, mitotic irregularities are infrequently met with, and Huettner (1927) states that such as are encountered are produced for the most part by "excessive polyspermy." He has described various mitotic abnormalities and aberrant meioses, which he noticed in preparations of eggs from different wild-type strains. In the eggs of hybrid females derived from crosses, made both ways, between *D. miranda* and *D. pseudoobscura*, Kaufmann (1940) regularly finds only a few cleavage divisions completed and subsequent embryonic degeneration. The failure of the eggs to develop is caused by abnormal cleavages as well as unusual proliferation of the polar body group of chromosomes.

Exposure of mature sperms and oöcytes of *D. melanogaster* to X-rays affects the gametes so that, after fertilization, orderly development becomes impossible in a high percentage of the eggs. The divisions are frequently asynchronous, and irregularities in the mechanics of mitoses are of varied kinds. Asymmetrical distribution of chromosomes to daughter nuclei, multipolar spindle configurations, and chromosomal clumping have been noticed by the writer (1940). Degenerate figures lie side by side within the same egg with apparently normal figures which are undergoing mitoses. Because of their minuteness in the egg, no attempt has been made to analyze the influence of the radiations on the treated chromosomes with respect to cytological details. Similar mitotic upsets are observed in Drosophila eggs placed soon after deposition in an atmosphere one-third ether by volume. Multipolar spindles and other cytological abnormalities are frequent in X-deficient eggs (Poulson, 1940).

FORMATION OF THE POLE CELLS

The pole cells of Drosophila are polynuclear in origin (Huettner, 1923). When nuclei are first noted in the posterior polar plasm, wherein are observed those deeply staining granules considered by some to be "germ cell determinants," their number is found to vary. Different workers have reported from two to eleven nuclei initially penetrating the polar region, but the significant fact is that more than one nucleus is always present in that area as the cleavage products pass to the periphery. The future germ cells are thus not linearly descended from a single, specific stem nucleus but rather from several, usually three to seven, nuclei which penetrate the posterior cytoplasm. While the ninth cleavage division is in progress, the several nuclei present in the polar area move into the protoplasmic outpocketings which are being pinched off from the cytosome. During the next division of the peripherally disposed cleavage nuclei, more of the nucleated pockets are constricted off from the posterior end of the egg. Once they are detached from the main egg mass, the cells with their nuclei may be called "pole cells." Drosophila pole cells are therefore precociously segregated before blastoderm formation and comprise the first obvious differentiation of the developing egg.

The pole cells have been regarded as germ cell primordia since, in various insects, they have been traced more or less directly to the formed gonad. Concerning Drosophila, however, a word of caution is necessary. Only those pole cells which ultimately reach the gonads ought, strictly speaking, be considered germ cell primordia. Many of the pole cells of Drosophila, however, do not enter through the posterior invagination but instead push between the elongating blastodermic nuclei and wander back into the central yolk mass, where their further history is still unclear. A number of those which do enter the body by way of the posterior invagination are never included in the gonad (Sonnenblick, 1941). Sonnenblick finds that, from the time the movements have begun which will end finally with the inclusion of the pole cells in the gonads, the pole cells do not divide. It is accurate to say, therefore, that the definitive germ cell primordia can be traced to and are included among the initially segregated group of pole cells, but it is not true that all the pole cells are to be germ cells. We cannot as yet determine precisely which of the pole cells are destined to function in the male or female gonads as germ cells and which are to fall by the wayside. All pole cells are only "potential germ cells," as Sonnenblick called them, and recently Poulson (1947) made the interesting observation that those pole cells which do not enter the gonads become a specialized section of midgut epithelium.

The pole cells appear spheroidal and, as a group, form a characteristic amicropylar polar cap in the area between the posterior blastodermal layer and the vitelline membrane (Fig. 24). Such amicropylar caps are well seen at a stage when furrows are pressing in from the egg surface to mark off the individual blastodermal cells. The number of pole cells present in the posterior clusters differs from egg to egg, as we shall see later, and is probably attributable to the varying quantity of cells initially budded forth. The number of times a single pole cell can reproduce while it is still in position outside the body of the organism is difficult to establish since the mitoses are asynchronous.

There is no general type of germ cell origin in the Diptera. In different forms the germ cells may be mononuclear, binuclear, or polynuclear in origin. In the fly, Miastor, a single pole cell, detached from the posterior pole before blastoderm formation, will eventually give rise to all the germ cells produced by the individual. Miastor development has been studied by Mecznikow (1866), who traced the primordia to the larval gonad and, most intensively, by Kahle (1908) and Hegner (1914). Of the four spindles present in the egg, in passing from the four- to the eight-cell stage, the most posterior spindle becomes oriented almost parallel to the long axis of the egg. The lower nucleus on this spindle remains in the polar plasm and is then cut off from the rest of the egg as the definitive germ cell primordium. The other daughter nucleus is left behind in the general unsegmented cytoplasm and gives rise to somatic cell nuclei. The germ primordium retains its full complement of chromosomes, whereas the somatic nuclei undergo chromatin diminution.

In Sciara (DuBois, 1932), the germ line is regularly initiated at fifth cleavage when two nuclei migrate into the posterior cytoplasm. Two protoplasmic buds containing the pair of nuclei appear at the back end of the egg, but, unlike Drosophila, these are not at any time completely separate from the main egg mass, remaining in direct contact with the yolk. The pole cell mitoses are not synchronous with those of the blastoderm. Neither does there appear to be any simultaneity of division among the original pole cells.

The phenomena in Drosophila and in Melophagus (Lassmann, 1936), and probably in Calliphora (Noack, 1901), typify those instances where the pole cells initially extruded from the caudal end of the egg are not derived from a single ancestral nucleus or regularly, as in Sciara, from two nuclei, but rather from several nuclei which chanced to migrate to the posterior polar plasm. In Drosophila the number varies but is typically more than two, while in Melophagus twelve nuclei simultaneously penetrate the polar plasm to give rise to the pole cells. The photomicrographs of Figs. 21–24 illustrate some of the events which occur during pole cell formation in Drosophila.

Figures 21–22 are high-power views of polar nuclei which have entered posterior protoplasmic pockets. The pockets have not as yet been severed from the general oöplasm, although constriction of the bud shown in Fig. 22 is fairly well advanced. As Rabinowitz (1941a) has indicated, the long axes of the protuberances are parallel to the egg periphery, with the spindles lying lengthwise across the buds. The deeply staining granules noticed in these buds are the polar granules which accompany the nuclei into the pole cells and which some investigators, especially Hegner (1914), have considered to be "germ cell determinants." Notice in Fig. 22 the comparative scarcity of these granules in the neighboring and still-connected egg cytoplasm. The distribution of the granules to the buds, and thus eventually to the primordia, is variable and indefinite, although Huettner maintains, as did Noack (1901), that the central pole cells receive a relatively larger amount of this material than do the more lateral buds. During the subsequent asynchronous pole cell divisions, the granules pass to the spindle poles and are so allotted to the daughter cells. Pole cells are sometimes noticed, however, which do not contain any of the "determinants." Such cells, completely constricted from the egg and devoid of the granules, have been figured by Rabinowitz (1941a), and this the writer can confirm. If these pole cells do contain the granules, the latter are either altered or too few in number to be discerned. We cannot go into any lengthy discussion here concerning the polar granules. There is no doubt that in many insects the region where the germ track develops does contain some visible cytoplasmic differentiations, such as the polar granules seen here in Figs. 21–22, whose origin

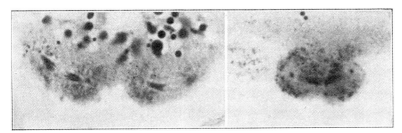

FIG. 21   FIG. 22

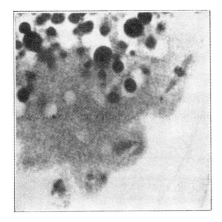

FIG. 23

FIGS. 21, 22, and 23. Posterior polar pockets with dividing nuclei which will give rise to pole cells. In Fig. 22 a bud is almost constricted off from the general oöplasm, while in Fig. 23 one cell is seen completely severed from the main cytoplasmic mass. Note the large number of deeply staining granules present in these polar pockets. From Huettner (1933).

and function are still problematical. That these cytoplasmic differentiations are organ-forming substances is, on the basis of the Drosophila investigations as well as from various experimental embryological studies, no longer tenable.

In the lower portion of Fig. 23, some protoplasmic pockets are in the process of constriction. Laterally and forward a metaphase figure is seen. There is a possibility here, upon the completion of the division, that one daughter nucleus will enter the pocket beneath it to become part of the germ line, while the daughter nucleus above will remain in the soma.   Figure 24 shows the posterior end of an egg, sectioned not quite sagittally, with segregated pole cells clustered in the posterodorsal region

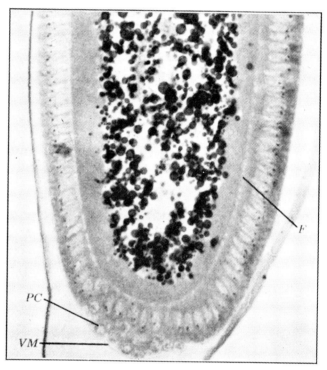

FIG. 24.   Section showing a group of segregated pole cells in the posterior polar cap.   Compare the size and shape of the pole cells with the blasto-dermic cells which are being delimited above.   The white streak in the cytoplasm indicates the bases of the furrows passing between and marking off the cell membranes.   600×.   *F*, lateral and inner cell limits of blasto-derm cells; *PC*, pole cells; *VM*, vitelline membrane.

between the vitelline membrane and the continuous peripheral layer above.   Blastodermal cell formation is in progress at this time but is as yet incomplete.   Note how the delicate vitelline membrane which is so closely applied to the egg (seen best at the dorsal, or left, side of the figure) has stretched in order to accommodate the mass of cells in the amicropylar cap.   The nuclei of the pole cells are small and round as compared with the elongate columnar nuclei of the somatic line, and polar granules are still evident in the cytoplasm of the cells. Earlier, the nuclei at egg periphery and in the pole cells were more comparable in size, but with the onset of blastoderm formation the surface nuclei elongate and appear columnar. Observations on living eggs indicate (cf. Child and Howland, 1933) that the pole cells are segregated 2¼ hours after development is initiated, and that they remain in the polar cap for 70 minutes more before they are passively carried into the interior of the organism.

There is no hard and fast distinction between somatic and germ cell in early ontogeny. The cleavage nuclei may be regarded as stem cells, random factors determining whether a specific cleavage product becomes a vitellophag (yolk nucleus) or whether it enters the pole cell or blastoderm cell lines. In Drosophila, as in Miastor and Calliphora, instances may occur whereby a daughter nucleus at one pole of the spindle becomes the nucleus of a pole cell which may, eventually, produce germ cells, while the other nucleus persists and functions in the soma. From an early stage the posterior protoplasm appears to be a differentiated germinal cytoplasm, in the sense that only those nuclei which happen to reach this area have an opportunity to become part of the germ line. Many of the pole cells are destined never to be included in the gonads to give rise to gametes; the unincluded cells have only recently been followed (Poulson, 1947).

BLASTODERM FORMATION

From the clear, nucleated layer of protoplasm which envelops the central yolk mass will be derived the greater part of the future somatic tissues and organs. This protoplasmic layer, during the early cleavages, is rather thin (Fig. 13), but it broadens when the cleavage nuclei have reached the periphery. The nuclei, when entering the cortical region and even after their first division there (Fig. 15), do not completely occupy the peripheral protoplasm. However, after several divisions at the egg surface the nuclei become closely packed (Figs. 18, 24).

Segmentation of the layer into separate cells begins with the appearance of furrows on the surface of the egg which gradually press down toward the centralized yolk mass. Observations on living dechorionated eggs indicate that the surface at this time is wavy, undulating in outline, while in prepared material (Figs.

16, 18, 19) slight protuberances are noted. Investigators studying the early developmental processes in the eggs of certain flies
(e.g., Pratt, 1900, in Melophagus; Child and Howland, 1933, in
Drosophila; and Auten, 1934, in Phormia) have stated that the
protuberances result from the action of the surface nuclei, which
actively push a little protoplasm ahead of them, giving the exterior of the egg the wavy contour mentioned. Rabinowitz
(1941a), however, maintains that the undulations are indicative
rather of furrow formation, since the rigidity of the vitelline membrane precludes the forcing out of the protoplasm as buds. The
writer agrees with Rabinowitz that it is the fissures pressing in
from the exterior which impart the wavelike character to the egg
surface, but the point made by him concerning the rigidity of the
vitelline membrane as a deterrent to protoplasmic outpocketing
is not understandable, since pole cells arise as posterior buds,
and upon the termination of the process of segregation (Fig. 24)
the membrane is no longer close to the egg but quite stretched.

The progress of the furrows as they press between adjacent
nuclei and traverse the protoplasmic sheet enveloping the yolk
can be followed. In Fig. 18 they have reached the outer edges
of the nuclei. The nuclei, originally spherical, have begun to
lengthen as the furrows push inward. As the inward progression
continues and the furrows make their appearance in the cytoplasm between the nuclei and the central yolk area (Fig. 24),
the bases of the fissures are indicated by what appears to be a
continuous streak extending about the egg. Actually, the innermost part of a single furrow can often be seen as a discrete, clear,
round spot. The seemingly continuous streak apparent in fixed
material represents the collective bases of the channels passing
inward between the more than three thousand nuclei which are
aligned at the surface of the organism. The furrows eventually
reach the inner edge of the protoplasmic layer (Fig. 19), where,
finally, the inner cell walls are marked off. Note the delimited
cells of the dorsal blastodermal wall of Fig. 25, high columnar in
appearance. Three hours after fertilization, blastoderm formation is completed, and a true cellular envelope covers the centralized yolk inclusions; the egg is no longer a syncytium. The
pole cells, we might mention, are still clustered in the posterior
cap, remaining quiescent during the process of blastodermic cell
formation.

Most of the clear protoplasmic material is taken up by the blastodermal cells as the separate cells are being defined. Except
under special circumstances, for example, absence of the X chromosome (Poulson, 1940) or X-ray-induced nuclear alterations
(Sonnenblick and Henshaw, 1941), little or no non-cellular cytoplasm is evident in the egg from this developmental stage on.
The writer cannot confirm those who have reported that upon
absorption of the peripheral substance the cells along the ventral
side of the organism are much larger than are those situated
dorsally. From the distribution of the extra yolk material in the
egg as the furrows are pressing inward (Fig. 18), it is apparent
that there is little or no difference in the sizes of the cells which
are in process of being delimited on the various surfaces of the
egg. It is true, though, that during the complex cellular movements occurring later in gastrulation the sizes and shapes of cells
in different regions become distinctly altered (Fig. 25).

Near the end of blastoderm formation, in a number of dipterous eggs, a separate inner layer of protoplasm, the so-called
"innere Keimhautblastem" of Weismann, can be observed. In
section it is a dense, granular region which is separated from the
outer oöplasm by a circlet of yolk granules and is eventually absorbed by the forming blastodermic cells. Such a distinct layer
has been reported present in the eggs of many flies, including
Musca, Calliphora, and Phormia. Although an "innere Blastema" has been mentioned in Drosophila, the writer's observations indicate that a second, separate, granular layer, as described
by investigators such as Blochmann, Noack, and Auten in the
material studied by them, is not apparent in the egg of the fruit
fly during the completion of the blastoderm.

It is only when the synchronous divisions of the nuclei at the
egg cortex have ceased that nucleoli are noted. During the intensely rapid mitoses which occur in early ontogeny, no nucleoli
are visible in the interkinetic phase, the stage when the chromosomes gradually become indefinable and the nucleus appears
virtually homogeneous. However, when furrows have barely
begun to move inward from the surface and while the cortically
disposed nuclei are elongating (Fig. 18), nucleoli become visible
(Sonnenblick, 1947). These elements are consistently located in
the portion of the nucleus closest to the egg exterior. In Figs.
18, 19, and 24, the nucleoli are seen oriented toward the outer

edge of the nuclei. Their appearance seems to be another manifestation of the synchronous, coördinated processes in the early egg since, when noticed for the first time in prepared material, they are found to be present in every nucleus. There is no evidence that they appear at different times; it seems rather that their presence is manifested simultaneously throughout the egg. In these and in all later stages of development, nucleoli and associated heteropycnotic chromosomes and portions of chromosomes are observed in interphase nuclei (Figs. 25, 26, 30, 31, and others). The remaining threads are poorly staining and constitute the vague reticulum of the nucleus.

The varying nuclear patterns found in larval cells (Kaufmann, 1934), involving the nucleolus, the associated sex chromosomes, and other pycnotic regions of the autosomes, can already be noticed in eggs which are a little more than 3 hours of age (Fig. 25).

FORMATION OF THE INNER LAYERS: GASTRULATION

The Inner Layers

Three hours after the beginning of development, the egg is covered by a columnar cellular layer, the yolk is centrally congregated, and posterodorsally the pole cell cap is evident. The pole cells are not dividing and are compacted in the perivitelline space, while the blastodermic layer below them has flattened. Between 190 and 210 minutes after deposition, a complicated series of cellular movements is initiated, almost but not quite simultaneously, in several regions of the seemingly quiescent organism. The time when these movements begin will vary a little from egg to egg at any one temperature and will be influenced by changes in temperature.

The initial indication of gastrulation is the appearance of a longitudinal ventral furrow which extends from an area approximately one-seventh of the distance from the anterior pole all along the bottom of the egg to the posterior pole (Figs. 25, 26). The furrow, as observed in living material, is open only momentarily. Thus, in order to prepare material at this stage for sectioning, it is advisable either to add hot fluid to the medium in which the eggs are developing and, through immediate coagula-

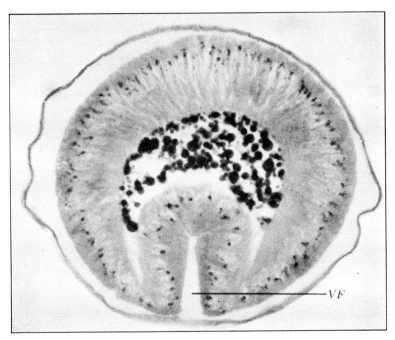

FIG. 25

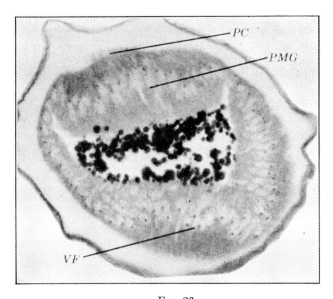

FIG. 26

FIGS. 25 and 26. Transverse sections through a 3½-hour embryo. In Fig. 25 the ventral furrow and inner layer are seen in the brief period before closure. Note different cell types dorsally and ventrally. Figure 26 is from the caudal end of the embryo. Pole cells are dorsal, while ventrally the furrow is shallow. 600×. *PC*, pole cells; *PMG*, posterior midgut rudiment; *VF*, ventral furrow.

111

tion, stop the processes then occurring or to remove the eggs quickly to cold fixative and puncture the envelopes. Forward, the furrow is shallow, but in transverse sections through the middle of the egg it is quite deep (Fig. 25). At the posterior end of the egg (Fig. 26) the ventral groove is again shallow. Such caudal sections have the peculiar flattened appearance evident in the illustrations because of the fact that the sections are cut dorsally through the flat blastoderm layer below the pole cells (especially see Fig. 26) and ventrally through the shallow part of the longitudinal furrow. The dorsal plate under the pole cells is the posterior midgut rudiment.

After closure of the ventral furrow, the inner tubelike portion will flatten and thus obliterate the cavity seen in Fig. 25. Briefly after closure the inner layer forms a distinct tube with a lumen, giving the egg a "tube within a tube" character. This is a decidedly fleeting stage, ending with the obliteration of the lumen. The columnar epithelial nature of the lower cells is lost, and they appear as flat, irregular polygons. Note the difference in cell size and shape of the dorsal blastodermic elements and of the cells which have invaginated, as shown in Fig. 25. In Fig. 29 attention is directed toward the flattened polygonal cells of the inner ventral layer. The thick ventral portion of the organism, comprising an outer layer of columnar cells, one cell in thickness, and an invaginated inner layer of irregular cells, several cells deep, will be referred to as the germ band.

In different insects there are variations in the manner of deriving the inner layer. Three general types have been described by Korschelt and Heider (1899, see pp. 309 et seq.). It is interesting to observe that some investigators consider the edges of the ventral furrow (Fig. 25) to be the lips of an unusually long blastopore. Such interpretations follow from the attempts of the earlier students to homologize structures in members of various animal phyla, Haeckel's biogenetic law serving as a stimulus to numerous investigations. Whatever it can be compared with, the ventral longitudinal furrow is simply a mechanism which the organism has derived whereby that cellular material which is subsequently to play a role in the development of internal structures is brought into the interior from the surface blastodermal envelope. The genetic factors governing this type, or any other types of inner layer formation such as are detailed by Korschelt

and Heider, are normally transmitted to the following generations.

The cells carried inward from the ventral blastoderm to form
the lower layer will not produce any external characters. Counts
made upon several embryos killed at the stage when the furrow
is still open in the midventral line indicate that about 20 per
cent of the cells are shifted to the interior. When we consider
further that additional cells are brought in from the outer blastodermic wall to form the proctodaeum, the stomodaeum, and
the paired salivary glands, we can estimate that some 30–35 per
cent of the cells comprising the original blastodermic epithelium
(Figs. 19, 24) are lost with respect to the surface expression of
characters. Were the blastoderm cells of diverse nuclear composition, because of some genetic alteration in an early cleavage
and because of the random penetrance of nuclei in the peripheral
protoplasm (Figs. 14–17), the cells which invaginate to form the
structures mentioned might also be of mixed mosaic composition
and give rise to tissues and organs with mosaic elements.[2]

Although the formation of the longitudinal ventral furrow and
the other gastrular events now to be described, appear in living
material to occur almost simultaneously, sectioned material offers
evidence that the ventral furrow is the first of the gastrular
phenomena. Transverse sections of a blastoderm show that the
cells are oriented toward the center of the organism, so that a
plane from the exterior traversing the length of a surface cell will
roughly reach a point at the center of the yolk mass. This can be
judged from early stages represented by Fig. 18, where the cell
membranes have begun to press in from the exterior, while such
an arrangement is still noted dorsally and laterally in Fig. 25.
Initial evidence of the ventral furrow is seen in preparations
which show about twelve cells to the right and left of the midventral line to have lost this orientation. The straight columnar
form is lost by these cells which show, instead, a decided curva-

[2] L. V. Morgan (1939) has described a spontaneous translocation which
she observed in a few cells of a salivary gland in *Drosophila melanogaster*
but which was not present in other cells analyzed. The time of origin of
the aberration has not been ascertained. Salivary glands of the F_1 larval
offspring of a cross between X-rayed normal males and untreated normal
females have been found by Helfer (1940, 1941) to be mosaic for certain
chromosomal aberrations.

ture; meanwhile, the ventral side of the egg flattens slightly. It is this group of curved cells which is brought inward. Increased mitotic activity is not among the forces, whatever they may be, which induce this inward movement of cells originally located on either side of the midventral line. There appears to be a downward pressure which initially causes the cells nearest the midline to bend and then to be squeezed inward; continuance of the pressure results in the closure of the temporary opening to the exterior. In preparations which give evidence of the beginning of the ventral furrow, there is as yet no sign of the anterior oblique cleft (the "cephalic furrow" or "head fold" of other investigators) or of the anterior and posterior invaginations, which appear almost at once after ventral furrow formation.

The Extension of the Germ Band

Laterally, and about one-third of the distance from the forward end of the embryo, cells buckle inward, and within 3–5 minutes the resulting cleft has deepened and reached the central yolk mass, where the movement appears to be impeded. The cleft is completed dorsally and ventrally, but since the indentations on these surfaces occur at different levels from the anterior pole, the cleft characteristically appears in profile view to be oblique to the long axis of the embryo. As determined from preparations, the cleft is shallow, barely discernible in fact, in the midventral area, but laterally it is quite marked (Figs. 27, 28). Figure 28 indicates a longitudinal section through the lateral portion of an embryo.

Concurrently with the appearance of the anterior oblique cleft, the posterodorsal region of the blastoderm just below the pole cell cluster invaginates to form, initially, a shallow trough, the posterior midgut rudiment. While the cell limits are earlier in process of being defined (Fig. 24), the flattening of the nuclear layer below the group of pole cells is already apparent. The pole cells in the concavity are passively carried along as the invaginating material is shunted dorsally (Fig. 29), along with the germ band as it extends around the caudal pole to the dorsal side. This obliterates the clear posterior perivitelline space in which the pole cells (Fig. 24) had been congregated. Fifteen minutes after the posterior plate appears, it has been forced an-

teriorly to a position about one-fifth of the distance from the
caudal end of the embryo, where it deepens to the extent noted

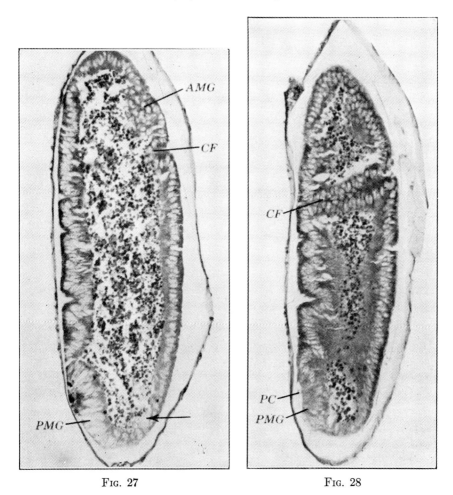

Fig. 27   Fig. 28

FIGS. 27 and 28. Median and lateral sagittal sections from the same embryo,
3½–3¾ hours old. Note dorsally the temporary buckling caused by the
forward movement of the posterior plate. The infolding occurs up to but
not beyond the anterior oblique cleft, readily seen in these sections. 300×.
AMG, anterior midgut rudiment; *CF,* anterior oblique cleft, cephalic furrow; *PC,* pole cells; *PMG,* posterior midgut rudiment.

in Fig. 29. In living eggs at this stage, the pole cells within the
concavity can hardly be seen, and as the dorsal progression of the
mouth of the invagination continues (Fig. 30) the cells within
the lengthening diverticulum disappear from view. The cells

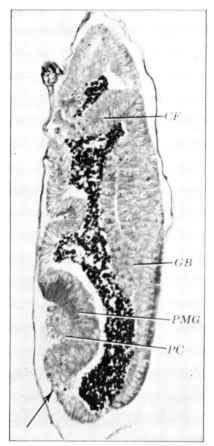

FIG. 29. Sagittal section of an embryo, a few minutes older than the embryo shown in Figs. 27 and 28. The posterior rudiment has both deepened and been extended dorsally. Pole cells are in the concavity. A single deep mid-dorsal fold is apparent between the invagination and the anterior oblique cleft. Note the thicker ventral germ band. The arrow indicates the region at the posterior lip of the invagination where much mitotic activity can be noted. 300×. *CF*, anterior oblique cleft, cephalic furrow; *GB*, germ band; *PC*, pole cells; *PMG*, posterior midgut rudiment.

of the dorsad-moving invagination are large and columnar, strikingly different from the smaller pole cells within the trough (Figs. 29, 30).

It has been stated above that mitotic activity is apparently not concerned in the midventral infolding which precedes the formation of the inner component of the germ band. Mitotic activity, however, does play a part in the dorsal extension of both the ventral blastodermic cells (outer layer) and of the inner layer. Just before and during the period when the posterior midgut rudiment becomes apparent, blastodermal cells in the extreme posteroventral region (Fig. 27, arrow) appear to thin out and taper. In living material as well, activity can be noted in this area. When the depression has moved forward one-fifth of the distance from the caudal end, preparations show that the cells in the neighborhood of the posterior lip are undergoing mitosis and are smaller in size and more numerous than are the cells located forward of the anterior lip of the invagination. The arrow in Fig. 29 indicates the area at the posterior lip of the invagination where the numerous cells are dividing. At the dorsal lip of the invagination, the cells are fewer, larger, still columnar in appearance, and relatively quiescent. No mitoses are evident in this region. It would seem, therefore, that extension of the ventral blastoderm is occasioned, at least in part, by the rapid proliferation of the cells behind the posterior invagination; this proliferation, in turn, forces the mouth of the invagination anteriorly. In Figs. 30 and 31, the lips of the invagination have approximated one another, and small dividing cells can also be seen near the posterior lip.

The inner layer is also extended around the hindpole of the egg, following, as it were, the forward progression of the aperture of the invagination (Figs. 30 and 31). When the concavity has advanced for only a slight distance from the caudal pole (Fig. 29), very few of the inner layer elements are cleaving, but when the mouth of the invagination has moved further forward, as in Fig. 30, intense mitotic activity is noted among these cells. Dividing cells are seen throughout the inner layer, from the anteroventral region where the stomodaeal invagination will soon appear to the posterior and dorsally flexed regions of the extended column. The rate of mitoses of the cells of the inner layer cannot be accurately determined since the divisions are asynchronous; adjacent cells are often not in the same phase of mitosis. The increase in number and the growth of these many active cells would seem to be the primary factor behind the anterior and dorsal flexure of the inner portion of the germ band. In Figs. 30 and 31 note that the outer layer, along the ventral surface, is comprised of columnar, blastodermlike cells, while posteriorly and dorsally the external cells are small and round, evidence of the divisions occurring earlier in this area (Fig. 29, arrow) which tend to assist in the elongation of the ventral blastoderm.

As the posterior midgut rudiment forms and begins to move anteriorly on the dorsal side of the embryo, the cells between the dorsal lip of the invagination and the anterior oblique cleft are seen to buckle in several places (Figs. 27, 28). These dorsal foldings can be observed clearly in dechorionated living embryos as well as in fixed material. At first, several clefts are apparent along the upper surface. The entire dorsal surface is not involved in these bucklings, and they are not to be confused with the true larval metameres which appear several hours later, in the 9-hour-old embryo (Figs. 43–45). The dorsal foldings are

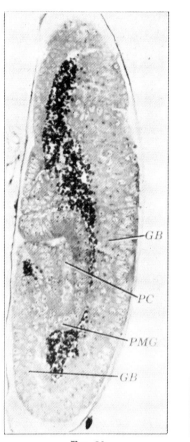 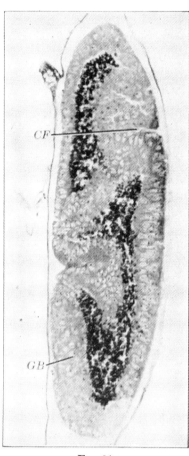

FIG. 30   FIG. 31

FIGS. 30 and 31. Two sections from the same embryo. The posterior midgut rudiment is still further forward dorsally. The diverticulum with its contained pole cells has also lengthened. The germ band is continuous ventrally and has been extended around the hind pole to the dorsal side. Note numerous mitoses in the germ band. In Fig. 31, a lateral section, the oblique cleft can be seen. 300×. *CF*, anterior oblique cleft, cephalic furrow; *GB*, germ band; *PC*, pole cells; *PMG*, posterior midgut rudiment.

irregular in outline and are temporary in character, lasting but a few minutes.

The orientation of the dorsal cells is obviously different, at this time, from that of the ventral surface cells. The impression gathered from observation of living eggs is of a vigorous movement of the posterior invagination which, of necessity, causes the dorsal blastoderm between the oblique cleft and the anterior lip of the invagination to quaver and then to give way, so to speak, resulting in the temporary dorsal foldings. The cells anterior to the oblique cleft do not buckle, retaining their original appearance and orientation. Finally, after the invagination has moved further cephalad, one deep mid-dorsal fold remains (Fig. 29). The cells in this region of the dorsal side are becoming slender and are thinning out; it is from this region that part of the embryonic membranes, the amnion and serosa, will be derived. Figure 38 represents a section across the dorsal opening of the proctodaeum with the double sheet, the serosa being external and the amnion internal, covering the aperture. The rudimentary double sheet of amnion and serosa is also clearly shown in Figs. 36 and 37. The embryonic folds of Drosophila remain rudimentary and never completely grow over the germ band as they do in certain other Diptera.

The mouth of the posterior invagination is forced anteriorly until it is found about one-third of the distance from the fore-end of the embryo (Figs. 32, 33). The stomodaeum, the rudiment of which is observed earlier, is decidedly more prominent at this time, as can be seen in these photographs. This is the most anterior point reached by the posterior end of the embryo. During this time the proctodaeal rudiment becomes apparent for the first time, and the proctodaeal-amniotic invagination is established. This follows rather than precedes the posterior midgut rudiment in time of origin. The relations between these are subsequently discussed by Poulson. The embryonic membranes, rudimentary and transitory features which are evident for only a few hours during the course of development, are more marked at this time. The deep mid-dorsal buckling has disappeared, and the columnar cells of that area move into the now-deepening proctodaeal-amniotic invagination. The undifferentiated cells between the proctodaeal-amniotic invagination and the oblique cleft which had earlier begun to thin out are decidedly attenuated when the stage shown by Fig. 30 is reached. These cells move inward with the backward growth of the tube, from the fore and lateral lips of which the double layer of amnion and serosa projects posteriorly (Figs. 32 and 33). The anterior cleft is less conspicuous at the time of maximal anterior extension of the germ band.

We venture to suggest that the anterior oblique cleft may be considered a stabilizing factor which tends to offset the stresses

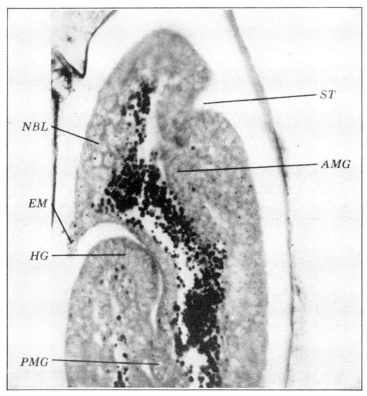

FIG. 32. Sagittal section of an embryo about 5½–6 hours old. The mouth of the posterior invagination is now one-third of the distance from the forward end of the organism. Rudimentary amnion and serosa cover the opening of the hindgut. The stomodaeum is forward on the ventral side. 525×. *AMG*, anterior midgut rudiment; *EM*, embryonic membranes; *HG*, hindgut; *NBL*, neuroblasts; *PMG*, posterior midgut rudiment; *ST*, stomodaeum.

resulting from the diverse surface changes and concomitant cellular translocations which take place in various areas of the organism. The egg appears quiescent until with the advent of the longitudinal ventral furrow (Fig. 25) the series of large-scale and relatively complex cellular movements is inaugurated. The various phenomena described in the preceding pages, including the derivation of the inner layer, the formation of the posterior midgut invagination and the entrance therein of the pole cells, the dorsal flexure of the germ band, the bucklings of the dorsal undifferentiated cells, the progression of the posterior invagination to one-third of the distance from the cephalic pole, and the appearance and caudal extension of the embryonic membranes, occur within a maximal period of 45 minutes, being completed by the fourth hour of development. As the posterior invagination is initiated and then extended dorsally, the cells ahead of the concavity "give" in several places, and the anterior cleft simultaneously appears, cutting in laterally, bringing in cells from the sides of the egg. Cellular dislocation and buckling do not occur forward of the anterior cleft but only in the region between the forward lip of the invagination and the anterior cleft. The cleft is then extended to the other surfaces and can be seen in living embryos to press in deeper as the large posterior invagination is thrust toward the fore-end of the organism.

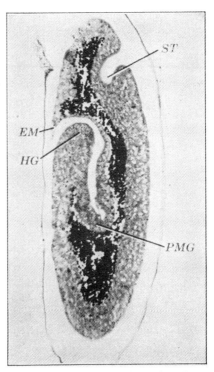

Fig. 33. Sagittal section of embryo of about similar age to that seen in Fig. 32. Mouth of the invagination is forward. Note the length of the diverticulum, which extends backward into the interior of the embryo. The stomodaeum has deepened somewhat. 235×. *EM*, embryonic membranes; *HG*, hindgut; *PMG*, posterior midgut rudiment; *ST*, stomodaeum.

When the mouth of the invagination is at its most anterior point (Figs. 32, 33), cells of the outer layer have been turned into the lengthening tube, and anteroventrally the stomodaeum begins to appear. At this period the anterior oblique cleft has flattened out and is difficult to discern in the preparations. The oblique cleft in Drosophila is regarded by the writer as a characteristic but transitory feature which may conceivably act as a buffer during a period of extensive and vigorous cellular movements. On the other hand, Poulson is convinced that it is the definitive boundary between the head and body regions of the germ band.

We may note here that the processes of gastrulation in Drosophila can be upset in various ways. Gastrulation does not occur in the absence of the X chromosome (YY eggs), for embryogenesis is disrupted quite early (Poulson, 1940). X-ray-induced dominant lethals can occasionally adversely affect gastrulation, although cellular proliferation is not inhibited (Sonnenblick and Henshaw, 1941).

The Question of Germ Layers in Drosophila

Interpreting the homologies of the germ layers in insects is unusually difficult, and many observations have been made and speculations proposed. Nelson (1915), quoting Weismann, says, "It becomes more and more evident that nowhere in the entire animal kingdom is the ontogeny so distorted and coenogenetically degenerate as in the insects, so that scarcely anywhere are the germ layers so difficult to recognize as here."

The book on general insect embryology by Johannsen and Butt (1941) contains a discussion of this controversial matter. From this book, as well as the review by Eastham (1930) on germ-layer formation in insects, it would appear that much of the confusion and discord concerning the layers which have arisen is the result, primarily, of too many attempts at generalization, based on weak or insufficient evidence, as well as the rather forced attempts at homologizing structures observed in members of different orders. Those working in insect embryology are severely restricted by their inability to alter the position of cells at will, with operative techniques, in order to ascertain the total as well as the actual potencies of the cells in various regions of the organism at the time separation into parts, that is, layers, occurs. Because of this lack, the insect embryologist is limited therefore to interpretations of the specificity or non-specificity of the germ layers based primarily on observation rather than on observation plus experiment.

Although all investigators are agreed on the manner of the origin of the mesodermal component of the inner layer of Drosophila, the origin of the rudiments of the midgut was not entirely clear until recently. Neither Parks (1936) nor Poulson (1937) was able to give a completely satisfactory account, and both interpreted the posterior invagination (which is now seen to be that of the posterior midgut rudiment) as proctodaeal in nature, Parks labeling it the proctodaeal-amniotic invagination while Poulson called it the proctodaeal invagination, though recognizing that cells at the tip became part of the midgut. Rabinowitz (1941a) also refers to the posterior invagination as the proctodaeal-amniotic invagination. Until recently the author too held this interpretation and regarded the anterior and posterior rudiments as derived from the two ends of the ventrally invaginated inner layer and thus different from the situation as described for Calliphora and other blowflies by Escherich (1902) and Noack (1901). A series of critical stages obtained by Poulson (see Chapter 3) since this account was first written makes it clear that, as in Calliphora, the original posterior invagination is that of the posterior midgut rudiment, and that the proctodaeum arises later during the dorsal extension of the germ band and the differentiation of the embryonic membranes. Poulson shows, furthermore, that the anterior rudiment of midgut has its origin in an anterior invagination at the tip of the ventral furrow, with the lumen of which it is momentarily confluent. From this time on the rudiments of the midgut are separate from the mesodermal tube. The situation in Drosophila is thus essentially the same as that in the blowflies.

This is not the place to discuss at length the question of whether the anterior and posterior invaginations are to be considered continuous with, or separate from, the initial midventral invagination and therefore represent either the formation of a single continuous inner layer, the mesentoderm, or separate entodermal and mesodermal rudiments. Although the former is the case in some insects, including such lower Diptera as Sciara (Butt, 1934), it is pretty clear that the latter situation prevails generally among higher Diptera. In Drosophila, Calliphora, and Lucilia the cell groups which give rise to the midgut are separate from the moment of their invagination from those which produce the mesodermal organs. It is logical then to refer to these rudiments as entodermal in the general sense of the term. It is true that without experimental analysis we know little enough about the potencies of particular regions of the blastoderm. Critical extirpation and marking experiments are much needed if our understanding of localization in the blastoderm of Drosophila is to be furthered. Slight differences in rates of embryological processes may easily be responsible for such variation as an apparently continuous single inner layer in one form as against clearly separate entoderm and mesoderm rudiments in another. Richards (1932) has expressed this succinctly: "The question is purely a matter of the time of determination of the parts involved."

THE DEVELOPMENT OF THE GONADS

The sex of the gonad is already determined and can be distinguished with certainty in the freshly hatched larva, as shown by Kerkis (1931), who studied the development and the rate of growth of the gonads in wild-type larvae and pupae of *Drosophila melanogaster*. The size of the gonad and its relation to the adjacent fat body were the characteristics used to determine whether a gonad was male or female. Female gonads were considerably smaller in size than were male gonads and were embedded in the fat body, whereas male gonads were only bordered by that tissue. These gonad characters, namely, size and relation to the fat body, are the primary morphologic evidences of phenotypic sex to become apparent. The indifferent stage in gonad development, that is, one in which the sex of the gonad cannot be distinguished either anatomically or by histological examination, is not found in newly emerged larvae, and it remained to be determined whether such a condition existed earlier, during the embryonic period. To establish this, embryos have to be fixed and sectioned, since it is almost impossible technically to find and to dissect out the gonads from embryos and measure their dimensions, as Kerkis did with the postembryonic stages.

In some late embryos Poulson (1937) has noted a difference in the size of the gonads and thus extended the observations of Kerkis. However, he does not give any counts of the number of primordial germ cells which are included in these gonads (such counts being the only reliable means of determining gonad size), nor does he refer to the existence, if any, of an indifferent stage in the development of the genital rudiment. Later Sonnenblick

(1941) made counts of the number of germ cells present in gonads of embryos at successively older developmental stages, from the time the gonads are formed to the newly emerged larva. Gonads which differ markedly in size are found even as early as the tenth hour of development, when the primitive gonads are just formed and in their definitive position four segments from the posterior end of the embryo. Thus, with size as a morphologic indicator of the sex of the gonad, Sonnenblick could find no evidence for the existence of an indifferent stage in the development of the gonad.

A detailed account of the pole cells from the time of their segregation to their inclusion in the gonads, during which period they undergo devious wanderings, diminish in number, and are never seen to divide, will now be given. The polynuclear origin of the pole cells and their seemingly passive inclusion in the posterior midgut invagination forming below them have been previously described. However, before their passage into the posterior invagination and while blastodermic cell formation is occurring, an interblastodermal migration, involving the passage of a variable number of free, fully detached pole cells from the polar cap to the large central yolk mass takes place; this has been reported by Rabinowitz (1941a). That these spheroidal cells which press in between the elongate blastoderm nuclei are actually pole cells is indicated not only by their appearance but also by counts of the pole cells during different phases of the migratory process. Before the interblastodermal movement has begun, Rabinowitz finds an average of 55 pole cells, with a range in different eggs from 36 to 73, present in the amicropylar cluster (see our Fig. 24). While the movement is occurring, the average number of cells in the polar cap decreases to 39. He asserts that the movement between the incipient blastodermic cells is a normal and regular phenomenon and that the migrating cells eventually reach the yolk, where they completely degenerate and disappear. Thus not all the pole cells of Drosophila enter the interior of the organism in the same manner. In Melophagus, Lassmann (1936) has reported that cells "with apparently insufficient amount of germ plasm" can leave the region of the germ cell mass at the caudal end of the egg to become part of an internal posterior proliferation. Nuclei may then wander into the yolk from the posterior proliferation to become vitellophags or yolk nuclei.

There are then two methods by which the pole cells of Drosophila enter the body of the developing organism. One is the passage between the posterior blastodermic nuclei, and the other

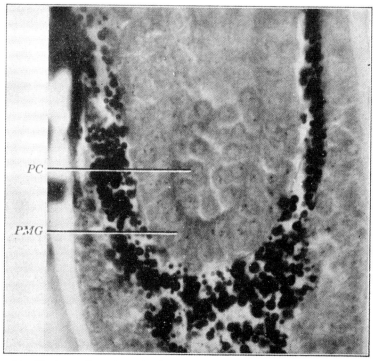

Fig. 34. Horizontal section of 6-hour-old embryo, showing pole cells in the interior of the embryo grouped at the blind end of the saclike posterior midgut rudiment. 600×. *PC*, pole cells; *PMG*, posterior midgut rudiment.

is by way of the posterior midgut invagination. The former method has been observed also in the eggs of several other flies, such as Sciara, Simulium, and Phormia by DuBois (1932), Gambrell (1933), and Auten (1934), respectively. Whereas in Drosophila both interblastodermal movement and inclusion in the posterior invagination occur, this early movement in the eggs of the other insects mentioned is the sole method by which incipient germ cells reach the interior of the animal. Reports indicate that the numbers of pole cells which are included in the gonadal rudiments of various insects differ markedly.

The data of Rabinowitz (1941a) show that from 15 to 44 pole cells, the average being 31, pass into the deepening posterior invagination. Counts made by another (Sonnenblick, 1941) in preparations of embryos of a different strain of Drosophila give

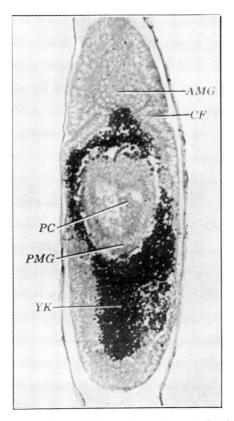

Fig. 35. Horizontal section of 6-hour-old embryo, indicating anterior rudiment below the stomodaeum with several mitoses, the yolk mass unenclosed, and pole cells loosely grouped in the posterior midgut invagination. 300×. *AMG*, anterior midgut rudiment; *CF*, anterior oblique cleft, cephalic furrow; *PC*, pole cells; *PMG*, posterior midgut rudiment; *YK*, yolk.

an average of 40 pole cells which enter the embryo (Figs. 29, 30), while in some organisms as many as 52 cells may be observed grouped at the end of the lengthened gut rudiment. It is possible that counts in additional preparations might bring these averages into closer approximation. At any rate, we can consider that about 31–40 pole cells usually make the initial passage into the depression forming below them. However, the study of embryonic gonad size, based on the number of enclosed primordia, made by the writer shows that many of the pole cells which enter by way of the posterior invagination are never incorporated in the formed gonads to function as primordial germ cells.

The average group of 31–40 pole cells which are in the midgut rudiment remains at the end of the diverticulum as it lengthens and grows caudally. In living material the cells can no longer be seen, but in preparations of 4- to 6-hour-old organisms they are observed loosely assembled in the posterior midgut rudiment, which appears saclike in horizontal sections (Figs. 34, 35). It has been reported (Poulson, 1937) that the pole cells which have passed to the inner end of the proctodaeum are readily recognizable by the presence in their cytoplasm of the polar granules originally found in the polar plasm and incorporated in the cells when they are segregated. The writer has not observed such granules in interiorly located pole cells. Fixation and staining procedures which earlier indicated the presence of the polar granules do not now show them. They may be present, of course, but in somewhat altered form which renders them undiscernible.

The pole cells remain quiescent at the inner end of the gut for 2 hours; sometime during the sixth hour of development varying numbers, but not all, of the cells pass through the wall of the gut rudiment into the body cavity. This migration through the gut wall, which cannot be seen in the living Drosophila embryo, has also been described for Melophagus by Lassmann (1936). The movements in Drosophila and Melophagus appear different from that described for Musca by Escherich (1902), for in Musca there is a definite canal in the gut wall through which the pole cells migrate. Such a canal does not appear to be present in Drosophila. In Calliphora, the pole cells pass through the gut rudiment while it is still being extended dorsally and is yet a shallow concavity (Noack, 1901). Parks (1936) believes that the phenomena in Drosophila are similar to that reported by Noack, whereas the views of the writer are more like that of Lassmann, reporting on Melophagus. The recent study of Aboim (1945) confirms the writer's account. The cells which Parks has figured between the invaginated blastoderm forming the "proctodaeal-amniotic invagination" and the yolk mass may possibly be some of the cells concerned in the interblastodermal migration.

From a study of embryos fixed with heated formol-alcohol-acetic acid mixture, in which the contours of individual cells stand out strikingly because of the alteration, if not destruction, of the intercellular cement substance (Figs. 36, 37), supplemented by preparations of embryos treated as usual with the same but cold fixative, one gathers the impression that the cells extend blunt pseudopodial-like protuberances and push their way singly through the hindgut wall. As the pole cells move between the columnar cells of the gut, they enlarge the spaces between the loosely aligned cells and distort the appearance of the gut components (Fig. 36). In 7- to 8-hour embryos the migration is completed. Not all the pole cells, however, penetrate into the body cavity. This has been observed in Drosophila by the writer and in Melophagus by Lassmann. The latter believes that the germ cells are able at one period only to pass between the cells of the gut rudiment, and those that fail to do so will disintegrate. He is apparently of the opinion that there are differences between pole cells, for he remarks that cells without a "sufficient quantity of germ plasm are unable to develop into germ cells." The significant factor determining whether a cell will become a functional primordium, he believes, is the amount of "germ plasm" accompanying the cleavage nuclei which first penetrate the posterior polar region. Just what is a sufficient amount of germ plasm?

As yet it has not been determined in Drosophila why certain of the originally segregated pole cells are not incorporated in the gonads but instead remain as part of the gut (Poulson, 1947). Differences in size or appearance of the pole cells are indiscernible to the observer. It may be presumed, furthermore, that the pole cells contain identical nuclear complexes, for the mitoses at the time of their formation appear in no way aberrant or different from any other divisions. Occasionally, in the enclosed yolk of the midgut of some older embryos, cells strongly resembling pole cells can be seen congregated (Fig. 57). These cells do not resemble and are unrelated to the gut epithelial cells. Since we do know that the number of pole cells included in both gonads of any one embryo is below the average number of cells entering the interior through the posterior invagination, it is possible that the trapped cells are some remaining members of those which never reached the gonads.

The wanderings of the pole cells are, even yet, incomplete. From their new position in the body cavity, the incipient germ cell primordia have to move or be moved to their final laterodorsal site four segments from the caudal end of the embryo. These

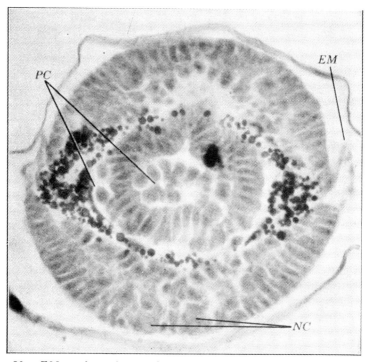

Fig. 36.   *EM,* embryonic membranes; *NC,* nerve cord; *PC,* pole cells.

Figs. 36 and 37.   Transverse sections of two embryos after heat fixation. Amoeboid nature of the pole cells indicated. The pole cells are passing through the gut wall, several having completed their migration. Simple embryonic membranes are evident laterally. The neural groove is seen in both figures, and the neuroblasts are prominent in Fig. 37. The embryos are approximately 6½ hours old.   600×.

movements are quite difficult to work out.   Parks (1936) makes no attempt to follow the pole cells into the gonad, and Poulson (1937) describes the final step in the process, namely, the enclosure of the primordia by elements of the inner layer.   The pole cells, the writer believes, do not migrate to their definitive location, but are forced or shunted there seemingly by the growing cellular strands of the midgut wall.

With the onset of metamerism (Figs. 44, 45, 62), the lengthy proctodaeal-amniotic invagination, opening on the dorsal surface and extending backward, begins to move caudally. Eventually, the proctodaeal opening reaches its final position (Figs. 46–50), thus reversing the movements started hours earlier when the lips

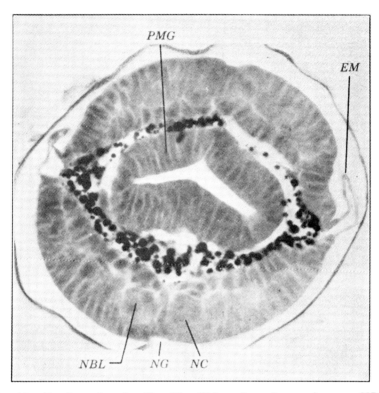

Fig. 37. See legend under Fig. 36. *EM*, embryonic membranes; *NBL*, neuroblasts; *NC*, nerve cord; *NG*, neural groove; *PMG*, posterior midgut rudiment.

of the newly formed diverticulum are carried anteriorly by the growth of the ventral blastoderm. As the proctodaeal opening retreats posteriorly, the other end of the lengthy tube turns forward, also reversing its position. The posterior midgut rudiment, closely united to the tip of the proctodaeum, becomes cleft. Strands extend out to encircle the large yolk mass and eventually unite with strands of the anterior rudiment coursing down from below the stomodaeum (Figs. 44, 50–52). These joined cellular ribbons form most of the midgut epithelium. In many preparations we now notice two groups of pole cells which appear at the sides of the posterior midgut in the lateral mesoderm. These free groups, as yet unsheathed, are laterally but somewhat dorsally situated below the yolk (Fig. 45). The primordial germ cells, for such they may now be called, are surrounded by a number of small cells derived from the mesodermal portion of the inner

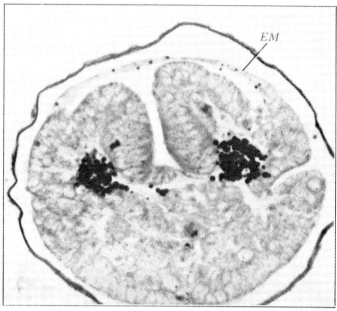

FIG. 38. Transverse section of embryo about 6 hours old. The embryonic membranes are seen as a double sheet, serosa external and amnion internal. Also note the membranes in the two previous figures. 600×. *EM*, embryonic membranes.

layer; from this material the sheath of the gonad and the associated fat body are differentiated. The movement of the pole cells, from the polar cap into the posterior invagination, to the blind end of the tube, through the wall of the gut, and the shunting to the final location four segments from the posterior pole are now (i.e., during the tenth hour of development) completed, and the encompassed cells are the gamete precursors (Figs. 55, 63).

In 14-hour-old embryos, the sheath or covering of the gonad has been differentiated (Fig. 56). It is quite distinct, comprising a circlet of minute cells about each group of primordia. The number of cells in the sheath has been determined for a number of gonads, with results indicating that approximately 26–37 cells make up any one sheath. The cells of the gonad cover are tiny, and care must be taken to distinguish them from the neighboring cells which are differentiating into the fat body. This latter structure, so prominent in larval sections, is composed at this time of aggregates of irregularly shaped cells which have as yet no noticeable fat or other secretory deposits or inclusions.

From the time the pole cells are segregated and clustered in the amicropylar polar cap and throughout all the wanderings just recounted, they are never observed to divide. Examination of these cells in many embryos at progressive levels of development gives no evidence of proliferation. Until the pole cells are incorporated into the gonads and have completed their wanderings, they are apparently unable to multiply. As a matter of fact, it is not until the sixteenth hour of development, that is, 6 hours after the rudimentary gonads have formed and about 14 hours after the pole cells are segregated, that germ cell mitoses are noted for the first time (Sonnenblick, 1941).

It was mentioned at the beginning of this section that sex differentiation of the gonad has already occurred in larvae newly emerged from the embryonic cases (Kerkis, 1931). In 6- to 10-hour-old larvae, the male gonads are developed enough to be visible through the larval integument, enabling one to separate the sexes of living larvae quite accurately without recourse to dissection and measurement. Gonads in female larvae are minute, and Kerkis remarks that "the young female gonads are so small that it is difficult to measure them as accurately as the male gonads or the older female gonads." The reason for this, as we shall see, is that so few cells make up a gonad, especially a female gonad, in young larvae.

To study the growth of the gonads in embryonic stages, material was fixed, and the germ cell primordia in 110 gonads were counted. On the basis of these counts certain comments can be made. A size difference is discernible as soon as the primordia have arrived at their permanent location. This disparity in volume can be traced from the 10-hour embryo on until the young larva hatches. The variations in volume result, primarily, from the number of apparently identical primordia enclosed in the sheath rather than any difference in the size of the individual germ cells. In 10- to 12-hour embryos, gonads are observed containing, for example, the following numbers of germ cells: 4–5, 6–6, 7–7, 9–10, 10–10, 10–11, 11–11, 12. In the last instance, the number of primordia in only one member of the gonad pair could be reliably determined. The numbers of primordia in the two members of any pair of gonads are virtually the same. In all the instances studied there were only two in which the members

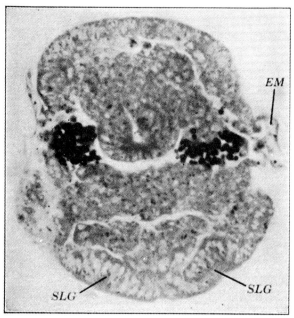

Fɪɢ. 39. An obliquely cut section of an 8–8½-hour-old embryo, showing early salivary glands as paired ectodermal plates. 600×. *EM*, embryonic membranes; *SLG*, salivary gland.

of a gonad pair gave counts which differed to some degree. In one embryo there were 5 and 9 germ cells in the gonad pair, while in another embryo the gonads had 8 and 13 of the cells. These are definitely exceptional counts, since the remaining counts gave identical, or almost identical, counts in the gonad pairs.

When one recalls that some 31–40 pole cells, and sometimes more as indicated above, enter the interior of the body through the posterior invagination but that gonads are formed containing but 5, 6, or 7 primordia, it is obvious that numbers of the pole cells of the amicropylar polar cap are frequently destined never to function as stem germ cells. An hereditary alteration in a pole cell, either spontaneous or induced (see Altenburg, 1933, 1934), may readily be eliminated with the pole cell containing it during the course of the pole cell movements.

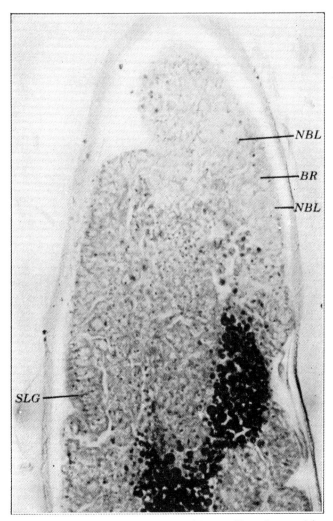

Fɪɢ. 40. Longitudinal section of an 8–8½-hour-old embryo with salivary glands. The supraoesophageal ganglion is here being delimited, and large, clear dividing neuroblasts, as well as smaller non-dividing ganglion cells, can be seen. 600×. *BR*, brain, supraoesophageal ganglion; *NBL*, neuroblasts; *SLG*, salivary gland.

After the initiation of the movements of the pole cells in the early embryo, no mitoses are ever observed among these cells, as we have noted above, until the embryo has incubated for 16 hours. While in the midgut invagination and in the body cavity, and also during the first few hours after their inclusion in the formed gonads, the primordia appear quiescent. Between the sixteenth hour of development and hatching, one and possibly two mitoses occur. The primordia in gonads of these later embryos range from 8 to 38. These figures are of interest, since Kerkis observed that the ratio of the volumes of the larger (male) gonads to the smaller (female) gonads in newly emerged larvae was of the order of $3\frac{1}{2}$ to 1. We can also understand why Kerkis had difficulty in determining, through measurements after their removal, the dimensions of female gonads in young larvae, since, after all, so very few cells are contained in a female gonad in a late embryo. Should even one or two further mitotic divisions take place during the first larval hours, such gonads would still be extremely small and most difficult to measure.

This account of embryonic gonad development is essentially that presented in 1941 by the writer. It is in accord with that of Aboim (1945), who has made a careful study of the origin and structure of normal gonads and of experimentally produced agametic gonads. He describes the pole cells as passing between the gut cells during the seventh and ninth hours as indicated above, although his evidence suggests purely passive movement. Subsequently some of the pole cells pass into the lateral mesoderm on both sides, and with the shortening of the embryo in the ninth to tenth hours these germinal pole cells become concentrated at the level which the definitive gonad is to occupy.

THE SALIVARY GLANDS

The salivary glands make their initial appearance about the eighth hour of development (Sonnenblick, 1939). In transverse sections they appear as lateroventral ectodermal plates immediately adjacent to the ventral nerve cord (Figs. 39–41). Poulson believed at first (1937) that the glands originated as invaginations from the ventral wall of the pharynx, but his later observations (1940) showed that the glands arise as ectodermal infoldings. As ingrowth of the paired plates continues and the early shallow trough deepens, the external aperture contracts, leaving a small orifice on the surface of the organism (Figs. 42, 43). It is not until some 3 hours after the invagination of the ectodermal plates that differentiation into duct and gland proper is noticed. Until this time all the cells comprising the gland appear to be similar in size and appearance, and no differentiation has occurred.

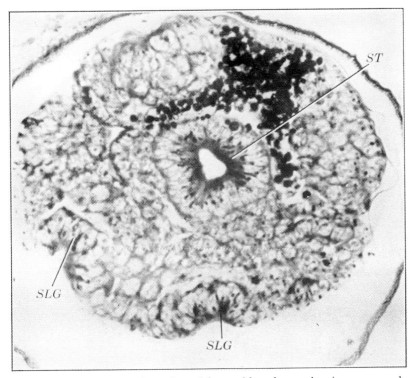

Fig. 41. Transverse section of an 8-hour-old embryo, showing very early paired salivary glands as shallow ventral plates. 760×. *SLG*, salivary gland; *ST*, stomodaeum.

In cross-section, seen in Fig. 59, the glands are noted extending into the body cavity lateral to the foregut from the second to the fifth body segment. By the twelfth hour of development, however, the anterior end of the gland, which is to become the duct, has thinned out considerably. The cells of the duct and their nuclei are decidedly smaller than are the cells and nuclei of the gland proper. The ducts lengthen and approach one another, so that in 16-hour embryos they have united medially (Fig. 60) to

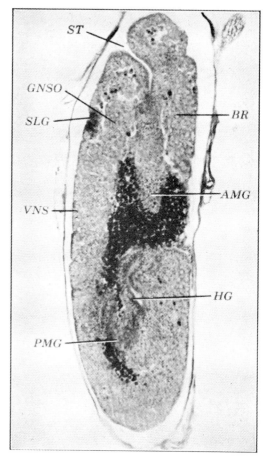

Fig. 42. *AMG*, anterior midgut rudiment; *BR*, brain, supraoesophageal ganglion; *GNSO*, suboesophageal ganglion; *HG*, hindgut; *PMG*, posterior midgut rudiment; *SLG*, salivary gland; *ST*, stomodaeum; *VNS*, ventral nervous system.

Figs. 42 and 43. Sagittal sections of a 9-hour-old embryo. Segmentation evident, especially along the ventral surface. The salivary glands have deepened, but an external orifice still remains. Cleft stomodaeal strands coursing posteriorly are seen in Fig. 43. Hindgut more anterior than posterior midgut rudiment at this time. Malpighian diverticula are noted in these figures. The supra- and suboesophageal ganglia are evident. 300×.

form a common duct which runs ventrally and passes into the floor of the pharynx (Fig. 61).

Beginning with the twelfth hour of development an intensive, deeply stained mass appears in the lumen of the gland. This

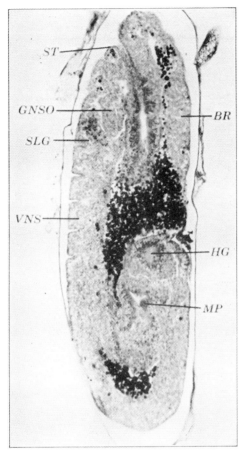

FIG. 43. See legend under Fig. 42. *BR*, brain, supraoesophageal ganglion; *GNSO*, suboesophageal ganglion; *HG*, hindgut; *MP*, Malpighian tubules; *SLG*, salivary gland; *ST*, stomodaeum; *VNS*, ventral nervous system.

streak, characteristic for and distinct in all glands 12 hours of age and older, can be observed in material sectioned in various ways. In sagittal and horizontal sections the pycnotic mass appears as a compact rod (Figs. 51–54, 58, 60), while in transverse section it is circular (Fig. 59). It is very likely that the glandular cells have begun to function, and that the accumulated secretion present in the lumen of the gland readily absorbs haematoxy-

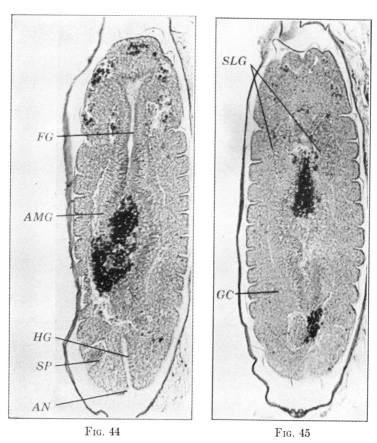

Fig. 44   Fig. 45

Figs. 44 and 45. Horizontal sections of an embryo 10½ hours old. These are ventral sections with segmentation clearly indicated. There are twelve segments, the first being the largest. Paired salivary glands are evident, and a posterior spiracle is noted in Fig. 44. Junction of the anterior and posterior midgut strands to form the ventral wall of the midgut, enclosing the yolk. Hindgut in definitive position. 300×. *AMG,* anterior midgut rudiment; *AN,* anus; *FG,* foregut; *GC,* germ-cell primordia; *HG,* hindgut; *SLG,* salivary gland; *SP,* spiracle.

lin.   In sections stained with toluidine blue the secretion stains metachromatically.

From the twelfth to sixteenth hours of development the cytoplasm of the gland cells becomes vacuolated; clear, vacuolar areas differing in size are noted in the cells. The presence of these vacuoles, together with the substance in the lumen of the gland, which readily accepts various stains, is further indication of secretory activity in these early glands.

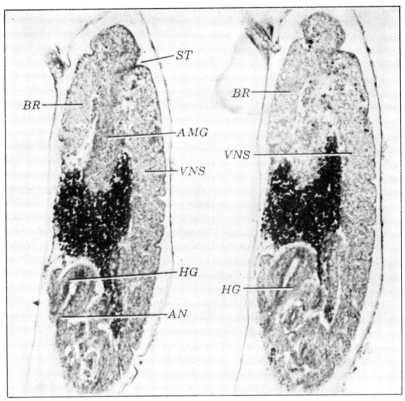

FIG. 46   FIG. 47

FIGS. 46 and 47. Successive sagittal sections from an embryo 9½ hours old. The opening of the proctodaeal tube is no longer anteroventrally but is moving posteriorly. Segmentation is becoming evident. The anterior rudiment below the stomodaeum is seen, as are the supraoesophageal (dorsal) ganglion and the ventral nerve cord. 300×. *AMG,* anterior midgut rudiment; *AN,* anus; *BR,* brain, supraoesophageal ganglion; *HG,* hindgut; *ST,* stomodaeum; *VNS,* ventral nervous system.

Examination of the glands in several hundred embryos permits the statement that no mitoses ever occur in salivary gland cells. From the time of origin of the glands as simple ectodermal plates through the period of differentiation in the later embryo, no cell division is noted. Increase in size of the glands is due solely to growth of the component cells.

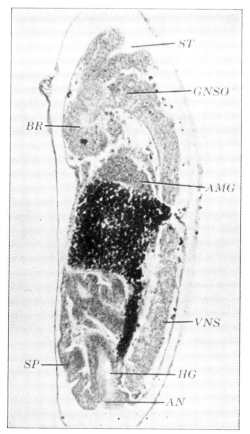

FIG. 48. *AMG*, anterior midgut rudiment; *AN*, anus; *BR*, brain, supraoesophageal ganglion; *GNSO*, suboesophageal ganglion; *HG*, hindgut; *SP*,
spiracle; *ST*, stomodaeum; *VNS*, ventral nervous system.

FIGS. 48, 49, and 50. Sagittal sections from a 10-hour-old embryo. The anal
opening is now at the posterior pole (Fig. 48). The hindgut passes forward,
then ventrally. Note the strands which will unite to form the midgut epithelium and enclose the yolk. The anterior strands are evident in Fig. 50,
and the posterior strands in Figs. 49 and 50. The nerve cord is separated
from the ventral ectoderm and is now restricted to the ventral side except
for the large dorsal ganglion. A tracheal opening is seen posterodorsally.
300×.

CESSATION OF MITOSES IN CERTAIN EMBRYONIC TISSUES

Since it is known that many larval tissues, especially those which are to undergo histolysis at metamorphosis, grow by in-

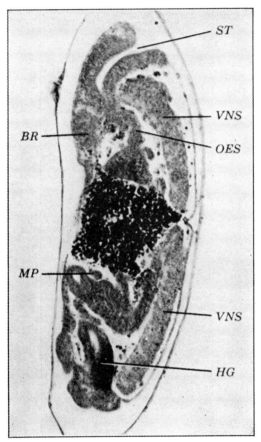

FIG. 49. See legend under Fig. 48. *BR*, brain, supraoesophageal ganglion; *HG*, hindgut, *MP*, Malpighian tubules; *OES*, oesophagus; *ST*, stomodaeum; *VNS*, ventral nervous system.

crease in cell size rather than by cell multiplication, the writer has attempted to find where and when in the developmental history of various tissues the mitoses cease. Some evidence has been found pertaining to the pole cells, the gut components, and the salivary glands.

It has already been noted that the pole cells do not divide from the period when the varied pole cell movements begin in eggs 3 hours of age until the embryo is 16 hours old. At this time the

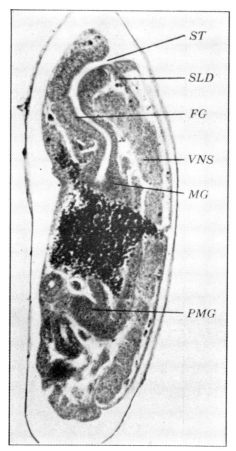

Fig. 50. See legend under Fig. 48. *FG*, foregut; *MG*, midgut; *PMG*, posterior midgut rudiment; *SLD*, salivary gland duct; *ST*, stomodaeum; *VNS*, ventral nervous system.

germ cell primordia have already been incorporated in the definitive gonad for 6 hours. Thus before the formation of the gonads and for some time afterward the pole cells and future germ cells do not reproduce. The events after the sixteenth hour of development are given in the section on gonad development.

With regard to the alimentary canal, it has been found that no gut cells can be observed dividing in embryos older than 12 hours, when the openings of the fore- and hindintestines are located at

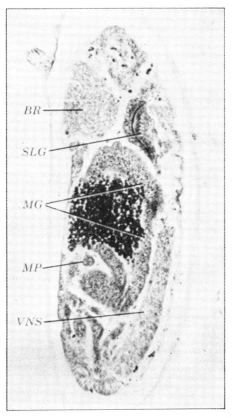

FIG. 51. Sagittal section of a 12-hour-old embryo. Anterior and posterior midgut rudiments are enclosing the yolk, especially evident ventrally. A salivary gland with deeply staining material in its lumen is noted. The large dorsal ganglion and the ventral nerve cord are seen. 300×. *BR*, brain, supraoesophageal ganglion; *MG*, midgut; *MP*, Malpighian tubules; *SLG*, salivary gland; *VNS*, ventral nervous system.

the anterior and posterior poles and the yolk is within the midgut epithelium. (Notice Figs. 56–58, 64–66.) Previously, mitoses could be noticed in the rudiments and in the cellular accumulations at the blind ends of the invaginations. During the latter part of embryogenesis, through the period of further differentiation of the alimentary canal while the embryo is still within the egg envelopes, and subsequently throughout the larval stages, the component cells do not proliferate. Growth of the gut tissues will therefore occur by increase in cell size throughout the

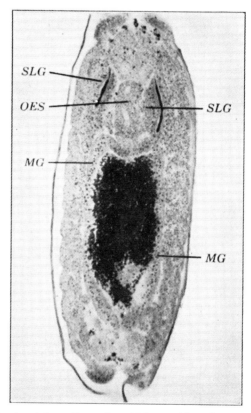

FIG. 52. A horizontal section of a 12-hour-old embryo, showing paired salivary glands with a secretion product in the lumen. The yolk is enclosed by the midgut. 300×. *MG*, midgut; *OES*, oesophagus; *SLG*, salivary gland.

embryonic-larval periods beginning about the twelfth hour of embryonic development. Since the nuclei of the gut cells in larvae of *D. melanogaster* (Cooper, 1938) and of *D. virilis* (Makino, 1938) have been reported as possessing polytene chromosomal complexes, it is possible that polyploidy can occur in the intestinal elements at any time during the latter half of the embryonic period.

In regard to the salivary glands, it has been noted that no mitoses are observed in these cells from the time of their origin and all through the remainder of embryonic development.

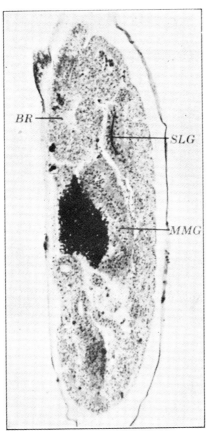

FIG. 53. Sagittal section of an embryo about 12 hours old. The thick ventral wall of the midgut, a salivary gland, and the supraoesophageal ganglion can be seen. 300✕. *BR*, brain, supraoesophageal ganglion; *MMG*, middle rudiment of midgut; *SLG*, salivary gland.

THE RATE OF EARLY DEVELOPMENT

The eggs of many of the higher Diptera develop at rates which are very rapid when compared with those of other insects. If one surveys the information available concerning the species of muscids and of Drosophilidae which have received any embryological attention, it is clear that they differ less in the rate

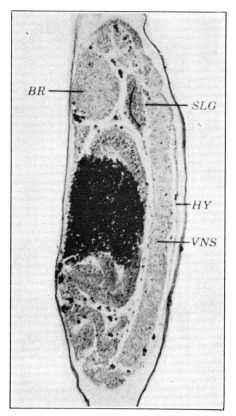

FIG. 54. Sagittal section of an organism which is slightly older than that in Fig. 53. A salivary gland with an attenuated tail is noted. 300×. *BR*, brain, supraoesophageal ganglion; *HY*, hypoderm; *SLG*, salivary gland; *VNS*, ventral nervous system.

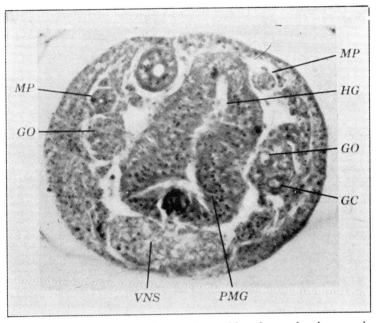

F<small>IG</small>. 55. Transverse section of an 11-hour-old embryo, showing a pair of gonads, with germ-cell primordia, near the site of junction of the hindgut and midgut. Sections of the Malpighian tubules, as well as the dual character of the ganglia of the ventral nerve cord, are evident in this figure. 600×. *GC*, germ-cell primordia; *GO*, gonad; *HG*, hindgut; *MP*, Malpighian tubules; *PMG*, posterior midgut rudiment; *VNS*, ventral nervous system.

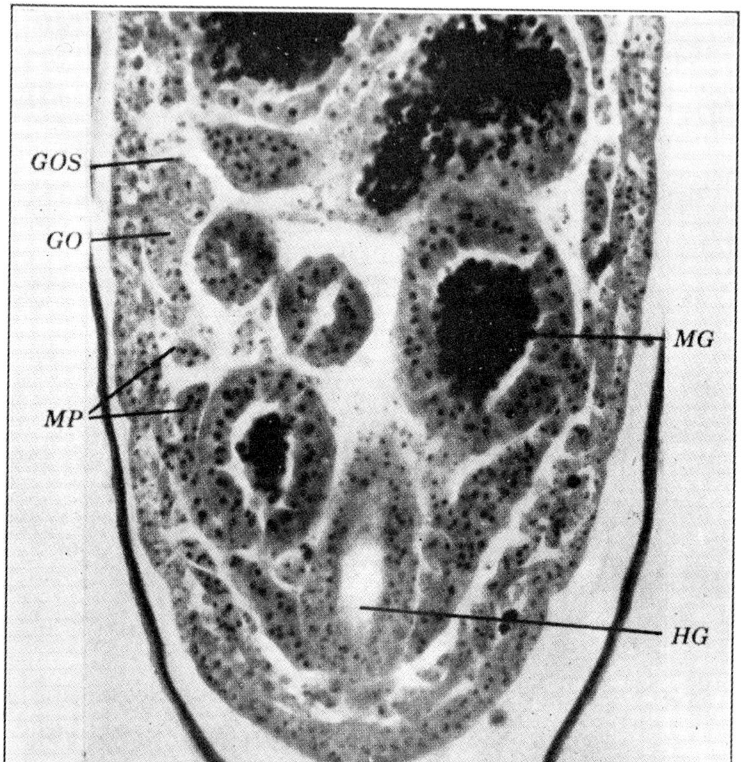

FIG. 56. Section of a 13–14-hour-old embryo to show a gonad in its definitive location four segments from the posterior end of the organism. Note the variety of cell types in the gonad, midgut, hindgut, wall of the embryo, and Malpighian tubule with no mitotic activity apparent. 600×. *GO*, gonad; *GOS*, gonad sheath; *HG*, hindgut; *MG*, midgut; *MP*, Malpighian tubules.

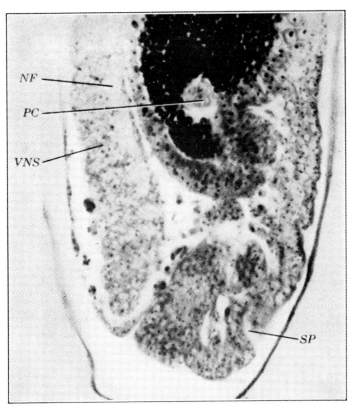

FIG. 57. Section from a 13-hour-old embryo with a cluster of small round cells, resembling pole cells, in the midgut near the region where the hindgut earlier contributed to formation of the midgut wall. The ventral epithelial wall has separated from the underlying nerve cord. Clear nerve fibrous substance in the cord is present. 600×. *NF*, nerve fibers and tracts; *PC*, pole cells; *SP*, spiracle; *VNS*, ventral nervous system.

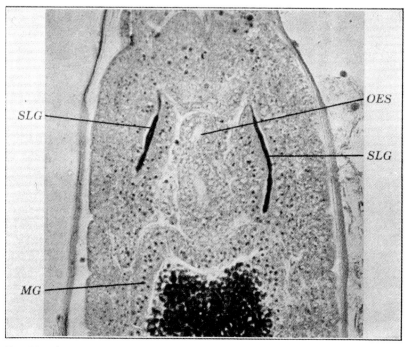

FIG. 58. A ventral horizontal section of a 12–13-hour-old embryo to show the position and arrangement of the salivary glands. 600×. *MG*, midgut; *OES*, oesophagus; *SLG*, salivary gland.

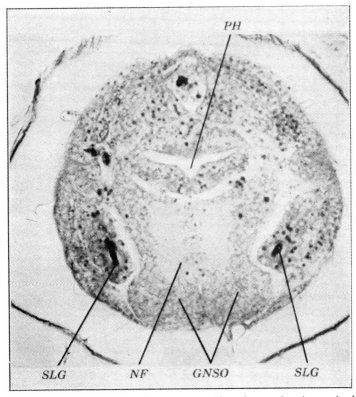

Fig. 59.   Transverse section of a 15-hour-old embryo, showing paired salivary glands and the suboesophageal ganglion of the nerve cord.   The lateral
nodes, the commissures, and the nerve fibrous substance are evident.   600×.
GNSO, suboesophageal ganglion; NF, nerve fibers and tracts; PH, pharynx;
SLG, salivary gland.

of early development than in the later stages. Since the total time of development ranges between 20 and 36 hours at ordinary temperatures, it is obvious that some precision in staging is essential. Likewise, the control of temperature is important, as indicated by the date of Powsner (1935) concerning total time of egg development and the data of Rabinowitz (1941a) concerning the earlier stages. Although no complete account of the effects of temperature at different stages is available, the studies of Rabinowitz (1941a) provide a model for further work.

Tables 2 and 3, which are taken from Rabinowitz's paper, provide the most accurate timetable of the events of early development and will be indispensable to anyone concerned with experimental work on early stages.

TABLE 2

TIMETABLE OF EVENTS AT VERY EARLY STAGES OF EGG DEVELOPMENT
(Rabinowitz, 1941a)

Temperature: 24°C; total number of eggs: 354

Stage of Development	Number of Eggs	Mean Age (in minutes)
Telophase of 2nd maturation division to conjugation of pronuclei	13	15 ± 1.21
1st division of cleavage nuclei	10	23 ± 1.72
2nd " " " "	14	34 ± 1.72
3rd " " " "	17	47 ± 1.47
4th " " " "	10	53 ± 1.05
5th " " " "	18	60 ± 0.57
6th " " " "	22	70 ± 0.58
7th " " " "	14	78 ± 1.28
8th " " " "	31	93 ± 1.13
1st " " blasteme "	82	99 ± 0.60
2nd " " " "	97	109 ± 0.51
3rd " " " "	24	120 ± 0.90

Finally, no attempt will be made to deal here with the further stages of development and of organogenesis, although considerable work has been done in this direction by the author. Interruption of the project by World War II made necessary a change in the original plan, and the later stages are dealt with by Poulson in Chapter 3 on organogenesis. The texts and illustrative materials of the two chapters comprise a rather thorough study of Drosophila embryonic development.

TABLE 3

TIMETABLE OF EVENTS AT EARLY STAGES OF EGG DEVELOPMENT
(Rabinowitz, 1941a)

Temperature: 25°C; total number of eggs: 172

Stage of Development	Number of Eggs	Mean Age (in minutes)
Blastoderm nuclei in early interkinesis, nuclei small	10	144 ± 1.70
Blastoderm nuclei increasing in size, large rounded nuclei with chromatic masses	19	153 ± 1.16
Blastoderm nuclei elongating; bases of furrows reach outer edges of nuclei	15	156 ± 1.48
Blastoderm nuclei elongated; bases of furrows extend from outer edges of nuclei to halfway down nuclei	10	171 ± 2.84
Blastoderm nuclei elongated; bases of furrows extend from halfway down nuclei to halfway between nuclei and yolk	32	182 ± 1.61
Furrows reach yolk on ventral side; gastrulation occurring; amnio-proctodaeal * invagination not formed	23	186 ± 1.69
Beginning of amnio-proctodaeal * invagination to deepening of it; migration of pole cells into amnio-proctodaeal * invagination	44	215 ± 1.00
Amnio-proctodaeal *invagination deep, extends posteriorly; tail fold has progressed forward; amnion and serosa formed; opening of amnio-proctodaeal invagination anterior; pole cells deep in invagination	19	234 ± 0.94

* Posterior invagination to form midgut rudiment. Actual proctodaeal invagination arises later; see text.

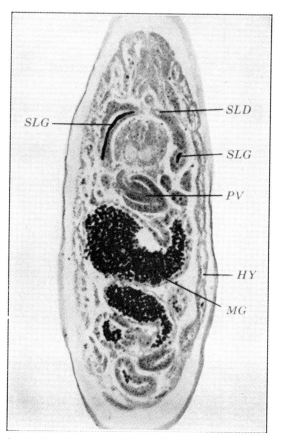

Fig. 60. Horizontal section of a 16-hour-old embryo, showing ducts of salivary glands approaching each other medially. The proventriculus is present. The midgut is no longer a simple yolk-filled reservoir as in Fig. 54 but is becoming longer and convoluted. The hypodermis has been delimited. 300×. *HY*, hypoderm; *MG*, midgut; *PV*, proventriculus; *SLD*, salivary gland duct; *SLG*, salivary gland.

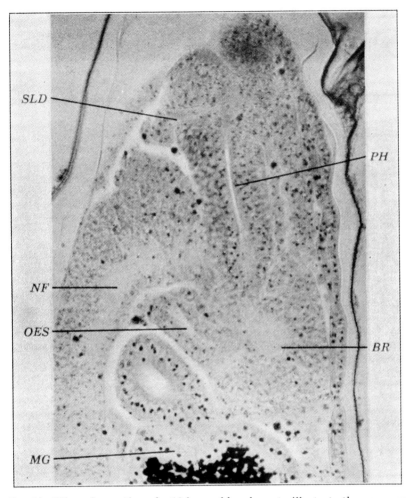

FIG. 61. View of a portion of a 16-hour-old embryo to illustrate the common duct of the salivary glands entering the floor of the pharynx. Nervous tissue with fibrous substance and elements of the fore- and midgut are clearly shown in this section. 600×. *BR*, brain, supraoesophageal ganglion; *MG*, midgut; *NF*, nerve fibers and tracts; *OES*, oesophagus; *PH*, pharynx; *SLD*, salivary gland duct.

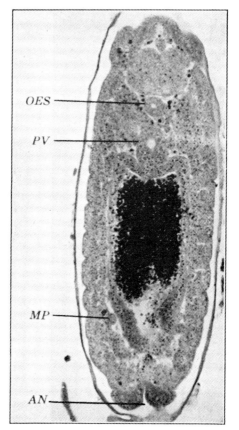

Fig. 62. *AN,* anus; *MP,* Malpighian tubules; *OES,* oesophagus; *PV,* proventriculus.

Figs. 62 and 63. Horizontal sections of an 11–12-hour-old embryo, showing fusion of ganglia around the oesophagus. Malpighian tubules are seen, as are the paired gonads in their permanent location. 300✕.

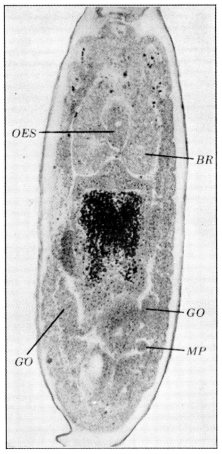

FIG. 63. See legend under Fig. 62. *BR*, brain, supraoesophageal ganglion; *GO*, gonad; *MP*, Malpighian tubules; *OES*, oesophagus.

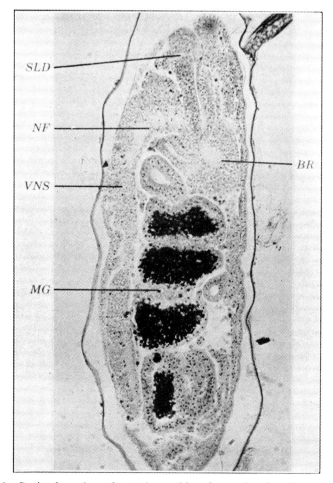

FIG. 64. Sagittal section of a 16-hour-old embryo, showing the contracting ventral nerve cord, the now convoluted midgut, and the common salivary gland duct entering the ventral wall of the pharynx. 300×. *BR*, brain, supraoesophageal ganglion; *MG*, midgut; *NF*, nerve fibers and tracts; *SLD*, salivary gland duct; *VNS*, ventral nervous system.

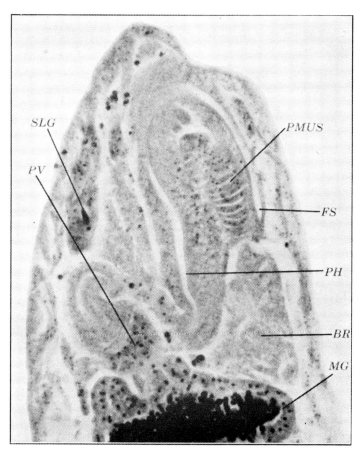

FIG. 65. Section of a 16–17-hour-old embryo to show elements of the fore-gut. Especially note the pharyngeal musculature, the skeletal bars of which contain syncytial nuclei, and the proventriculus at the site of junction with the midgut. 600×. *BR*, brain, supraoesophageal ganglion; *FS*, frontal sac; *MG*, midgut; *PH*, pharynx; *PMUS*, pharyngeal musculature; *PV*, proventriculus; *SLG*, salivary gland.

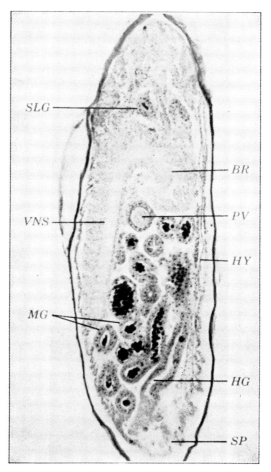

FIG. 66. In this sagittal section of an 18-hour-old embryo, note the contracted ventral nerve cord and dorsal ganglion, the salivary gland, the highly convoluted and still yolk-filled midgut, and the larval hypodermis. 300×. *BR*, brain, supraoesophageal ganglion; *HG*, hindgut; *HY*, hypoderm; *MG*, midgut; *PV*, proventriculus; *SLG*, salivary gland; *SP*, spiracle; *VNS*, ventral nervous system.

LITERATURE CITED

ABOIM, A. N. 1945. Développement embryonnaire et post-embryonnaire des gonades normales et agamétiques de Drosophila melanogaster. Rev. suisse zool. **52**:53–154.

ALTENBURG, E. 1933. The production of mutations by ultra-violet light. Science **78**:587.

ALTENBURG, E. 1934. The artificial production of mutations by ultra-violet light. Am. Naturalist **68**:491–507.

AUTEN, M. 1934. The embryological development of Phormia regina: Diptera (Calliphoridae). Ann. Entomol. Soc. Am. **27**:481–506.

BUTT, F. H. 1934. Embryology of Sciara. Ann. Entomol. Soc. Am. **27**:565–579.

CHILD, G. P., and R. B. HOWLAND. 1933. Observations on early developmental processes in the livîng egg of Drosophila. Science **77**:396.

COOPER, K. W. 1938. Concerning the origin of the polytene chromosomes of Diptera. Proc. Natl. Acad. Sci. **24**:452–458.

DARLINGTON, C. D. 1937. Recent Advances in Cytology. 2nd ed. Blakiston, Philadelphia.

DELBRÜCK, M. 1941. A theory of autocatalytic synthesis of polypeptides and its application to the problem of chromosome reproduction. Cold Spring Harbor Symposia Quant. Biol. **9**:122–126.

DEMEREC, M. 1941. Unstable genes in Drosophila. Cold Spring Harbor Symposia Quant. Biol. **9**:145–150.

DOBZHANSKY, TH. 1934. Studies on hybrid sterility. I. Spermatogenesis in pure and hybrid Drosophila pseudoobscura. Z. Zellforsch. u. mikroskop. Anat. **21**:169–223.

DUBOIS, A. M. 1932. A contribution to the embryology of Sciara (Diptera). J. Morphol. **54**:161–195.

EASTHAM, L. E. S. 1930. The formation of germ layers in insects. Biol. Rev. **5**:1–29.

ESCHERICH, K. 1902. Zur Entwicklung des Nervensystems der Musciden, mit besonderer Berücksichtigung des sog. Mittelstranges. Z. wiss. Zoöl. **71**:525–549.

GAMBRELL, L. H. 1933. The embryology of the black fly Simulium pictipes (Hagen). Ann. Entomol. Soc. Am. **26**:641–671.

GOWEN, J. W. 1933. Meiosis as a genetic character in Drosophila melanogaster. J. Exptl. Zoöl. **65**:83–106.

HALLEZ, P. 1886. Sur la loi de l'orientation de l'embryon chez les insectes. Compt. rend. **103**:606–608.

HEGNER, R. W. 1914. The Germ-cell Cycle in Animals. Macmillan, New York.

HELFER, R. G. 1940. Two X-ray induced mosaics in Drosophila pseudoobscura. Proc. Natl. Acad. Sci. **26**:3–7.

HELFER, R. G. 1941. A comparison of X-ray induced and naturally occurring chromosomal variations in Drosophila pseudoobscura. Genetics **26**:1–22.

HILL, D. L. 1945. Chemical removal of the chorion from Drosophila eggs. *Drosophila Information Service* 19:62.

HOWLAND, R. B., and G. P. CHILD. 1935. Experimental studies on development in *Drosophila melanogaster*. I. Removal of protoplasmic materials during late cleavage and early embryonic stages. *J. Exptl. Zoöl.* 70:415–424.

HOWLAND, R. B., and B. P. SONNENBLICK. 1936. Experimental studies on development in *Drosophila melanogaster*. II. Regulation in the early egg. *J. Exptl. Zoöl.* 73:109–125.

HUETTNER, A. F. 1923. The origin of the germ cells in *Drosophila melanogaster*. *J. Morphol.* 37:385–423.

HUETTNER, A. F. 1924. Maturation and fertilization in *Drosophila Melanogaster*. *J. Morphol.* 39:249–265.

HUETTNER, A. F. 1927. Irregularities in the early development of the *Drosophila melanogaster* egg. *Z. Zellforsch. u. mikroskop. Anat.* 4:599–610.

HUETTNER, A. F. 1933. Continuity of the centrioles in *Drosophila melanogaster*. *Z. Zellforsch. u. mikroskop. Anat.* 19:119–134.

HUETTNER, A. F., and M. RABINOWITZ. 1933. Demonstration of the central body in the living cell. *Science* 78:367–368.

JOHANNSEN, O. A., and F. H. BUTT. 1941. *Embryology of Insects and Myriopods.* McGraw-Hill, New York.

KAHLE, W. 1908. Die Paedogenese der Cecidomyiden. *Zoologica* (Stuttgart) 21:1–80.

KALISS, NATHAN. 1939. The effect on development of a lethal deficiency in *Drosophila melanogaster:* with a description of the normal embryo at the time of hatching. *Genetics* 24:244–270.

KAUFMANN, B. P. 1934. Somatic mitoses of *Drosophila melanogaster*. *J. Morphol.* 56:125–155.

KAUFMANN, B. P. 1940. The nature of hybrid sterility—abnormal development in eggs of hybrids between *Drosophila miranda* and *Drosophila pseudoobscura*. *J. Morphol.* 66:197–213.

KAUFMANN, B. P. 1941. Induced chromosomal breaks in Drosophila. *Coid Spring Harbor Symposia Quant. Biol.* 9:82–92.

KAUFMANN, B. P., and M. DEMEREC. 1942. Utilization of sperm by the female *Drosophila melanogaster*. *Am. Naturalist* 76:445–469.

KERKIS, J. 1931. The growth of the gonads in *Drosophila melanogaster*. *Genetics* 16:212–224.

KORSCHELT, E., and K. HEIDER. 1899. *Textbook of the Embryology of Invertebrates.* Vol. 3. Macmillan, New York.

LASSMANN, G. W. P. 1936. The embryological development of *Melophagus ovinus* L., with special reference to the development of the germ cells. *Ann. Entomol. Soc. Am.* 29:397–413.

LILLIE, F. R. 1906. Observations and experiments concerning the elementary phenomena of development in *Chaetopterus*. *J. Exptl. Zoöl.* 3:153–268.

LOWNE, B. T. 1890–1895. *The Anatomy, Physiology, Morphology and Development of the Blow-fly.* Vols. 1–2. R. H. Porter, London.

MAKINO, S. 1938. A morphological study of the nucleus in various kinds of somatic cells of *Drosophila virilis. Cytologia* **9**:272–282.

MECZNIKOW, E. 1866. Embryologische Studien an Insekten. *Z. wiss. Zoöl.* **16**:389–500.

MORGAN, L. V. 1939. A spontaneous somatic exchange between nonhomologous chromosomes in *Drosophila melanogaster. Genetics* **24**:747–752.

MULLER, H. J. 1941. Résumé and perspectives of the symposium on genes and chromosomes. *Cold Spring Harbor Symposia Quant. Biol.* **9**:290–308.

NELSON, J. A. 1915. *The Embryology of the Honey Bee.* Princeton University Press.

NOACK, W. 1901. Beiträge zur Entwicklungsgeschichte der Musciden. *Z. wiss. Zoöl.* **70**:1–57.

NONIDEZ, J. F. 1920. The internal phenomena of reproduction in Drosophila. *Biol. Bull.* **39**:207–230.

PAINTER, T. S. 1940. On the synthesis of cleavage chromosomes. *Proc. Natl. Acad. Sci.* **26**:95–100.

PAINTER, T. S., and E. REINDORP. 1939. Endomitosis in the nurse cells of the ovary of *Drosophila melanogaster. Chromosoma* **1**:276–283.

PARKS, H. B. 1936. Cleavage patterns in Drosophila and mosaic formation. *Ann. Entomol. Soc. Am.* **29**:350–392.

PATTEN, W. 1884. The development of phryganids with a preliminary note of *Blatta germanica. Quart. J. Microscop. Sci.* **24**:549–602.

PATTERSON, J. T. 1929. The production of mutations in somatic cells of *Drosophila melanogaster* by means of X-rays. *J. Exptl. Zoöl.* **53**:327–372.

PATTERSON, J. T. 1931. The production of gynandromorphs in *Drosophila melanogaster* by X-rays. *J. Exptl. Zoöl.* **60**:173–211.

PATTERSON, J. T., and W. STONE. 1938. Gynandromorphs in *Drosophila melanogaster. Univ. Texas Publ.* 3825.

POULSON, D. F. 1937. The embryonic development of *Drosophila melanogaster. Actualités sci. et ind.* 498. Hermann et Cie., Paris.

POULSON, D. F. 1940. The effects of certain X-chromosome deficiencies on the embryonic development of *Drosophila melanogaster. J. Exptl. Zoöl.* **83**:271–325.

POULSON, D. F. 1945. On the origin and nature of the ring gland (Weismann's ring) of the higher Diptera. *Trans. Conn. Acad. Arts and Sci.* **36**:449–487.

POULSON, D. F. 1947. The pole cells of Diptera, their fate and significance. *Proc. Natl. Acad. Sci.* **33**:182–184.

POWSNER, L. 1935. The effects of temperature on the durations of the developmental stages of *Drosophila melanogaster. Physiol. Zoöl.* **8**:474–520.

PRATT, H. S. 1900. The embryonic history of imaginal discs in *Melophagus ovinus* L., together with an account of the earlier stages in the development of the insect. *Proc. Boston Soc. Natl. Hist.* **29**:241–272.

RABINOWITZ, M. 1941a. Studies on the cytology and early embryology of the egg of *Drosophila melanogaster*. *J. Morphol.* **69**:1–49.

RABINOWITZ, M. 1941b. Yolk nuclei in the egg of *Drosophila melanogaster*. *Anat. Record* **81**, Suppl. 2:80–81.

REITH, F. 1925. Die Entwicklung des Musca-Eis nach Ausschaltung verschiedener Eibereiche. *Z. wiss. Zoöl.* **126**:181–238.

RICHARDS, A. G. 1932. Comments on the origin of the midgut in insects. *J. Morphol.* **5**:433–441.

RICHARDS, A. G., and A. MILLER. 1937. Insect development analyzed by experimental methods: a review. *J. N. Y. Entomol. Soc.* **45**:1–60.

SCHRADER, F. 1944. *Mitosis.* Columbia University Press, New York.

SCHULTZ, J. 1941. The evidence of the nucleoprotein nature of the gene. *Cold Spring Harbor Symposia Quant. Biol.* **9**:55–65.

SEHL, A. 1931. Furchung und Bildung der Keimanlage bei der Mehlmotte *Ephestia Kühniella* nebst einer allgemeinen Übersicht über den Verlauf der Embryonalentwicklung. *Z. Morphol. Ökol. Tiere* **20**:535–598.

SHARP, L. W. 1934. *Introduction to Cytology.* 3rd ed. McGraw-Hill, New York.

SONNENBLICK, B. P. 1939. Salivary glands in the embryo of *Drosophila melanogaster*. *Rec. Gen. Soc. Am.* **8**:137.

SONNENBLICK, B. P. 1940. Cytology and development of the embryos of X-rayed adult *Drosophila melanogaster*. *Proc. Natl. Acad. Sci.* **26**:373–381.

SONNENBLICK, B. P. 1941. Germ cell movements and sex differentiation of the gonads in the Drosophila embryo. *Proc. Natl. Acad. Sci.* **27**:484–489.

SONNENBLICK, B. P. 1947. Synchronous mitoses in Drosophila, their intensely rapid rate, and the sudden appearance of the nucleolus. *Rec. Gen. Soc. Am.* **16**:52.

SONNENBLICK, B. P., and P. S. HENSHAW. 1941. Influence on development of certain dominant lethals induced by X-rays in Drosophila germ cells. *Proc. Soc. Exptl. Biol. Med.* **48**:74–79.

STERN, C. 1936. Somatic crossing-over and segregation in *Drosophila melanogaster*. *Genetics* **21**:625–730.

STUART, R. R. 1935. The development of the mid-intestine in *Melanoplus differentialis*. *J. Morphol.* **58**:419–438.

STURTEVANT, A. H. 1921. The North American species of Drosophila. *Carnegie Inst. Wash. Pub.* 301.

STURTEVANT, A. H. 1929. The claret mutant type of *Drosophila simulans*: a study of chromosome elimination and cell-lineage. *Z. wiss. Zoöl.* **135**:323–356.

TIEGS, O. W., and F. V. MURRAY. 1938. The embryonic development of *Calandra oryzae*. *Quart. J. Microscop. Sci.* **80**:159–271.

WALD, H. 1936. Cytologic studies on the abnormal development of the eggs of the claret mutant type of *Drosophila simulans*. *Genetics* **21**:264–281.

WEISMANN, A. 1863. Die Entwicklung der Dipteren im Ei nach Beobachtungen an Chironomus sp., *Musca vomitoria,* und *Pulex canis. Z. wiss. Zoöl.* **13**:107–220.

WILSON, E. B. 1928. *The Cell in Development and Heredity.* 3rd ed. Macmillan, New York.

WILSON, E. B., and A. F. HUETTNER. 1931. The central bodies again. *Science* **73**:447–448.

3

Histogenesis, Organogenesis, and Differentiation in the Embryo of Drosophila Melanogaster Meigen

D. F. POULSON*

INTRODUCTION

It can be said without exaggeration that there is no single insect whose descriptive embryology is established in all its details —this in spite of a rather large literature. In the case of the Diptera a number of studies, beginning in the second half of the last century, have established the general features as well as many details in a number of forms ranging from Chironomus, Miastor, Sciara, and Simulium among the lower Diptera to higher forms, such as Calliphora, Drosophila, Lucilia, Melophagus, Musca, and Phormia, but again completeness is lacking and the picture is fragmentary. The earlier stages are better known than the later in all cases. In *Drosophila melanogaster* these have been subjected to more detailed and critical study than in any of the other forms, chiefly by Huettner and his students, as elaborated by Sonnenblick in Chapter 2 on the early embryology of Drosophila. The present article is an attempt to set forth in reasonably brief form analytical accounts of the development and differentiation of individual tissues and organs up to the time of hatching of the larva and so to answer the numerous questions of "What happens when?" with regard to the principal systems. It does not pretend to be complete, and gaps in our knowledge will be indicated. Since this professes to be an analytical treatment,

* Osborn Zoological Laboratory, Yale University.

it will first be well to consider briefly certain fundamental features in the embryogenesis of Drosophila before proceeding with the details.

The earliest morphological changes in the fertilized egg are nuclear division and movement and cytoplasmic movements, leading to the establishment of at least three principal lines of nuclei: blastoderm, pole cell, and yolk cell nuclei. Together with the accompanying cytoplasm the first of these give rise to the major portion of embryonic cells; the second, together with the polar plasm, become either germ cells or midgut cells, depending on the position they later attain within the embryo; and the third, together with the inner cytoplasm, become the yolk cells which form a primitive gut enclosing the yolk up to the time of formation of the midgut and then may have other fates (see below). The subsequent period of morphogenetic movements, "gastrulation," leads to the establishment of cell groups, the so-called "germ layers." This is seen to be a less simple process than generally envisioned in textbook accounts, involving not only invagination and folding, but migration of cells as well. The inclusion of the external pole cells and the completion of the germ band inaugurate the period of regional differentiation which precedes organogenesis and histological differentiation. Visible signs of this, such as mesoderm segmentation, separation of neuroblasts and dermatoblasts, tracheal plates, and salivary gland plates, precede the appearance of body segmentation. A second period of morphogenetic movements begins in the approach of the midgut rudiments and the invagination and consolidation of the tracheal tubes. It is strikingly evidenced in the shortening of the germ band and culminates in the dorsal extension of the mesoderm and hypoderm and the involution of the head. Further movements are involved in the later differentiation of particular regions. An important feature of dipteran development is the establishment of two systems, the larval and the imaginal. Whether they are as separate as once thought is doubtful (Henson, 1946), but just how they are interrelated is not clear. The present account is confined principally to the origin and differentiation of larval structures, for only a few of the imaginal disc rudiments are morphologically distinguishable before the time of hatching of the larva. An attempt will be made in a subsequent section to derive from the separate accounts of the organ systems a cell lineage and a general plan of embryogenesis within the egg of *D. melanogaster*. The order of the succeeding sections reflects in a general way the sequence in which the rudiments of the several systems make their appearance.

Except for the line drawings, all the figures are from original photomicrographs made with a Leitz Makam camera on a Spencer microscope with apochromatic objectives, using Wratten M plates and appropriate Wratten filters. Initial magnifications were 200X, 600X, or 900X. In certain cases, chiefly the frontal and sagittal sections of Figs. 40–71, enlargements were necessary to keep all figures on the same scale. Except where indicated in the legends, all figures are oriented with the anterior end to the left and dorsal side up. The photographs of Figs. 72–83 are of dechorionated eggs mounted while living in salt solution. Dechorionation is easily accomplished by treatment for one minute with a solution of 3 per cent sodium hypochlorite.

Attention is called to the many excellent figures of later stages of development included in Sonnenblick's Chapter 2. These, which are iron haematoxylin preparations, should be compared with the Bodian preparations that provide the majority of the illustrations of this chapter.

The writer is deeply indebted to a number of people for their aid, criticisms, and suggestions during the course of this investigation. He is grateful to his colleagues, especially Dr. Maxwell Power, Dr. G. E. Pickford, Professor G. E. Hutchinson, Professor A. Petrunkevitch, and Professor J. S. Nicholas, for many helpful discussions, to Dr. M. Demerec for the extension of time and his editorial forbearance, to Dr. B. P. Sonnenblick and Mr. Dietrich Bodenstein for suggestions and criticisms, to Miss Regina Spignesi for technical assistance, and to Miss Lisbeth Krause for the drawings. Grateful acknowledgment is made to the Committee on Grants of the George Sheffield Fund of Yale University for grants which provided for technical assistance and a part of the photomicrographic equipment.

DEVELOPMENT OF MESODERMAL TISSUES AND ORGANS

The mesoderm arises in Drosophila, as in other higher Diptera, from midventral cells which first change shape (Figs. 1 and 2) and then invaginate along the ventral midline between 3 and $3\frac{1}{2}$

hours.   The ventral furrow closes quickly, producing an elongate mesodermal tube.   With the extension of the germ band, the tube flattens and loses its lumen.   At these early stages it shows no segmentation.   The flattened mesoderm remains associated with the ectoderm along the midventral line through a small number of median cells which arose by mitosis at the time of closing of the furrow.   They form a thin vertical layer joining the two. This median group of slender mesoectodermal cells separates the

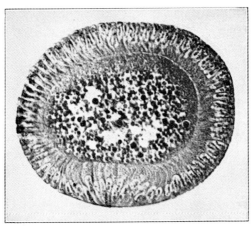

Fig. 1.   Changes in cell shape which precede the inturning of the mesoderm along the ventral midline.   Transverse section of a 3-hour embryo stained with iron haematoxylin.   ca. 400✕.

developing neuroblasts of the two sides and later becomes incorporated in the median portion of the ventral nerve cord, where it appears to contribute the neuroglial elements of the nervous system.   During the extension of the germ band the mesodermal cells undergo frequent mitoses.   The divisions continue, more or less randomly distributed, until the stomodaeum is definitively formed (6 hours at 22–23°C), after which their frequency declines.   No divisions have been observed later than the period just before the shortening of the germ band (8–9 hours at 22–23°C). By this time all the cells which will produce the mesodermal tissues and organs are present, and the period of visible differentiation of mesoderm is reached.

Certain features in the early development of the mesoderm are not readily followed, in particular the establishment of its segmental structure, which becomes visible by the fifth hour.   A primary regional differentiation of the mesoderm which is early apparent is the separation of the head mesoderm from that of the thorax and abdomen. The portion of the mesoderm anterior to the cephalic furrow (oblique crossfurrow) becomes the mesoderm of the head, the remainder that of the trunk. The most anterior part of the inturning mesodermal tube is in close association with another invagination which begins to form slightly later and at right angles to the former. The lumen of the mesodermal tube is confluent with that of the latter. This gives a configuration which in frontal section has the appearance of a hollow T (Fig. 43). The hollow condition persists only a short time, for the lumen of the crossbar of the T is rapidly obliterated and its cells form a nearly solid mass while undergoing several mitoses. This cell mass, which remains closely associated with the superficial ectoderm, is the anterior rudiment of the midgut. It is soon as clearly separate from the nearly mesodermal cells as is the posterior rudiment of the midgut at the other end of the germ band.

The writer's observations on Drosophila are in essential agreement with those of Escherich (1901) and Noack (1901) on muscoid flies. See also Henson (1945).

The cephalic mesoderm gives rise to the connective tissues and musculature of the head and foregut, including the special buccopharyngeal musculature. Within the head it is extremely difficult, if not impossible, to distinguish segmentation of the mesoderm. Any segmentation is much more fleeting in nature than that of the external parts of the head.

Since the cavity in the mesoderm is obliterated early (Fig. 2), it is not entirely clear whether the inner and outer layers of mesodermal cells retain their relative positions into the later stages. Roughly it appears that they do. Before the time of shortening of the germ band (9–10 hours) the mesoderm becomes fully separated into inner layer (splanchnopleure) and outer layer (somatopleure) in both head and body regions. Of these the former gives rise to the visceral musculature, while the latter produces the more conspicuous mesodermal organs and tissues. The separation of these portions of the mesoderm is not evident before the time of invagination of the tracheal pits (6–7 hours) but becomes clearly visible in the next hour, during which the segmentation of the body mesoderm is a fairly conspicuous fea-

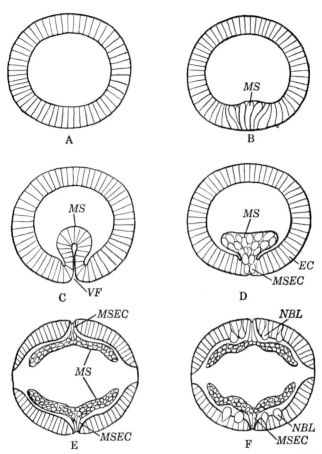

Fig. 2.   Diagrams illustrating origin and early development of the mesoderm as seen in transverse section.   A: Blastoderm before "gastrulation."
B: Changes in cell shape preceding gastrulation.   C: Invagination of
mesoderm, open ventral furrow.   D: Closed furrow and flattened mesodermal tube.   Mesectodermal cells which subsequently become included
in the nervous system maintain a connection between mesoderm and ectoderm.   E: Mesoderm after the onset of mitotic activity, showing increased
number of smaller cells.   F: Mesoderm at the time of neuroblast separation, showing lateral neuroblasts separated by mesectoderm cells.   *EC*,
ectoderm; *MS*, mesoderm; *MSEC*, mesectoderm; *NBL*, neuroblast; *VF*,
ventral furrow.

ture of longitudinal sections. The rudiments of the visceral mesoderm are clearly demarcated along the anterior portion of

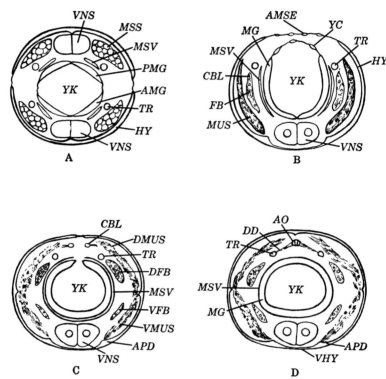

FIG. 3. Diagrams of transverse sections to illustrate the further history of the mesoderm and the dorsal closure of the embryo. A: The midregion of an embryo just before shortening (8–9 hours), showing separation of somatic and visceral mesoderm. B: The midregion of a recently shortened embryo (10 hours), showing dorsal extension and separation of rudiments of fat bodies. C: Dorsal closure nearly completed; approach of cardioblasts (11 hours). D: Closure completed, aorta formed, gut epithelium enclosed by muscles (12 hours). *AMG*, anterior midgut rudiment; *AMSE*, amnioserosa; *AO*, aorta; *APD*, apodeme; *CBL*, cardioblast; *DD*, dorsal diaphragm; *DFB*, dorsal fat body; *DMUS*, dorsal musculature; *FB*, fat body; *HY*, hypoderm; *MG*, midgut; *MSS*, somatic mesoderm; *MSV*, visceral mesoderm; *MUS*, muscle; *PMG*, posterior midgut rudiment; *TR*, trachea; *VFB*, ventral fat body; *VHY*, ventral hypoderm; *VMUS*, ventral musculature; *VNS*, ventral nervous system; *YK*, yolk.

the developing foregut and the posterior part of the hindgut in 8-hour embryos, while those associated with the two major parts of the midgut become easily distinguishable within the next hour.

The fore- and hindguts have their muscular layers more or less complete before the midgut fully closes over on the dorsal side.

Subdivision of the somatic mesoderm sets aside the cell groups which will make up the external muscles and the fat bodies, circulatory system, and associated elements. These come to lie in their definitive positions only after the shortening of the embryo (between the ninth and tenth hours) and the dorsal extension of the body wall lead to the union of the originally separate dorsal rudiments (Fig. 3) and thus to the completion of the dorsal aspect of the larva.

The development of the mesoderm is illustrated in the diagrams of Figs. 2 and 3.

Somatic Musculature

With the separation into somatic and splanchnic mesoderm the former becomes divided into larger, more ventrolateral and smaller, more dorsolateral portions in each segment (Figs. 3, 54, and 58). The ventrolateral portions give rise to the body musculature, the dorsolateral portions to the fat bodies, gonadial mesoderm, and the circulatory system and associated elements. The segmented ventrolateral portions, or myotomes, remain in close association with the adjacent hypoderm (Figs. 50, 51, 54, and 55), and, with the shortening of the embryo, this association is seen to be limited to the intersegmental regions. Here the ends of the now elongated myogenic elements establish their attachments to the developing apodemes (Figs. 4, 59, and 63). These points of attachment are highly localized. The individual myogenic cells appear to become fused, first at their attached ends and then throughout their length, to produce the multinucleate muscle fibers. No clear evidence has been obtained for any other origin of the multinucleate condition than by fusion. Although some mitoses have been found in myogenic cells just before the shortening of the embryo, no critical figures indicating origin of the multinucleate condition through failure of cytokinesis have been observed. Myofibrils make their first appearance in the elongating cells and are distinguishable in sections by 10 to 11 hours. The details of myogenesis and the arrangement of the various muscles will not be presented here. It is sufficient to note that the larval muscular system is complete in all its details by 13–14 hours and that coördinated muscular activity which now begins increases from this time through hatching. Figures 4 and 13 illustrate the structure of segmental muscles and the mode of

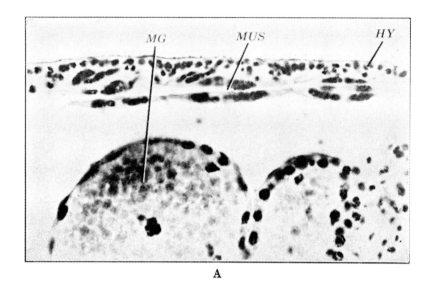

A

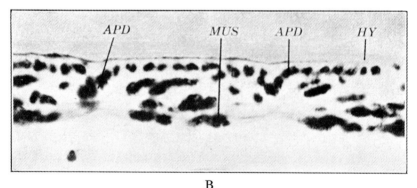

B

FIG. 4. Body musculature, showing multinucleate fibers and attachments to apodemes of body wall in frontal section. A: Section of body wall of embryo of 13 hours. Bodian preparation. 600×. B: Enlargement of same region of next section of this embryo. 1000×. *APD*, apodeme; *HY*, hypoderm; *MG*, midgut; *MUS*, muscle.

attachment to the apodemes. The largest and most conspicuous muscles are the anterior longitudinal muscles (Figs. 20B and 71) and those associated with the cephalopharyngeal apparatus (Figs. 60, 64, and 68). The dorsal pharyngeal muscles are arranged vertically in two rows between the dorsal wings of the cephalopharyngeal skeleton and the dorsal wall of the buccopharyngeal passage. They serve to dilate the passage and are an indispensable part of the larval feeding mechanism.

Fat Bodies and the Gonad Envelopes

The fat bodies and the mesodermal portions of the gonads separate from the somatic mesoderm during the shortening of the embryo but are not clearly distinguishable as such until the tenth to the eleventh hours. The fat bodies are small and inconspicuous and are most readily identified in sections by locating the germ cells of the developing gonads, which lie embedded in the posterior portions of the larger fat bodies. The cells and nuclei are small at this time, and little differentiation is evident until later stages. Although the methods used in making the preparations did not allow determination of the time of appearance of fat within the cells, there is no striking evidence of vacuolization in the cytoplasm much before the time of hatching. Small fat globules can be distinguished within the fat body cells of living larvae at the time of hatching.

The development of the gonad has been described by Sonnenblick (1941; and this volume) and has been the object of a detailed study by Aboim (1945), who has confirmed and extended the earlier observations of Geigy (1931b) that in the absence of germ cells the mesodermal components differentiate into agametic gonads complete in every other respect. Thus the differentiation of this part of the mesoderm is not dependent on the presence of germ cells. Instead, the pole cells are themselves dependent on the mesoderm for their differentiation as germ cells, for their developmental fate is very different when they do not reach the mesoderm (see below; and Poulson, 1947). The germ cells of the gonads are clearly distinguishable from the mesodermal cells at all stages in normal gonads, from the time they arrive in the lateral mesoderm and their aggregation with the mesoderm to form the gonads in their definitive position in the ninth segment (Figs. 51, 59, and 66).

The mesoderm contributes not only the envelope of the gonads, but the internal supporting and accessory cells as well. Thus in larvae, or later stages, from eggs whose pole cells were extirpated by the action of ultraviolet light at an early stage, all the nongerminal elements are fully and completely developed, as illustrated in Aboim's (1945) figures. For further information concerning the germ cells within the gonad see Sonnenblick's Chapter 2.

The Dorsal Vessel and Associated Tissues

The rudiments of the dorsal vessel are two rows, or lines, of inconspicuous cardioblasts atop the lateral crests of the mesoderm advancing toward the dorsal midline during the period of dorsal closure of the embryo (11–12 hours). At earlier stages these cells are indistinguishable among those dorsolateral mesoderm cells from which they, the pericardial cells, and the alary muscles of the dorsal diaphragm are derived. This portion of the dorsolateral mesoderm is separated by the developing tracheal trunks from those which give rise to the fat bodies and the mesodermal elements of the gonads. The two lines of cardioblasts, which extend posteriorly from the cerebral commissure to the last segment, meet at the dorsal midline in the eleventh hour and remain approximated to each other, achieving actual contact only at their dorsomedial and ventromedial surfaces. The space between them becomes the lumen of the vessel. Their union is completed first in the more posterior, or heart, region. The cardioblasts are flanked by rows of smaller paracardial and dorsal diaphragm cells. With the completion of the dorsal closure of the embryo the dorsal vessel and the diaphragm lie between and slightly above the tracheal trunks (Fig. 5).

The lumen of the dorsal vessel is largest in the more posterior, or heart region, and tapers down anteriorly to the dorsal aorta (Figs. 6A, B). At its anterior end the dorsal vessel passes through the ring gland (Weismann, 1864; Pantel, 1898) and is anchored to the cephalopharyngeal apparatus by several pairs of large ventrolateral cells (Figs. 33 and 34). The termination of the dorsal vessel at this point is characteristic of the Diptera. For a comparative account of the heart of insects, see the paper of Gerould (1938). There are reasons to suppose that the mesoderm producing the dorsal pharyngeal musculature (see p. 176) in these forms is serially homologous with that from which the dorsal vessel is derived (Poulson, 1945b).

The heart and aorta are contractile by the sixteenth hour; the pulsations become increasingly regular as hatching approaches. The flow of blood is anteriorly in the dorsal vessel through the suprapharyngeal space, thence posteriorly through the haemocoele, and back into the heart through the ostia. The origin of the ostia has not been followed, and it is not clear when they first appear. Large uninucleate pericardial nephrocytes lie be-

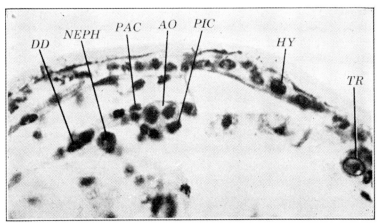

FIG. 5. Anterior end of dorsal aorta in transverse section just posterior to brain in embryo of 15 hours. Bodian preparation. 1000×. *AO*, aorta; *DD*, dorsal diaphragm; *HY*, hypoderm; *NEPH*, nephrocytes; *PAC*, paracardial cells; *PIC*, pericardial cells; *TR*, trachea.

neath and lateral to the dorsal vessel, closely associated with it and the dorsal diaphragm. They are similar in appearance to the garland cells ("Guirlandenzellen" of Weismann) which encircle the oesophagus just anterior to the proventriculus. However, garland cells are typically binucleate, although this is not always apparent in sections.

Two pericardial bodies, or glands, lying on either side of the anterior portion of the dorsal aorta just behind the ring gland, but independent of it in origin, are to be found in embryos 11–12 hours of age or older. These constitute the first pair of the so-called "blood-forming organs" of Stark and Marshall (1930), the remainder of which arise some time after the hatching of the larva. The full nature of their function is unknown. While not all the steps in their origin are clear, it is fairly certain that they come from the dorsolateral mesoderm just posterior to the lobes of the brain (Poulson, 1945b). An outer thin connective tissue

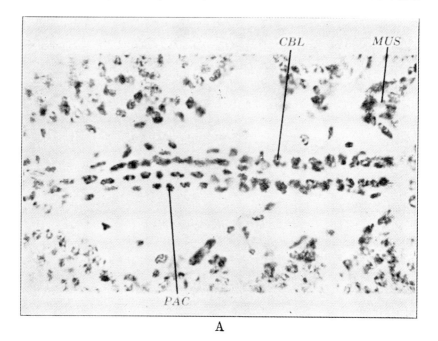

A

B

FIG. 6. Dorsal aorta and heart in frontal section in embryos of 20 hours. A: Anterior end of aorta with parallel strands of paracardial cells (see also Fig. 34). Bodian preparation. 1200×. B: Posterior end of dorsal vessel, or heart, with large lumen. Bodian preparation. 1200×. *CBL*, cardioblast; *HT*, heart (posterior aorta); *MUS*, muscle; *PAC*, paracardial cells.

coating encloses about a dozen cells somewhat larger than any of the cells of the ring gland and with nuclei large in proportion to the amount of cytoplasm (Fig. 34B).

The origin of embryonic and larval blood cells has not been adequately studied, and although the organs described above may have something to do with later larval blood cell production, they are clearly not the source of the early embryonic blood cells. There is evidence in the sections on which this study is based that some of them arise from yolk cells which fail, with certain parts of the yolk, to become enclosed in the midgut. These are found (Fig. 17A) in the most anterior and posterior regions of the embryo at the time that shortening is being completed (11–12 hours). Unenclosed yolk spheres are readily seen in the haemolymph of late embryos and larvae about ready to hatch. These provide a ready means of following the movements of the haemolymph (Poulson, 1937). Most blood cells probably are derived from the thin midventral portion of mesoderm, as in many other insects (cf. Hirschler, 1928; Johannsen and Butt, 1941).

Visceral Musculature

The visceral musculature is derived from the splanchnic or inner of the two layers into which the mesoderm separates between the sixth and seventh hours. The rudiments become recognizable as such in the next hour as they approach and become applied to the outside of the several gut rudiments. They are best followed in transverse and frontal sections and are illustrated in successive stages in Figs. 49, 50, 51, 54, 55, and 59, where the splanchnic layer is spreading over the foregut and anterior midgut rudiments. The gut mesoderm cells become flattened and more elongate with time and eventually become arranged into two layers, the inner giving rise to the circular muscles and the outer to the longitudinal muscles around the midgut and hindgut, while the reverse is true of the foregut. The separation and differentiation of the two layers are completed first in the oesophageal portion of the foregut and along the length of the hindgut before the midgut is fully closed dorsally. The two muscular layers of the midgut are distinguishable, in the main, before the saclike gut (12 hours) becomes transformed into the long convoluted tube characteristic of the larva at hatching. Study of the differentiation of the gut musculature is facilitated in Bodian-type preparations in which the individual circular and longitudinal muscle fibers are clearly discernible. It is possible in such preparations to learn much about the innervation and tracheation of the gut, although no detailed studies have been directed toward these ends.

DEVELOPMENT AND DIFFERENTIATION OF THE GUT

The gut of insects, its origin much disputed, is clearly compound in its derivation. The situation in Drosophila, as in other Diptera, is complex. In the preparation of this paper the whole subject had to be restudied and previous work reinterpreted. The principal findings provide the basis of this account. A definitive presentation cannot be given until the morphological studies receive experimental verification and extension with the aid of techniques of vital staining, extirpation, etc.

Although, as will be shown below, primary yolk cells which never migrated to the surface play an important part in the history of the gut the first distinguishable active step in the complex of processes leading to the formation of the gut is the entry of certain of the pole cells between the posterior blastoderm cells to the interior of the egg, as described by Rabinowitz (1941), but not so interpreted by him or by those who observed similar movements of pole cells in other Diptera (cf. Escherich, 1901; Noack, 1901; as well as others). The entry of these pole cells is followed by the more familiar processes of "gastrulation," the infolding of cells along the ventral midline to form the mesodermal tube, and the invagination, at the two ends of the "gastrular" furrow, of cell groups which give rise to the anterior and posterior sections of the midgut, the ventriculus, or stomach of insects. As will be elaborated below, three principal sections, or divisions, of the midgut can be distinguished on the basis of their origin. These may be referred to as the anterior midgut, the middle midgut, and the posterior midgut, which arise respectively from the anterior endoderm rudiment, certain of the pole cells and yolk cells, and the posterior endoderm rudiment. On the basis of his own studies and those of others on the metamorphosis of the insect gut, Henson (1945, 1946) distinguishes three principal divisions of the midgut of Calliphora, which he designates respectively as the pro-, the mes-, and the metenterons. Since he describes the midgut as originating not only from anterior and posterior endoderm rudiments but also by transformation of the ends of the stomodaeum and proctodaeum, it seems best not to apply his terms in this instance and to use instead the terms widely used by insect physiologists (Wigglesworth, 1939; Waterhouse, 1940, 1945) for the principal functional divisions of the larval midgut.

The foregut arises from the stomodaeal invagination, which makes its appearance only after "gastrulation" is complete. The hindgut takes its origin in parts of the so-called proctodaeal-amniotic invagination anterior and lateral to the posterior endoderm rudiment and not fully understood by most previous workers, including the writer. From the time of its origin the foregut is closely associated with the anterior midgut rudiment; a similar relation holds for the hindgut and the posterior midgut rudiment. So close, indeed, is the latter association that it is not easy to distinguish their common boundary, a condition which has furthered the view that the Malpighian tubules are of ectodermal origin. That the Malpighian tubules probably arise in Drosophila from the posterior midgut rudiment will be demonstrated below. The union of the several rudiments into a continuous structure and the differentiation of the numerous regions to give the definitive larval gut require about 8 hours of development (i.e., eighth to sixteenth hours). The origin of the salivary glands will be dealt with in connection with the foregut.

The regions of the gut will be treated in sequence according to their complexity and time of origin: midgut, hindgut, and foregut.

The Midgut: Origin and Early Development

The midgut, or ventriculus, of the Diptera has been described as arising from an anterior and a posterior rudiment whose fusion eventually completes the larval intestine (Kowalevsky, 1886; Graber, 1889; Escherich, 1901; Noack, 1901; DuBois, 1932; Gambrell, 1933; Butt, 1934; Lassmann, 1936; Parks, 1936; Poulson, 1937). The yolk cells have been considered solely as vitellophags which eventually degenerate and could therefore have no part in the formation of the gut. That this is altogether an oversimplified and incomplete picture is evident from a careful study of the development of the midgut in *Drosophila melanogaster*. The over-all similarity of the embryologies of Drosophila and the muscoid flies suggests that the situation is much the same through-out the higher Diptera. A comparative study of representative Diptera with the aid of the techniques here employed would be most enlightening.

An understanding of the development of the midgut requires a knowledge of the nature and behavior of the yolk cells. Although the mode and time of origin of the yolk cells in other Diptera are subject to some variation (cf. Johannsen and Butt, 1941), these arise in *D. melanogaster* primarily from the non-migrating, or syncytial, nuclei which remain behind in the yolk at the time of blastoderm formation (1½–2 hours). There is no evidence of re-entry from the blastoderm itself in this material. These cells are, therefore, primary yolk cells. Other yolk cells which appear subsequently in the yolk are secondary yolk cells. The only source of these which has been established in *D. melanogaster* is from the migration of certain of the pole cells through the blasto-derm, as described in detail by Rabinowitz (1941) and confirmed in this material. This re-entry of pole cells which may be inter-preted as the first phase in gastrulation is followed by the in-turning of the median ventral cells to form the mesodermal tube (3½ hours).[1] This in turn is followed closely by the anterior and posterior invaginations, which produce the anterior and pos-terior midgut rudiments at the two ends of the mesodermal tube.

The primary yolk cells remain in the interior of the yolk mass until a slightly later stage. The secondary yolk cells (of pole cell origin) tend to remain on the surface of the yolk at the pos-terior end and then move anteriorly over the dorsal surface of the yolk as the germ band elongates (4 hours). They tend to remain in the vicinity of and associated with the invaginating posterior midgut rudiment. Their cytoplasm, like that of the other yolk cells, appears to become fused with the thin layer of cytoplasm which bounds the outer surface of the yolk mass. The cell bodies are thus spread out over the yolk, and the nuclei be-come somewhat flattened. During the later phases of "gastrula-

[1] Another interpretation is that these entering pole cells provide the source of the future germ cells and therefore cannot be considered as secondary yolk cells. All the remaining pole cells (external) would then be concerned in the formation of the middle midgut. Extirpations, even more closely timed than those of Geigy (1931b) and Aboim (1945), will be necessary to establish the fates of these pole cells.

tion" and the elongation of the germ band the primary yolk cells move to the surface, and by the fifth hour few, if any, remain in the interior. The yolk mass is thus entirely enclosed by the yolk cells in a kind of "primitive gut" or yolk sac, which persists until subsequent movements of the anterior and posterior midgut rudiments and the shortening of the embryo lead to their final disposition as described below. This is a very different role from that previously ascribed to the yolk cells in Diptera and is in better agreement with the situation in many other insects (cf. Stuart, 1935; Snodgrass, 1935; Johannsen and Butt, 1941). That this has not been described before is probably due to the fact that so much use has been made of iron haematoxylin preparations, in which the yolk spheres are intensely stained and the yolk cells correspondingly difficult to follow. The yolk cells have been observed and traced most advantageously in Bodian, Feulgen, Harris haematoxylin, and toluidin blue preparations. In the last the cytoplasm of the yolk cells can be traced continuously around the yolk mass, and it is quite clear that almost no cytoplasm is left among the central yolk spheres. It is not easy to distinguish primary and secondary yolk cells in this period, for, although there is a difference in size of the nuclei, it is not always apparent in sectioned material. The nuclei of the primary yolk cells, though showing some size variation, are larger than those of the secondary yolk cells, which remain about the size of other pole cell nuclei. Precise studies of this difference are yet to be made, but the fact that a certain number of yolk cells become incorporated into the middle midgut makes such information highly desirable.

The anterior midgut rudiment arises in gastrulation as a short broad invagination (Fig. 7B) which stands in relation to the mesodermal tube as the crossbar to the base of the letter T (Fig. 43). With the closure of the lumen of the ventral tube, the walls of the anterior invagination become closely approximated, so that only a slit persists. Numerous mitoses in the anterior midgut rudiment result in a mass of cells from which even the slit disappears. From this time on, the anterior midgut rudiment is fully separated from the mesoderm. When between 5½ and 6 hours the stomodaeal plate begins to push in to form the stomodaeal invagination, the anterior midgut rudiment is already intimately associated with its tip. This relation persists, and eventually

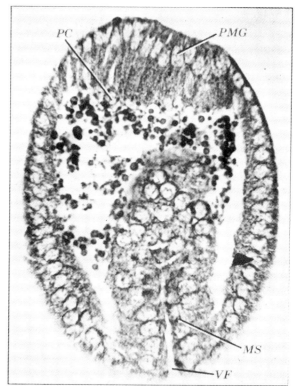

FIG. 7A.

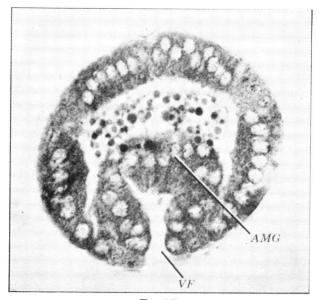

FIG. 7B.
186

the lumen of the foregut becomes confluent with that developing in the anterior midgut.

Mitoses continue in the anterior rudiment, which begins to grow posteriorly along the ventral side, pushing out two processes, or bands, which make the whole rudiment a somewhat Y-shaped unit between the yolk and the now flattened mesoderm. Although there is no strict localization of mitoses, they are more frequent in the proximal regions of the lateral bands. The originally rather cuboidal cells become more columnar with time and show a definite polarization with the nuclei lateral (peripheral) and a deep inner (medial) cytoplasmic border (Figs. 54, 55, and 59). A similar polarization is characteristic of the cells of the posterior midgut rudiment as well as those of the fore- and hindguts. In their progress beneath the yolk mass the lateral bands come into intimate contact with the adjacent yolk cells. These yolk cells become smaller and smaller with time and eventually disappear, evidently digested and absorbed by the cytoplasmic border of the anterior midgut cells. The lateral bands approach one another with time at their ventral edges until, with the shortening of the embryo, they support the anterior half of the yolk mass as they fuse with the other parts of the midgut. It is clear that the arms of the Y-shaped anterior rudiment give rise to the epithelium of the anterior midgut and the caeca, while the base of the Y, fused with the stomodaeum, becomes the epithelial wall of the "proventriculus."

The posterior midgut rudiment is a readily recognizable region of the blastoderm immediately before its invagination in gastrulation because of the way in which its deep columnar cells are separated from the thinner, more nearly cuboidal bounding cells, which appear to owe their shape, at least partly, to the mechanical consequences of the infolding of the ventral mesodermal tube (Fig. 7A). The invagination of the posterior midgut rudiment follows shortly and is a conspicuous feature of the $3\frac{1}{2}$-hour em-

FIG. 7. Origin of the midgut. A: Posterior rudiment of midgut before invagination, as seen in semifrontal section at posterior end of embryo of $3\frac{3}{4}$ hours. One of the early migrating pole cells lies internally to it. Iron haematoxylin preparation. 600×. B: Portion of anterior rudiment in transverse section at anterior end of ventral furrow. Harris haematoxylin preparation. 600×. *AMG*, anterior midgut rudiment; *MS*, mesoderm; *PC*, pole cells; *PMG*, posterior midgut rudiment; *VF*, ventral furrow.

bryo (Figs. 76 and 77). By 4 hours the invagination is very deep, contains the formerly external pole cells, and is separated from the mesodermal tube. As the germ band elongates, it is carried by the tail end of the embryo more and more anteriorly along the dorsal side until it reaches the limiting position just posterior

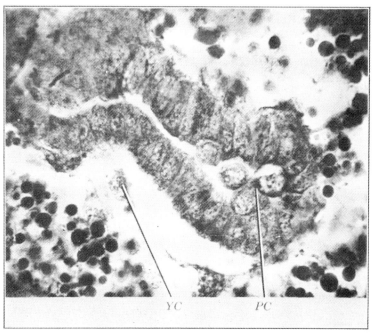

FIG. 8. Posterior midgut rudiment with included pole cells in sagittal section of an embryo of 5½ hours (see also Figs. 44 and 46). Yolk cells surround yolk forming primitive gut. Borax carmine preparation. 900×. From Poulson (1937), courtesy Hermann et Cie. PC, pole cells; YC, yolk cells.

to the brain level (Figs. 78 and 79). During this course the adjacent superficial cells become the cells of the amnion, and the aperture of the amniotic cavity is easily mistaken for the proctodaeum, which is only just beginning to be formed. The pole cells lie deep in the lumen (Fig. 8) near the tip of the invagination, whence according to Sonnenblick (1941) and Aboim (1945) some of them later pass between the midgut cells and into the lateral mesoderm to provide the germinal elements of the gonads.[2]

[2] If the interpretation suggested in Footnote 1 (p. 184) proves to be correct then none of the pole cells which enter the posterior midgut invagi-

Although there are occasional cell divisions before this time, it is in the fifth and sixth hours that mitotic activity becomes marked in the region just anterior to the tip of the invagination. This mitotic activity is more or less concurrent with the invagination of the proctodaeum and leads shortly to the outpocketing of the midgut rudiment, where its lumen is confluent with that of the developing proctodacum. These outpocketings will be shown below to give rise to the Malpighian tubules. In the seventh and eighth hours mitoses are frequent in cells at the tip of the midgut rudiment. From the tip there gradually extend two proliferating cell masses, or bands, morphologically very like those of the anterior midgut rudiment. These pass along the dorsal surface of the yolk and appear to carry with them many of the dorsal yolk cells. With them also move those pole cells which did not reach the gonads. The latter, together with some of the yolk cells, become the middle rudiment of midgut (Figs. 9, 55, and 59). There is no evidence of digestion and absorption of yolk cells by the posterior rudiment. A detailed account of this is to be published elsewhere.

Between the eighth and ninth hours the inner wall (yolk side) of the posterior rudiment becomes thinner in the vicinity of the tip and eventually ruptures, or opens out, so that the posterior rudiment is no longer a blind invagination bearing proliferations. As the rudiment extends back along the yolk mass and around it, the latter is literally scooped up in the arms of the rudiment. When, between the ninth and tenth hour, the embryo shortens, the posterior rudiment comes to lie posteriorly and ventral to the yolk mass. Its cells become so compacted during this movement that they then have a deep columnar form. The rudiment continues to move anteriorly with its two branches, or arms, touching along the midventral line, but not yet fused. Fusion takes place only when all the gut rudiments are united. As a consequence a small amount of posterior yolk and associated yolk

nation ever pass into the mesoderm to become the germinal elements of the gonads, and the question as to the mechanism of pole cell passage between the midgut cells is eliminated. The best evidence for actual passage of pole cells between the midgut cells is purely circumstantial (Sonnenblick and Aboim), and the writer has never found any unequivocal instances in his own preparations. The question of the fates of the pole cells and their relation to the middle midgut is being subjected to detailed investigation.

cells appear to be cut off regularly and left outside the gut. These yolk cells may be the source of some of the embryonic blood

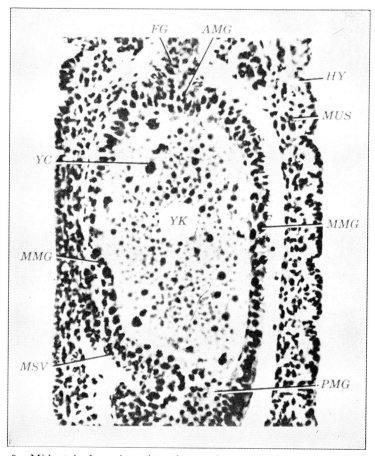

FIG. 9. Midgut in frontal section of an embryo of 11 hours, showing large cells of the middle midgut between the smaller cells of the anterior and posterior portions (see also Fig. 59). Bodian preparation. 600×. *AMG*, anterior midgut rudiment; *FG*, foregut; *HY*, hypoderm; *MMG*, middle midgut; *MSV*, visceral mesoderm; *MUS*, muscle; *PMG*, posterior midgut rudiment; *YC*, yolk cell; *YK*, yolk.

cells, as suggested above (p. 181). The cells of the middle rudiment are not so compacted or columnar. As they are also larger than the posterior rudiment cells, they are readily recognized. They compose a solid group at the anterior edge of each lateral part of the posterior rudiment.

As the anterior and posterior rudiments approach, it is the middle rudiment which makes contact and fuses with the pos-

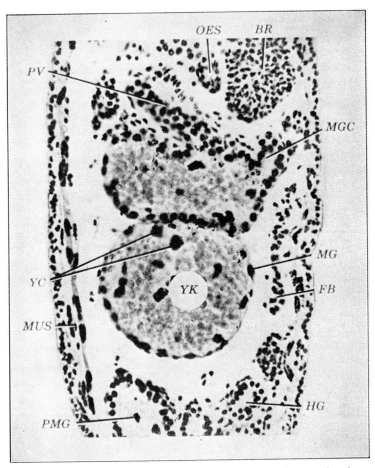

FIG. 10. Midgut in frontal section of an embryo of 13 hours, showing constriction in the middle region of large cells and the beginnings of midgut caeca (see also Fig. 4A and Fig. 63). Bodian preparation. 600×. *BR,* brain; *FB,* fat body; *HG,* hindgut; *MG,* midgut; *MGC,* midgut caecum; *MUS,* muscle; *OES,* oesophagus; *PMG,* posterior midgut rudiment; *PV,* proventriculus; *YC,* yolk cell; *YK,* yolk.

terior edge of the anterior rudiment on each side of the midline (Fig. 55). Thus the continuity of the midgut is established. The lateral parts of the three rudiments then fuse along the ventral midline and begin their spread dorsolaterally over the yolk until the main mass is enclosed. During this time (tenth and eleventh hours) the remaining yolk cells begin to leave the surface of the yolk for the interior, where they degenerate. The cells of the middle rudiment show no such changes and are fully incorporated as a characteristic middle section of the midgut epithelium. While the dorsal extension of the midgut wall proceeds the cells of the anterior and posterior sections become less and less columnar until eventually they are essentially cuboidal in form. By the time of closure (eleventh hour) the muscular layer is closely applied to the exterior of the epithelial wall. No further mitoses are found in the embryonic midgut during or after the shortening of the embryo, and the rudiments are complete in cell numbers.

The Midgut: Regional Differentiation

The subsequent development of the midgut involves a number of form changes, the most conspicuous of which are the conversion of the large central sac (Figs. 9, 56, and 59) into an elongate tube (Figs. 10, 11, 64, and 67) and regional differentiation throughout the length of the midintestine. The form change from sac to tube begins by the cutting in of a number of furrows or constrictions in the twelfth hour. Two marked constrictions appear equally spaced in the main body of the sac, one anterior to the level of the middle rudiment, the other at that level. A lesser constriction appears between the sac and the posteriormost part of the midgut, which bears the Malpighian tubes. The most anterior of these constrictions separates the future "proventricular" (cardiac) region from the body of the midgut. The end result of these constrictions and of the movements involved in them is the long convoluted larval midgut. Other movements than constrictive ones are involved, as seen during the formation of the four caeca just posterior to the "proventriculus" (cardia). These develop during this same period (12–16 hours) as anteriorly directed outward pockets. All these form changes are probably intimately related to the differentiation of the muscular layers of the gut, in which the oriented fibers of the circular and longitudinal layers are appearing.

The characteristic features of this regional differentiation may be summarized briefly. The anteriormost part of the anterior rudiment of midgut becomes the "proventriculus" (cardia). The misapplication of the term "proventriculus" to the cardia, or cardiac region of the ventriculus, of Diptera is so deeply rooted in

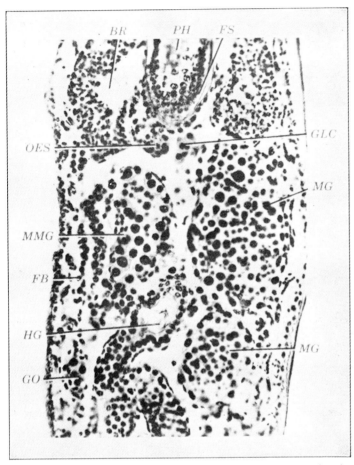

FIG. 11. Midgut in frontal section of an embryo of 16 hours, showing convolutions and middle region of large cells which has now become small in diameter and much longer (enlargement of part of Fig. 67). Bodian preparation. 600×. *BR*, brain; *FB*, fat body; *FS*, frontal sac; *GLC*, Guirlanden cells; *GO*, gonad; *HG*, hindgut; *MG*, midgut; *MMG*, middle midgut; *OES*, oesophagus; *PH*, pharynx.

the literature that it seems best to retain the term here, enclosed in quotation marks. The "proventriculus" is conical in shape, with the extremity of the foregut inserted into the base of the cone as part of the cardiac, or stomodaeal, valve (Fig. 13). At the junction with the foregut the midgut epithelium is reflected posteriorly over the surface of the inserted oesophageal wall, thus forming the cardiac valve. Within this annulus of midgut cells are those which later secrete the peritrophic membrane

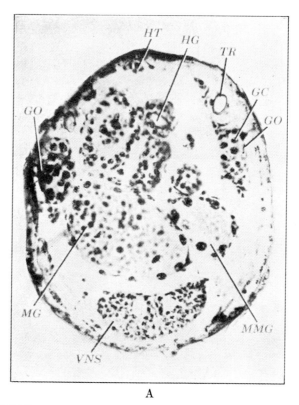

A

Fig. 12. Middle midgut region of large cells in transverse sections of embryos of 16 hours. A: Entire section at the level of the gonads to show comparative cell size and smaller lumen of middle region. Bodian preparation. 600×. B and C (p. 195): Adjacent sections of the same level in another embryo for comparison of nuclear size of middle midgut cells and germ cells in the gonad. Bodian preparations. 900×. *GC*, germ cells; *GO*, gonad; *HG*, hindgut; *HT*, heart (posterior aorta); *MG*, midgut, *MMG*, middle midgut; *MMGC*, middle midgut cells; *MUS*, muscle; *TR*, trachea; *VNS*, ventral nervous system.

(Wigglesworth, 1930), as well as those which will produce the anterior imaginal ring. The tip of the conelike "proventriculus" opens into the succeeding wider section of the ventriculus at the level of the four gastric caeca. The cells of the "proventricular wall" are more compacted and columnar than those of the other

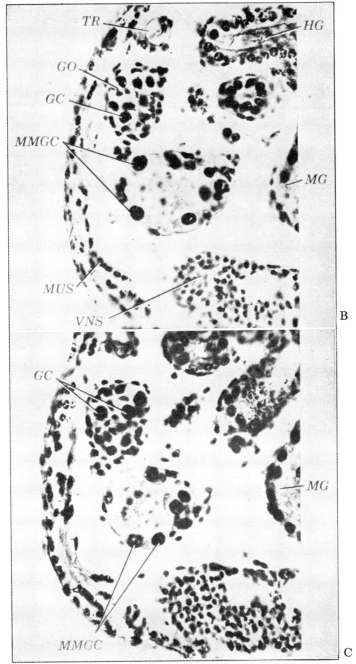

FIG. 12B and C.

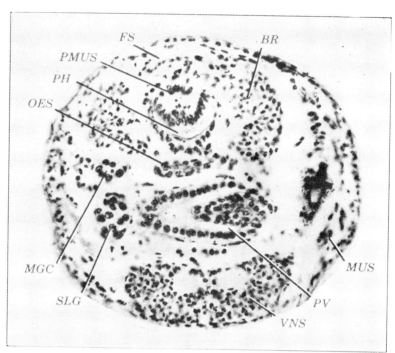

FIG. 13A.

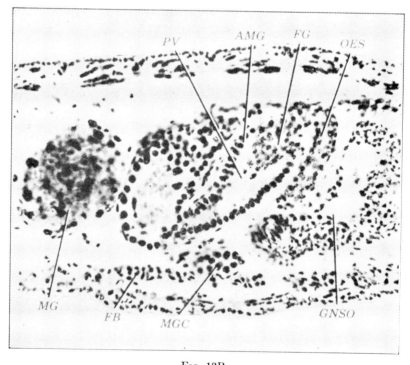

FIG. 13B.

regions of the ventriculus. Those of the caeca and other anterior
parts of the ventriculus are of the low cuboidal type, whose nuclei
later develop salivary-gland-type chromosomes. The diameter
of the ventriculus is largest in the caecal region and tapers down
to the middle region of large cells derived from the middle rudiment. In this middle section the large, rounded epithelial cells
protrude into the lumen and are its characteristic feature (Figs.
11 and 12). Posterior to this the ventricular epithelium is considerably flattened, consisting of low cuboidal cells which continue past the Malpighian tubes to the junction with the hindgut.
The same cell type characterizes the Malpighian tube cells.
The cells of the most posterior part of the midgut later give rise
to the posterior imaginal ring, which lies between mid- and hindgut in the same relative position as the posterior lip of the blastopore, to which Henson (1946) holds it is homologous. No transformations of the tips of the fore- and hindguts to give midgut,
such as Henson believes to occur in Calliphora, are evident in
Drosophila. If the above account is correct, Henson's interpretation of metamorphosis as a repetition of embryogenesis still
holds without the necessity of auxiliary assumptions.

The Malpighian Tubes

The origin of the Malpighian tubes of the higher Diptera has
not previously had adequate attention. They have been considered to originate as outpocketings of the proctodaeum (cf.
Snodgrass, 1935; Poulson, 1937) and were so treated by Marie
Strasburger (1932) in her account of the gut of Drosophila
melanogaster, although she presented some grounds for considering them as endodermal in origin. Henson (1945, 1946), from a

Fig. 13. Cardiac region of midgut in embryos of 13–14 hours. A: Transverse section at the level of the proventriculus. Bodian preparation.
600×. B: Frontal section, showing the relation of foregut and midgut
at the proventriculus. Note especially the size differences between the
nuclei of these two parts of the gut. Bodian preparation. 600×. AMG,
anterior midgut rudiment; BR, brain; FB, fat body; FG, foregut; FS,
frontal sac; GNSO, suboesophageal ganglion; MG, midgut; MGC, midgut
caecum; MUS, muscle; OES, oesophagus; PH, pharynx; PMUS, pharyngeal muscles; PV, proventriculus; SLG, salivary gland; VNS, ventral nervous system. In B anterior end is at right.

study of their position in the larval gut of Calliphora and from embryological observations, considers Malpighian tubes as midgut derivatives, but by transformation of the tip of the proctodaeum, as he has previously demonstrated to be the case in Lepidoptera (Henson, 1932).

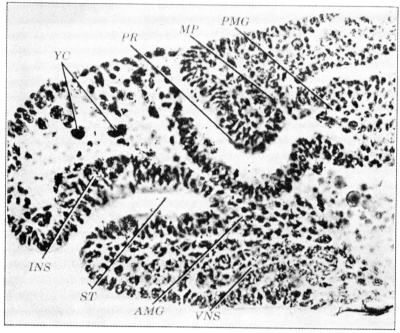

FIG. 14. Stomodaeum and proctodaeum in sagittal section of an embryo of 8 hours. Stomodaeum left, showing invaginations which give rise to stomodaeal nervous system and corpora cardiaca. Proctodaeum and posterior midgut rudiment above. Malpighian tubes at the point of junction. Bodian preparation. 600×. *AMG*, anterior midgut rudiment; *INS*, invaginations in roof of stomodaeum; *MP*, Malpighian tube; *PMG*, posterior midgut rudiment; *PR*, proctodaeum; *ST*, stomodaeum; *VNS*, ventral nervous system; *YC*, yolk cell.

The appearance of outpocketings in the wall of the posterior midgut rudiment close to its junction with the proctodaeal invagination (in the sixth to seventh hour) is the first indication of the Malpighian tubes in the embryo of *D. melanogaster* (Fig. 14). Although the line of demarcation between these two rudiments of the gut is not sharply defined, the evidence makes it probable that this is posterior to the level of the Malpighian tubes. The previous interpretation of the Malpighian tubes of Drosophila as ectodermal in origin (Poulson, 1937) was based on the mistaken identification of a large part of the posterior midgut rudiment as part of the proctodaeum (see below). The development of these outpocketings into the four definitive Malpighian tubes progresses at first slowly and then more rapidly during the period of visible differentiation of the gut (Fig. 15). The walls of the tubes, at first thick and composed of tightly packed columnar cells (Figs. 14 and 51), become thinner and their cells more cuboidal (Figs. 15A, B) as the tubes elongate until their definitive larval form is achieved between the sixteenth and eighteenth hours of development. By this time there is every evidence that the tubes are fully functional, and crystalline excretory products (probably carbonates) are to be seen within them in living embryos. Pigment is normally present by this time in the tube cells of wild-type strains. The bases of each lateral pair of tubes become united in the later stages. Thus each lateral pair shares a common opening into the posterior midgut. This common portion receives a muscular coat continuous with that of the midgut and is thus contractile.

The Hindgut

In previous work on the embryology of Drosophila the relationships between the posterior endoderm rudiment and the proctodaeum were incompletely understood. The posterior endodermal invagination was referred to as the proctodaeal or proctodaeal-amniotic invagination (Parks, 1936; Poulson, 1937; Rabinowitz, 1941; Sonnenblick, 1941). It is now clear that the posterior plate of cells immediately beneath the pole cell cap in the late blastoderm stage is the first evidence of the posterior midgut. Only after this has invaginated, carrying the external pole cells in its lumen, as the embryo elongates on the dorsal side of the egg, do the first signs of the proctodaeal rudiment become apparent. This rudiment is seen to consist of a region of deeply columnar cells, which begins to invaginate at the tail end of the embryo (Fig. 44) between the fifth and sixth hours. The manner in which the invagination takes place and the proctodaeal tube is formed is difficult to visualize without the aid of diagrams. Those of Figs. 16 and 17 illustrate a stage of the proctodaeal invagination and clarify its relation to the previously inturned posterior midgut rudiment and to the embryonic membranes. The succeeding stages

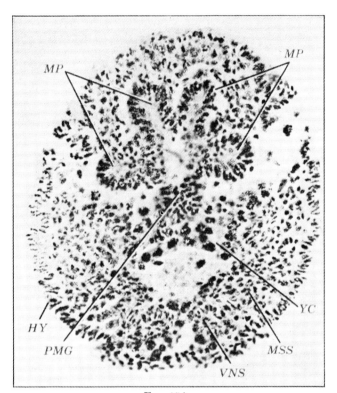

FIG. 15A.

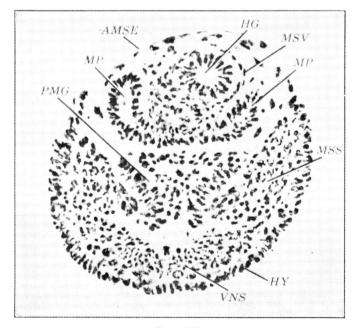

FIG. 15B.

Fig. 15. Malpighian tubes in 9- and 10½-hour embryos. A: At the time of shortening of the embryo (9 hours) cut slightly off from transverse, so that all four of the Malpighian tubule rudiments are evident at the posterior end of the midgut. Bodian preparation. 600×. B: In transverse section in a 10½-hour embryo only two of the tubes are visible with the hindgut above and the posterior midgut rudiment below. Bodian preparation. 600×. *AMSE*, amnioserosa; *HG*, hindgut; *HY*, hypoderm; *MP*, Malpighian tube; *MSS*, somatic mesoderm; *MSV*, visceral mesoderm; *PMG*, posterior midgut rudiment; *VNS*, ventral nervous system; *YC*, yolk cell.

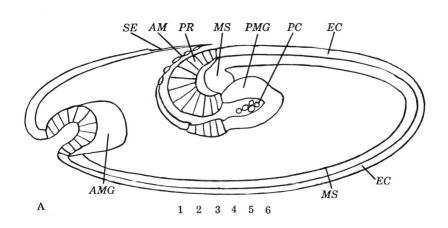

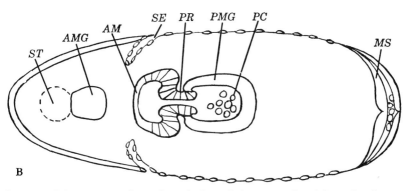

Fig. 16. Diagrams to show the relation of the proctodaeal invagination to the posterior midgut rudiment in an embryo of 6 hours. A: Sagittal section. B: Frontal section. For transverse sections see Fig. 17. *AM*, amnion; *AMG*, anterior midgut rudiment; *EC*, ectoderm; *MS*, mesoderm; *PC*, pole cells; *PMG*, posterior midgut rudiment; *PR*, proctodaeum; *SE*, serosa; *ST*, stomodaeum.

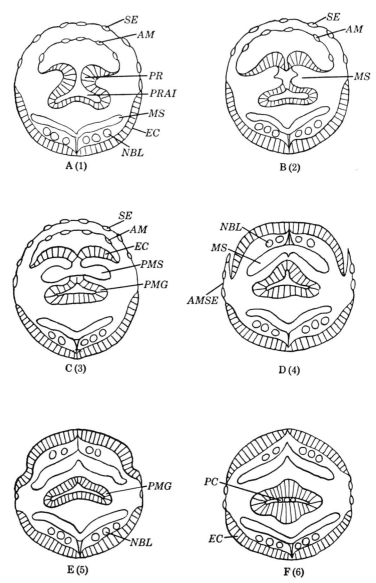

FIG. 17. Diagrams to show the relation of proctodaeum to the posterior
midgut rudiment. A–F: Transverse sections at the levels 1–6 in Fig. 16.
AM, amnion; *AMSE*, amnioserosa; *EC*, ectoderm; *MS*, mesoderm; *NBL*,
neuroblast; *PC*, pole cells; *PMG*, posterior midgut rudiment; *PMS*, posterior mesoderm; *PR*, proctodaeum; *PRAI*, proctodaeal-amniotic invagination; *SE*, serosa.

can be followed in the photographs of Figs. 44, 48, 51, 52, and 55, which show the progressive transformation of the deepening troughlike proctodaeum into a complete tube by the time that the embryo shortens between the ninth and tenth hours. The precise boundary between the proctodaeum and the posterior

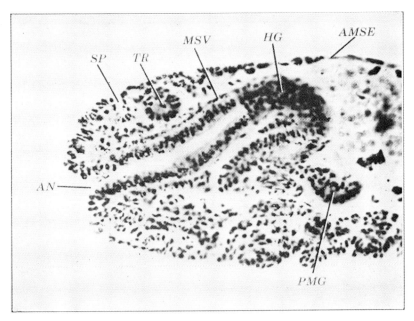

FIG. 18. Hindgut in near sagittal section, showing large loop and rudiment of gut musculature. Bodian preparation. 600×. *AMSE*, amnioserosa; *AN*, anus; *HG*, hindgut; *MSV*, visceral mesoderm; *PMG*, posterior midgut rudiment; *SP*, spiracle; *TR*, trachea. Posterior at left.

endoderm has not been established with certainty, but it appears to be just posterior to the region from which the Malpighian tubes develop. This is clearly shown in sagittal section in Fig. 14 and in frontal section in Fig. 51.

With the shortening of the embryo between the ninth and tenth hours the posterior section of the proctodaeum is pulled to the posterior end of the egg. This movement produces the characteristic great loop of the hindgut, which now courses from the posterior end of the embryo over the posterior midgut and thence back down and under the latter to the point of junction. This S-shaped course is best seen in Fig. 18, which also shows very

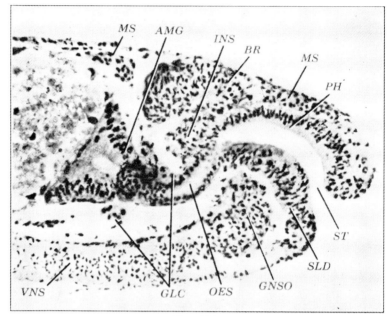

Fig. 19A.

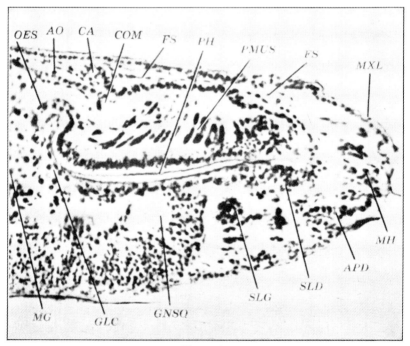

Fig. 19B.

clearly the muscular layer of the hindgut. The musculature of the hindgut is derived from the most posterior section of the mesoderm, which is sufficiently conspicuous to be followed in Figs. 44, 51, and 55.

The cells of the hindgut wall, which are still deeply columnar at the eleventh to twelfth hours, gradually become more cuboidal as the hindgut lengthens in the later stages of differentiation. With this change the diameter of the hindgut decreases. By the time of hatching the epithelial lining is typically low cuboidal with very little or no regional differentiation.

The Foregut

The rudiment of the foregut is first visible between the fourth and fifth hours of development as a stomodaeal plate on the ventral side just back from the anterior tip of the egg. It lies in the position from which the cells which form the anterior rudiment of the midgut were previously turned inward. The center of this plate of columnar cells invaginates between $5\frac{1}{2}$ and 6 hours (Fig. 44) and continues moving inward and posteriorly following (or pushing with it) the anterior midgut rudiment to which its tip is closely applied. Shortly after the beginning of the invagination numerous mitoses begin in the distal part of the stomodaeum. The axes of the spindles are normally parallel to the lumen. (During this same period there are also mitoses in the proctodaeal rudiment.) Subsequent mitoses are less frequent; practically none are to be observed after the period of shortening of the embryo.

FIG. 19. Stomodaeum and anterior end of midgut. A: Nearly median sagittal section, showing larval pharynx and oesophagus in embryo of $10\frac{1}{2}$ hours. Bodian preparation. 550×. B: Median sagittal section of an embryo at 16 hours, showing foregut after involution of head is complete. Bodian preparation. 550×. *AMG*, anterior midgut rudiment; *AO*, aorta; *APD*, apodeme; *BR*, brain; *CA*, corpus allatum; *COM*, cerebral commissure; *FS*, frontal sac; *GLC*, Guirlanden cells; *GNSO*, suboesophageal ganglion; *INS*, invaginations in roof of stomodaeum; *MG*, midgut; *MH*, mouth hook; *MS*, mesoderm; *MXL*, maxillary sensory rudiment; *OES*, oesophagus; *PH*, pharynx; *PMUS*, pharyngeal muscles; *SLD*, salivary duct; *SLG*, salivary gland; *ST*, stomodaeum; *VNS*, ventral nervous system.

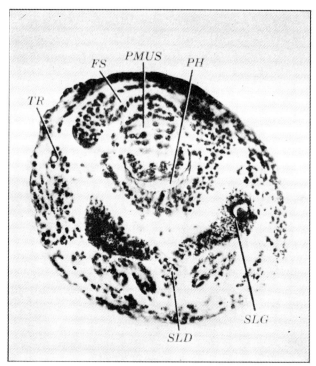

Fig. 20A.

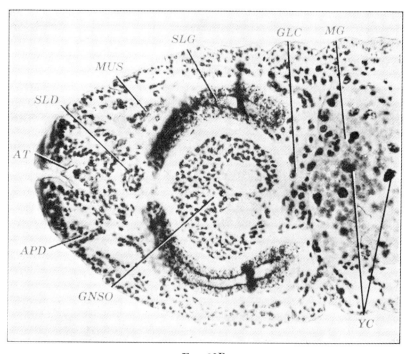

Fig. 20B.

In the anterior roof of the stomodaeum, between the seventh and eighth hours, there are a few divisions perpendicular to the lumen in cells which appear to be neuroblasts. They appear to be closely associated with three small invaginations which develop in the same region in the eighth hour. These represent the origin of the stomodaeal nervous system and one of the components of the ring gland, which are discussed below (see pp. 222 and 234). As the stomodaeal invagination deepens, more and more material from the outer ventrolateral surface is moved to the interior through the aperture, so that by the tenth hour the salivary glands have also passed in and open through their newly developed ducts on the ventral wall of the buccal cavity. At this time (Figs. 19A and 56) the orifice represents the true larval mouth aperture; the region just posterior is the buccal cavity, which is continued into the pharnyx; and the terminal section attached to the anterior midgut rudiment is the oesophageal portion. At this time there is little morphological differentiation between any of these regions. Practically all the characteristic regional features develop during the eleventh to sixteenth hours, when the larval head has been involuted and the period of major differentiation is under way. The involution of the head is dealt with on pp. 230–234. Only the differentiation of the stomodaeal derivatives themselves and the salivary glands will be treated here.

After the ninth hour the roof of the buccopharyngeal region becomes progressively thickened, and the floor thins out (Figs. 56, 60, 61, and 63). Concommittantly the whole region becomes dorsoventrally flattened, and the lumen increasingly slitlike, reaching a climax by the time the involution of the head is completed (Figs. 64, 65, and 67). The oesophageal region lengthens

---

FIG. 20. Salivary glands in 14-hour embryos. A: Transverse section of anterior end, showing common duct of glands, and anterior end of pharyngeal apparatus and frontal sac. Bodian preparation. 600×. B: Frontal section, showing position of glands, silver-blackening cytoplasmic elements, and lumen with secretion. No silver-blackening elements in the duct cells in A. Bodian preparation. 600×. *APD*, apodeme; *AT*, atrium; *FS*, frontal sac; *GLC*, Guirlanden cells; *GNSO*, suboesophageal ganglion; *MG*, midgut; *MUS*, muscle; *PH*, pharynx; *PMUS*, pharyngeal muscles; *SLD*, salivary duct; *SLG*, salivary gland; *TR*, trachea; *YC*, yolk cell.

and makes a partial loop (Figs. 60, 63, 64, and 67) before the junction with the anterior midgut. It decreases in diameter, and its cells become low cuboidal. The whole foregut develops a chitinous lining during and following the involution of the head.

The salivary glands originate as described by Sonnenblick (1939; and this book) from paired lateral thickenings which become platelike between the seventh and eighth hours (Fig. 48). These are sharply differentiated from this time on by the silver-blackening cytoplasmic elements characteristic of the salivary gland cells of most insects (Poulson, 1945b). By the eighth hour central depressions appear in the plates and deepen into invaginations (Fig. 49), which are completed by the time that these gland rudiments have passed in through the stomodaeal opening. They become united by a common region (ninth to tenth hours), which differentiates into the common duct. The anteriormost cells of each glandular mass become the duct cells, the remainder of the main body of secretory cells. No mitoses have been observed in the salivary glands from the time that they are first distinguishable as ventrolateral thickenings. The definitive number of cells is thus present from the time of their inception. This information is important in any evaluation of the complexity of structure of the salivary gland chromosomes. Evidences of secretory activity, such as stainable secretion in the ducts, are found to parallel the invagination of the plates and the development of the silver-blackening cytoplasmic apparatus, which becomes a conspicuous feature in all embryos of 10 hours and older. The position of the salivary gland plates at the time at which they are first distinguishable is important as a regional marker in the embryo, for the salivary glands of Diptera are clearly labial glands and so indicate the position of the labium, which is formed from the second maxillary segment of the larval head (see p. 231). The final position of the salivary glands is rather far back from the head alongside the anterior portion of ventral nerve cord, where they are seen in Figs. 20, 64, and 65. The nuclear structure of the glands is not at first visibly different from that of cells of the anterior part of the alimentary tract, and no special study of their chromosome structure has been undertaken, as yet, in embryos of Drosophila.

DEVELOPMENT OF THE PRINCIPAL ECTODERMAL ORGANS

The Nervous System

As in other insects, the nervous system of the Diptera originates
from neuroblasts which become differentiated from the rest of the
ectoderm at a comparatively early stage in the embryos (cf.
Graber, 1889; Wheeler, 1889, 1891, 1893; Escherich, 1902). Parks
(1936) did not deal with the origin of the nervous system of Drosophila, and Poulson (1937) gave only the briefest account of it.
The first signs of neuroblasts in embryos of *D. melanogaster* are
found in the enlargement of certain cells in the anterior dorsolateral ectoderm, from which the lobes of the brain are derived,
and shortly after among cells along the length of the embryo on
either side of the midventral line. The subsequent history and
behavior of neuroblasts are quite uniform throughout the prospective neural regions, except among certain cells of the anterior
dorsolateral ectoderm, as will be indicated below. The history
of a typical ventral neuroblast will be briefly described.

Neuroblasts are first distinguishable from other ectodermal
cells, in 4–4½ hour embryos, by their increasing size, their
shape, and their slightly more basophilic cytoplasm. As the
neuroblast enlarges, its nucleus moves, or is pushed, toward the
inner end of the cell, which bulges out to give the whole a rather
characteristic flask shape. The cytoplasm of the outer end of
the developing neuroblast is squeezed between neighboring cells
and has the appearance of the neck of a flask. This neck, or
column, of cytoplasm does not retract from between the neighboring cells until the first neuroblast mitosis begins. It does so
then as the cell rounds up during the prophase of that division.
These steps are illustrated in the diagrams of Fig. 21, which
include the course of the first neuroblast mitosis. In this and subsequent neuroblast divisions the *spindle is always oriented perpendicular to the plane of the embryonic surface,* in contrast to
other ectodermal cell divisions all of which are in the plane of the
surface. As neuroblasts move inward from the surface, their
places are taken by the division products of dermatoblast and
other ectoderm cells.

The early pattern of distribution of neuroblasts has not been established in detail, although it is clear that the neuroblasts are grouped and that in the thoracic and abdominal regions these groups are segmental and represent the Anlagen of the ganglia. It appears to be relative to the latter that the segmental structure

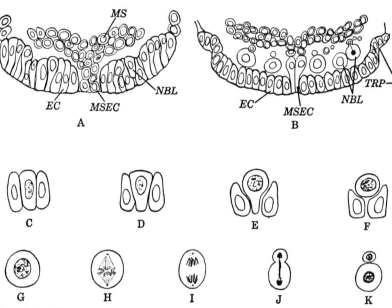

Fig. 21. Diagrams to illustrate the origin and differentiation of neuroblasts. A: Enlarging neuroblasts in ventral ectoderm of 4- to 5-hour embryo, transverse section. B: Separated neuroblasts in mitosis in 5- to 6-hour embryo, transverse section. C–F: Stages in the separation of a ventral neuroblast. G–K: Stages in neuroblast mitosis, which is characterized by unequal cytoplasmic division, producing small preganglion cell. *EC,* ectoderm; *MS,* mesoderm; *MSEC,* mesectoderm; *NBL,* neuroblast; *TRP,* tracheal pit.

of the mesoderm becomes established. Although it is of considerable theoretical interest, the neuromeric structure of the head has not been investigated, for it appears difficult in this material.

The differentiation of cells as neuroblasts appears to be limited to the period between the fourth and the seventh hours in the ventral nervous system, as none of the flask-shaped, undivided neuroblasts has been observed in older embryos. At the level of the brain the period may be somewhat longer.

The nuclear phenomena of neuroblast mitosis are entirely regular. The spindle is large, the centrioles are distinctly present, and the chromosome number is always diploid. Since these are the largest mitotic cells in the embryo, they provide the best available material for the study of chromosomes during embryonic development. Even these chromosomes are none too large for cytological study. Typical ventral neuroblasts, when rounded up before the first division, have a diameter of about 12.5 μ, as compared with 7.5 μ of adjacent mesoderm cells in 6-hour embryos fixed in F.A.A. Nuclear diameters of these neuroblasts average about 7.5 μ, those of mesoderm cells about 5.0 μ, and those of the small ganglion cells about 3.3 μ. Because of their shape the size of other ectoderm cells cannot be readily expressed in terms of a single measurement. It is evident that none of them attains the size of neuroblasts and that they are intermediate between the latter and the mesoderm cells. The characteristic feature of neuroblast divisions is the unequal cytoplasmic division illustrated in Figs. 21G–K. The cell furrow, instead of forming equatorially, is eccentric, and the result of cytokinesis is a large neuroblast cell, which will continue to divide in this fashion, and a small bud, which becomes a ganglion cell. This bud is proportionately smaller in the first than in later divisions, but the absolute size of ganglion cells and nuclei appears to remain about the same. This suggests that there is a decrease in neuroblast size during the later divisions. Such measurements as have been made on neuroblasts in later embryos are in complete agreement with this hypothesis. By an appropriate set of measurements at successive stages it should be possible to establish the number of divisions through which a given neuroblast has passed and so determine the age distribution of neuroblasts within the nervous system. Mitotic activity of neuroblasts is most intense between the fifth and ninth hours (Fig. 44), after which it drops off until, with the completion of shortening of the embryo, relatively few mitoses are to be observed. Most of these are to be found in the posterior cortical regions of the brain lobes, where they have been observed to continue through the eighteenth hour. Rarely, in these late stages, are any mitoses present in the ventral nervous system.

Although neuroblasts divide repeatedly, it is not known how many times each undergoes mitosis; nor is the original number of neuroblasts established. At present it cannot be said exactly how many neuroblasts comprise a segmental ganglion (or neuromere). An estimate can be made from counts of the numbers observed per section at different ages up to the cessation of mitotic activity and the distribution of frequency along the length of the embryo. At the time of writing only a small amount of such data is available. This indicates roughly 16 neuroblasts per segment in a 6-hour embryo. Since the bilateral arrangement of neuroblasts is remarkably symmetrical, this means about 8 neuroblasts per segment on either side of the midline at this time. Exact data of this type are now being sought for the brain and suboesophageal ganglion and for the rest of the ventral nervous system. The numbers of ganglion cells in successive stages remain to be determined. The number is very large in the fully developed nervous system.

Until recently it was thought that the small cells, when "budded off" by the neuroblasts, underwent no further divisions, as seems to be the case in several insects (cf. Johannsen and Butt, 1941), and became at once ganglion cells. Very clear evidence is now available for at least one mitosis before these cells become ganglion cells (*sensu structu*), and they might better be referred to as preganglion cells. Nelson (1915) describes a similar situation in the honeybee. As these are small cells, good mitotic figures are not readily observed, but it is clear that in the prophase of this preganglionic mitosis somatic synapsis, such as described by Kaufmann (1934) in later larval neuroblast and ganglion cell mitoses, is at a maximum and that it persists into the metaphase of this division. The few good metaphase figures which have been found show only the haploid number of paired chromosomes. Evidence is accumulating that this division may be of the nature of a somatic reduction to give haploid ganglion cells. This cannot be considered established until critical anaphase figures are found. Thus the small size of the early larval ganglion cell nuclei may prove to be due to haploidy. Whether all or only certain of the preganglonic mitoses may be of such a nature remains to be determined.

The length of time required to complete the neuroblast mitotic cycle is as yet undetermined, but an estimate can be made for the ventral nervous system where, in a period of 4–5 hours, a comparatively small number of neuroblasts (roughly 16 per segment) produces a large number of ganglion cells (roughly 300 per segment, on the basis of some counts before the condensation of the nervous system). If one takes into account the preganglionic mitosis, this gives, very roughly, about 26–32 minutes per neuroblast mitotic cycle. This rate of division, while it does not approach that determined by Rabinowitz (1941) for the preblastoderm mitoses, is of the same order of magnitude as determined for rapidly dividing cells in other organisms. It is not surprising, therefore, that almost no section of an embryo is devoid of dividing neuroblasts during the period of 5–9 hours inclusive.

There is a remarkable synchrony of division between cells in corresponding positions on the two sides of the embryo. Some synchrony between ganglia is also apparent, but it is less striking than the bilateral correlation.

Any really satisfactory account of the development of the nervous system must include data of the type indicated, which are at present fragmentary.

The neuroblasts of the brain originate (Fig. 22A) in the same way as those of the ventral nervous system, except that in one region on either side some of the first neuroblasts, instead of moving to the interior, round up and divide at the surface. In some sections these actually protrude externally beyond the neighboring cells. In these regions of the brain some of these and subsequent neuroblasts remain dividing at the surface and are only later covered over by neighboring ectoderm cells. All indications are that these neuroblasts represent the beginnings of the cephalic sensory organs and sensory regions of the brain.

The central nervous system of insects (Hanström, 1928; Snodgrass, 1935) is composed of a series of segmental ganglia derived from the neuroblasts, of which those of the ventral portion are nearly always distinctly recognizable. These consist of the suboesophageal ganglion, which belongs to the head, three thoracic ganglia, and eight abdominal ganglia. The brain consists of a series of three fused ganglia which in some instances may be separately recognizable, but not readily so in the Diptera. In Drosophila (Hertweck, 1931) the ganglia are closely approximated to one another and are indicated externally only by the segmental nerves, except in the middle stages of embryonic development before the ventral cord begins its characteristic condensation. The

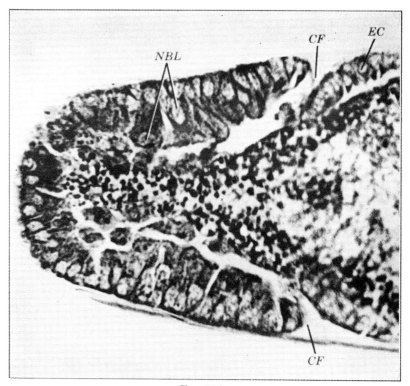

Fig. 22A.

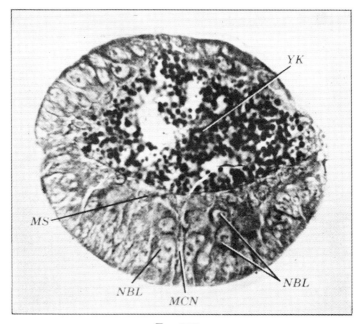

Fig. 22B.
214

interganglionic portions arise by fusion. The internal connections between the ganglia are derived by the outgrowth of processes, the nerve fibers, from the ganglion cells. These connections are transverse and longitudinal within the ganglia and longitudinal between the ganglia. From each of the segmental ganglia paired nerves extend out to innervate the corresponding segments of the body.

Up to the time of shortening of the embryo (9–10 hours) the ventral nervous system is a rather loose aggregate of segmentally arranged neuroblasts and ganglion cells lying between the outer ectoderm and the mesoderm and hedged in laterally by the rapidly developing tracheal invaginations. The neuroblasts lie beneath the ganglion cells, which accumulate above them. The bilaterality of the nervous system is conspicuous from the beginning, the cells of the two sides remaining separated by median cells (see p. 210) which are continuous with the mesoderm and maintain a connection with the ectoderm of the ventral midline. These are the cells described by Escherich (1902) as the middle cord and believed by him to be the source of a median nerve in blowflies. The history of the median cells in Drosophila is similar in many ways to that given by Escherich, but there are differences. No evidence has been found of an actual median nerve, and it appears that these cells contribute supporting elements and sheath cells to the ventral cord and the segmental nerves. It has not been determined whether some of these cells may have a purely neural or conducting function in the ventral cord.

The median cell structure is originally sheetlike and uniform, uniting ectoderm and mesoderm. As a result of cell division and perhaps of cell shifts, it comes to show regular variations in thickness along its length (as seen in transverse and frontal sections), which are correlated with the segmental structure of both ventral cord and the overlying mesoderm (Fig. 23). With the completion of shortening of the embryo this sheet is thickest

---

FIG. 22. Neuroblasts in the brain and ventral nervous system. A: Separation of the neuroblasts from the anterior lateral ectoderm in an embryo of 4–4½ hours. Frontal section of iron haematoxylin preparation. 600×. B: Transverse section just posterior to head of embryo of 5–6 hours, showing separated neuroblasts (see also Figs. 44, 45, and 46). Iron haematoxylin preparation. 600×. CF, cephalic furrow; EC, ectoderm; MCN, median cells of nervous system; MS, mesoderm; NBL, neuroblast; YK, yolk.

at the dorsal level in the interganglionic regions and thinnest within the ganglia. As the lateral mesodermal masses separate

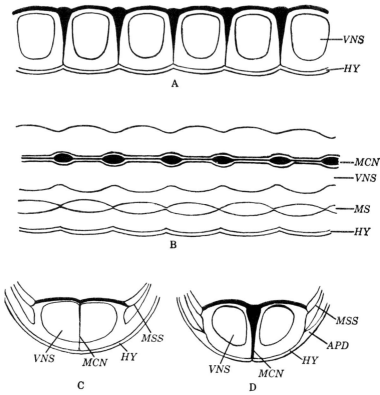

Fig. 23. Diagrams to illustrate the relations of the ventral ganglia and the segmental divisions of the mesoderm. A: Sagittal section to show ganglionic and interganglionic regions and the median cells of mesectodermal origin at the time of shortening of the embryo. B: Frontal section of the same, showing continuity of median cell strand. C: Transverse section in a ganglionic region to show thinness of median strand and bilateral nature of nerve cord. Median cells still associated with ventral mesoderm. D: Transverse section in an interganglionic region, showing median cells still in contact with midventral hypoderm and ventral mesoderm. APD, apodeme; HY, hypoderm; MCN, median cells of nervous system; MS, mesoderm; MSS, somatic mesoderm; VNS, ventral nervous system.

and move apart, the more dorsal median cells remain in contact with them, spreading out in a thin sheet atop the ventral nerve cord. This sheet is continuous with the vertical portion, which separates the two halves of the nerve cord. The median cells thus come to form a structure which is T-shaped in cross-section with the base set upon the ectoderm of the ventral midline, the partition separating the lateral parts of the nerve cord, and the arms spread over the dorsal surface. With the shortening of the embryo the contact with the lateral mesoderm remains and is more conspicuous in the interganglionic regions. The central partition becomes very thin within the ganglionic regions, and it is there that its connections with the outer ectoderm are broken as the nervous system begins its further condensation. It would not be difficult to interpret this structure as a median nerve at this time, but its subsequent history makes clear that it is not. Instead the outer, upper sheet of cells spreads over the surface of the cord to form the sheath cells or neurolemma. The larger cells of the interganglionic regions become the characteristic columnar cells of these regions, and some of the smaller cells appear to surround the developing nerve fiber bundles between the ganglia. Such observations as have been made on the outgrowth of nerve fibers to form the segmental nerves suggest that their course is guided, or aided, to the lateral mesoderm by the thin dorsal connections described above, and that the latter supply the sheath cells of the nerves.

The first clear signs of nerve fibers are to be found at the time the embryo has shortened (tenth hour). Both longitudinal and transverse fibers (Fig. 25) make their appearance and become more and more marked with time as the ventral nerve cord continues its condensation, the steps in which may be followed in Figs. 56, 60, 64, and 68. It is at this time that the outgrowth of segmental nerves is noted. These are somewhat difficult to follow and will not be dealt with further here, except to indicate that from this time on twelve pairs of ventral nerves are to be found, including the pair originating in the suboesophageal ganglion. By the time of hatching the ventral nervous system is highly condensed with its ganglia closely approximated. The interganglionic regions are marked by the tall columnar median cells (Figs. 21A, B, and 24). The paired nerves extend from the segmental ganglia to the corresponding segments.

Before the shortening of the embryo (ninth hour) the brain lobes consist of lateral aggregates of neuroblasts and ganglion cells, with no sign of segmental structure, becoming somewhat

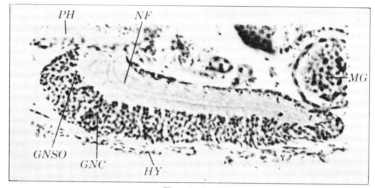

FIG. 24A.

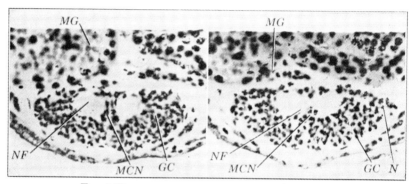

FIG. 24B.   FIG. 24C.

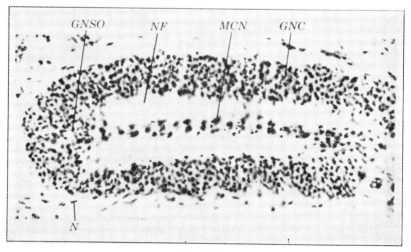

FIG. 24D.

FIG. 24. Structure of the ventral nervous system in later embryos. Bodian preparations. A: Sagittal section of an embryo of 20 hours just to one side of median cells, so that longitudinal nerve fiber tracts are visible. 265×. The nervous system is almost completely condensed. B: Transverse section at interganglionic level in an embryo of 16 hours, showing median cells. 400×. C: Transverse (next but one) section at ganglionic level of same embryo as B, showing transverse commissure and part of a segmental nerve. 400×. D: Frontal section through upper level of the ventral nervous system in an embryo shortly before hatching. Segmental arrangement of median cells very striking. Two of segmental nerves near anterior end below. 400×. GC, germ cells; GNC, ganglion cell; GNSO, suboesophageal ganglion; HY, hypoderm; MCN, median cells of nervous system; MG, midgut; N, nerve; NF, nerve fibers; PH, pharynx.

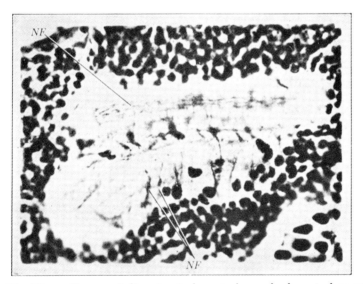

FIG. 25. Nerve fibers and fiber tracts in anterior end of ventral nervous system as seen in near frontal section of larva at hatching. Both longitudinal and transverse fibers are evident. Bodian preparation. 1200×.
NF, nerve fibers.

more separated from the rest of the ectoderm than at earlier stages. Just before the shortening, lateral ectodermal thickenings on either side at the posterior level of the brain lobes are invaginated (Figs. 27 and 53) against the brain and subsequently become incorporated into it. These persist as ball-like cell groups for some hours before they become indistinguishable from other cells in this region. It is probable that these are the

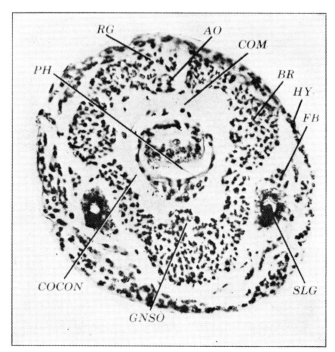

Fig. 26A.

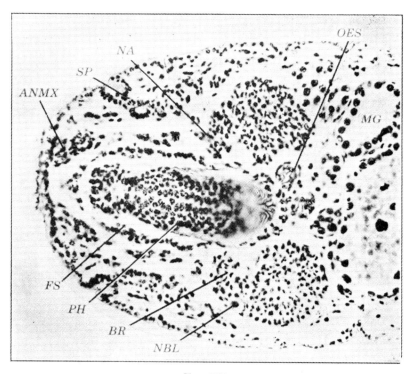

Fig. 26B.

rudiments of the optic lobes and the source of the large neuroblasts later observed dividing in the posterior region of the brain.

During the shortening of the embryo the brain becomes sharply
separated from the remaining ectoderm except in two lateral regions on either side, where cellular connections are retained with
ectodermal thickenings. These thickenings, or plaques, subsequently move anteriorly (though retaining connections with the
brain) and give rise to certain of the larval cephalic sensory
organs of the antennal-maxillary complex (Keilin, 1927; Hertweck, 1931).

In the brain there are no median cells of the type characteristic
of the median cord. The lobes are separated in the early stages
by the anterior end of the yolk mass. During the movements
in the shortening of the embryo the bases of the lobes come into
contact with the suboesophageal ganglion, where connections
are soon established. Nerve fibers extending in both directions
on either side produce the circumoesophageal commissures. The
brain lobes make contact in the midline above the oesophagus,
where nerve fibers establish the principal transverse connections
between the lobes. Thus the oesophagus becomes encircled by
nervous tissue. The brain (Fig. 26) now has a central fibrous
structure, which becomes increasingly complex, surrounded by
small ganglion cells. The larger neuroblasts, which continue their
divisions into the later stages, are confined to the posterior cortical regions.

With the involution of the head the whole central nervous system is shifted posteriorly and is separated from the extreme anterior end of the larva by the highly developed cephalopharyngeal apparatus, as illustrated in Figs. 60, 63, 64, and 67. The

FIG. 26. Brain and associated structures in embryos of 16 hours. A: Transverse section at the level of the cerebral commissure and circumoesophageal
connectives. Bodian preparation. 600×. B: Frontal section at the level
of the cephalic sense organs and anterior spiracles, including parts of
antennal maxillary complex and the antennal nerve. Bodian preparation.
600×. *ANMX*, antennal-maxillary complex; *AO*, aorta; *BR*, brain;
COCON, circumoesophageal connective; *COM*, cerebral commissure; *FB*,
fat body; *FS*, frontal sac; *GNSO*, suboesophageal ganglion; *HY*, hypoderm; *MG*, midgut; *NA*, antennal nerve; *NBL*, neuroblast; *OES*, oesophagus; *PH*, pharynx; *RG*, ring gland; *SLG*, salivary gland; *SP*, spiracle.

characteristic position of the brain is within the third thoracic and first abdominal segments from this time on.

The stomodaeal, or stomatogastric, nervous system (Bickley, 1942) arises from neuroblasts which move internally along with three invaginations in the roof of the stomodaeum (Fig. 14) between the eighth and tenth hours (Poulson, 1945b). These give

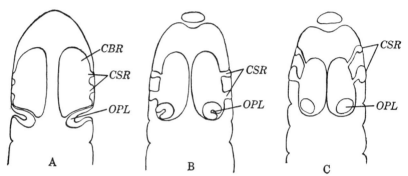

Fig. 27. Diagrams of ırontal sections to illustrate origin of optic lobes of brain and the sensory rudiments which will give rise to the photoreceptor organs and the antennal parts of the antennal-maxillary complex. A: Just before the shortening of the embryo (8–9 hours) showing invagination of optic lobes and the lateral regions which remain associated with the hypoderm. B: Just after shortening (10 hours) at the beginning of involution of the head. C: Before involution of head is complete. The connections between the two sensory rudiments and the brain subsequently fuse to form the antennal nerve. Their distance apart is exaggerated in these diagrams. *CBR*, cerebrum; *CSR*, cerebral sensory rudiment; *OPL*, optic lobe of brain.

rise to three ganglia: the frontal, which lies within the cephalopharyngeal apparatus above the pharynx; the hypocerebral, which lies just behind the transverse commissure of the brain, but beneath the ring gland and the anterior end of the dorsal aorta; and the ventricular, on the forepart of the midgut. Between the frontal and hypocerebral ganglia the recurrent nerve appears. This is found in later embryos to extend posteriorly from the hypocerebral ganglion to the ventricular ganglion of the midgut. Connectives between the brain and the frontal ganglion may also be distinguished in late embryos, as are nerves joining the hypocerebral ganglion with the brain. The latter are the nervi corpori cardiaci of Hanström (1942).

No attempt has been made to determine the origin of sensory organs, except to follow the ectodermal regions (Fig. 27) associated with the lateral parts of the brain. In addition to the two regions described above there is a region of close association between the anterolateral surface of the suboesophageal ganglion and the lateral ectoderm. With the involution of the head these three regions move anteriorly (Fig. 28), retaining their connec-

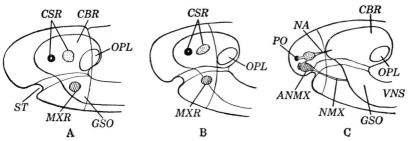

Fig. 28. Brain and larval sense organs during the involution of the head: origin of the antennal maxillary complex. Diagrams of whole mounts of anterior end seen in side view at successive stages in the involution of the head. A: At 10 hours before involution is under way. B: During involution. C: After the completion of involution and differentiation of the sense organs, showing antennal and maxillary nerves to the central nervous system. *ANMX*, antennal-maxillary complex; *CBR*, cerebrum; *CSR*, cerebral sensory rudiment; *GSO*, suboesophageal ganglion; *MXR*, maxillary sensory rudiment; *NA*, antennal nerve; *NMX*, maxillary nerve; *OPL*, optic lobe of brain; *PO*, photoreceptive organ; *ST*, stomodaeum; *VNS*, ventral nervous system.

tions with the central nervous system, and finally lie at the anteriormost tip of the larva, above and lateral to the mouth hooks, on the maxillary lobes. They form the so-called "antennal-maxillary complex" (cf. Weismann, 1864; Wandolleck, 1898; Keilin, 1927; Hertweck, 1931) of cyclorraphous Diptera. The morphology of this in the larva of *D. melanogaster* is described by Keilin and Hertweck. The antennal-maxillary complex is almost fully differentiated between the fourteenth and sixteenth hours. From this time on it consists (on each side) of a large outer and more ventral ganglion (maxillary ganglion) with a broad end-organ, and an inner and more dorsal ganglion (antennal ganglion) centrally constricted from which a branch (antennal-maxillary ganglion) runs to the broad maxillary end-organs. The anterior termination of the antennal ganglion is in an ocellus-like end-organ which has previously been referred to as the larval antenna. In view of its structure, the studies of Ellsworth (1933) on the photoreceptive properties of the corresponding organ in *Lucilia sericata*, and the light responses of larvae of Drosophila, it is probable that this is a photoreceptor, the larval ocellus. If this is correct, the larval antenna is then incorporated into the maxillary end-organ, and the latter ought more properly be referred to as the antennal-maxillary end-organ. Keilin (1927) and Hertweck (1931) have described the chordotonal organs of the maxillary complex. Their origin has not been studied here, nor has that of larval chordotonal organs generally, the structure and distribution of which have been described by Hertweck (1931). However, the origin of the ventral ganglion of the pharynx and its end-organ (hypophyseal ganglion and end-organ) has been followed from an association of the suboesophageal ganglion with ventral ectodermal thickenings.

The functional activity of the nervous system may be inferred by the beginnings of muscular movements after the dorsal closure of the embryo in the twelfth hour. A more objective criterion is provided by measurements which have been made on cholinesterase activity in embryos at different stages in development (Poulson and Boell, 1946a, b). No appreciable cholinesterase activity has been found before 10 hours. After that it is present and increases rapidly between the thirteenth hour and the time of hatching. The higher enzyme activities correspond to the period of increasing movement on the part of the larva. That the cholinesterase activity is localized in the nervous system has been shown by determinations on isolated embryonic nervous systems.

The Tracheal System

During the period of neuroblast formation the superficial ectoderm cells undergo mitosis. From the resulting cells are derived the tracheal system, the hypoderm, and, much later, the imaginal discs. Although the establishment of the segmental structure of the embryo is first detectable in the nervous system and then in the mesoderm, the first external evidence of segmentation is the appearance in the sixth and seventh hours of eleven pairs of lateral tracheal pits distributed at equal intervals along the

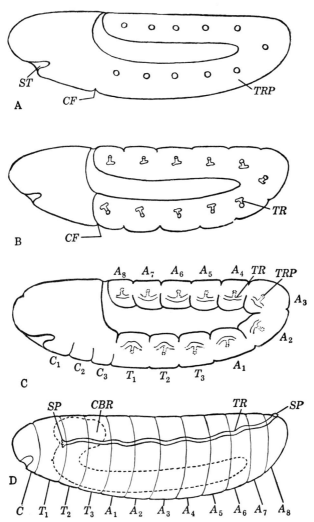

FIG. 29. Diagrams to illustrate the development of the tracheal system as seen in lateral views of whole mounts. A: At the time of formation of tracheal pits (6–7 hours). B: Tracheal invaginations T-shaped (7–8 hours). C: Beginning of consolidation of tracheal rudiments, disappearance of pits (8–9 hours). Maximum head segmentation. D: Tracheal trunks and anterior and posterior spiracles after dorsal closure of embryo (12–14 hours). Segmentation of head and body indicated. *CBR*, cerebrum; *CF*, cephalic furrow; *SP*, spiracle; *ST*, stomodaeum; *TR*, trachea; *TRP*, tracheal pit.

trunk (Fig. 29). None has been observed in the head region. These begin in the sixth hour as slightly thickened lateral plates with very shallow central depressions, which deepen into pitlike invaginations in the seventh hour, as can be seen in Figs. 30, 46,

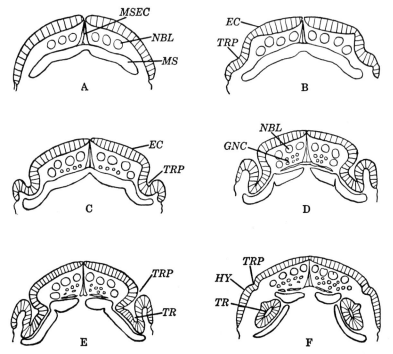

Fig. 30. Diagrams of transverse sections near the posterior end of the embryo while it is still extended before shortening. A: Before the appearance of tracheal pits. B: Tracheal pits (6 hours). C: Tracheal pits deepening. D: Tracheal invaginations (7–8 hours). E: Later stage. F: Tracheal pit remnants and tracheal rudiments at the time of shortening. *EC*, ectoderm; *GNC*, ganglion cell; *HY*, hypoderm; *MS*, mesoderm; *MSEC*, mesectoderm; *NBL*, neuroblast; *TR*, trachea; *TRP*, tracheal pit.

and 50. These invaginations branch and become somewhat T-shaped at the blind ends, the cells numbers increasing by mitosis. Subsequently few or no mitoses are to be found among tracheal cells. In the eighth hour the walls of the invaginations become closely approximated (Fig. 31C) so that the openings to the surface are greatly reduced. During the ninth hour the shortening of the embryo brings about their complete closure, and these external evidences of the tracheal system are lost.

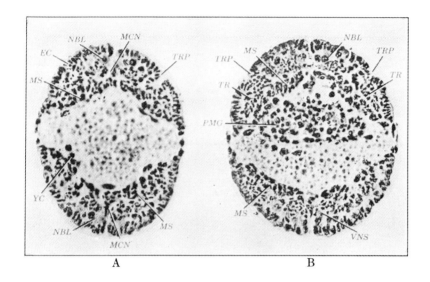

A   B

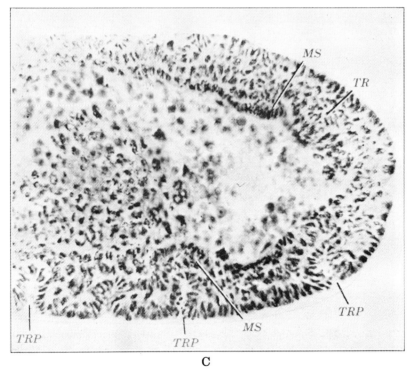

C

Fig. 31. Tracheal invaginations. A and B: Transverse sections of an embryo of 7–8 hours. Bodian preparations. 400×. C: Frontal section of slightly later embryo, showing position of pits in relation to segmentation of mesoderm. Bodian preparation. 600×. *EC*, ectoderm; *MCN*, median cells of nervous system; *MS*, mesoderm; *NBL*, neuroblast; *PMG*, posterior midgut rudiment; *TR*, trachea; *TRP*, tracheal pit; *VNS*, ventral nervous system; *YC*, yolk cells.

227

Meanwhile the branches of the invaginations become unequal, the more lateral one becoming longer. As the embryo undergoes shortening, these become oriented longitudinally, and the rudiments join those of adjoining segments. In this way the two longitudinal tracheal trunks arise. The lumen remains almost indistinguishable for the next few hours, becoming readily recognizable only when the chitinous intima begins to appear after the twelfth hour. The establishment of the branches and their extensions is not readily followed as the outgrowing tracheocytes are very slender and elongate. The courses of these finer branches and the tracheoles are more readily followed after the chitinous intima has been secreted. In the later stages spiral taenidia are clearly present in the trunks and the larger branches. During differentiation the cells change from cuboidal to flattened epithelial. The anterior and posterior crossconnections between the trunks are established at the time of dorsal closure of the embryo (eleventh to twelfth hour). By the sixteenth to eighteenth hours the tracheal system, as described by Ruhle (1932) for the first-instar larva, is fully established. For details of the courses of the branches the reader is referred to his paper.

The spiracles originate separately from the tracheal vessels. The posterior pair arise from separate posterior invaginations in the terminal segment between the ninth and tenth hours and are clearly present when the embryo has fully shortened. They join with the posterior ends of the tracheal trunks and provide the only connection to the outside. Very shortly the posterior spiracles are raised from the dorsal surface posterolateral to the anus. Their characteristic sculpture, setae, etc., appear somewhat later at the time of differentiation of the larval integument and setae.

The anterior spiracles make their appearance during the time of involution of the head as ventrolateral invaginations near the posterior edge of the prothoracic segment. Into the tips of these rather deep invaginations grow the anterior tips of the tracheal trunks. The anterior spiracles remain closed and internal until a much later period in the life history.

The appearance of air in the trachea is noted in living embryos from about the eighteenth hour on. The entrance of air from the tissues and its penetration into the finer branches and tracheoles proceed in the manner described by Kaliss (1939) for Drosophila and for other forms by Sikes and Wigglesworth (1931).

The Oenocytes

The origin of the oenocytes has not so far been traced in this material. Oenocytes are to be found in later embryos before hatching in the characteristic lateral positions in each abdominal segment between the lateral longitudinal muscles and the hypoderm. This is the point at which the original tracheal invaginations were formed and closed off. It is thus clear that in Drosophila, as in other insects, the oenocytes arise in close association with the tracheal system and probably from material invaginated at the same time. However, this remains to be definitely established. The situation in the late embryo with regard to numbers and positions of oenocytes is in full agreement with the descriptions of Koch (1945) and Bodenstein (p. 275) for later larval stages.

The Hypoderm

The hypoderm arises largely from the ventral and ventrolateral cells which remain at the surface after the differentiation of the neuroblasts and the invagination of the tracheal rudiments, except in a few special regions, such as those which become the salivary glands. The characteristic appearance of these dermatoblasts may be seen in Figs. 45, 46, 49, and 50, where they still retain something of their earlier columnar shape. With time the hypoderm cells become more cuboidal and finally nearly flattened. In the extreme lateral regions before the dorsal closure of the embryo the ectoderm changes abruptly into the much-flattened cells of the embryonic membranes. At the time of dorsal closure of the embryo some of these cells appear to be incorporated into the dorsal hypoderm, where they soon become indistinguishable from neighboring cells. There is evidence that many of the larger amnioserosa cells, especially those just posterior to the head, are replaced by lateral cells and are absorbed by the yolk. This may be true of such cells in other regions, but crucial instances of this are lacking.

Except for the dorsal and lateral coverings of the head region, the hypoderm is more or less uniform in thickness and structure up to the time of shortening of the embryo (9–10 hours). During and after the shortening, regional differences become apparent, for the hypoderm is then sharply folded in the intersegmental regions. The folds are deepest laterally and almost completely absent on the ventral midline. The hypoderm beneath the nervous system thins out in this flattened region (Figs. 49, 50, 53, 54, 57, and 58). The transition between this and the more lateral hypoderm is sharp. From the intersegmental folds develop the apodemes, to which the segmental muscles attach (Fig. 4). This attachment appears to begin just before and during the shortening of the embryo. With the dorsal closure the lateral hypoderm thins out in the segmental regions as it spreads dorsally. The closure is completed in the tenth to eleventh hour before the involution of the head is very far under way. From this time on the structure of the hypoderm is essentially uniform in all regions of the embryo.

In the twelfth hour, when the involution of the head has been completed, the first signs of the chitinous cuticle appear. The cuticle develops more or less uniformly on the original outer surface of all the non-neural ectoderm derivatives, thus forming not only the outer cuticle of the integument and the intima of the trachea, but also the linings of the hind- and foreguts and the framework of the cephalopharyngeal skeleton. From the time of first appearance of the cuticle, a thin inner basement membrane can be distinguished sharply, bounding the inner edges of the hypoderm cells. Chitinization proceeds rapidly after this time, and setae and other special cuticular differentiations begin to appear about the fourteenth hour. The larval cuticle is very thin and flexible. It has not so far been possible to distinguish any separate layers within it.

The Involution of the Head

The larvae of cyclorraphous Diptera are characteristically acephalic and apodous. The absence of an external larval head is a consequence of the involution of the head regions in middle embryonic development. Although studied from a comparative morphological point of view by a number of workers (Wandolleck, 1899; Holmgren, 1904a, b; Becker, 1910; Wahl, 1914; Keilin, 1915, 1917; deMeijere, 1916; Bischoff, 1922; Imms, 1924; Snodgrass, 1924, 1935), no account of the embryological events of involution is available for any Diptera except *Melophagus ovinus* (Pratt, 1900) and *D. melanogaster* (Poulson, 1945).

The early movements of the anterior ectoderm to the interior begin with the stomodaeal invagination (5½–6 hours). These movements continue during the subsequent period and do not completely cease until the head has involuted. The stomodaeal invagination proceeds at first rapidly and then more slowly for some hours (Figs. 44, 14, 48, and 52). Most of the cells turned in during this period appear to be of ventral and ventrolateral origin. Between the eighth and ninth hours more rapid movements begin, and at this time appear the three invaginations in the roof of the stomodaeum which give rise to the stomodaeal nervous system and contribute to the lateral parts of the ring gland. The now-invaginating salivary glands move anteriorly and by 10½ hours open through their ducts into the hypopharynx (Fig. 56). Immediately after this begins the invagination of the frontal sac, into which go the dorsal and the remaining lateral and ventral cells of the head region, except those which form the integument of the tiny external head, or neck region. With the inmoving frontal sac go the groups of cells from the mandibular region, which form the corpus allatum, or central portion of the ring gland. The frontal sac passes in and then dorsally over and laterally around the buccopharyngeal region. Beneath it differentiate the large vertically arranged buccopharyngeal muscles (Figs. 32, 60, and 64). The larval cephalic primordia (Figs. 27 and 28) move anteriorly to their final positions, retaining connection with the brain lobes by nerves. The abdominal and thoracic segments expand so that the latter occupy the original position of the head segments. This very active period in the involution of the head is of short duration, beginning between 10½ and 11 hours and lasting 30–45 minutes. Although this involution has been followed very closely in living embryos, it is clear that the point of origin in the outer ectoderm of each portion of the frontal sac can be determined with certainty only by the aid of marking and other experimental procedures.

The derivatives of the frontal sac, as described here, are the mouth hooks and the skeletal plates of the cephalopharyngeal apparatus. According to previous authors, the posterior portion of the frontal sac later gives rise to the cephalic complex of imaginal discs (Wahl, 1914; Snodgrass, 1924; Chen, 1929). Although that

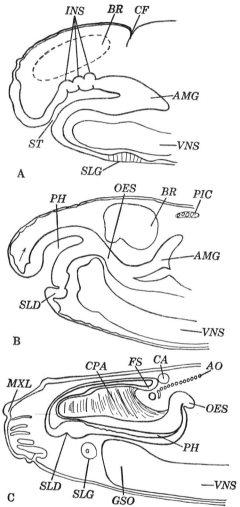

FIG. 32. Involution of the head illustrated by *camera lucida* outlines of sagittal sections. A: Anterior end of embryo of Fig. 48, at 8 hours. B: Anterior end of embryo of Fig. 56, at 10 hours. C: Anterior end of embryo of Fig. 64, at 16 hours. *AMG*, anterior midgut rudiment; *AO*, aorta; *BR*, brain; *CA*, corpus allatum; *CF*, cephalic furrow; *CPA*, cephalopharyngeal apparatus; *FS*, frontal sac; *GSO*, suboesophageal ganglion; *INS*, invaginations in roof of stomodaeum; *MXL*, maxillary lobe; *OES*, oesophagus; *PH*, pharynx; *PIC*, pericardial cells; *SLD*, salivary duct; *SLG*, salivary gland; *ST*, stomodaeum; *VNS*, ventral nervous system.

is clearly the case in the embryo of *Melophagus ovinus* (Pratt, 1900), there has been no thorough early embryological study of their origin in other higher Diptera. The origin of imaginal discs will be dealt with in a separate section. The mouth hooks and pharyngeal bars are secreted as chitinous structures by cells of the inner or more ventral wall of the frontal sac. They darken and become increasingly visible, until by the time of hatching they are conspicuous features of the anterior end. The more dorsal portions of the frontal sac, especially along the midline, thin out and form the so-called "neck membrane." The only posterior or lateral thickenings that have been found before the time of hatching are the points of attachment of the protractor muscles of the pharynx. There are no clear indications here of antennal-optic rudiments.

During the observations on living embryos one anomalous case was found which throws considerable light on the origin of parts and the movements concerned in the involution. In this embryo dorsal closure was not complete and the midgut herniated dorsally over the head region. Involution stopped, and the frontal sac was not invaginated. The head ectoderm remained external. When first observed, the embryo was 18 hours old. It was followed closely until it was 31 hours old and then fixed. The normals in this batch of eggs all hatched between 22 and 24 hours. Active movements of body muscles and peristalsis of the gut continued up to the time of fixation, and differentiation of all the externally visible parts was essentially normal. On the surface of the ventral ectoderm of the head of the abnormal embryo there developed a number of chitinized structures which clearly represented the mouth hooks and the cephalopharyngeal plates and skeleton. The structures which appeared to be the mouth hooks lay posterior and lateral to the cephalopharyngeal skeletal parts. Thus these derivatives of the frontal sac come from ventral and ventrolateral ectoderm of the mandibular and maxillary segments. This case cannot be reported more fully until a study of sections is correlated with the observations which have been made on the living embryo and the subsequent whole mount preparation. Cases of this sort are very rare; this is the only one among the thousands of eggs on which this study is based.

In addition to the shifts in the outer layers of the embryo internal shifts are initiated as a consequence of involution. The most conspicuous of these is the posterior displacement of the brain, which appears to be pushed backward against and into the saclike midgut. The gut buckles in at this point around the back of the brain. Shortly afterward appears the first of the constricting furrows (Figs. 60 and 63) which lead to the elongation of the midgut. What the relations may be between involution of the head and the shortening and condensation of the ventral nervous system is unknown. At this time the ventral nervous system extends to the posterior tip of the embryo just beneath the termination of the hindgut. As the brain moves backward, it pulls the suboesophageal region of the ventral nervous system slightly forward and up from beneath it.

The mechanisms of the shortening of the embryo, the dorsal closure, and the involution of the head are very incompletely understood. From a consideration of normal development it might be concluded that shortening is a consequence of the attachment and early contraction of the longitudinal muscles. However, shortening is a regular feature in Notch-deficient embryos (Poulson, 1940, 1945a), in which the ventral hypoderm is lacking and musculature is only partially differentiated. Thus something other than muscle contraction as such is involved. The experimental results of Reith (1925) and Pauli (1927) on muscid eggs are in agreement with this. Given the shortened condition of the embryo, dorsal closure is a simple mechanical consequence. Likewise involution of the head is a further consequence of both shortening and closure in an embryo in which the differentiation of the head region is incomplete or retarded. A complete analysis of these movements in the embryo by experimental means is much needed.

The Ring Gland

The ring gland is complex in its origin as well as in its structure (Poulson, 1945b) and can best be understood in relation to the movements and changes in the head region which culminate in involution. The lateral parts of the ring gland and the hypocerebral ganglion appear to arise from the largest of the three invaginations which form in the roof of the stomodaeum just before the shortening of the embryo (see p. 207 and Figs. 14 and 48). Some of these invaginated cells are small neuroblasts and undergo typical neuroblast mitoses to produce the ganglion cells of the hypo-

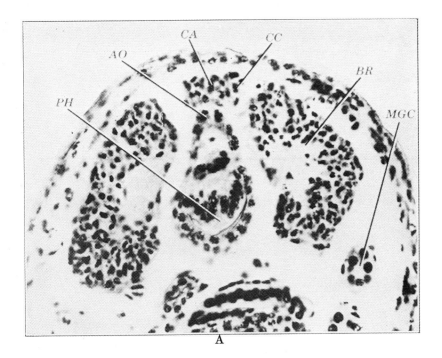

A

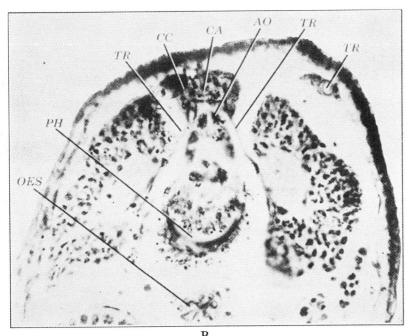

B

Fig. 33. The ring gland in transverse section in Bodian preparations. A:
At 15–16 hours. 900×. B: At 17–18 hours, showing tracheal branches
through ring gland to brain. 900×. *AO*, aorta; *BR*, brain; *CA*, corpus
allatum; *CC*, corpus cardiacum; *MGC*, midgut caecum; *OES*, oesophagus;
PH, pharynx; *TR*, trachea.

cerebral ganglion. The other cells form two groups which move dorsally and in the twelfth hour approach the anterior end of the aorta, where they remain lateral to its terminus. During this movement they become invested with thin connective tissue cells.

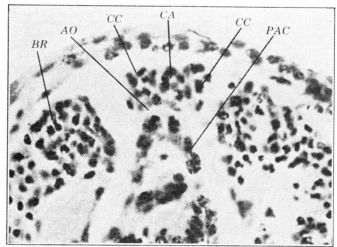

FIG. 34A.

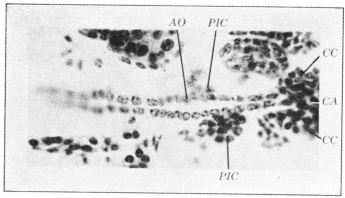

FIG. 34B.

Ventrally they are intimately associated with the cells of the hypocerebral ganglion.

During the invagination of the frontal sac in the eleventh hour a compact group of cells is to be found at its extreme tip. These cells become separate from the other cells of the frontal sac, although they remain associated with them. This group of cells is pushed between the lobes of the brain to lie over the anterior termination of the dorsal aorta, where it is flanked by the two lateral groups of cells described above. This group is invested also by a thin coat of connective tissue and bound to the lateral cells. This central group of small cells, clearly derived from the

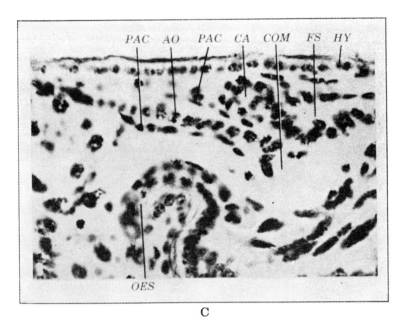

C

FIG. 34. Ring gland and dorsal aorta as seen in three principal planes. Bodian preparations. A: Transverse, detail of Fig. 33A. 1200×. B: Frontal section of 20-hour embryo, showing position of groups of large pericardial cells posterior to ring gland. 900×. C: Sagittal section of 16-hour embryo, showing relations of corpus allatum, aorta, frontal sac, paracardial cells. 900×. This portion of the frontal sac will later give rise to optic-antennal imaginal discs. AO, aorta; BR, brain; CA, corpus allatum; CC, corpus cardiacum; COM, cerebral commissure; FS, frontal sac; HY, hypoderm; OES, oesophagus; PAC, paracardial cells; PIC, pericardial cells.

ectoderm of the mandibular segment, appears to represent the fused corpora allata of lower Diptera. The larger lateral cells, on the basis of their origin from the roof of the stomodaeum, appear to be the corpora cardiaca. Together with the hypocerebral ganglion these cell groups form a ring (Figs. 33 and 34) around the anterior end of the aorta, referred to variously as "Weismann's ring," "anneau de soutien," and, more currently, the "ring gland." In the fourteenth hour the ring gland lies in its characteristic position above the cerebral commissure, held in place by the tracheal branches which pass through the lateral parts and thence to the brain. A pair of short nerves (nervi corpori cardiaca I of Hanström), which join the hypocerebral ganglion to the brain at the level of the commissure, may be readily demonstrated in favorably oriented sections. The presence of these nerves supports the homology of the lateral parts of the ring gland with the corpora cardiaca. Whether all the cells of the lateral parts of the ring gland have the same origin cannot be determined without further and more detailed study. Certainly the more dorsal lateral cells are larger than the ventral ones (Poulson, 1945b), but there seems to be no gradation between larger cells and the smaller cells of the hypocerebral ganglion.

As the time of hatching approaches, the ring gland becomes more and more tilted anteriorly over the brain so that its appearance in section is somewhat changed. No special staining techniques have as yet been tried to determine when secretory activity begins. The appearance of the ring gland at 15 and 17 hours is illustrated in Figs. 33 and 34.

The Imaginal Discs

The studies which have been made on the early development of the imaginal discs in *D. melanogaster* by Chen (1929), Medvedev (1935), Auerbach (1936), and Steinberg (1941, 1943) have made it clear that most of these first become morphologically recognizable in early larval development. Although the rudiments of the thoracic discs are present and recognizable in section at the time of hatching (Auerbach), it is some hours before they are large enough to be discerned in dissections. The cephalic complex of antennal-optic discs was found by Chen to be first recognizable about 16 hours after hatching. Steinberg (1941) has shown that these discs are attached to the brain through the optic stalk 24 hours after hatching. The rudiments of the abdominal discs have not so far been reported in the early larval stages of Drosophila. They are described by Snodgrass (1924) simply as hypodermal thickenings in the larva of Rhagoletis. Kaliss (1939) has figured the newly formed genital disc at the posterior end of the larva at the time of hatching.

The lack of satisfactory information concerning the earliest stages in the origin of the cephalic complex of imaginal discs has

been a stumbling block in the study of the relations of the eye
and brain and of the development of eye mutants. A careful
search was therefore made for rudiments of the antennal-optic

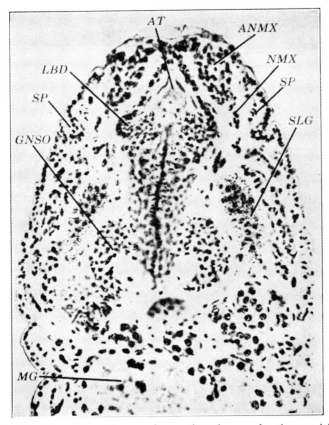

FIG. 35. Frontal section of an embryo of 16 hours, showing positions of
labial discs, antennal-maxillary complex, and other anterior structures.
Bodian preparation. 600×. *ANMX*, antennal-maxillary complex; *GNSO*,
suboesophageal ganglion; *LBD*, labial imaginal disc; *MG*, midgut; *NMX*,
maxillary nerve; *SLG*, salivary gland; *SP*, spiracle.

discs within the embryo. In view of the findings of Pratt (1900)
concerning the early origin of the cephalic complex in *Melophagus
ovinus* it has been generally accepted that the antennal-optic discs
of the higher Diptera have their origin in the deep posterior parts
of the frontal sac (cf. Snodgrass, 1924, 1935) as lateral pouches.
No such pouches or deeper invaginations of the posterior portions of the frontal sac are to be found in *D. melanogaster* either

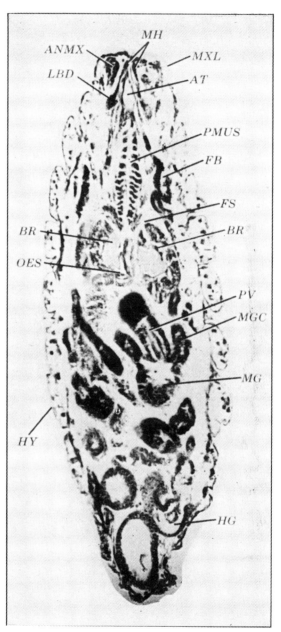

Fig. 36. Frontal section of newly hatched larva to show relationships of organs at time of hatching. Iron haematoxylin preparation. 300×. *ANMX*, antennal-maxillary complex; *AT*, atrium; *BR*, brain; *FB*, fat body; *FS*, frontal sac; *HG*, hindgut; *HY*, hypoderm; *LBD*, labial imaginal disc; *MG*, midgut; *MGC*, midgut caecum; *MH*, mouth hook; *MXL*, maxillary lobe; *OES*, oesophagus; *PMUS*, pharyngeal muscles; *PV*, pro-ventriculus.

before or at the time of hatching. All that can be found is a posterior thickened region (Fig. 34C) of slightly larger cells which shows signs of bilaterality. Although this thickening increases and becomes clearly separated to two, there are no signs of antennal-optic pouches up to 6 hours after hatching, the oldest stages observed in this material. During this time the region is clearly separated from the brain, and no connection is established until sometime later. The period between the sixth and the twenty-fourth hours of larval life should receive careful attention if the relations between the developing discs and the brain are to be fully elucidated.

The only imaginal rudiments clearly recognizable well before the time of hatching are the labial buds, which later give rise to the proboscis. These are first noticed as thickened ventrolateral regions of the atrium just in front of the opening of the frontal sac and anterior to the termination of the common salivary duct (Fig. 35). By the time of hatching these are round lateral buds with a definite lumen opening into the atrium (Fig. 36).

The thoracic rudiments at the time of hatching are, as described by Auerbach (1936), hypodermal thickenings in the case of the three pairs of leg discs and small invaginated cell groups in the case of the three pairs of dorsal discs. The steps in the invagination of the dorsal discs have not been followed, and little can be added here to Auerbach's account. No evidence of the single genital disc has been noticed in this material.

CELL LINEAGE AND THE GROUND PLAN OF THE EMBRYO

It is of some interest to summarize our knowledge of the general plan of Drosophila development so that it can be compared with what is known of other insects (Seidel, 1936; Richards and Miller, 1937; Krause, 1939). Our present conceptions of the plan of development within the egg of Drosophila have been based mainly on the ingenious analysis of mosaics and gynandromorphs initiated by Sturtevant (1929) and continued by Parks (1936) and Patterson and Stone (1938). Although some direct experiments have been carried out, with the aim of determining the pattern of development, by Geigy (1931a) and especially by Howland and her coworkers (Howland and Child, 1935; Howland and Sonnenblick, 1936; Howland, 1941), a really thorough-going experimental analysis remains to be done. In the absence of detailed defect and marking experiments an attempt will be made to derive from the existing data on normal and abnormal development a general picture of localization and the Anlagenplan in the blastoderm and preblastoderm stages. This is most readily done by tracing the cell groups which comprise the various larval organs backward to the regions of origin in the blastoderm. It is recognized that this approach has its pitfalls and gives only a provisional picture, but it may serve as a stimulus for further investigations along these lines.

A chronological summary of developmental events such as is given in Table 1 is of considerable practical value for many pur-

TABLE 1

Timetable for Development of the Egg of D. Melanogaster

[Based principally on the data of Rabinowitz (1941) and the writer]

Time (in hours)	General Features of Stages at 23–25°C
0–1½	Premigration; synchronous mitoses; no pole cells
1½	Migration; preblastoderm; pole cell formation
2	Syncytial blastoderm; synchronous blastoderm mitoses
2½	Cleavage furrows; blastoderm cell formation
3	Cellular blastoderm; pregastrular cell movements
3½	Ventral furrow; posterior plate; early gastrulation
3¾	Ventral furrow deep; cephalic furrow; extension begins
4	Extending germ band; anterior and posterior invaginations
4½	Extending germ band; dorsal folds; flattening mesoderm
5	Extended germ band; large neuroblasts; primitive gut
5½	Beginning stomodaeum and proctodaeum; neuroblast mitoses
6	Stomodaeum and proctodaeum deep; mesoderm segmented
7	Tracheal invaginations; salivary gland plates
8	Segmentation of head and trunk; muscle attachments
9	Shortening of embryo begins; salivary glands internal
10	Beginning involution of head and dorsal closure
11	Involution of head and dorsal closure; gonads compact
12	Frontal sac deep; midgut saclike, unconstricted
13	Midgut constrictions; beginning chitinization
14	Condensing ventral nervous system; first muscular movements
16	Further condensation of nervous system; regular muscular movements
18	Larval differentiation nearly complete; active movements
20	Continued active movements; first air in tracheae
22–24	Hatching from the egg

poses but, being purely descriptive, does not indicate relationships. Information on cell lineage has been brought together in Fig. 37, which represents an attempt to present the relationships between the cell groups which contribute to the various tissues

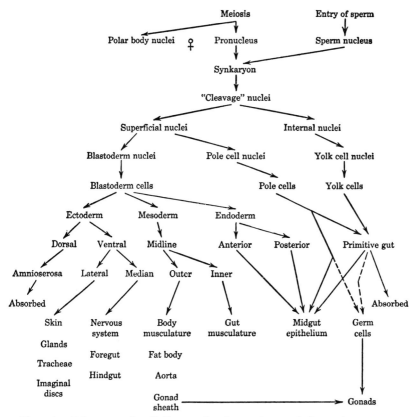

FIG. 37. Diagram of cell lineage in the embryo of *D. melanogaster.*

and organs. Although it is not possible to extract more than goes into such a diagram, it is useful in visualizing these relationships and helpful in making a projection of the principal organs back on the blastoderm. To construct an accurate map of localization would require complete information concerning cell movements as well as the distribution and numbers of postblastoderm mitoses. A small amount of such information is available and has been combined with that given in the diagram of cell lineage to prepare the provisional fate-map, or Anlagen-plan, of Fig. 39.

It is clear from the accounts of Huettner (1923) and Sonnenblick in Chapter 2 and from Parks' (1936) study that the early distribution of nuclei is random, or nearly so. The migration to the surface results in the three principal cell lines, those of the blastoderm, pole cells, and yolk cells. Rabinowitz (1941) established the occurrence of three blastoderm mitoses preceding cell formation, the lack of synchrony in the early pole cell divisions, and the absence of mitoses in the early yolk cells. The nuclei of the early yolk cells increase in size, evidently undergoing endomitosis, and show every resemblance to the vitellophag nuclei of Chironomus described by Melland (1942). Although the actual numbers of yolk nuclei have not been determined, Feulgen whole mounts show well over a hundred of them at the time the blastoderm is completed. Thus a very substantial proportion of the early cleavage nuclei do not reach the surface at all.

As there are few mitoses during the period of "gastrulation," and none at that time along the ventral midline, counts of numbers of cells turned under in the ventral mesodermal tube enable the limits of the mesoderm to be rather accurately represented within a region of eight to ten cells on either side of the midline along the ventral side extending from the level of the anterior endoderm to the posterior endoderm plate. It can be seen in Fig. 38 that this represents about one-sixth of the circumference of the blastoderm. As the mesoderm is turned in, the remaining cells spread over the surface to form a slightly thinner outer layer. At the ends of the mesodermal tube the anterior and posterior midgut rudiments are then turned in and remove from the surface an even more substantial proportion of blastoderm cells per section. Of the cells of the ventrolateral sixths of the original blastoderm about six to eight per section move in to become neuroblasts in the regions posterior to the brain. On the basis of their size it is inferred that neuroblasts are derived directly from blastoderm cells without an intervening mitosis. The remaining ventrolateral cells of the thoracic and abdominal regions give rise to the hypoderm and the tracheal system, while the most dorsal cells become embryonic membrane. It is not easy to estimate the fraction of original blastoderm cells which goes into the tracheal system, as some mitoses precede the formation of the tracheal pits. It is clear, however, that the tracheal regions are located high up on the lateral blastoderm and rather close together. These, like other regions of the blastoderm, become elongated and more widely separated as the embryo lengthens around onto the dorsal side. The exact pattern of these movements remains to be determined here and in the stomodaeal and proctodaeal rudiments. The early location of the salivary glands

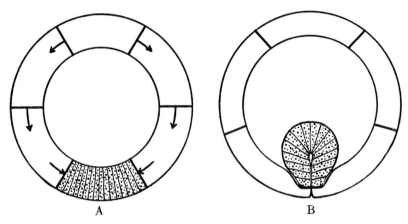

A   B

Fɪɢ. 38. Diagram to illustrate cell movements and extent of inturned mesoderm. A: Blastoderm just before movements, showing region which will turn in as mesoderm. B: Mesodermal tube just formed, showing spread of remaining blastoderm cells over surface as suggested by arrows.

is clear, and their subsequent movements can be readily followed in the head region. The imaginal discs are derived from the ventrolateral ectoderm, but adequate localization studies remain to be made.

Mitoses are numerous in the inturned mesoderm and in the midgut rudiments. Whether all inturned cells divide and how many divisions each undergoes remain to be determined. As noted elsewhere, mitotic activity ceases in all parts of the embryo by the tenth to eleventh hours except in the nervous system, where certain neuroblasts continue to divide. There are no mitoses in the salivary glands from the time that these are first recognizable. It is probable, as Sonnenblick suggests, that the salivary gland cells are derived directly from the blastoderm without any intervening mitoses. With regard to the pole cells, Sonnenblick has pointed out that there are no mitoses in them between the time of gastrulation and the sixteenth hour, when the first germ cell mitoses begin in the gonads.

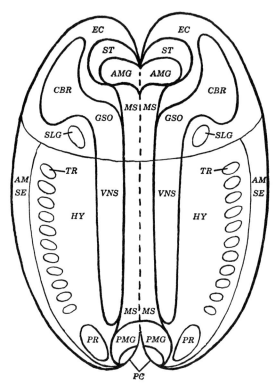

Fig. 39. Tentative diagram of preblastoderm localization in the egg of *D. melanogaster*. Ventral view of an egg split open along dorsal midline and spread out nearly flat. *AM*, amnion; *AMG*, anterior midgut rudiment; *CBR*, cerebrum; *EC*, ectoderm; *GSO*, suboesophageal ganglion; *HY*, hypoderm; *MS*, mesoderm; *PC*, pole cells; *PMG*, posterior midgut rudiment; *PR*, proctodaeum; *SE*, serosa; *SLG*, salivary gland; *ST*, stomodaeum; *TR*, trachea; *VNS*, ventral nervous system.

In the main the largest nuclei and cells within the later embryo are those which have undergone the fewest mitoses since the blastoderm stage. The largest nuclei are those of the yolk cells; next come those of the pole cells (both in the midgut and in the gonads), the large neuroblasts, and the salivary gland cells. The large binucleate garland cells appear to arise from failure of cytokinesis at the last mitosis. The multinucleate muscle fibers appear to arise by fusion, although failure of cytokinesis in later mitoses is not excluded. It is thus tempting to postulate that the size of cells at the time they appear to be morphologically differentiated in embryonic development is inversely related to the number of mitoses they have passed through. Exact data concerning nuclear size and cell size in different tissues and organs throughout embryonic development are not yet available, nor has the nuclear structure of non-mitotic cells been adequately investigated. These are indispensable in any attempt to relate differentiation to the mitotic and endomitotic cycles.

The general plan of the embryo of *D. melanogaster* (Fig. 39) and, presumably, that of the higher Diptera generally show many resemblances to those established for other insects (Seidel, 1936; Richards and Miller, 1937; Krause, 1939) but differ in that almost the whole of the blastoderm is embryonic and only a small proportion becomes membranes. That this is correlated with the reduction of the larval head in the cyclorrhaphous Diptera is suggested by the condition in the lower Diptera (e.g., Chironomus: Weismann, 1863; Miall and Hammond, 1900; Sciara: DuBois, 1932; Butt, 1934; Simulium: Gambrell, 1933), in which the embryonic area is restricted, and a full set of embryonic membranes is formed.

The only other species of Drosophila which has been used for embryological study is *D. funebris*. The figures of early and late stages published by Strasburger and Körner (1939) show a very close resemblance to corresponding stages in *D. melanogaster*. These authors give a brief account of the normal embryology of this species, including a timetable of stages of the second half of development, from which it is clear that the chief difference between the two species lies in rate of development. *Drosophila funebris* requires 30–32 hours from egg laying to hatching at 25°C, whereas *D. melanogaster* completes its embryological development in 20–22 hours at the same temperature. The few observations made by the writer on other species (*D. gibberosa, D. pseudoobscura*) show that these forms are likewise very similar to *D. melanogaster*.

FIG. 40. Nearly median sagittal section of 4-hour embryo at the time of gastrulation and early extension of the germ band. Section includes only a lateral part of the posterior endodermal invagination, which is shown in transverse section in Fig. 42. Harris haematoxylin preparation. 240✕. *AEN*, anterior endoderm; *CF*, cephalic furrow; *DEC*, dorsal ectoderm; *MS*, mesoderm; *PEN*, posterior endoderm; *VEC*, ventral ectoderm; *YC*, yolk cells.

FIG. 41. Transverse section through the anterior endodermal invagination with furrow still open. Harris haematoxylin preparation. 240✕. *VF*, ventral furrow; for other abbreviations, see Fig. 40.

FIG. 42. Transverse section through the posterior endodermal invagination. Harris haematoxylin preparation. 240✕. *MSEC*, mesectoderm; for other abbreviations, see Fig. 40.

FIG. 43. Frontal section at level of the anterior endodermal invagination, showing its relation to the anterior end of the mesodermal tube. Mitotic activity in many regions. Note great size of yolk-cell nuclei. Sperm tail visible near anterior end. Bodian preparation. 240✕. *AMS*, anterior mesoderm; *EC*, ectoderm; *SPM*, sperm tail; *YK*, yolk; for other abbreviations, see Fig. 40.

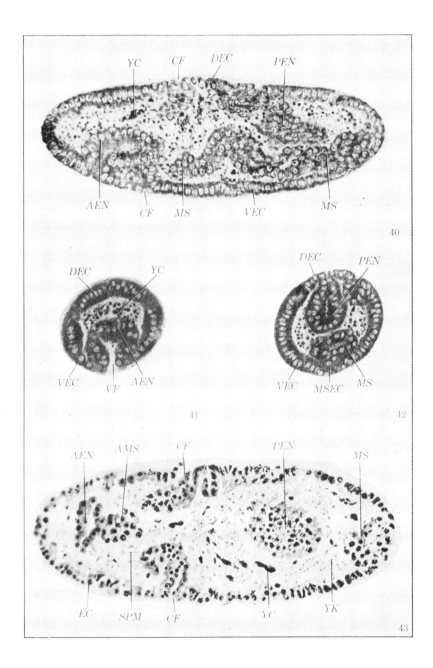

Fig. 44. Median sagittal section of a 6-hour embryo with deep stomodaeal and proctodaeal invaginations and actively dividing neuroblasts. Bodian preparation. 300×. *AEN*, anterior endoderm; *AM*, amnion; *EC*, ectoderm; *MS*, mesoderm; *NBL*, neuroblast; *PC*, pole cells; *PEN*, posterior endoderm; *PR*, proctodaeum; *SPM*, sperm tail; *ST*, stomodaeum; *YC*, yolk cells.

Fig. 45. Transverse section of an embryo of same age at level of the stomodaeum and brain lobes with dividing neuroblasts in the latter. Bodian preparation. 300×. *VEC*, ventral ectoderm; for other abbreviations, see Fig. 44.

Fig. 46. Transverse section of embryo of same age at the level of the hindgut rudiment, showing pole cells within latter, and dividing neuroblasts in ventral nervous system. Bodian preparation. 300×. *AMSE*, amnioserosa; *MSEC*, mesectoderm; for other abbreviations, see Fig. 44.

Fig. 47. Frontal section of embryo of the same age, showing relation of stomodaeal and proctodaeal invaginations to the anterior and posterior rudiments of midgut. Lower side damaged by puncture indicated by arrow. Harris haematoxylin preparation. 300×. *AMSE*, amnioserosa; *CF*, cephalic furrow; for other abbreviations, see Fig. 44.

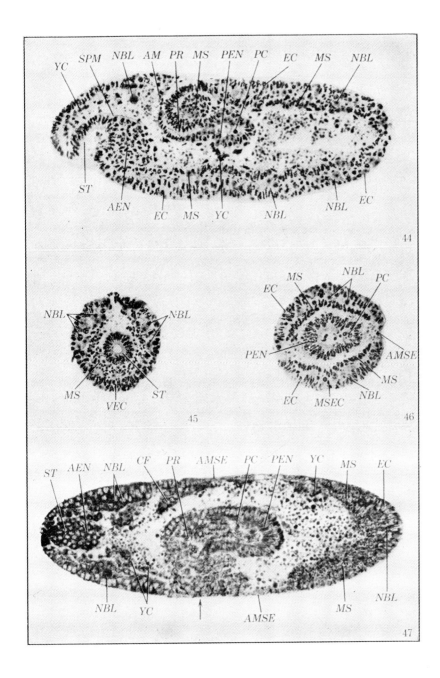

Fɪɢ. 48. Near sagittal section of an 8-hour embryo before the beginning of shortening. Bodian preparation. 300×. *AMG*, anterior midgut rudiment; *EC*, ectoderm; *INS*, invaginations in roof of stomodaeum; *MP*, Malpighian tube; *MS*, mesoderm; *NBL*, neuroblast; *PC*, pole cells; *PMG*, posterior midgut rudiment; *PR*, proctodaeum; *SLG*, salivary gland; *SPM*, sperm tail; *ST*, stomodaeum; *VNS*, ventral nervous system; *YC*, yolk cells.

Fɪɢ. 49. Transverse section at posterior end of brain, showing the invagination of the optic lobes and the invaginated tips of the salivary glands. Bodian preparation. 300×. *OPL*, optic lobe of brain; for other abbreviations, see Fig. 48.

Fɪɢ. 50. Transverse section of an embryo slightly younger than the above near the end of the posterior midgut rudiment, showing especially well a pair of tracheal invaginations. Bodian preparation. 300×. *AMSE*, amnioserosa; *MCN,* median cells of nervous system; *TR,* trachea; for other abbreviations, see Fig. 48.

Fɪɢ. 51. Frontal section of an embryo of 8 hours, showing the great development of the posterior midgut rudiment and the Malpighian tubes. Germ cells in the lateral mesoderm not yet compacted into gonads. Pole cells and yolk cells associated with the tip of the posterior midgut rudiment to form middle rudiment of midgut. Bodian preparation. 300×. *AMSE*, amnioserosa; *CF*, cephalic furrow; *GC*, germ cells; *MCN,* median cells of nervous system; *MSS*, somatic mesoderm; *MSV,* visceral mesoderm; *OPL*, optic lobe of brain; *TR,* trachea; for other abbreviations, see Fig. 48.

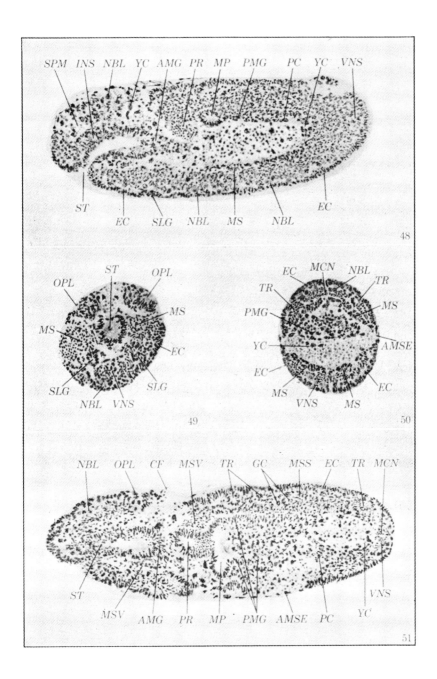

FIG. 52. Nearly median sagittal section of a 9-hour embryo during the shortening. Bodian preparation. 240✕. *AMG*, anterior midgut rudiment; *AMSE*, amnioserosa; *BR*, brain; *GNSO*, suboesophageal ganglion; *MMG*, middle midgut; *MSS*, somatic mesoderm; *NBL*, neuroblast; *PH*, pharynx; *PMG*, posterior midgut rudiment; *PMUS*, pharyngeal muscles; *PR*, procto-daeum; *ST*, stomodaeum; *VNS*, ventral nervous system; *YC*, yolk cells; *YK*, yolk.

FIG. 53. Transverse section at posterior region of brain, showing continuing neuroblast mitoses and invaginated optic lobe on one side. Bodian prepara-tion. 240✕. *HY*, hypoderm; *MSV*, visceral mesoderm; *OPL*, optic lobe of brain; for other abbreviations, see Fig. 52.

FIG. 54. Transverse section through embryo of same age as those above at level of anterior midgut rudiment, showing its bipartite nature and relation to yolk cells and visceral mesoderm. Splitting up of somatic mesoderm is also evident. Bodian preparation. 240✕. *DMSS*, dorsal somatic meso-derm; *HY*, hypoderm; *MSV*, visceral mesoderm; *VMSS*, ventral somatic mesoderm; for other abbreviations, see Fig. 52.

FIG. 55. Frontal section through a 9-hour embryo slightly less advanced than others, showing position of proctodaeum and union of midgut rudi-ments. Pole cells in middle rudiment clearly recognizable. Bodian prepa-ration. 240✕. *MS*, mesoderm; *MSV*, visceral mesoderm; *PC*, pole cells; for other abbreviations, see Fig. 52.

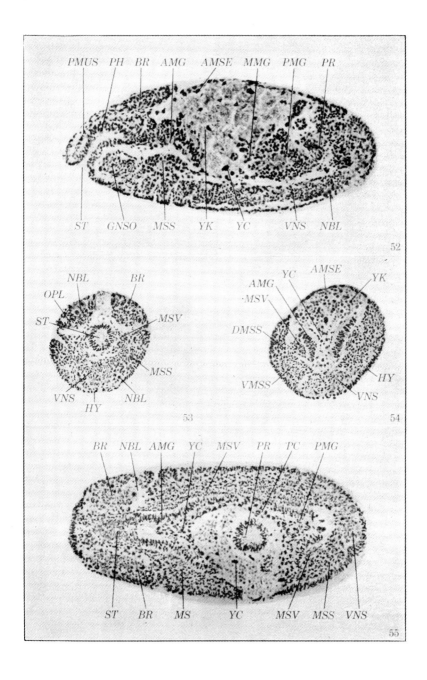

Fig. 56. Median sagittal section of a 10½-hour embryo after completion of shortening, but before frontal sac invagination. Bodian preparation. 240×. *AMG*, anterior midgut rudiment; *AN*, anus; *AO*, aorta; *BR*, brain; *GLC*, Guirlanden cells; *GNSO*, suboesophageal ganglion; *HG*, hindgut; *HY*, hypoderm; *MMG*, middle midgut; *MP*, Malpighian tube; *NF*, nerve fibers; *OES*, oesophagus; *PH*, pharynx; *PIC*, pericardial cells; PMG, posterior midgut rudiment; *PMUS*, pharyngeal muscles; *SLD*, salivary duct; *TR*, trachea; *VNS*, ventral nervous system; *YK*, yolk.

Fig. 57. Transverse section at brain level of embryo of the same age. Bodian preparation. 240×. *FB*, fat body; *MS*, mesoderm; *MSV*, visceral mesoderm; *SLG*, salivary gland; for other abbreviations, see Fig. 56.

Fig. 58. Transverse section through anterior end of midgut of embryo of same age before dorsal closure is complete. Large amnioserosa cells in yolk. Nervous system clearly separate from rest of ectoderm in this and other sections. Bodian preparation. 240×. *AMSE*, amnioserosa; *DMS*, dorsal mesoderm; *MCN*, median cells of nervous system; *MS*, mesoderm; *MSV,* visceral mesoderm; *VNS*, ventral nervous system; for other abbreviations, see Fig. 56.

Fig. 59. Frontal section of an embryo of the same age, showing union of midgut rudiments and middle section on upper side with pole cells. Segmentation indicated along lower side. Bodian preparation. 240×. *CSR*, cerebral sensory rudiment; *GC*, germ cells; *GO*, gonad; *MSV*, visceral mesoderm; *MUS*, muscle; *N*, nerve; *NBL*, neuroblast; *PC*, pole cells; *ST*, stomodaeum; for other abbreviations, see Fig. 56.

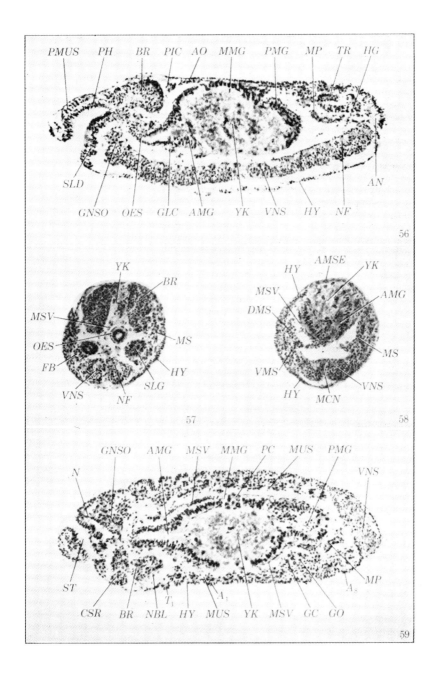

FIG. 60. Median sagittal section of a 13-hour embryo, showing frontal sac invaginated and midgut constrictions. Ventral nervous system has begun to shorten. Bodian preparation. 240×. *AMG*, anterior midgut rudiment; *FS*, frontal sac; *GNSO*, suboesophageal ganglion; *HG*, hindgut; *HY*, hypoderm; *MCN*, median cells of nervous system; *MMG*, middle midgut; *MXL*, maxillary lobe; *NF*, nerve fibers; *OES*, oesophagus; *PH*, pharynx; *PIC*, pericardial cells; *PMG*, posterior midgut rudiment; *PMUS*, pharyngeal muscles; *PV*, proventriculus; *RG*, ring gland; *SLD*, salivary duct; *TR*, trachea; *VNS*, ventral nervous system; *YK*, yolk.

FIG. 61. Transverse section (tilted slightly forward) of embryo of the same age, showing tip of frontal sac above and anterior end of midgut centrally. Bodian preparation. 240×. *BR*, brain; *GLC*, Guirlanden cells; *MUS*, muscle; for other abbreviations, see Fig. 60.

FIG. 62. Transverse section through anterior midgut just back of brain in embryo of same age as above. Bodian preparation. 240×. *AO*, aorta; *BR*, brain; *DHY*, dorsal hypoderm; *FB*, fat body; *MUS*, muscle; *VHY*, ventral hypoderm; *YC*, yolk cell; for other abbreviations, see Fig. 60.

FIG. 63. Frontal section (nearly median) of a 13-hour embryo, showing brain pushed back against the midgut and frontal sac and cephalopharyngeal apparatus between lobes of brain. Maxillary sense organ and ganglion at anterior end; large cells of middle midgut near center. Bodian preparation. 240×. *AT*, atrium; *BR*, brain; *CPA*, cephalopharyngeal apparatus; *FB*, fat body; *GLC*, Guirlanden cells; *GNMX*, maxillary ganglion; *LBD*, labial imaginal disc; *MP*, Malpighian tube; *MUS*, muscle; *MXR*, maxillary sensory rudiment; *NMX*, maxillary nerve; *SLG*, salivary gland; *YC*, yolk cells; for other abbreviations, see Fig. 60.

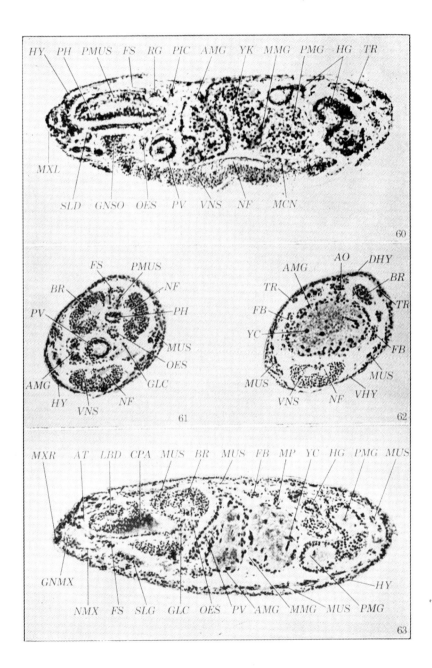

Fig. 64. Median sagittal section of a 16-hour embryo, showing cephalo-pharyngeal apparatus and coiling of gut. Bodian preparation. 240×. *AMG*, anterior midgut rudiment; *AO*, aorta; *APD*, apodeme; *CA*, corpus allatum; *FS*, frontal sac; *GNSO*, suboesophageal ganglion; *HG*, hindgut; *HY*, hypoderm; *MH*, mouth hook; *MMG*, middle midgut; *MXL*, maxillary lobe; *NF*, nerve fibers; *OES*, oesophagus; *PH*, pharynx; *PMG*, posterior midgut rudiment; *PMUS*, pharyngeal muscles; *SLD*, salivary duct; *SLG*, salivary gland; *TR*, trachea; *VNS*, ventral nervous system; *YK*, yolk.

Fig. 65. Transverse section of an embryo of the same age, through brain and suboesophageal ganglion. Nerve tracts well developed, salivary glands full of secretion. Bodian preparation. 240×. *BR*, brain; *COCON*, circum-oesophageal connective; *MUS*, muscle; for other abbreviations, see Fig. 64.

Fig. 66. Transverse section of an embryo of the same age, through the level of the gonads and middle midgut. Bodian preparation. 240×. *CT*, cuticle; *GO*, gonad; *MUS*, muscle; for other abbreviations, see Fig. 64.

Fig. 67. Frontal section (median) of an embryo of the same age, showing convolutions of the gut. In this and the other sections cuticle is clearly present over the hypoderm. Bodian preparation. 240×. *BR*, brain; *CT*, cuticle; *GNMX*, maxillary ganglion; *GO*, gonad; *MP*, Malpighian tube; *MUS*, muscle; *SP*, spiracle; for other abbreviations, see Fig. 64.

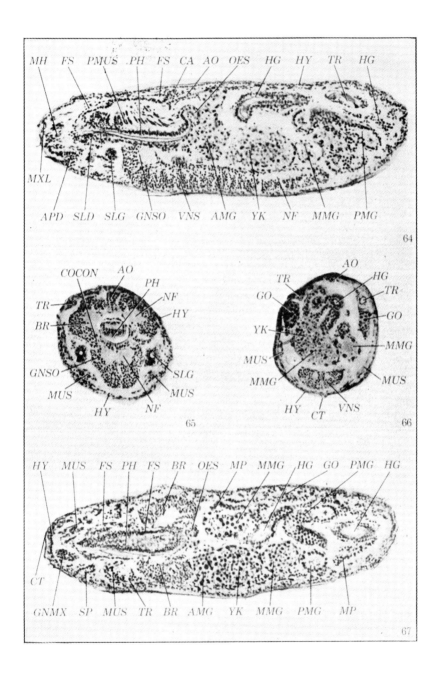

FIG. 68. Sagittal section (nearly median) of an embryo at the time of hatching, showing characteristic features of the larva. Bodian preparation. 240×. *AMG*, anterior midgut rudiment; *APD*, apodene; *FS*, frontal sac; *GNMX*, maxillary ganglion; *GNSO*, suboesophageal ganglion; *HY*, hypoderm; *MGC*, midgut caecum; *MUS*, muscle; *MXR*, maxillary sensory rudiment; *NF*, nerve fibers; *OES*, oesophagus; *PH*, pharynx; *PMG*, posterior midgut rudiment; *PMUS*, pharyngeal muscle; *PO*, photoreceptive organ; *PV*, proventriculus; *TR*, trachea; *VNS*, ventral nervous system.

FIG. 69. Transverse section of an embryo at the time of hatching, through the brain and ring gland, showing chitinous pharyngeal skeleton. Bodian preparation. 240×. *AO*, aorta; *BR*, brain; *COCON*, circumoesophageal connective; *COM*, cerebral commissure; *CPA*, cephalopharyngeal apparatus; *CT*, cuticle; *RG*, ring gland; *SLG*, salivary gland; for other abbreviations, see Fig. 68.

FIG. 70. Transverse section of embryo at the time of hatching, through the midregion. Bodian preparation. 240×. *AO*, aorta; *CT*, cuticle; *DD*, dorsal diaphragm; *FB*, fat body; *MG*, midgut; for other abbreviations, see Fig. 68.

FIG. 71. Frontal section at the time of hatching, freed from membranes so that maxillary lobes protrude in front. Setae present on cuticle at this time. Bodian preparation. 240×. *BR*, brain; *CPA*, cephalopharyngeal apparatus; *MH*, mouth hook; *MMG*, middle midgut; *MP*, Malpighian tube; *MXL*, maxillary lobe; *SLG*, salivary gland; for other abbreviations, see Fig. 68.

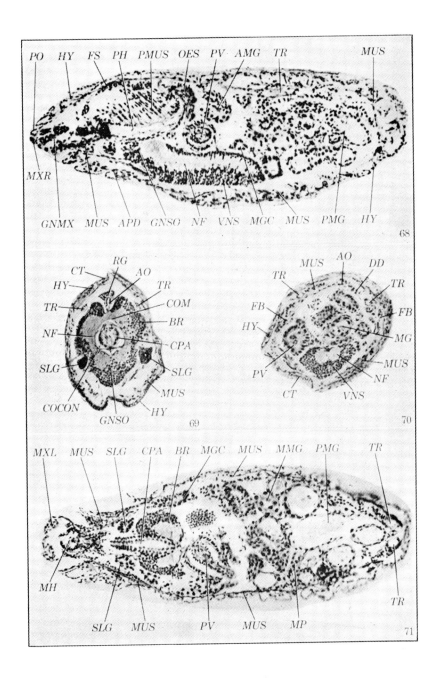

FIG. 72. Photomicrograph of living egg at the time of migration of nuclei to the surface (1½ hours). Anterior end above. ca. 175×.

FIG. 73. Photomicrograph of living egg during cell formation (2½ hours). Pole cells at posterior end below. ca. 175×.

FIG. 74. Photomicrograph of living egg near completion of cell furrows. Pole cells clearly separate at posterior end. ca. 175×.

FIG. 75. Photomicrograph of living egg at the time of completion of cellular blastoderm (3 hours), seen from the side, pole cells at posterior. ca. 175×.

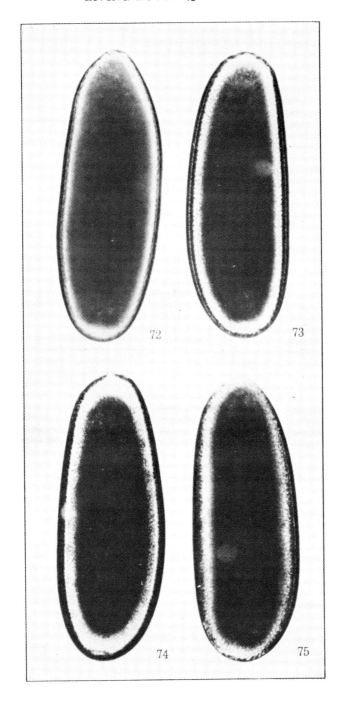

FIG. 76. Photomicrograph of living egg at the beginning of gastrulation (3–3½ hours), seen from the side. ca. 175×.

FIG. 77. Photomicrograph of living egg during gastrulation and early extension. Cephalic furrow visible (3¾ hours). *CF*, cephalic furrow. ca. 175×.

FIG. 78. Photomicrograph of living egg with germ band fully extended but before invagination of stomodaeum (5 hours). *CF*, cephalic furrow. ca. 175×.

FIG. 79. Photomicrograph of living egg during dorsal extension (4–4½ hours). *CF*, cephalic furrow. ca. 175×.

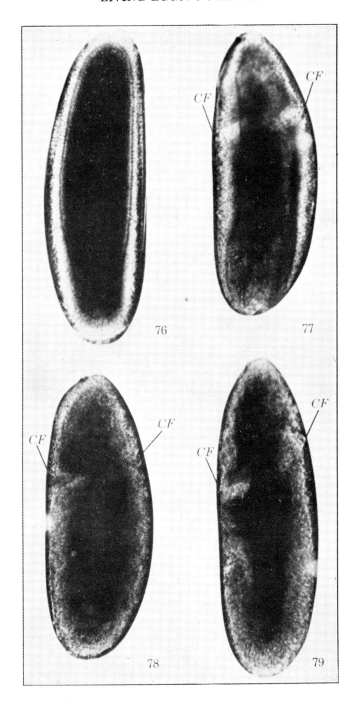

Fig. 80. Photomicrograph of living egg at the completion of shortening of the embryo but before involution of the head (10 hours). Side view, showing brain and ventral nervous system and the gut. ca. 175×.

Fig. 81. Photomicrograph of living egg at the same age as Fig. 80, seen from the ventral side, showing gut and segmentation. ca. 175×.

Fig. 82. Photomicrograph of living egg at the time of beginning of active muscular movements (16 hours), seen from side. ca. 175×.

Fig. 83. Photomicrograph of living egg with larva about ready to hatch (22–24 hours), showing air-filled tracheae, seen from dorsal side. ca. 175×.

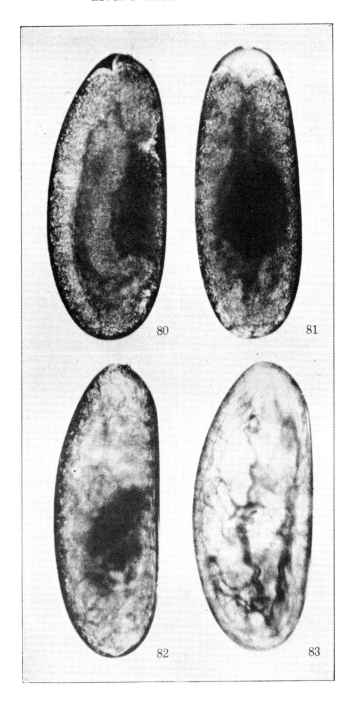

80   81

82   83

LITERATURE CITED

ABOIM, A. N. 1945. Développement embryonnaire et post-embryonnaire des gonades normales et agamétiques de *Drosophila melanogaster*. *Rev. suisse zool.* **52**:53–154.

AUERBACH, CHARLOTTE. 1936. The development of the legs, wings, and halteres in wild type and some mutant strains of *Drosophila melanogaster*. *Trans. Royal Soc. Edinburgh* **58**:787–816.

BECKER, R. 1910. Zur Kenntnis der Mundteile und des Kopfes der Dipterenlarven. *Zool. Jahrb. Anat. u. Ont.* **29**:281–314.

BICKLEY, WILLIAM E. 1942. On the stomodeal nervous system of insects. *Ann. Entomol. Soc. Am.* **35**:343–354.

BISCHOFF, W. 1922. Über die Kopfbildung der Dipterenlarven: Über die Deutung der Mundhaken der Cyclorhaphalarven. *Arch. Naturgeschichte A:* **88**.

BUTT, F. H. 1934. Embryology of Sciara. *Ann. Entomol. Soc. Am.* **27**:565–579.

CHEN, TSE-YIN. 1929. On the development of imaginal buds in normal and mutant *Drosophila melanogaster*. *J. Morphol.* **47**:135–199.

DuBOIS, ANNE MARIE. 1932. A contribution to the embryology of Sciara (Diptera). *J. Morphol.* **54**:161–192.

ELLSWORTH, J. K. 1933. The photoreceptive organs of a flesh fly larva, *Lucilia sericata* (Meigen): an experimental and anatomical study. *Ann. Entomol. Soc. Am.* **26**:203–215.

ESCHERICH, K. 1901. Ueber die Bildung der Keimblatter bei den Musciden. *Nova Acta Leopoldina* **77**:303–367.

ESCHERICH, K. 1902. Zur Entwicklung des Nervensystems der Musciden mit besonderer Berucksichtigung des sog. Mittelstranges. *Z. wiss. Zoöl.* **71**:525–549.

GAMBRELL, MRS. F. L. 1933. The embryology of the black fly, *Simulium pictipes* Hagen. *Ann. Entomol. Soc. Am.* **26**:641–676.

GEIGY, RUDOLF. 1931a. Erzeugung rein imaginaler Defekte durch ultraviolette Eibestrahlung bei *Drosophila melanogaster*. *Arch. Entwicklungsmech. Organ.* **125**:406–447.

GEIGY, RUDOLF. 1931b. Action de l'ultra-violet sur le pôle germinal dans l'oeuf de *Drosophila melanogaster*. *Rev. suisse zool.* **38**:187–288.

GEROULD, JOHN H. 1938. Structure and action of the heart of *Bombyx mori* and other insects. *Sep. Acta Zoologica* **19**:297–352.

GRABER, VEIT. 1889. Vergleichende Studien über der Embryologie der Insecten und Insbesondere der Musciden. *Denkschn. Kaiserl. Akad. Wiss. Wein Math. Nat. Cl.* **56**:257–314.

HANSTRÖM, B. 1928. *Vergleichende Anatomie des Nervensystems der wirbellosen Tiere unter Berucksichtigung seiner Funktion.* Springer, Berlin.

HANSTRÖM, B. 1942. Die Corpora cardiaca und Corpora allata der Insekten. *Biol. Generalis* **15**:485–531.

HENSON, H. 1932. The development of the alimentary canal in *Pieris brassicae* and the endodermal origin of the Malpighian tubules of insects. *Quart. J. Microscop. Sci.* **75**:283–309.

HENSON, H. 1945. The embryology and metamorphosis of the alimentary canal of the blowfly (Calliphora). *Proc. Leeds Phil. Soc.* **4**:184–190.

HENSON, H. 1946. The theoretical aspect of insect metamorphosis. *Biol. Rev.* **21**:1–14.

HERTWECK, HEINRICH. 1931. Anatomie und Variabilität des Nervesystems und der Sinnesorgane von *Drosophila melanogaster* (Meigen). *Z. wiss. Zoöl.* **139**:559–663.

HIRSCHLER, JAN. 1928. Embryogenese der Insekten. *Schroder Handb. Entomol.* **1**:570–824.

HOLMGREN, N. 1904a. Zur Morphologie des Insektenkopfes. I. Zum metameren Aufbau des Kopfes der Chironomuslarve. *Z. wiss. Zoöl.* **76**:439–477.

HOLMGREN, N. 1904b. Zur Morphologie des Insektenkopfes. II. Einiges über die Reduktion des Kopfes der Dipterenlarven. *Zool. Anz.* **27**:343–355.

HOWLAND, RUTH B. 1941. Structure and development of centrifuged eggs and early embryos of *Drosophila melanogaster*. *Proc. Am. Phil. Soc.* **84**:605–616.

HOWLAND, RUTH B., and GEORGE PERCY CHILD. 1935. Experimental studies on development in *Drosophila melanogaster*. I. Removal of protoplasmic materials during late cleavage and early embryonic stages. *J. Exptl. Zoöl.* **70**:415–427.

HOWLAND, RUTH B., and B. P. SONNENBLICK. 1936. Experimental studies on development in *Drosophila melanogaster*. II. Regulation in the early egg. *J. Exptl. Zoöl.* **73**:109–125.

HUETTNER, A. F. 1923. The origin of the germ cells in *Drosophila melanogaster*. *J. Morphol.* **37**:385–423.

IMMS, A. D. 1924. *A General Textbook of Entomology*. E. P. Dutton, New York.

JOHANNSEN, O. A., and F. H. BUTT. 1941. *Embryology of Insects and Myriapods*. McGraw-Hill, New York.

KALISS, NATHAN. 1939. The effect on development of a lethal deficiency in *Drosophila melanogaster:* with a description of the normal embryo at the time of hatching. *Genetics* **24**:244–270.

KAUFMANN, BERWIND P. 1934. Somatic mitoses of *Drosophila melanogaster*. *J. Morphol.* **56**:125–155.

KEILIN, D. 1915. Recherches sur les larves de Dipteres Cyclorhaphes. *Bull. sci. France Belg.* **49**:15–198.

KEILIN, D. 1917. Recherches sur les Anthomyides alarves carnivores. *Parasitology* **9**:325–450.

KEILIN, D. 1927. The chordotonal organs of the antenno-maxillary complex in the larvae of cyclorrhaphous Diptera. *Ann. Nat. Hist.* **20**:334–336.

KOCH, J. 1945. Die Ocnocyten von *Drosophila melanogaster*. *Rev. suisse zool.* **52**:415–420.

KOWALEVSKY, A. 1886. Zur Embryonalentwicklung der Musciden. *Biol. Zentr.* **6**:49–54.

KRAUSE, G. 1939. Die Eitypen der Insekten. *Biol. Zentr.* **59**:495–536.

LASSMANN, G. W. P. 1936. The early embryological development of *Melophagus ovinus* L., with special reference to the development of the germ cells. *Ann. Entomol. Soc. Am.* **29**:397–413.

MEDVEDEV, N. N. 1935. Genes and the development of characters. *Z. indukt. Abst. Vererbl.* **70**:55–72.

MEIJERE, J. C. H. DE. 1916. Beiträge zur Kenntnis der Dipteren-Larven und -Puppen. *Zoolog. Jahrb. Abt. Syst.* **40**:177–322.

MELLAND, A. M. 1942. Types of development of polytene chromosomes. *Proc. Roy. Soc. Edinburgh* **61B**:316–327.

MIALL, L. C., and HAMMOND, A. R. 1900. *The Harlequin Fly.* Clarendon Press, Oxford.

NELSON, J. A. 1915. *Embryology of the Honey Bee.* Princeton University Press, Princeton.

NOACK, W. 1901. Beiträge zur Entwicklungsgeschichte der Musciden. *Z. wiss. Zoöl.* **70**:1–57.

PANTEL, J. 1898. Essai monographique sur les caractères extérieurs, la biologie et l'anatomie d'une larve parasite du groupe des Tachinaires. *Cellule* **15**:1–290.

PARKS, HAL B. 1936. Cleavage patterns in Drosophila and mosaic formation. *Ann. Entomol. Soc. Am.* **29**:350–392.

PATTERSON, J. T., and WILSON STONE. 1938. Gynandromorphs in *Drosophila melanogaster. Univ. Tex. Pub.* 3825:1–67.

PAULI, MARGARETE E. 1927. Die Entwicklung geschnurter und centrifugierter Eier von *Calliphora erythrocephala* und *Musca domestica. Z. wiss. Zoöl.* **129**:481–540.

PEREZ, CHARLES. 1910. Recherches histologiques sur la metamorphose des Muscides, *Calliphora erythrocephala* Mg. *Arch. zool. exp. et gén.* **4**:1–274.

POULSON, D. F. 1937. The embryonic development of *Drosophila melanogaster. Actualités sci. et. ind.* **498**:1–51. Hermann et Cie, Paris.

POULSON, D. F. 1940. The effects of certain X-chromosome deficiencies on the embryonic development of *Drosophila melanogaster. J. Exptl. Zoöl.* **83**:271–325.

POULSON, D. F. 1945a. Chromosomal control of embryogenesis in Drosophila. *Am. Naturalist* **79**:340–363.

POULSON, D. F. 1945b. On the origin and nature of the ring gland (Weismann's ring) of the higher Diptera. *Trans. Conn. Acad. Arts and Sci.* **36**:449–487.

POULSON, D. F. 1947. The pole cells of Diptera, their fate and significance. *Proc. Natl. Acad. Sci.* **6**:182–184.

POULSON, D. F., and E. J. BOELL. 1946a. The development of cholinesterase activity in embryos of normal and genetically deficient strains of *Drosophila melanogaster. Anat. Record* **96**:12.

POULSON, D. F., and E. J. BOELL. 1946b. A comparative study of cholinesterase activity in normal and genetically deficient strains of *Drosophila melanogaster. Biol. Bull.* **91**:228.

PRATT, H. S. 1900. Embryonic history of imaginal discs in *Melophagus ovinus* L. *Proc. Bost. Soc. Nat. Hist.* **29**:241–272.

RABINOWITZ, MORRIS. 1941. Studies on the cytology and early embryology of the egg of *Drosophila melanogaster*. *J. Morphol.* **69**:1–49.

REITH, FERDINAND. 1925. Die Entwicklung des Musca-eies nach Ausschaltung verschiedener Eibereiche. *Z. wiss. Zoöl.* **126**:181–238.

RICHARDS, A. GLENN, JR., and ALBERT MILLER. 1937. Insect development analyzed by experimental methods: a review. *J. N. Y. Entomol. Soc.* **45**:1–60.

RUHLE, HERMANN. 1932. Das larvale tracheensystem von *Drosophila melanogaster* Meigen und seine variabilität. *Z. wiss Zoöl.* **141**:159–245.

SEHL, ALFRED. 1931. Furchung und Bildung der Keimanlage bei der Mehlmotte *Ephestia kuehniella* Zell. Nebst einer allgemeinen übersicht über den Verlauf der Embryonalentwicklung. *Z. Morphol. Ökol. Tiere* **20**:535–598.

SEIDEL, F. 1936. Entwicklungsphysiologie des Insekten-Keims. *Verhandl. deut. zoöl. Ges.,* 291–336.

SIKES, ENID K., and V. B. WIGGLESWORTH. 1931. The hatching of insects from the egg, and the appearance of air in the tracheal system. *Quart. J. Microscop. Sci.* **74**:166–192.

SNODGRASS, R. E. 1924. Anatomy and metamorphosis of the apple-maggot, *Rhagoletis pomonella* Walsh. *J. Agr. Research* **28**:1–35.

SNODGRASS, R. E. 1935. *Principles of Insect Morphology.* McGraw-Hill, New York.

SONNENBLICK, B. P. 1939. The salivary glands in the embryo of *Drosophila melanogaster. Rec. Gen. Soc. Am.* **8**:137.

SONNENBLICK, B. P. 1941. Germ cell movements and sex differentiation of the gonads in the Drosophila embryo. *Proc. Natl. Acad. Sci.* **27**:484–489.

STARK, M. B., and A. K. MARSHALL. 1930. The blood-forming organ of the larva of *Drosophila melanogaster. J. Am. Inst. Homeopathy* **23**:1204–1206.

STEINBERG, ARTHUR G. 1941. A reconsideration of the mode of development of the bar eye of *Drosophila melanogaster. Genetics* **26**:325–346.

STEINBERG, ARTHUR G. 1943. The development of the wild type and bar eyes of *Drosophila melanogaster. Can. J. Research* **21**:277–283.

STRASBURGER, E. H., and LONTA KÖRNER. 1939. Untersuchungen über die Wirkung des Polyphaen-Gens in der Entwicklung von *Drosophila funebris. Biol. Zentr.* **59**:366–387.

STRASBURGER, M. 1932. Bau, Funktion, und Variabilität des Darmtraktus von *Drosophila melanogaster* Meigen. *Z. wiss. Zoöl.* **140**:539–649.

STUART, R. R. 1935. The development of the mid-intestine in *Melanoplus differentialis* (Acrididae: Orthoptera). *J. Morphol.* **58**:419–437.

STURTEVANT, A. H. 1929. The claret mutant type of *Drosophila simulans:* a study of chromosome elimination and of cell-lineage. *Z. wiss Zoöl.* **135**:323–356.

WAHL, B. 1914. Über die Kopfbildung Cycloraphen Dipterenlarven und Postembryonale Entwicklung des Fliegenkopfes. *Arb. Zool. Inst. Univ. Wien* **20**:159–272.

WANDOLLECK, B. 1898. Die Fuhler der cyclorhaphen Dipterenlarven. *Zool. Anz.* **21**:283–294.

WANDOLLECK, B. 1899. Zur Anatomie der cyclorhaphen Dipterenlarven, Anatomie der Larve von *Platycephala planifrons* (F). *Abhandl. Ber. k. zool. anthrop. Mus. Dresden* **7**.

WATERHOUSE, D. F. 1940. Studies of the physiology and toxicology of blowflies. 5. The hydrogen-ion concentration in the alimentary canal. 6. The absorption and distribution of iron. 7. A quantitative examination of the iron content of *Lucilia cuprina*. *Council for Sci. and Ind. Research Pamph.* **102**:1–67.

WATERHOUSE, D. F. 1945. Studies of the physiology and toxicology of blowflies. 10. A histochemical examination of the distribution of copper in *Lucilia cuprina*. 11. A quantitative investigation of the copper content of *Lucilia cuprina*. *Council for Sci. and Ind. Research Bull.* **191**:1–39.

WEISMANN, AUGUST. 1863. Die Entwicklung der Dipteren im Ei. *Z. wiss. Zoöl.* **13**:107–220.

WEISMANN, AUGUST. 1864. Die nachembryonale Entwicklung der Musciden nach Beobachtungen an *Musca vomitoria* und *Sarcophaga carnaria*. *Z. wiss. Zoöl.* **14**:187–336.

WHEELER, WM. M. 1889. The embryology of *Blatta germanica* and *Doryphora decemlineata*. *J. Morphol.* **3**:291–386.

WHEELER, WM. M. 1891. Neuroblasts in the arthropod embryo. *J. Morphol.* **4**:337–344.

WHEELER, WM. M. 1893. A contribution to insect embryology. *J. Morphol.* **8**:1–160.

WIGGLESWORTH, V. B. 1930. The formation of the peritrophic membrane in insects, with special references to the larvae of mosquitoes. *Quart. J. Microscop. Sci.* **73**:593–616.

WIGGLESWORTH, V. B. 1939. *The Principles of Insect Physiology*. Methuen, London.

ZALOKAR, MARKO. 1943. L'ablation des disques imaginaux chez la larve de Drosophile. *Rev. suisse zool.* **50**:232–237.

ZALOKAR, MARKO. 1947. Anatomie du thorax de *Drosophila melanogaster*. *Rev. suisse zool.* **54**:17–53.

4

The Postembryonic Development of Drosophila

DIETRICH BODENSTEIN

INTRODUCTION

The developmental changes in the larva and pupa of Drosophila have not been systematically described heretofore, and indeed are not yet completely understood. Strasburger in 1935 provided the most useful guide to the structure and development of Drosophila which was possible at that time, but this work was intended only as an introductory survey. Since then the increased interest in the developmental mechanics and genetics of Drosophila has augmented considerably our knowledge of larval and pupal development. The various investigations of the origin and changes in the developmental sequence of the imaginal discs, Robertson's study of metamorphosis, and the discovery of the metamorphosis hormones have made possible a more connected account of the relations between the different stages of postembryonic development.

The main purpose in assembling this present account has been to provide a useful and practical guide for those who will use Drosophila as scientific material. Most of the facts presented have been verified from personal observation of the writer's own material, but, in order not to delay completion of this study, in a number of instances the findings of other workers have simply been reported. A timetable of developmental events taking place in each of the main parts of the animal has been devised, which should enable any student not only to recognize the part but also to diagnose the stage of development reached. All data given in this paper are based on observations at 25°C and refer, if not otherwise mentioned, to *Drosophila melanogaster*.

The author is especially indebted to Dr. B. P. Sonnenblick for invaluable aid in the preparation of some of the parts and for permission to use his sectioned material, from which some of the illustrations were prepared. For Fig. 19, the author is indebted to Dr. M. Power, who gave him those photographs from some of his unpublished works.

It also gives the author great pleasure to acknowledge the helpful criticism and suggestions offered by Drs. L. C. Dunn and M. Demerec during the preparation of this manuscript. The expert assistance of Miss A. M. Hellmer, who is responsible for most of the drawings, is also gratefully acknowledged.

GENERAL SURVEY OF POSTEMBRYONIC DEVELOPMENT

The life cycle of *Drosophila melanogaster* from egg-hatching to the emergence of the imago is about 192 hours at 25°C. The larvae pass through three instars and two larval molts. The first molt occurs at about 25 hours, the second molt at about 48 hours, and puparium formation at about 96 hours after hatching.

The Larva

The larva (Figs. 1A–C) has twelve segments: the head segment, three thoracic segments, and eight abdominal segments. The mouth aperture is located ventrally in the head segment. The body wall is soft and flexible and consists of the outer non-cellular cuticula and the inner cellular epidermis. The latter is internally confined by a fine basement membrane. The cuticula is made up of two layers: an outer one, the exocuticle, and a thicker lamellar one, the endocuticle. The cuticle of each segment carries around its anterior border a multiple ring of small chitinous hooks (Fig. 3D). Around the mouth there are also set a number of chitinous hooks. A great number of sense organs are spread regularly over the whole body.

The larvae are quite transparent. Their fat bodies, in the form of long whitish sheets (Figs. 1C and 2), the coiled intestine, and the yellowish Malpighian tubes, as well as the gonads embedded in the fat body, can easily be distinguished in the living larva when observed in transmitted light. The dorsal blood vessel is

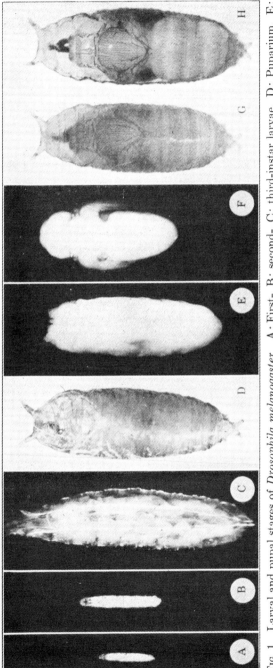

FIG. 1. Larval and pupal stages of *Drosophila melanogaster*. A: First-, B: second-, C: third-instar larvae. D: Puparium. E: Prepupa removed from pupal case. F: Young pupa removed from pupal case. G: Older pupa visible through pupal case. H: Dorsal surface of pupal case removed to show pupa clearly. Note the mouth armature in the anterior part of the case.

the circulatory organ of the larva. It is a fine muscular tube running along the dorsal midline of the body. Its anterior section is known as the aorta, and the posterior wider part, the heart, is in life easily visible by its pulsation. The two great lateral tracheal trunks passing from end to end through the entire organism are the most prominent features of the larva. The larval muscles, segmentally arranged, are transparent but can be made visible when the larva is fixed in hot water. Apart from its purely larval structures, the larva contains a number of primitive cell

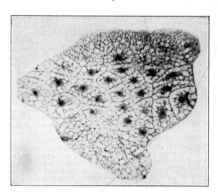

complexes called imaginal discs, which are the primordia for later imaginal structures. The primary mechanism by which the larva grows is molting. At each molt the entire cuticle of the insect, including many specialized cuticular structures, as well as the mouth armature and the spiracles, is shed and has to be rebuilt again. During each molt, therefore, many recon-

FIG. 2.   Section through fat body of mature larva.

struction processes occur, leading to the formation of structures characteristic of the ensuing instar. The growth of the internal organs proceeds gradually and seems to be rather independent of the molting process, which mainly affects the body wall. Organs such as Malpighian tubes, muscles, fat body, and intestine grow by an increase in cell size; the number of cells in the organ remains constant. The organ discs, on the other hand, grow chiefly by cell multiplication; the size of the individual cells remains about the same. In general one might say that purely larval organs grow by an increase in cell size, whereas the presumptive imaginal organs grow by cell multiplication. In Fig. 3 are given cross-sections through different regions of a mature larva, showing the various larval organs.

The Pupa

A series of developmental steps by means of which the insect passes from the larval into the adult organism is called "meta-

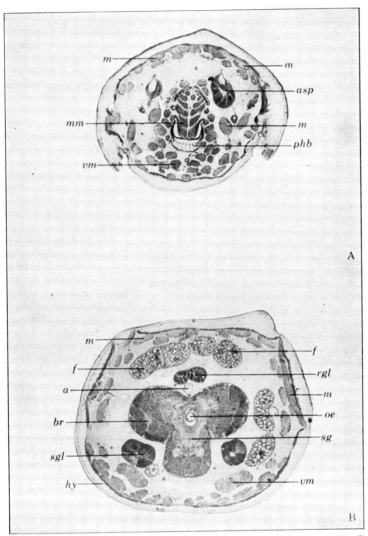

Fig. 3. Cross-section through various levels of mature larvae. A: Cross-section through spiracle region. B: Cross-section through brain region. *a,* aorta; *asp,* anterior spiracle; *br,* brain; *f,* fat body; *hy,* hypoderm; *m,* muscles; *mm,* muscle of pharynx; *oe,* oesophagus; *phb,* pharyngeal bars; *rgl,* ring gland; *sg,* suboesophageal ganglion; *sgl,* salivary gland; *vm,* ventral muscles.

FIG. 3 (cont'd). C: Cross-section through thoracic region. D: Cross-section through region of stomach. a, antenna disc; c, cuticular processes; dv, dorsal vessel; f, fat body; gc, gastric caecae; h, haltere disc; hy, hypodermis; lgl, lymph glands; L1–L3, first, second, third leg discs; m, muscles; mt, Malpighian tubes; oe, oesophagus; pc, peripodial cavity; pm, peripodial membrane; sgl, salivary gland; st, stomach; tr, trachea; vm, ventral muscles; w, wing disc.

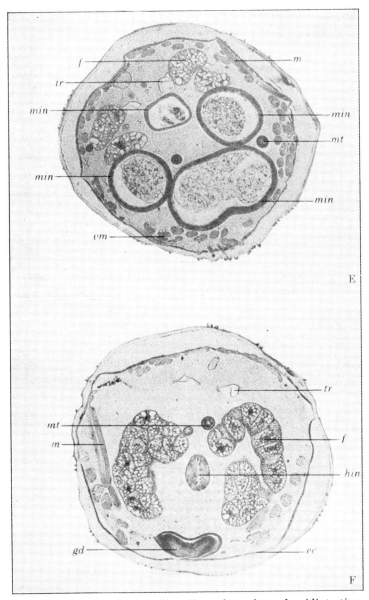

FIG. 3 (cont'd). E: Cross-section through region of midintestine. F: Cross-section through last abdominal segment. *ec*, epidermis; *f*, fat body; *gd*, genital disc; *hin*, hindintestine; *m*, muscles; *min*, midintestine; *mt*, Malpighian tubes; *tr*, trachea; *vm*, ventral muscles.

morphosis." The most drastic changes in this transformation process take place during the pupal stage.

Shortly before pupation the larva leaves the food and usually crawls onto the sides of the culture bottle, seeking a suitable place for pupation, and finally comes to rest. The larva is now very sluggish, everts its anterior spiracles, and becomes motionless. Soon the larva shortens and appears to be somewhat broader, thus gradually acquiring its pupal shape. The shortening of the larval cuticle, which forms the case of the puparium, is caused by muscular action (Fraenkel and Rudall, 1940). The puparium, which is the outer pupal case, is thus identical with the cuticle of the last larval instar. When the shaping of the puparium is completed, the larval segmentation is obliterated, but the cuticle is still white. This stage lasts only a few minutes and is thus an accurate time mark from which to date the age of the pupa. Immediately after the cuticle reaches the white prepupal stage, the hardening and darkening of the cuticle begin and proceed very quickly. About $3\frac{1}{2}$ hours later the puparium is fully colored (Fig. 1D). Pigmentation apparently starts in the external surface of the cuticle and proceeds inward.

About 2 hours after the puparium is formed, a fine seam appears ventrally to the anterior spiracles. It runs across the front of the puparium, curves posteriorly, where it proceeds along the flanks, and bends upward in the fourth segment, where the left and right flank seams finally meet dorsally. The dorsal surface of the puparium circumscribed by the seams is known as the "operculum." With the help of an eversible sac located just above the base of the antennae, the so-called "ptilinum," the insect separates the seams anteriorly and laterally, thus opening the operculum, and emerges from the pupal case. Four hours after the formation of the puparium, the animal within it has separated its epidermis from the puparium and has become a headless individual having no external wings or legs and known as the "prepupa" (Fig. 1E). A very fine prepupal cuticle has been secreted and surrounds the prepupa.

Pupation takes place about 12 hours after puparium formation. By muscular contraction the prepupa draws back from the anterior end of the puparium and everts its head structures. This movement also ejects the larval mouth armature, which until now was attached to the anterior end of the prepupa. The wings, halteres, and legs are also everted. The prepupal cuticle, together with tracheal linings and spiracles, is completely shed, while a new pupal cuticle is formed. A typical pupa with head, thorax, and abdomen is thus shaped (Fig. 1F). In section it is seen that the pupa now lies within three membranes: an outer membrane, the puparium; an intermediate membrane, the prepupal cuticle; and an inner membrane, the newly secreted pupal cuticle.

Now metamorphosis involves the destruction of certain larval tissues and organs (histolysis) and the organization of adult structures from primitive cell complexes, the imaginal discs. It must, however, be realized that some larval organs are transformed into their imaginal state without any very drastic change in their structure. The duration and extent of these transformation processes vary greatly for the different organs involved. Larval organs which are completely histolyzed during metamorphosis are the salivary glands, the fat bodies, the intestine, and apparently the muscles. All these organs are formed anew, either from imaginal disc cells already present in the larva or from cells which come visibly into being in the course of pupal reorganization. The Malpighian tubes are relatively little altered during metamorphosis but nevertheless undergo some change in their structural composition. The same situation seems to prevail in the brain, which is not completely histolyzed. The details of its metamorphotic changes are, however, not known. The extremities, eyes, antennae, mouthparts, and genital apparatus differentiate from their appropriate imaginal discs, which were already present in the larval stage and which undergo histogenesis during pupal development. The body wall of the imaginal head, thorax, and abdomen is also formed from imaginal disc cells. The body wall of head and thorax is formed by the combined effort of all the imaginal discs in this region, each of which contributes its part. The hypoderm of the abdomen is formed by segmentally arranged imaginal cells which first become visible in young prepupae.

THE THREE LARVAL INSTARS

For those using Drosophila larvae in any experimental work, it often becomes necessary to know with some precision the age of the material—more specifically, the separation of the larvae into the respective instars. There are three characteristics, one of which requires sectioned material, by means of which one can determine with relative ease the various larval instars. These characteristics are concerned with the structure of the mouth apparatus, the appearance and presence of the anterior and posterior spiracles, and the number and morphology of the pharyngeal bars, which are consistently located along the ventral wall of the pharynx. With the aid of the material presented in Figs. 4 and 5, consisting of drawings and of photomicrographs of whole mounts and of histological sections, together with a brief discussion of the several distinguishing characteristics, the reader will be enabled readily to separate larvae of the three stages.

The Larval Mouthparts

A detailed analysis of the morphology of the mouth apparatus in the three larval instars will not be given here, since the discussion of the parts by M. Strasburger (1932) is sufficiently clear and complete for most purposes. Furthermore, in her paper will be found a comparative analysis, including terminology, of the mouth armature of various cyclorrhaphic Diptera. The forward portion consists of a pair of hooks which articulate behind with a chitinized sclerite having the form of an H, the so-called "H-piece," posterior to which is a pair of large, vertical, cephalopharyngeal plates. There are also a number of smaller parts to the rather complicated mouth armature, as pictured by Strasburger. In the first instar, and only in this instar, there is an unpaired dorsal element, the "Medianzahn." The median tooth is apparently utilized by the young larva during the period when it is attempting to rupture the embryonic cases at eclosion.

Apart from the mouth hooks, the other elements of the apparatus are practically identical in all the instars. In Figs. 4A-C are shown drawings of mouth hooks from larvae of the three instars. (These are *camera lucida* representations taken at the same magnification.) The cephalopharyngeal skeleton is shed periodically at the two molts between the three instars and can be observed in the food medium. It is reconstituted at the next stage.

The size differences between the various mouth hooks are apparent; at each molt a larger structure replaces the former hook.

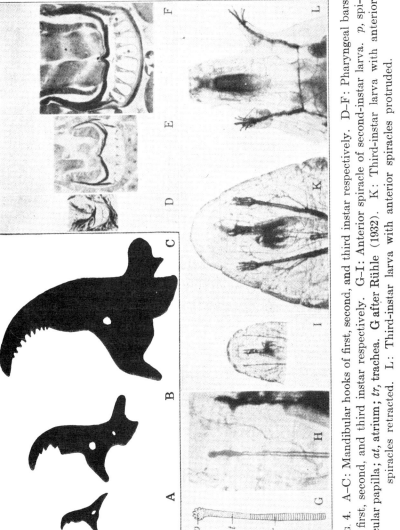

FIG. 4. A–C: Mandibular hooks of first, second, and third instar respectively. D–F: Pharyngeal bars of first, second, and third instar respectively. G–I: Anterior spiracle of second-instar larva. *p*, spiracular papilla; *at*, atrium; *tr*, trachea. G after Rühle (1932). K: Third-instar larva with anterior spiracles retracted. L: Third-instar larva with anterior spiracles protruded.

The number of teeth on the mandibular hooks in individuals representative of the three stages differs sufficiently to enable one thereby to separate the instars. This was noted by Alpatov (1929). The first-instar hooks usually have one tooth, whereas, after molting, there are two or three teeth on the sclerites of second-instar larvae. A row of approximately nine to twelve teeth is present in larvae of the third instar.

Although the differences in teeth number are sufficiently striking to provide a clue to the identity of the instar of the larva to which they belong, it must be noted that there is some variability with regard to this character. The number of teeth in any one individual (i.e., on the right and left mandibular hooks) will be found to vary. Furthermore, the number of teeth present in individuals of different species of Drosophila, as well as in different wild-type races which we have studied, is dissimilar. The variability between individuals and among races and species, however, does not invalidate the use of this character as a means for the separation of larvae of the three instars.

The Anterior and Posterior Spiracles

The presence or absence and the appearance of the spiracular openings at the fore- and hindend of the organism provide other means of identification of the larval instars. The anterior spiracles are found at the ends of the great lateral tracheal trunks, opening in the first thoracic (prothoracic) segment (i.e., the segment next to the head segment).

In the first instar there are no anterior spiracles, although some investigators have maintained that rudiments of the anterior spiracles are present as hypodermal thickenings of the first thoracic segment. The absence of distinct anterior spiracles thus immediately distinguishes first-instar larvae from older organisms. In larvae of the second stage (Figs. 4G–I) anterior spiracles are present as enlargements at the end of the great trunks, but the openings appear closed. Figure 4 shows an enlarged view of one club-shaped anterior spiracle, and in Fig. 4I the pair of anterior spiracles is observed terminating at the posterior end of the first thoracic segment. The fingerlike papillae, characteristic of only third-instar larvae, are evident as minute buds on the surface of the second-instar spiracles. This is indicated in Fig. 4G. Larvae of the final instar are represented in Figs. 4K and L. The spiracles appear as small hypodermal mounds from which protrude some seven to nine fingerlike processes with open ends (Figs. 4L and 5E). The processes and the spiracular chamber can be retracted within the organism, as in Fig. 4K, or relaxed and exposed, as seen in Figs. 4L and 5E.

Thus the identity of the larval instars can be ascertained from the absence of any forespiracles, indicative of the youngest larvae,

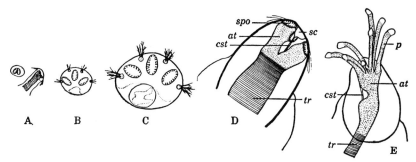

FIG. 5. A–C: Front view of posterior spiracular openings of first-, second-, and third-instar larvae respectively. D: Posterior spiracle of third-instar larva. E: Anterior spiracle of third-instar larva. *at,* atrium; *cst,* cuticular strand, i.e., remnant of cuticular spiracle; *p,* spiracular papilla; *sc,* scar; *spo,* spiracular orifice; *tr,* trachea. After Rühle (1932).

or from the appearance of the anterior spiracles, which are present after the first molt. Further detailed comments concerning the structure of the respiratory system of the larvae, especially well studied by Rühle (1932), will be given in a later section.

Posterior spiracles are present in larvae of all stages, opening on the dorsal side of the last (eighth abdominal) segment. The apertures of the hindspiracles are rimmed by stigmatic plates which differ in larvae of different ages. Chitinous rods protrude into the ovoid apertures from the inner surface of the plates. The posterior spiracles of the first-instar larvae have only two openings (Fig. 5A), whereas those of second- and third-instar larvae have three (Figs. 5B and C). Four groups of hairs radiate from the cuticle of the spiracle arranged around the stigmatic plates. The hairs are small and apparently unbranched in young larvae but are larger and branched in third-instar larvae (cf. Figs. 5B and C).

The Pharyngeal Bars

There is yet a third means of determining larval instars, but for this, sectioned material is required. The pharynx is the passageway between the oral hooks extending from the mouth to the oesophagus. In larvae of all stages a group of longitudinal ridges extends along the floor of the pharynx. In transverse section these ridges or folds appear as a series of riblike bars projecting upward into the lumen of the cavity. The appearance of the pharyngeal bars is characteristic for each of the instars, and their number appears to be constant, at least in the second and third instars. Auerbach (1936) has presented diagrams of these riblike folds in Drosophila and noted the relationship of pharyngeal bar morphology to larval instar. They have also been figured in the apple maggot, Rhagoletis, by Snodgrass (1924).

The material is difficult to prepare, but in Figs. 4D–F we have given photomicrographs of sections through the pharyngeal skeleton, showing the appearance of the bars in larvae of the three stages. All material was cut at the same thickness (8 μ), and the photographs were taken at the same magnification. Figure 4D is from a larva fixed at 2 hours after hatching from the egg. The bars are very delicate, with their upper ends branched at right angles to the vertical portion of the bar. In larvae of the second and third instar (Figs. 4E and F), the bars are progressively larger and sturdier, but the upper right-angled bend is absent. In the second stage (Fig. 4E) there are seven ribs, while in third-instar larvae (Fig. 4F) they number nine. The dilator muscles of the pharynx can be seen in the figures. Auerbach (1936) states that during each larval stage the pharyngeal bars, the cuticle, and the cuticular organs of the succeeding instar can be noted under the structures already present, and that the degree of development of these underlying structures indicates the time interval elapsed since the previous molt.

LOCATION AND IDENTIFICATION OF THE LARVAL ORGAN DISCS AND GONADS IN A MATURE LARVA

An attempt was made (Fig. 6A) to indicate as accurately as possible, in ventral aspect, the position and identification of certain organs and imaginal discs of a mature larva. The pharynx, brain, and ventral ganglia and a portion of one main tracheal

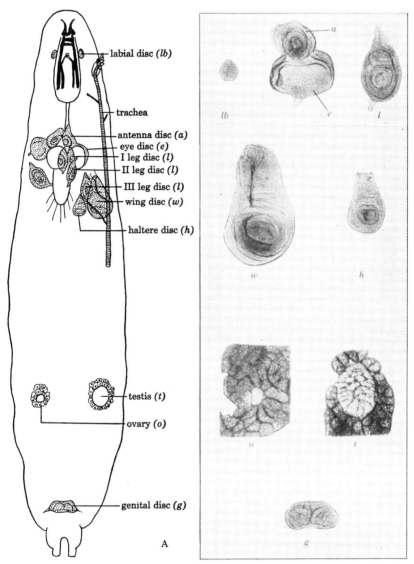

labial disc *(lb)*

trachea

antenna disc *(a)*
eye disc *(e)*
I leg disc *(l)*
II leg disc *(l)*

III leg disc *(l)*
wing disc *(w)*

haltere disc *(h)*

testis *(t)*

ovary *(o)*

genital disc *(g)*

A

Fig. 6. Location and identification of imaginal discs in the mature larva. (See text.)

trunk are shown, and the location of the organ discs can be determined by using these structures as markers. A few of the structures are shown on only one side, for example, the relation of certain discs to the chief tracheal trunk; and one male and one female gonad are presented in the region where the gonad pair is typically found.

The purpose in presenting this diagram is simply to assist the student or investigator who employs Drosophila as a research organism to locate and identify the desired structure or imaginal disc, since it has been found that sectioned material alone will not aid sufficiently in determining specific structures. With the aid of this chart and with some practice in dissection, it is believed that the various parts shown in the figure can be located with reasonable speed and accuracy. The remainder of Fig. 6 is occupied with photomicrographs of whole mounts of the discs and gonads. All these photographs were taken at the same magnification, and thus the comparative dimensions of the structures in the mature larva can be estimated. One of each disc (labial, antennal, and eye, leg, wing, haltere, and genital) is presented, as well as a testis and an ovary with the neighbouring fat body. The labeling is the same as that given in the drawing.

The Organ Rudiments

The paired labial discs are seen in a position lateral to the pharynx, and, from ventral view, the eye discs proper appear as flat cups over the brain. In the diagram the eye disc on the left side has been drawn with its neural connection severed, and the disc is pushed slightly aside to show its shape and appearance. A dissected eye disc is seen in Fig. 6, e, together with the anterior antennal rudiment (Fig. 6, a).

The first and second leg discs with a portion of their neural processes which proceed to the ventral ganglionic mass are evident. Not shown here are the stalks which connect the other ends of the discs to the hypodermis of the larva. The second leg disc on the left of the diagram has also been somewhat displaced with its neural connection cut. The discs of the third leg and of the wings and halteres are more laterally situated and have connection with the main tracheal trunk. None of this group of rudiments is innervated by nerves from the central ganglia, although all are in contact with the hypodermis by means of stalks.

The genital disc in the mature larva is a single structure lying near the posterior end of the organism, slightly cephal to the

anus. It is this rudiment which in the pupa will be transformed
and give rise to the genital ducts, the oviducts, and the vasa deferentia, as well as to the other accessory apparatus of the internal genitalia and to the external genitalia. This independent
origin of the genital ducts and of the gonads proper was first
clearly demonstrated for Drosophila by Dobzhansky (1930).

The Gonads

The larval gonads are situated in the fifth abdominal segment,
four segments from the caudal end of the organism. A pair of
ovaries or testes is found at this site, but only one of each is pictured in the drawing and is also shown in the whole mount (Fig.
6, *o* and *t*). The gonads are transparent, vesicular bodies lying
in the body cavity in close approximation to the fat body. They
are bordered by that tissue, but the minuteness of the female
gonad usually makes it appear as if it were almost entirely surrounded by fat cells. Figures 6, *o* and 6, *t* show well the gonad-fat
body relationship. The testis is considerably larger than the
ovary, the dimensions across the latter being no more than two
or three fat cells. This size difference of the gonads is characteristic for the sexes from the period in embryonic development when
the gonads are first formed (Sonnenblick, 1941) and can also be
used to determine the sex of larvae of all three instars without
recourse to dissection or to sectioning of the material (Kerkis,
1931). During pupation the gonads undergo differentiation, and
junction with the genital ducts will occur (see p. 302).

THE DEVELOPMENT OF THE EYE

A pair of saclike structures, originating as ectodermal evagi-
nations from the dorsal wall of the pharyngeal cavity, forming
the so-called "frontal sacs," is the Anlage of the eye and antenna
discs. The posterior parts of these frontal sacs or cephalic complex, as they have been termed, will, in the course of development,
give rise to the eye discs proper, while the anterior parts will form
the antennal buds. From the eye disc will develop not only the
ommatidia but also parts of the head hypodermis and perhaps
even the outer eye ganglion. This latter structure might arise
from the optic stalk (Pilkington, 1942). The antennal-eye com-

plex is already present at the time the larva hatches from the egg (Kaliss, 1939). Like the other organ discs, the eye-antennal discs grow during the larval period by cell multiplication, and the size of the individual cells remains approximately constant. In 1-day-old larvae the compound disc has increased in size; it has become oval-shaped and is made up of several cell layers. A well-developed optic stalk is present and is attached to the brain (Steinberg, 1941). As development proceeds, the anterior portion of the complex becomes somewhat elongated and widened, while the posterior portion broadens. At about 40 hours the disc has become leaf-shaped, and its inner wall has thickened, especially in the middle of the anterior portion and posteriorly. A thin cell layer makes up the outer wall of the compound disc, enclosing a narrow lumen. The anterior and posterior thickenings in the inner wall of the complex represent the Anlagen of the antennal and eye discs proper, which are, however, still connected by a broad cell layer at this stage.

By 48 hours a distinct anterior and posterior disc portion can be distinguished, which represents the antennal and eye discs, respectively (Fig. 7A). Although the antennal and eye disc cells are still very similar, the inner wall at the point where the two discs meet has become more constricted, thus setting off more clearly the antennal disc portions from the posterior eye disc portions. Ten hours later the antennal and eye discs are well formed. Both have increased in size and are connected by a thin layer of cells. The antennal disc becomes folded (Fig. 7B), indicating the first signs of segmentation. Probably about this time also the cells in the eye disc begin to organise into units, for such units have been observed in 72-hour-old pupae, where they are already well formed (Steinberg, 1943). Each of these units consists of a cluster of four cells. Only a few of these clusters are present as yet, but they increase in number as the eye grows older (Steinberg, 1941). Now it is doubtful whether these cell clusters divide into other similar clusters, or whether the newly added clusters are made by new formative material in the more peripheral eye regions. Because of this uncertainty, it is questionable whether each early cell cluster represents a single imaginal ommatidium or more than one. However, Steinberg (1943), who has studied the eye development in Drosophila very carefully, believes that each cluster is the precursor of a single ommatidium.

If this is the case, the mode of ommatidia increase would indeed resemble that observed in some hemimetabolous insects, where an addition of new ommatidia takes place from instar to instar from the eye border, the so-called "budding zone." In the mature

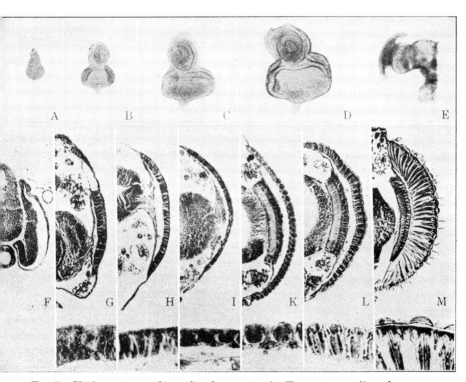

FIG. 7. Various stages of eye development. A: Eye antenna disc of a very young third-instar larva. B and C: Eye antenna discs of successively older third-instar larva. D: Eye antenna disc of mature larva. E: Eye disc of a pupa about 2 days old. F: Section through eye disc of mature larva. G–M: Successively older stages of pupal eye development. G–M after Bodenstein (1938).

larva, the cell clusters are arranged in regular rows, and their number approximates that of the final ommatidia. At this time each cluster is still composed of four cells (Steinberg, 1941). This shows that the cells which make up the component parts of an ommatidium, such as the cells which form the dioptic apparatus, the retinula, the pigment, and the hair cells, occur first in early pupal development. Each ommatidia precursor is innervated by a single nerve fiber ramification of the optic nerve. During the prepupal period the eye and antennal discs prepare for the process of evagination, which is quite complicated and which has been discussed by Robertson (1939) and Steinberg (1943). At pupation (i.e., about 11–12 hours after puparium formation) the eye and antennal discs evaginate. The differentiation and specialization of the eye disc cells into their final imaginal state take place during pupal life. In about 1-day-old pupae the eye is a relatively thick-walled shallow cup where the individual ommatidia are clearly visible as broad columnlike structures (Fig. 7G). As the eye cup flattens out, the ommatidia become thinner and shorter (Fig. 7H). By the middle of the second day, at about 132 hours, the eye cup is relatively thin-walled and has increased somewhat in area. The ommatidia have continued to shorten and are now rounded ball-like structures (Fig. 7I). At the end of the second day the formation of the cornea lenses begins, and hairs develop from special hair-forming cells. It is also at this stage that the first pigmentation begins (Fig. 7K). The ommatidia now begin to elongate and at 2½ days (156 hours) after puparium formation have reached a typical mushroom stage, where the cornea lenses and hairs are clearly formed. Pigmentation has progressed, and the eye has now a brownish color (Fig. 7L). During the remaining 36 hours of pupal life the ommatidia grow rapidly in length, and their differentiation is completed (Fig. 7M).

THE DEVELOPMENT OF THE WING

The wings of Drosophila, as of other Diptera, arise from buds which originally appear as thickenings of the hypodermal wall. The buds, which are masses of small embryonic cells, invaginate from the ectoderm and form pocketlike sacs or peripodial cavities enclosed in thin-walled peripodial membranes (Fig. 3C). In older larval stages the wing rudiments can be seen in the lateral group of discs (Fig. 6A) in the neighborhood of the main tracheal trunk, with which they are in close contact. The wing discs have been called the dorsal mesothoracic discs, since they were originally situated dorsally and laterally in the second thoracic segment; however, we shall refer to them in this study simply as wing discs. No innervation or neural connection can be noted in the wing rudiments. The stalk connecting the discs with the surface of the organism is evident in third-instar larvae although Robertson (1936) could first locate the hypodermal connections only in the 4-hour prepupa.

A number of investigators, including Chen (1929), Goldschmidt (1935), Auerbach (1936), Robertson (1936), and Waddington (1939), have studied the development of wings of normal Drosophila in various stages of ontogeny. Auerbach's report has been found especially helpful. She has followed wing development from the invaginated bud in the newly hatched larva to the fully developed pupal wing, describing histological features and illustrating her study with drawings and photomicrographs. The present author has also prepared sectioned material of all larval instars and finds the story is in most essentials in agreement with Auerbach's account. In Fig. 8 is shown a consecutive series of whole mounts of discs, beginning with the undifferentiated rudiment of the early third-instar larva and terminating with the folded imaginal wing; all photographs were taken at the same magnification. In this developmental series can be seen the emergence of the thoracic and wing-forming material, the elongation of the wing pouch, the appearance of the veins, and the final folding of the appendage. Figures 8A–D are discs from young to mature third-instar larvae, and Figs. 8E–I are rudiments from prepupae 0 to 10 hours after the larvae have become immobile and have everted the anterior spiracles. Figures 8K–N are progressively older stages in wing development.

Wing buds are already present in the interior of newly emerged larvae (Auerbach, 1936) and thus undoubtedly originate in the embryo. These observations do not confirm those of Chen (1929), who found the earliest signs of the buds in the 16-hour larva. A connection with the respiratory system is evident even in the early rudiments. The buds grow during the first two instars, and furthermore the cells of the lower wall elongate to form the thick epithelial layer. The buds now appear as sacs with a distinct lumen, the cavity being the peripodial cavity, the outer wall being quite thin and representing the peripodial membrane, and the thick inner wall being the wing rudiment proper (see Fig. 3C).

An early third-instar disc in ventral aspect (Fig. 8A) has as yet no visible differentiation. The stalk arises at the anterior pointed end. The visible dark threads are tracheoles. In older

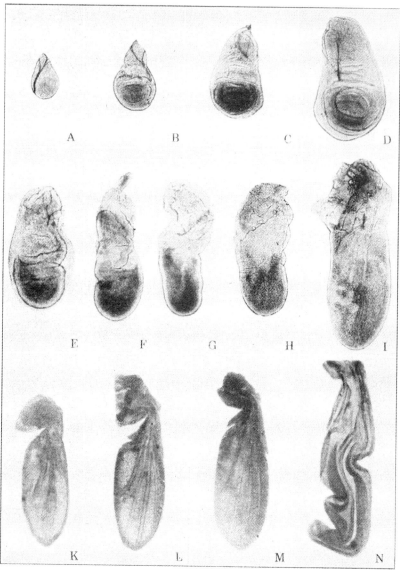

FIG. 8.   Various stages in the development of the wing.   A–D: Successively older stages of third-instar wing discs.   E–I: Successively older wing discs between puparium formation and pupation.   K: Wing of 18-hour-old pupa.   L: Wing of 28-hour-old pupa.   M: Wing of 38-hour-old pupa. N: Wing of 48-hour-old pupa.

third-instar larvae (Figs. 8B and C) the disc has enlarged, becoming more rectangular. Median crossridges appear, separating the forward portion which gives rise to the imaginal thorax from the posterior region which will form the wing. A distinct, clear, peripheral fold extends about the margin of the disc. In the mature larva (Fig. 8D) the disc is visibly differentiated. The rather smooth anterior region is separated by the ridges from the hind portion. A posterior circular epithelial mass has pushed inward and back, forming a pouch whose opening is indicated by the heavy concentric folds noted in the lower half of Fig. 8D. The area which invaginates is delimited in earlier stages.

In the early prepupal hours the wing pouch has lengthened considerably. (Figures 8E–H show several later stages found after the eversion of the anterior spiracles.) The pupal wing is everted about this time with the proximal portion forming the hinge or basal articulation with the thorax and the distal portion destined to form the wing proper. Figure 8I is a whole mount of a wing from a 10-hour prepupa with the wing evaginated, and the veins, indicated by the clear lengthy areas, already noticeable. These veins are the remnants of the wing cavity; they are apparently not identical with the later imaginal veins (Waddington, 1939). The pupal wing of this and slightly older stages apparently accumulates fluid, forcing the two surface layers, each one cell in thickness, to expand.

Toward the end of the first day of pupal life and throughout the second day, the saclike wing contracts, becoming thinner and platelike, and the venation becomes more marked. In a pupa about 20 hours old, veins are present in the tip and the base of the wing; about 3 hours later the posterior crossvein is clearly visible. Figures 8K–N are pictures of wings taken from 18-, 28-, 38-, and 48-hour-old pupae. Anteriorly, the area for articulation with the thorax has been hollowed out and can be seen in these four figures. The increasing age of the wing can be noted by a reduction in the width of the veins. The vein width diminishes in older wings (cf. Figs. 8L–N). Forty-eight hours after puparium formation the flattened wing has undergone a characteristic folding, and in the newly emerged imago the appendage is in just such a condition. The wing is straightened within about an hour after emergence of the young fly through pressure of the fluid accumulating between the two epithelial layers and also, as stated by Robertson (1936), through the brushing movements of the hindlegs.

The haltere buds (dorsal metathoracic buds) are also present; they are the lateral group of buds associated with the main tracheal trunk (Fig. 6A). These, as Auerbach has shown, have a developmental history similar to that followed by the wing buds and therefore do not require any further special or detailed report.

THE DEVELOPMENT OF THE LEG

The three pairs of legs of the fly are the ventral appendages of the three thoracic segments next to the head. The leg of the adult insect consists of a definite number of joints or divisions, specifically the coxa, which is the basal or proximal joint, followed by the trochanter, the femur, the tibia, and the tarsus, the tarsus composed normally of five small segments. As with so many other imaginal structures, the origin of the adult legs can be traced back to certain buds which originate in the embryo as ectodermal thickenings and later invaginate. From these simple formative centers, composed of small embryonic cells, the appendages are differentiated.

In Fig. 9 is presented a series of photographs of whole mounts of the rudiments which will eventually form the first (prothoracic) pair of legs of the imago. These have been taken from *Drosophila virilis* rather than *D. melanogaster,* for although a similar series of *melanogaster* material has been prepared, the *virilis* Anlagen were found to be superior for illustrative purposes. The findings are in all aspects similar in the two groups. The earliest buds shown (Fig. 9A) are from an early third-instar larva. The successively older stages of leg development presented in the figure are from individuals of the middle and late third instar, the prepupa, and the pupa. The progressive transformation of the buds from a rather primitive condition to the formed appendages is apparent in this group of pictures. All the photographs of these mounts were taken at the same magnification.

In the newly hatched larva, the leg buds are present as thickenings of the ventral hypoderm and can, in fact, be observed in the late embryo. Kaliss (1939) has reported the presence in

D. melanogaster of all three pairs of the ventral thoracic discs, confirming the description of them recorded by Auerbach (1936) for the young larvae. The derivation and initial appearance of. the rudiments in the embryo of Drosophila have not been reported as yet. In the literature concerned with the ontogeny of the Diptera, the leg Anlagen are known as the ventral prothoracic, ventral mesothoracic, and ventral metathoracic discs, corresponding to the thoracic segment wherein they are situated in the larva. We shall refer to them rather as first, second, and third leg discs (Fig. 6A).

Tracheal ramifications join all the leg buds very early. The first and second leg discs, but apparently not the third pair, are innervated by processes from the ventral ganglion. As in the case of the wing discs discussed earlier, the buds have passed into the interior of the organism, appearing as sacs, the lumen of each being the peripodial cavity (Fig. 3C). The external sheath or peripodial membrane is continuous with the larval hypodermis by means of a connecting stalk which very likely is the long neck of the invagination. The invagination of the buds occurs before the second larval instar has terminated; all the leg Anlagen have strands connecting them with the hypodermis. During these early instars the epithelial cells of the upper or dorsal wall of the buds elongate, and it is from these primary thickened regions, noted in sections of the buds, that the legs proper as well as a portion of the adult thorax will be derived.

A pair of first leg discs from an early third-instar larva is shown in Fig. 9A. It was preferable to demonstrate the rudiments of the first legs rather than of the other two pairs, since these buds are intimately associated in the early stages as well as during the period when differentiation is proceeding, and the changes occurring simultaneously in both members of the pair could more readily be demonstrated. There is as yet no visible differentiation in early third-instar rudiments. The tracheal connections which proceed laterally and the severed posterior neural processes are evident in Fig. 9A. During the progress of the third instar (Figs. 9B and C) not only have the discs increased in size but also morphological differentiation has taken place. The rings apparent in the discs represent the primitive leg segments. The thickened cellular wall of the rudiments has invaginated into the peripodial cavity, thereby diminishing the volume of the cavity, while a new cavity is forming. This can be seen in Fig. •3C, which is a transverse section of a mature larva. The lumina of the peripodial cavity and of the primitive leg are visible. The medial walls of the peripodial sacs are continuous, with a com-

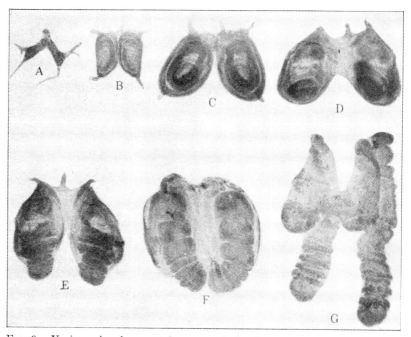

Fig. 9. Various developmental stages of the first pair of leg discs (*Drosophila virilis*). A: Leg disc pair of a very young third-instar larva. B: Same leg disc pair of a somewhat older third-instar larva. C: Leg disc pair of a mature larva. D: Leg disc pair of a white prepupa. E and F: Leg disc pairs of older prepupal stages. G: Leg disc pair after evagination, about 16 hours after puparium formation.

mon membrane separating the cavities of the first leg discs. The lumina can also be seen in the whole mount indicated in Fig. 9C. From the newly invaginated material will be derived, it may be repeated, not only the leg segments, but also a portion of the imaginal thorax (see page 313). The initial leg segment to be delimited is the tarsus, that is, the distal division of the adult leg.

Figures 9D–F are pairs of rudiments removed from progressively older prepupal stages. The first is from a larva which has just entered the prepupal period (larva immobile, anterior spiracles everted, larval skin still white); the discs presented in Figs. 9E and F are from prepupae which have developed for 4 and 8 hours respectively, after larval immobility. In Figs. 9D and E the traces of the later leg segmentation are now much more clearly visible. Furthermore, the peripodial sacs, whose medial walls were apposed earlier (Fig. 9B) appear to have fused for a portion of their length. The segments of the leg are oriented in the peripodial cavity in such a manner that the various divisions of the appendages slide into one another, in telescopic fashion (see Korschelt and Heider, 1899). In Fig. 9F the segments of the limb are becoming exposed. In this photograph the five tarsal joints, as well as tibia and femur, are seen. The stalks joining the peripodial sacs with the hypodermis have shrunken, and the lumina of the stalks, which represent the necks of the original invaginations, widen. The legs are gradually uncovered in a proximal-distal direction, as the peripodial sac slowly shrinks and disappears. The hypodermis of the larva undergoes histolysis, and a new hypodermal structure is developed from the anterior portion of the disc. The present writer has not studied this process in any detail, and the reader is referred to the several existing accounts in the literature. From a study of such literature we may say, in passing, that the phenomena concerned with the formation of the imaginal integument in Diptera need further careful investigation.

Figure 9G shows the appendages from a pupa 15 hours after larval quiescence and eversion of the spiracles. The legs have unfolded and elongated caudally and are enclosed in the secreted pupal cuticle. In the first hour of pupal life they elongate posteriorly until they extend almost to the top of the abdomen. At this stage the leg is a very thin-walled, inflated, saclike structure, which fills completely the pupal leg sheath. As development proceeds, however, the leg gradually contracts, bringing the leg epithelia together. In the 2-day-old pupa the leg has become quite reduced in width, resulting in the formation of a large space between the leg and the chitinous sheath. The segmentation of the leg is by now clearly visible in total mounts, and its histological differentiation is completed during the following 2 days.

THE DEVELOPMENT OF THE GENITAL DISC AND THE GONADS

The reproductive system consists of the paired gonads, the genital ducts with their accessory structures, and the external genitalia. Now the development of the gonads is quite independent from that of the genital ducts and external genitalia; these latter structures have their origin in the genital imaginal disc, which is situated median-ventrally in the posterior end of the larva, just below the intestine (Fig. 6A); during the entire larval life this disc has no contact with the gonads. It is in the pupal stage, when the transformation of the imaginal disc into genital ducts and external genitalia takes place, that the connection between gonads and ducts is accomplished.

The gonads, already present when the larva hatches from the egg, are roundish transparent bodies. They lie in the body cavity just below the body wall, one on each side in the fifth abdominal segment. The gonads are surrounded by the fat body, to which they have a definite spatial relationship (Fig. 6, o and t). The ovaries are much more embedded in the fat body than are the testes, which are but bordered by it. The testes are from the beginning much larger than the ovaries. This size difference is maintained throughout the larval period (Fig. 6, o and t) and is, as pointed out on p. 291, a good means of determining the sex of the larva. The ovary of the newly hatched larva is extremely small and contains only 8–12 germ cells, whereas in the larger testis 36–38 germ cells are found (Sonnenblick, 1941). The testes of first-instar larvae are, in fact, large enough to be recognised through the body wall of living larvae (Kerkis, 1931), but the ovaries at this time can be detected only in sectioned material. As larval development proceeds, the gonads grow in size. The testes become somewhat more ellipsoidal in older larval stages; the ovaries remain round.

The testes of 24-hour-old larvae contain only spermatogonia, which have large nuclei and but little cytoplasm (Fig. 10, 24 hours). Numerous division figures can be found at this time, indicating that the spermatogonia multiply rapidly (Kerkis, 1933). In 50-hour-old testes first spermatocytes are found. The anterior part of the testes contains the spermatogonia, while the posterior part contains the spermatocytes. There is no sharp division between the regions occupied by spermatogonia and first

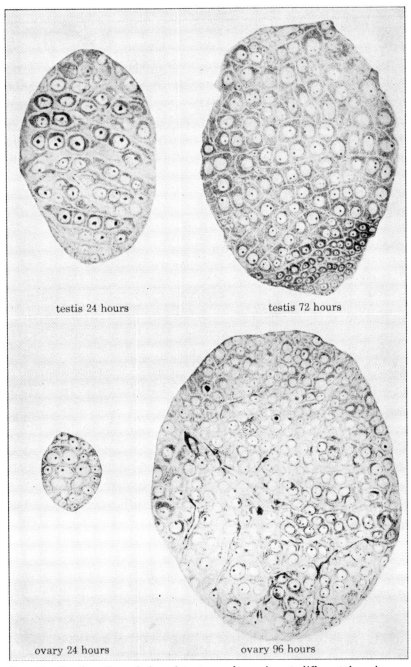

testis 24 hours   testis 72 hours

ovary 24 hours   ovary 96 hours

FIG. 10. Section through larval testes and ovaries at different larval ages. 72-hour-old testis about ½ as much enlarged as the other figures. After Kerkis (1933).

spermatocytes, for the transformation occurs rather gradually in an anterior-posterior direction. The cells of equal growth stages are arranged in more or less regular layers in such a way that each subsequent layer carries larger cells (Fig. 10, 24 hours). The largest cells are those located most posteriorly. This graded arrangement of the cells has become very pronounced in 72-hour-old testes. Here spermatogonia are found only in the anterior end of the testes; the greater part of it is filled with first spermatocytes (Fig. 10, 72 hours). The spermatocytes can easily be distinguished from the spermatogonia by their larger nuclei and the greater amount of the cytoplasm (Kerkis, 1933). Apart from the germ cells, a number of small cells, probably mesodermal in origin, are found in the extreme posterior end of the testes, in close relation to the testicular wall; these cells apparently play an important part in the pupal elongation of the testes (see p. 306). In mature larvae (96 hours) the spermatogonia are restricted to the extreme anterior tip of the testis, the rest being filled with spermatocytes, which have reached their maximal size and are ready for the meiotic divisions (Kerkis, 1933). Twenty-four hours after puparium formation nearly mature spermatozoa, as well as all the other stages of spermatogenesis, are found in the testes. At emergence of the fly the testes contain mostly mature spermatozoa, and only a few cells in the earlier stages of spermatogenesis.

Apart from an increase in size, the ovary changes relatively little during larval life. In Fig. 10 is shown the ovary of a 24-hour-old larva. At this stage the ovary contains only a small number of oögonia. The central part of the young ovary contains large cells, with nuclei and little cytoplasm, while the periphery is occupied by smaller oögonial cells, which are apparently formed by the large cells (Kerkis, 1933). As development proceeds, the number of small oögonia increases, but their size becomes progressively smaller. After a certain number of small cells have been formed, the large cells disappear. When this has happened, the division rate of the small oögonia is slowed down, and they now begin to grow in volume. During the fourth day of larval life they have reached the size of the larger cells present in the center region of the younger ovaries (cf. Fig. 10, 24 hours, with Fig. 10, 96 hours). Throughout the entire larval life the ovaries contain only oögonia, no oöcytes. Now the larval ovary carries a

number of small cells besides oögonia, cells which apparently give
rise to the follicular epithelium, the walls of the egg tubes (ovarioles, etc.). The differentiation of the egg tubes begins, according
to Kerkis (1933), at the end of the third day of larval life. This
process begins at one pole of the ovary and proceeds gradually
inward.

Sections through ovaries at the time of puparium formation reveal already quite well-differentiated egg tubes. It is, however,
not much before the twentieth hour after puparium formation
that these tubes can also be recognised in total dissections. During pupal development the ovary grows considerably and changes
its form, while at the same time its internal differentiation is completed. About 36 hours after puparium formation the ovaries
become attached to the oviducts (Fig. 13). At this time, however,
the ovaries still contain only oögonia. The first oöcytes appear
much later in development, namely, shortly before the emergence
of the fly (Kerkis, 1933). Meiosis takes place in the female at
the emergence of the adult, thus considerably later than in the
male.

The ovaries of newly emerged flies are not yet completely mature and contain no ripe eggs. They are relatively small and
abundantly supplied by tracheae. Each ovary is made up of
about 15–20 egg tubes or ovarioles, the exact number of which depends on the larval food condition (Saveliev, 1928). At this
relatively immature state the ovarioles show but little differentiation. They are filled with germ cells (some of which have already formed oöcytes, whereas others are destined to become
nurse cells) and with a number of smaller cells, which will give
rise to the follicle cells. Each ovariole is enclosed in a delicate
transparent membrane. As development proceeds, and the oöcytes
multiply and mature, the ovarioles increase rapidly in size. Each
ovariole now contains a chain of developing ova. The follicular
epithelium has been formed and divides the ovariole into 5 or 6
egg chambers, each of which encloses the egg and its 15 nurse
cells. Only the chamber at the apex of the ovariole, the germarium, consists of densely packed cells from which the oögonia become differentiated into oöcytes and nurse cells. This region represents the primitive egg tube of the young ovary. The greater
part of the somewhat older ovariole is known as the vitellarium;
it holds the developing eggs in a single chain, one behind the

other. The oldest egg is the one nearest the posterior end. At the anterior end the wall of the ovariole forms a threadlike prolongation known as the terminal filament, while proximally it forms a thin tube, the ovarial stalk, which leads to the oviduct. The ovary of 1-day-old flies contains eggs at various stages of development. The amount of yolk deposited in each egg indicates its stage of development. The egg is fully developed when the yolk fills the whole egg chamber and the nurse cells have degenerated. When the leading egg is ripe, it ruptures an epithelial plug, which closes the vitellarium from the ovariole stalk, and passes through the stalk into the oviduct. About two days after emergence of the fly the ovary contains fully developed eggs ready to be laid.

The growth rate of *D. melanogaster* male and female gonads during the larval and early pupal life has been studied by Kerkis (1931). He finds that during the first part of larval life the testes grow at a higher rate than the ovaries, whereas in the second half of larval life and in the young pupa the conditions are reversed. Now it is well known that the growth rate of animals or organs generally decreases with an increase in age. The growth rate of the testes is no exception to this. Yet the ovaries behave differently; their rate of growth increases with age, at least during larval life. After puparium formation, however, the rate of growth decreases also in the ovary (Kerkis, 1931).

During the larval and early pupal growth period the testis maintains its ellipsoidal form. However, when the testis becomes attached to the vas deferens, this situation changes, for the testis now begins to elongate and gradually assumes a spiral shape (Figs. 11–12). The factors determining this spiral growth have been analyzed experimentally by Stern (1941a and b), whose investigations revealed the following facts. The wall of the unattached pupal testis is made up of two distinct layers of cells, an outer and an inner. The outer layer is composed of large flat cells with large bulging nuclei. The inner layer is a fine membrane without visible structure and has apparently extremely flat nuclei. Now this inner layer increases in thickness toward the posterior end of the testis, and the nuclei, which are rather widely spaced anteriorly, are closely packed and larger in the thickened posterior region. The cells of the outer layer become pigmented at the end of pupal life and are thus responsible for the pigment coat of the adult testes, which is characteristic for many Drosophila species. During pupal life these presumptive pigment cells are able to migrate and may move on to the vasa efferentia, which originate from the imaginal disc, and which do not possess these pigment-forming cells (Stern and Hadorn, 1939).

It is the inner layer which gives the testis its typical shape. When the vas deferens becomes attached to the testis, it joins the testis at its posterior tip, at that region, therefore, where the cells of the inner layer are grouped closely together. Now the vas deferens apparently releases a substance which stimulates the growth of the testes at this point of contact, for the testes fail to

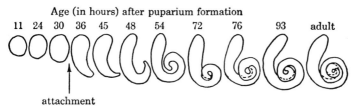

Age (in hours) after puparium formation

11   24   30   36   45   48   54   72   76   93   adult

attachment

Fig. 11. Successive growth stages in the pupal development of the testis. After Stern (1941).

elongate and coil if the vas does not become attached to them. After attachment has taken place, the closely grouped cells at the posterior end of the testes become more and more widely spaced as elongation and later coiling of the testes continue. The growth of the testes is thus mainly the result of an increase in the membrane area and is accomplished without an increase in the number of cells. The coiling of the testes is explained by Stern (1941a) on the assumption that the growth-promoting substance is released in different amounts to opposite sides of the testes, resulting in a different growth response of the different regions. As a consequence the elongation of the testes takes place unequally, which finally leads to coiling. It is clear that the interaction between testis and vas is to a great extent responsible for the final imaginal form of the testes. The important function of the vas in determining the shape of the testes is very clearly demonstrated in one further experiment by Stern. He was able to show that the vasa deferentia from species with non-coiling testes are unable to induce coiling in testes which normally have coils, and, vice versa, that the vasa deferentia of "coiling species" will cause normally uncoiled testes to coil (Stern, 1941b).

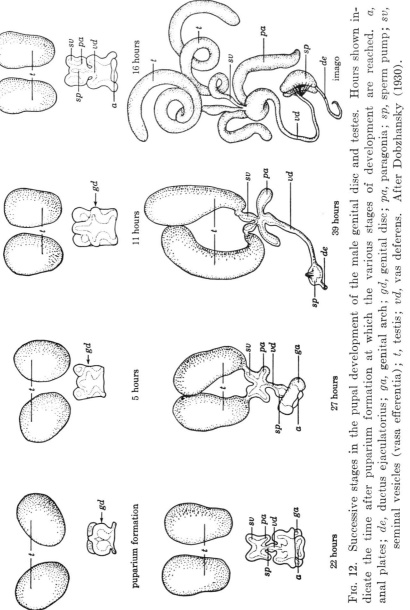

FIG. 12. Successive stages in the pupal development of the male genital disc and testes. Hours shown indicate the time after puparium formation at which the various stages of development are reached. *a*, anal plates; *de*, ductus ejaculatorius; *ga*, genital arch; *gd*, genital disc; *pa*, paragonia; *sp*, sperm pump; *sv*, seminal vesicles (vasa efferentia); *t*, testis; *vd*, vas deferens. After Dobzhansky (1930).

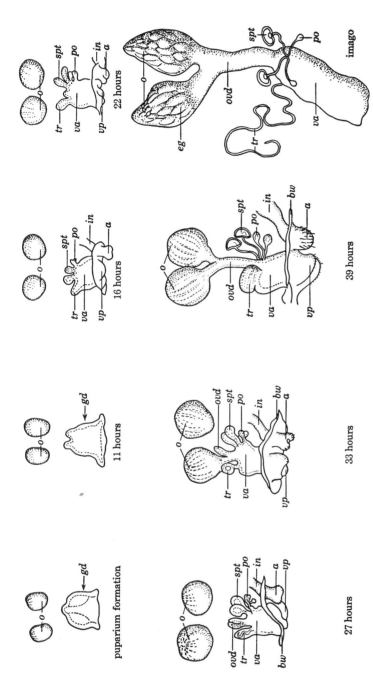

FIG. 13.   Successive stages in the pupal development of the female genital disc and the ovaries.   Hours shown indicate the time after puparium formation at which the various stages of development are reached.   *a*, anal plates; *bw*, body wall; *eg*, eggs; *gd*, genital disc; *in*, intestine; *o*, ovary; *ovd*, oviduct; *po*, parovaria; *spt*, spermatheca; *tr*, tubular receptacle; *va*, vagina; *vp*, vaginal plates.   After Dobzhansky (1930).

The origin of the genital ducts, accessory glands, and external genitalia from the imaginal genital disc was first clearly demonstrated by Dobzhansky (1930). Dobzhansky followed the development of the genital disc of male and female from the mature larva throughout the entire pupal life by dissecting the successive developmental stages of the discs. Figures 12 and 13 illustrate the pupal development of the genital discs of the two sexes. These drawings are self-explanatory; they show clearly how, during the course of pupal life, the internal genital ducts and accessory structures arise from the anterior disc regions, while the posterior regions give rise to the external genitalia. It may be pointed out here that Dobzhansky's observations were made on material cultured at 27°C. In the drawings presented in Figs. 12 and 13, his timing has been converted into that for 25°C for convenient comparison with other events. Such a procedure is not very satisfactory, for it gives only an approximation of the exact time. This fact becomes very obvious when one compares the converted timing in Figs. 12 and 13 with the timing of the normal growth stages of the testes taken from Stern (Fig. 11), who based his observations on pupae reared at 25°C. The observed difference may be partly due to differences in the races of *melanogaster* used by the two authors (both do not state them specifically), yet, even granting this uncertainty, the time data given in Figs. 12 and 13 seem somewhat too low.

The development of the genital disc during the earlier periods of larval life is but little known. The only information available on this subject is an investigation by Newby (1942) on *Drosophila virilis*. Newby finds that the genital disc in *D. virilis* begins to develop shortly before the third day of larval life. Taking into consideration the relatively long larval period of *D. virilis*, one might assume that the genital disc makes its initial appearance early in the second larval instar. Now in young third-instar *D. melanogaster* larvae the genital discs are present, but they are still extremely small and difficult to dissect. Thus also in *melanogaster* the beginning of development of the genital disc takes place before the third instar. How much earlier we do not know, but it seems reasonable to assume that it may be at the same time as in *virilis*, namely, between the first and second molts. Whatever the exact time may be, the genital disc develops apparently later than the gonads, which are already present when the larvae hatch from the egg.

The larval development of the genital disc in *virilis* has been described by Newby (1942) as follows. By the end of the second larval day, the genital disc is made up of a plate of cells, which have apparently arisen from the ectoderm in the midventral line (Fig. 14). A narrow band of elastic fibers, probably mesodermal in origin, extends anteriorly from either side of the disc and is attached to muscles. At the region where these fibers join the disc are located a number of mesodermal cells, which in the course of development will give rise to the connective tissue and muscle of the ducts and accessory structures. Ventrally the disc is connected with the hypoderm by a short stalk. Within the next 24 hours the disc increases rapidly in size, mainly by an addition of new cells from the hypodermis. The hypoderm cells which become transformed into the disc cells appear to enter the disc from the sides through the stalk. As these cells enter the stalk, which connects the epidermis with the disc proper, they become greatly reduced in size, they flatten out, and their nuclei become smaller. In 4-day-old larvae (Fig. 14) the cells at the surface of the disc have become arranged into an epithelial plate, which separates from the underlying cell layers, thus forming a narrow cavity (later the lumen of the reproductive tract). During the next 2 days the disc grows considerably. Growth at this time, however, is mainly caused by cell multiplication; only a few cells seem to be added from the hypoderm. Although it is difficult to establish this fact by histological study, it can be clearly demonstrated by transplantation experiments. It was found (Bodenstein, unpublished) that small isolated genital discs of 4- or 5-day-old larvae grew normally when transplanted into the body cavity of host larvae of the same age. From this it would seem that growth takes place normally, although the stalk connecting the disc with the hypoderm was severed at dissection before transplantation and had established no new connection with the hypoderm of the host. It may be mentioned that these experiments also confirm the observations of other investigators as to the origin of the internal genital ducts, accessory glands, and external genitalia from the genital disc, for all these structures were found to develop from the transplanted disc.

The genital discs of early 6-day-old male and female larvae
are essentially alike (Fig. 14). They are now relatively large

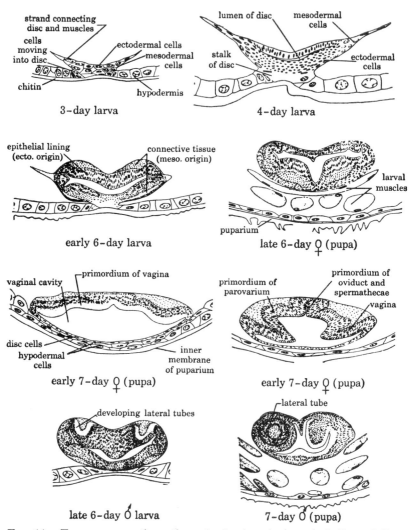

FIG. 14. Transverse sections through the imaginal genital disc of *Dro-sophila virilis* in various stages of larval and early pupal development. After Newby (1942).

and possess a wider lumen than younger discs. The epithelial
cells which surround the lumen are arranged into a stratified
columnar epithelium. This epithelial layer is covered at the outside by a compact layer of irregularly placed connective tissue

cells. In the mature larva (late sixth day) the male and female discs are externally about the same shape, but they differ internally (cf. Figs. 14, ♂ and ♀). A female disc at this stage is characterized by its large lumen and the thinness of the roof epithelium (Fig. 14). The male disc, on the other hand, is typified by the appearance of two grooves on the dorsolateral part of the disc (Fig. 14), which are the primordia of the vasa efferentia and accessory glands. The first visible differentiation of the reproductive structures within the compound disc thus takes place earlier in the male than in the female. The further pupal development of the disc is illustrated for both sexes in Figs. 12 and 13. Further details concerning these events, based mainly on histological examinations, can be found in Newby's (1942) instructive account.

THE DEVELOPMENT OF THE IMAGINAL HYPODERM OF THE THORAX AND ABDOMEN

The entire hypoderm of the imago is formed anew during the earlier part of pupal life, while simultaneously the larval hypoderm is histolyzed. The cells at the anterior end of each of the imaginal discs are the presumptive cells for the imaginal hypodermis of head and thorax. Shortly after puparium formation, when the stalks connecting the imaginal discs to the larval hypoderm have shortened considerably in preparation for the evagination of the discs, the anterior disc regions become closely connected with the larval epidermis. At this time the peripheral disc cells grow into the neighboring larval hypoderm, replacing it gradually, while the larval epidermis cells are cast off into the body cavity, where they are phagocytized. These replacement centers start at each disc, and, as the different patches of imaginal hypoderm grow in area, they meet finally and fuse with one another. Fusion takes place first in the lateral-ventral region of the thorax, about 6 hours after puparium formation. Within the next two hours the larval hypoderm of the thorax is entirely replaced by the imaginal hypodermis.

The imaginal hypoderm of the abdomen is formed by the so-called "hypodermal histoblasts." These histoblasts are small cell nests, consisting of tiny cells which appear between the larval epidermis cells about 4 hours after puparium formation. Within each segment, except the last, are one dorsal and one ventral pair of histoblasts. As development proceeds, the histoblasts proliferate, and their cells spread out along the larval hypodermis. In this process the larval hypoderm cells are gradually displaced and cast off into the body, where they are taken up by phagocytes. Cells from the lateral abdominal spiracles located half-way between the dorsal and ventral histoblasts also contribute apparently to the imaginal hypoderm (Robertson, 1936). The imaginal hypoderm of the last abdominal segment, including the outer genitalia of the fly, is formed in a manner similar to the thoracic hypoderm, namely, by cells which originate from the caudal regions of the genital disc. According to Robertson (1936), 2 hours after puparium formation some of the larval hypoderm has already been replaced by the imaginal hypodermal cells. The abdomen is completely covered by the imaginal epidermis in 34-hour-old pupae. At about 50 hours these cells begin to secrete the imaginal cuticle.

Although the shape and size of the various bristles in the skin of adult Drosophila vary greatly, they can be classified into two main types, the large macrochaetae and the small microchaetae. These bristles are innervated (Stern, 1938) and apparently function as tangoreceptors. Since the formation of macrochaetae and microchaetae is very similar, we shall treat in the following account only the development of the macrochaetae, as described by Lees and Waddington (1942), who studied the development of certain bristles in normal and mutant types of Drosophila in some detail. Each bristle arises by the outgrowth of a long cytoplasmic process from a large cell known as the "trichogen cell," while the socket in which the bristle articulates is formed by a second cell, the so-called "tormogen cell." Some typical stages of bristle development are shown in Fig. 15. About 16 hours after puparium formation, the bristle-forming cells become distinct from the other epidermis cells by their larger size. These cells are now arranged in groups of two. One of them is smaller than the other and is located near the surface of the skin. This represents the tormogen cell; the other, larger one, the trichogen cell, is placed somewhat below the tormogen cell (Fig. 15, 22 hours). Both cells increase rapidly in size between 16 and 40 hours after puparium formation and reach about ten times the linear dimension of a normal epidermis cell (Fig. 15, 40 hours).

Their nuclei also become very large and contain polytene chromosomes. The formation of the bristles begins about 30 hours after puparium formation. By 36 hours the bristles have reached about one-half their full length, and the tormogen cells have secreted the socket material, somewhat later than the outgrowth of the bristles. The outer surface of the bristles, which until now has been smooth, forms small elevated ridges, which run from the tip to the base of the bristles. By 48 hours the socket around the

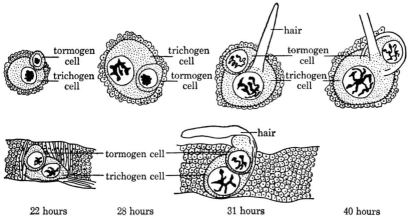

22 hours   28 hours   31 hours   40 hours

FIG. 15. Successive stages of bristle development. Upper row: surface view of whole mounts. Lower row: sections. After Lees and Waddington (1942).

bristle is completed, and the bristle has attained its full length. At this stage the bristle is a solid structure, but in the imago it is a hollow cone. This hollowing-out process begins immediately after the bristle has reached its full length. At the same time the tormogen and trichogen cells gradually begin to reduce in size. In 70-hour-old pupae, when bristle pigmentation begins, the tormogen and trichogen cells have become so small that it is difficult to distinguish them from the other cells of the epidermis.

The time of bristle pigmentation varies within the different regions of the pupa and proceeds, as observed in living pupae, as follows. In 70-hour-old pupae little black dots representing the first indication of bristle pigmentation appear dorsomedially on the thorax. Within the next 2 or 3 hours the surface of the head and thorax is covered with these little black dots, whereas the rest of the body is still white. Very fine dark bristles can first be recognized in the head, thorax, wings, and legs in pupae of 78 hours, while little black dots begin to appear dorsally in the first and second abdominal segments. About 82 hours after puparium formation the abdomen has fully colored bristles. At this time the pigment bands on the dorsal abdominal plates begin to show up, but their pigmentation is not yet complete. The pigmentation of the body wall is completed during the rest of pupal life.

Apart from the various-sized bristles many epidermal cells have minute hairlike projections, the so-called "cell hairs." These cell hairs, in contrast to the bristles, are formed by single cells and can be recognized as early as 50 hours after puparium formation.

THE OENOCYTES

The oenocytes are cells of ectodermal origin. Their function is as yet not fully understood. The imaginal oenocytes of chironomids (Zavrel, 1935) are said to produce a hormone needed for the development of the gonads. In Diptera, however, they were found to respond to a hormone released by the corpus allatum (Day, 1943). This latter observation seems to indicate that they are the reactor organs and do not produce a gonadotropic hormone. Yet their behavior under the influence of the hormone seems to indicate either that they liberate substances that are necessary as building materials for other organs, or that they function as metabolic regulators, which play an important part in the intermediate metabolism.

The oenocytes of Drosophila have been investigated by Koch (1945), who found that there are two generations of oenocytes, the larval oenocytes and the imaginal oenocytes. The larval oenocytes degenerate during metamorphosis, when a new crop of imaginal oenocytes is formed.

The Larval Oenocytes

The larval oenocytes are large cells which possess characteristically a dense homogeneous cytoplasm, which contains some yellowish granules. These granules increase in number as the larvae grow older. The round oenocyte nucleus is located centrally in the cell and contains chromosomes in polytene condition.

The oenocytes are grouped together into a cluster of six or seven cells which adhere closely to each other (Fig. 16). One such cell cluster is situated bilaterally in each of the abdominal segments with the exception of the last one. At this site the oenocytes are

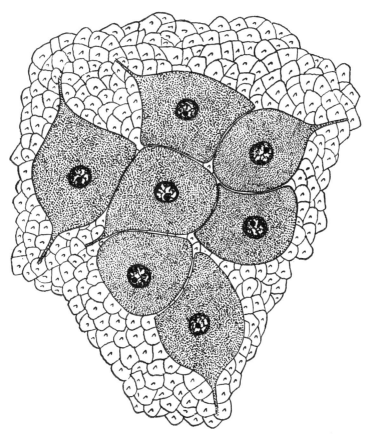

FIG. 16. Cell cluster of oenocytes of a last-instar larva. Note the characteristic grouping of these cells and the fine protoplasmic strands that connect the cells with the hypoderm.

found in the space between the dorsoventral muscles and the body wall. The cells are attached to the hypoderm by fine elastic plasma strands, which extend and contract as the larvae move about. Each cell group is supplied by a special tracheal branch.

In first-instar larvae the oenocytes are difficult to find, since they are at this stage not much different in size from the cells of the ectoderm. In second-instar larvae, however, they are easily detected by their larger size, their typical grouping, and their characteristic cytological appearance. Although there is only one generation of oenocytes throughout larval life, the cells increase progressively in size as they become older. In the fully grown larvae the oenocytes reach a cell diameter of about 80 μ, and a nuclear diameter of about 10 μ.

At the time of pupation the degeneration of the oenocytes begins and is completed in about 2 days. The events leading to the final destruction are as follows. The oenocytes round up, and the cell boundaries become irregular. The tracheal supply is severed, and the attachment to the hypoderm breaks off. The oenocytes can now be found scattered among the fat cells in the body cavity. Vacuoles appear, especially near the nucleus, and the cytoplasm takes up stains very irregularly. Finally the chromatin in the nucleus clumps; the cell membrane breaks down, and the cell fragments are absorbed.

The Imaginal Oenocytes

The imaginal oenocytes apparently arise from the imaginal discs that form the hypoderm of the adult (Koch, 1945). Before the different Anlagen of the adult hypoderm have united to form the continuous cover of the adult abdomen, cells are pinched off at the inner side of this growing sheet of cells. Some of these cells differentiate into the oenocytes while others give rise to fat cells. The fat cells grow more rapidly than the presumptive oenocytes and soon surpass them in size. As development proceeds, the oenocytes become segmentally arranged into a band of cells that occupies dorsally a space between two bands of fat body cells that have developed at the same time. Some of the oenocytes invade the fat body and are hence found at its inner surface. It must here be pointed out that the close spatial association of fat body cells and oenocytes is a common occurrence in many other forms of insects. Oenocytes appear also near the ventral midline of the abdomen; here they are found in small groups and again in close connection with the fat body.

The adult oenocytes are much smaller than the larval ones, being only 12–17 μ in diameter. There are, however, many more oenocytes in the adult than in the larva. By their dense cytoplasm, and especially by their yellowish green color, the imaginal oenocytes are easily recognizable and distinct from the whitish cells of the fat body. The color of the oenocytes, caused by granular inclusions in the cytoplasm, is more pronounced in older flies than in newly hatched individuals.

THE NERVOUS SYSTEM

The only information available on the development and meta-morphosis of the nervous system in Drosophila is the work of

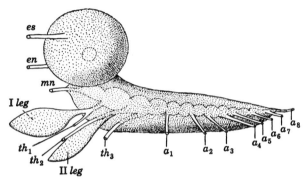

FIG. 17. Lateral view of the larval nervous system. a_1–a_8, abdominal nerves; *en*, antenna nerve; *es*, eye stalk; *mn*, maxillary nerve; th_1–th_3, thoracic nerves. After Hertweck (1931)

Hertweck (1931), on which the following description is based. Hertweck also gives some additional details on the metamorphosis of the nerves. Further literature on this subject may also be found in Hertweck's paper.

The larval central nervous system (Fig. 17) consists of the two brain hemispheres and the compound ventral ganglion. It is located in the third thoracic segment and reaches to about the end of the first abdominal segment. The two hemispheres are connected ventrally; this place of connection is pierced by the oesophagus, which runs backward dorsally above the ventral ganglion. The latter is anteriorly connected to the ventral point of fusion of the two hemispheres and consists anteriorly and pos-teriorly of the suboesophageal ganglion, the three pairs of thoracic ganglia, and the eight pairs of abdominal ganglia. Two nerves— the eye and antenna nerve, the latter innervating the larval an-tennal organ (see p. 329)—arise from each hemisphere. The maxillary nerves, which innervate the sense organs and muscles of the head, come from the most anterior region of the ventral ganglion. From the first and second thoracic ganglia arise the

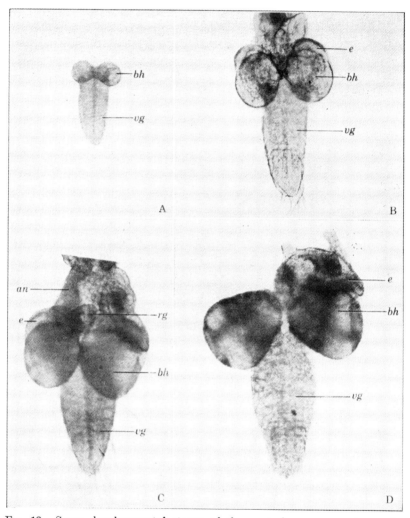

FIG. 18. Some developmental stages of the central nervous system. A: Brain and ventral ganglion of a second-instar larva, B: of a younger third-instar larva, C: of a mature larva, D: of a young prepupa. *an,* antenna disc; *bh,* brain hemisphere; *e,* eye disc; *rg,* ring gland; *vg,* ventral ganglion.

nerves for the first and second leg pair. Each of these nerves sends off a branch which innervates the muscles and sense organs of the first and second thoracic segments. The third thoracic ganglion has no nerves directly connected to the leg disc; thus the thoracic nerve innervating the third segment arises from the third thoracic ganglion directly. Each of the abdominal ganglia sends off a pair of nerves which innervate the corresponding segments. A small ganglion, the hypocerebral ganglion, located somewhat anteriorly to the proventriculus, represents the center of the visceral nervous system. The main morphological change

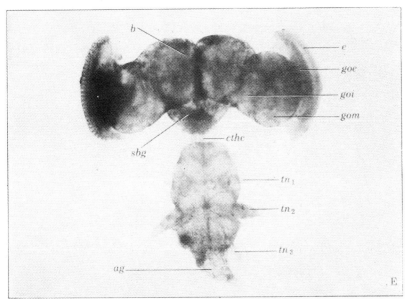

FIG. 18 (*cont'd*). E: The central nervous system of an about-two-days-old pupa. *ag*, abdominal ganglion; *b*, brain; *cthc*, cephalothoracic cord; *e*, eye disc; *goe*, outer eye ganglion; *goi*, inner eye ganglion; *gom*, middle eye ganglion; *sbg*, suboesophageal ganglion; *tn₁–tn₃*, thoracic nerves.

the brain and ventral ganglion complex undergo during larval life is an increase in size (Figs. 18A–C).

Cross-sections of the larval brain hemispheres (Fig. 19A) reveal that they are composed of a cellular cortex and a fibrous core. The core, organized into aggregates of fibers or glomeruli, discloses in the larva some of the characteristic centers of the insect brain, such as the central complex, the corpora pedunculata, and the antennal glomeruli. The largest part of the brain, however, is occupied by the larval ganglion cells and the cells of the cortex. The cortex cells, which are smaller than the other ganglion cells, are arranged into a regular epithelium and represent the formation center (Fig. 19A, 10) of the outer imaginal ganglion. Some good although indirect evidence is now available which indicates that the outer eye ganglion is derived

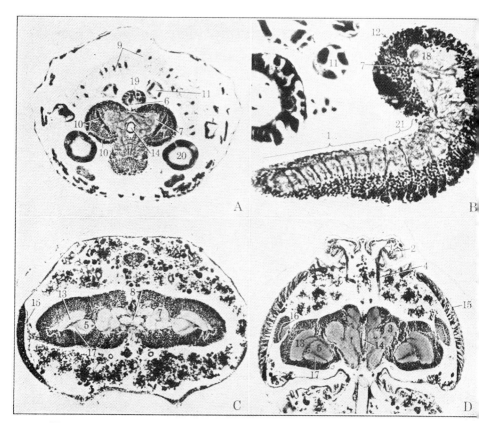

Fig. 19. Boudin-stained sections through various parts of the central nervous system of larvae and pupae. A: Cross-section through the brain region of a 60-hour-old larva. B: Sagittal section through the nervous system of a 78-hour-old larva. C: Cross-section through the brain region of a 3-hour-old pupa. D: Frontal section through the head of a 45-hour-old pupa. 1, abdominal ganglia; 2, antenna; 3, antennal ganglion; 4, antennal nerve; 5, anterior inner eye ganglion; 6, aorta; 7, brain; 8, central complex; 9, fat body; 10, formation centers; 11, gastric caecum; 12, large ganglion cells; 13, middle eye ganglion; 14, oesophagus; 15, ommatidia; 16, outer eye ganglion; 17, posterior inner eye ganglion; 18, posterior root of corpora pedunculata; 19, ring gland; 20, salivary gland; 21, thoracic ganglia.

from the eye disc and not from the brain (Power, 1943). The hollow cup-shaped formation center of the eye ganglia is divided by a thin wall of cells, as illustrated diagrammatically in

Fig. 20. These cells will later form the inner and middle imaginal eye ganglia. The fibrous cores of the inner and middle eye ganglia and the medulla externa and interna are located one on each side of the cell wall. The eye ganglia are, however, not complete in the larva.

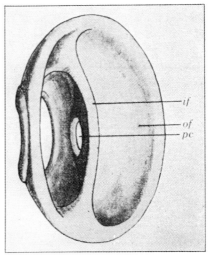

FIG. 20. Model of the inner and outer formation centers. *if*, inner formation center; *of*, outer formation center; *pc*, point of connection between outer and inner medullar masses. After Hertweck (1931).

The histological picture of the ventral ganglion is very similar to that of the hemispheres, although there are of course no cells present which correspond to the formation centers of the optic ganglia. The ganglion cells are larger than the cells of the cortex and are located mainly ventrally and laterally in the ventral ganglion. The segmentation of the ventral ganglion is especially clear in the region of the abdominal ganglia (Figs. 21 and 19B). In the nervous system of the larva and of the pupa one frequently finds giant ganglion cells (Fig.

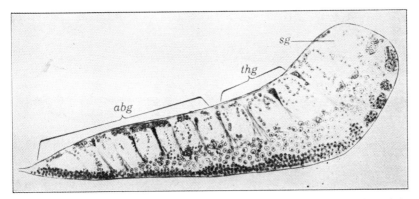

FIG. 21. Sagittal section through the larval ventral ganglia. *abg*, eight abdominal ganglia; *sg*, suboesophageal ganglion; *thg*, three thoracic ganglia. After Hertweck (1931).

19B, 12) which often show very clear mitotic pictures; these cells have been used for cytological studies.

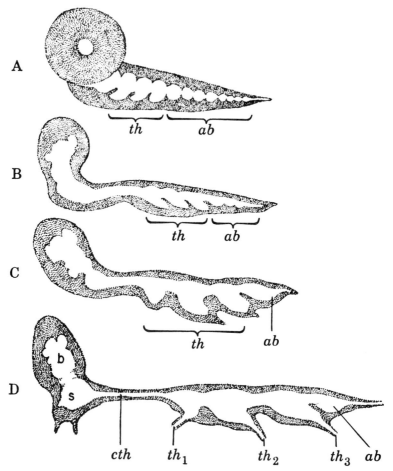

Fig. 22. Diagrammatic representation of the metamorphosis of the central nervous system. A: Larval central nervous system. B and C: Successively older stages of the pupal central nervous system. D: The metamorphosed imaginal central nervous system. *ab,* abdominal ganglia; *b,* brain; *cth,* cephalothoracic cord; *s,* suboesophageal ganglion; *th,* thoracic ganglia; *th₁-th₃,* thoracic nerves. Modified after Hertweck (1931).

In the mature larva the brain shows histologically some indication of its future metamorphosis, for, apart from the structures connected with the optic ganglia, one finds various types of ganglion cells often grouped together into clusters. Whether these cells are newly developed primordia for the formation of the imaginal brain or whether they are just larval ganglion cells which themselves have begun to undergo metamorphosis is not known. In any case the finer histological changes which take place during the transformation of the larval into the imaginal brain are not as yet fully understood. At the time of puparium formation the brain hemispheres become pear-shaped (Fig. 18D), because of the formation and development of the eye ganglia from the special optic centers described above. The outer and inner formation centers characteristic for the larval brain disappear as the ganglia are formed. As metamorphosis proceeds, the region between the suboesophageal ganglia and the first thoracic ganglia becomes constricted (Figs. 22B and C), forming the cephalothoracic cord connecting the suboesophageal ganglion with the thoracic ganglia (Figs. 18 and 22D). The two brain hemispheres unite also mediodorsally. In the hemispheres the development of the three eye ganglia continues. The outer eye ganglion comes to lie directly below the eye cup (Fig. 18E). The other two eye ganglia form the largest part of the brain, which is small compared with the immense eye ganglia (Fig. 19D). Some additional details concerning the histology of the brain are illustrated in Figs. 19A–D. The three thoracic ganglia remain separated while the eight pairs of abdominal ganglia unite into one abdominal ganglion (Figs. 18E and 22D). The morphological changes in the nervous system during metamorphosis are shown in Fig. 22.

THE RING GLAND

The problem of the hormone control of insect metamorphosis is one of the foremost problems in the developmental physiology of insects. It is thus necessary to discuss in some detail the development of a small organ of internal secretion, the so-called "ring gland," for it is this organ which controls, by hormonal action, molting, metamorphosis, organ growth, and apparently also general metabolic activities in Drosophila (Hadorn and Neel, 1938; Bodenstein, 1943a and b, 1944; Vogt, 1942 to 1943). The ring gland is located dorsocephally between the two brain hemispheres of the larva (Fig. 23A). Its lateral extremities encircle the aorta like a ring, and for this reason the structure has been given the name ring gland. The ring gland is easily found in dissection by the characteristic arrangement of the tracheae which supply it. Two larger tracheae, which arise from the hemispheres, border the ring gland laterally and pass on forward; these lateral tracheae are connected by a smaller transverse trachea which passes through the ring gland itself (Figs. 23A and B). The ring gland is already present when the larva hatches from the egg. At this time the ring gland is exceedingly small, but it grows as development proceeds. Growth apparently takes place by an increase in cell size rather than by cell multiplication.

The ring gland is a compound gland. In the ring gland of mature larvae three types of cells may be easily distinguished (Figs. 23B–D). Most of the wider anterior gland portion and both lateral extremities of the gland are made up of very large cells with large nuclei containing polytene chromosomes (Fig. 23, *lc*). A group of smaller cells is located in the anterior median portion of the gland (Fig. 23, *ca*). The proximal end of the lateral extremities of the gland and their ventral connection are made up of very small cells (Fig. 23B, *cc*). This region of the gland is closely connected with the aorta wall and the hypocerebral ganglion. Finally there are the tracheal cells. The epithelium of the tracheae, especially that of the lateral tracheal trunks, is extremely thick and consists of very small cells (Fig. 23, *tc*), which apparently represent the primordial cells for the imaginal trachea.

Now the ring gland, like many other larval structures, undergoes metamorphosis. During pupal development it gradually moves posteriorly, away from its anterior brain position, and becomes finally situated in front of the proventriculus in the anterior region of the thorax. While this movement takes place, the morphology and histology of the larval ring gland change drastically, as it is made over into the imaginal ring gland complex. The large ring gland cells which made up the bulk of the larval ring gland degenerate gradually, while the small cell group located anteromedially in the larval gland becomes what is known as the corpus allatum in the imago. The small cells in the posterior extremities of the larval gland apparently form all or part of an elongated cell structure which represents the hypocerebral ganglion and the corpus cardiacum of the imago. Figure 23E shows a total mount of the metamorphosed imaginal ring gland complex, and Fig. 23F shows the

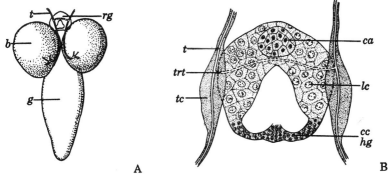

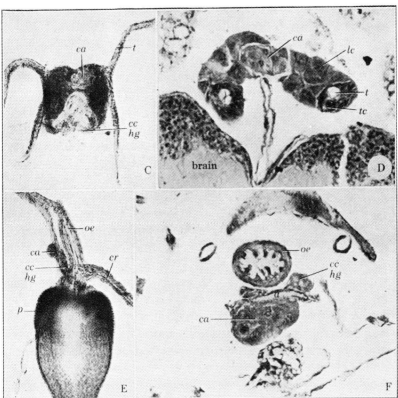

FIG. 23. Metamorphosis of the ring gland complex in *Drosophila virilis*.
A: Larval central nervous system, showing the position of the ring gland.
B: The ring gland of the mature larva, semidiagrammatically shown. C:
The ring gland of a mature larva, total mount. D: Cross-section through the
ring gland of a mature larva. E: The metamorphosed ring gland complex
in the imago, total mount. F: Cross-section through the imaginal ring
gland complex. *a,* aorta; *b,* brain; *ca,* corpus allatum; *cc,* corpus cardiacum;
cr, stalk of crop; *g,* ventral ganglion; *hg,* hypocerebral ganglion; *lc,* large
larval ring gland cells; *oe,* oesophagus; *p,* proventriculus; *rg,* ring gland;
t, trachea; *tc,* tracheal cells; *trt,* transverse trachea. B–F after Bodenstein
(1947).

same in cross-section. Anteriorly the small round cell body is the corpus allatum (*ca*). Slightly posterior to it is a larger elongated compound structure comprised of ganglionlike cells of two types. The structure apparently is the fused corpus cardiacum and the hypocerebral ganglion (*cc, hg*).

The large ring gland cells which degenerate during metamorphosis produce the hormone for molting and metamorphosis, while the imaginal corpus allatum releases a different hormone which regulates egg maturation (Hadorn and Neel, 1938; Bodenstein, 1942, 1943a and b, 1944, 1947; Vogt, 1942c).

THE LARVAL SENSE ORGANS

Hertweck (1931) has described the larval sense organs of Drosophila. The body wall of the larva is provided with a large number of very small sense organs. The external cuticular part of these organs (Figs. 24A–E) may take the form of hairs, bristles of various shapes, little cones fastened to a round platelet, or cones projecting from a minute chitinous funnel. The cones of the latter

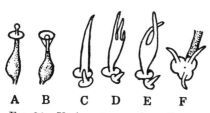

A   B   C   D   E   F

Fig. 24. Various types of small sense organs in the body wall of the larva. After Hertweck (1931).

type may be drawn into the funnel when stimulated. These sensilla are located in the thoracic and abdominal segments, where they are arranged in rows along the flank in a typical pattern characteristic for each segment. They function apparently as tactile organs or as chemoreceptors.

Ventrally in the thoracic segments, where the ligaments of the leg discs attach to the hypoderm, one other type of sense organ, consisting of three small sense hairs fastened to a rounded basal plate, is found (Fig. 24F).

Seven larger sense organs marked externally by wartlike cuticular protuberances are located on each side of the abdominal segment (Fig. 25). Structurally these sense organs are not all alike; three types may be distinguished (Figs. 25A–C). Type A is a cone-shaped stump carrying three rows of bristles. The distal portion of the sense cell protrudes as a small clublike process from the apex of the cone. The sense organs A_1–A_4 in Fig. 25 are of this type. Type B is a somewhat flatter cone which has at its end three slightly curved bristles. This type is represented in the sense organs B_1 and B_2 in Fig. 25. Type C is a smooth cone carrying at its tip two or three setae set into a

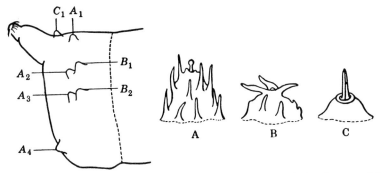

FIG. 25. Sense cones in the last abdominal segment of the larva. After Hertweck (1931). (For explanation, see text.)

circular chitinous ring. The sense organ located dorsally on the stigma process (Fig. 25C) belongs to this group.

In the chitinous intima of the pharynx are grouped together four small sense organs carrying slightly bent chitinous rods and two larger chitinous sense cones (Fig. 26) which are apparently chemoreceptors. One pair of sense organs, supposedly tangoreceptive in function, is located laterally on each side of the head segment; it is called the antenna-maxillary sense complex and consists externally of two chitinous plates covered with small papillalike protrusions. The antenna and maxillary nerves which innervate these organs form a slender ganglion just below them (Hertweck, 1931).

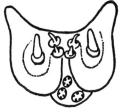

FIG. 26. Sense organ in the larval pharynx. After Hertweck (1931).

The scolopophorus organs (chordotonal organs) in Drosophila are of the cordlike type. They are small compound sense organs placed between the body wall and the muscle layers and are attached at both ends to the hypoderm (Fig. 27). Their function is primarily to control muscle tension. Each scolopophorus sensillum is made up of the following elements (Fig. 27). A basal sense cell (*bsc*) with an elongated distal process carries at its apex a small sense rod or scolops (*sco*). The axial fiber of the sense cell can be followed into the rod, where it ends in a knot. The elongated distal projection of the sense cell is embraced by an enveloping cell (*ec*). A long cap cell (*cc*) is attached at one end to the hypoderm, while the other end reaches the sense rod. The rod is connected to the hypoderm by a terminal fiber (*tf*). A ligament cell (*l*) attaches the basic part of the organ to the hypoderm.

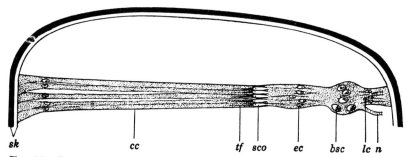

Fig. 27. Larval scolopophorus organs. *bsc*, basal sense cell; *cc*, cap cell; *ec*, enveloping cell; *lc*, ligament cell; *n*, nerve; *sco*, scolops; *sk*, body wall; *tf*, terminal fiber. After Hertweck (1931).

Now these scolopophorus organs consist either of a single sensillum of the type discussed above or of 3 or 5 combined sensilla (Fig. 27). They are present in all the larval segments, where they are arranged in a definite pattern (see Hertweck, 1931). Altogether there are 90 scolopophorus organs; 64 of them are made up of 3, and 14 of 5, sensilla each. Whether their number or distribution is the same in all the larval instars is not known. During pupation these larval scolopophorus organs are apparently histolyzed. Robertson (1936) found them in 12-hour-old pupae in a state of degeneration and could not detect them in later pupal stages. The scolopophorus organs of the imago appear relatively late in pupal development.

THE TRACHEAL SYSTEM AND ITS METAMORPHOSIS

The tracheal tubes originate as invaginations of the ectoderm, and their own wall thus contains the same structure as the body wall, only in reverse order. The outer epithelium of the spiracle continues with the epidermis and possesses a basement membrane. The inner wall of the trachea is the cuticular intima, which is very characteristic; it forms minute spiral ridges, the taeniola. Now the most prominent features of the larval tracheal

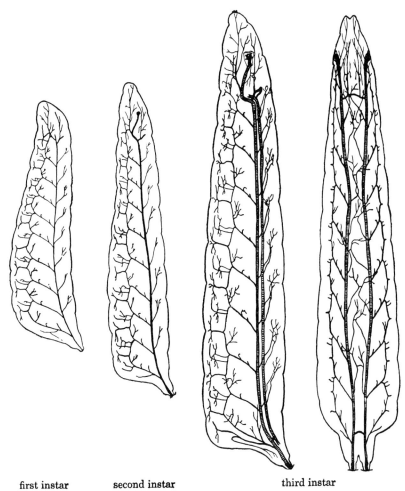

first instar   second instar   third instar

FIG. 28. The tracheal system of the larva in the first, second, and third instar. After Rühle (1932).

system are the two great lateral trunks (Figs. 1 and 28), which connect at each end with the anterior and posterior spiracles. A tracheal crossbranch connects the two lateral trunks anteriorly in the third thoracic segment and posteriorly in the eighth abdominal segment. Laterally each main trunk gives off, in the third thoracic segment and in all the abdominal segments, a branch which passes forward diagonally and ventrally. From each of these branches a visceral branch passes into the body. The lateral segmental tracheae also give off finer ramifications, the pattern of which is quite constant in each segment. The tracheal pattern in the head and in the two anterior thoracic segments is less regular. Figure 28 illustrates the tracheal system of the three larval instars; it can be seen that the tracheal pattern is quite similar in the three instars, but the branching becomes progressively more developed in the older instars.

The spiracle apertures open into a large chamber known as the atrium or atrial chamber (Figs. 5D–E), which connects with the dorsal tracheal trunks. At each molt except the last the entire spiracle is formed anew and takes the characteristic form typical for the ensuing instar (Figs. 4 and 5). The old atrial chamber serves for the discharge of the tracheal intima and is closed off when the new chamber becomes the functional new breathing orifice. The site of the earlier spiracle is marked by a scar on the surface of the integument (Fig. 5, sc) and remains connected with the base of the new atrial cavity by a strand of cuticular tissue (Fig. 5, c st). This rudiment of the preceding spiracle is present in the last instar in the anterior spiracle and in the second and third instar in the posterior spiracle. The anterior spiracles of the first two instars are not functional, whereas the posterior spiracles function throughout larval life. The anterior spiracles open to the outside only in the third instar by seven to nine fingerlike papillae which are open at their ends. Each of the posterior spiracles has two spiracular openings in the first-instar larvae and three in the second- and third-instar larvae. Details concerning the larval tracheal system in Drosophila can be found in the paper by Rühle (1932).

The following description of the metamorphosis of the tracheal system is based on the work of Robertson (1936). The metamorphosis of the tracheal system involves extensive changes, since the imaginal tracheal system is quite different from that of the larval organism. In the mature larva the abdominal tracheae of the imago begin to form. They originate from imaginal cell clusters located in each segment on the abdominal longitudinal tracheal trunk. These imaginal cells begin to grow out toward the hypodermis. At puparium formation the larva everts its anterior and posterior spiracles, and 1 hour later the abdominal tracheal cells have established connection with the abdominal hypoderm, forming a solid cord. Two hours after puparium formation this cord has become hollow, and chitin is secreted into the lumen. The larva now begins to contract away from the anterior end of the pupal case in preparation for the prepupal molt, thus partly withdrawing the anterior spiracle. At the prepupal molt, that is, 4 hours after puparium formation, the posterior spiracles are also partly withdrawn, and a new intima is secreted. The newly formed abdominal tracheae have now a continuous lumen which runs from the tracheal trunks to a pore in the prepupal membrane. The stigmata of these tracheae are apparently, however, not functional at this stage of development. Shortly before pupation the new anterior spiracles, later the prothoracic spiracles, which have developed just medially to the larval spiracles, become connected with the anterior longitudinal trunks of the larva and are filled with air just before the anterior spiracles are shed at pupation, that is, 12 hours after puparium formation. A continuous air supply is thus guaranteed during the entire process of metamorphosis. The posterior tracheal linings, as well as the intima of the newly formed abdominal tracheae, are also completely withdrawn at pupation. The tracheal system already formed in the head from the imaginal tracheal cells is everted together with the head and rapidly fills with air. The pupal spiracles now open into the space between the pupal cuticle and the prepupal membrane. The removal of the tracheal linings is accomplished by the contraction of the prepupa during pupation and by body movements of the young pupa. A gas bubble formed at the time of pupation is also helpful in this process.

In pupae 15 hours of age the tracheal cells in the abdominal region are still larval in character. By 18 hours, however, the intima of the abdominal lateral trunks, formed after the larval intima was shed at the prepupal molt, breaks near the posterior end. The caudal fragment soon degenerates, finally forming a small knot on the pupal cuticle. The intima of the posterior tracheal trunks anterior to the breaking point loosens from the epithelium and is shed into the lumen. After being shed, the intima shrinks, and, being connected with the intima of the anterior trunks, the posterior shrunken intima is pulled into the anterior trunk region. Now the posterior trunks are histolyzed.

The anterior trunks undergo marked changes, especially in the thoracic region, where they form the large thoracic air sacs. Posteriorly from the air sacs the trunks become gradually thinner and disappear in the region of the first and second abdominal segments, where they supply the abdominal spiracles. The spiracles of the other abdominal segments are apparently supplied by finer tracheal branches, which connect with the larger trunks in the two anterior abdominal segments. The closing device of the adult spiracles is first visible in pupae 72 hours of age; the intima of the anterior tracheal trunks, together with the remaining intima of the posterior trunks, which was pulled into the anterior tracheal region, as already described, is shed through the prothoracic spiracles at emergence of the fly.

THE METAMORPHOSIS OF THE MUSCLES

The histolysis of the larval muscles and the formation of the new imaginal muscles take place during pupal development. Although all the larval muscles are apparently histolyzed, it has been found that the time of their destruction varies in the different regions of the body and for the different kinds of muscles involved, for certain larval muscles may be used relatively late in pupal development to carry out definite functions at the time when other larval muscles are already in an advanced state of degeneration. The velocity with which the various imaginal muscles develop and differentiate also differs greatly. Thus, especially in the earlier stages of pupal development, one may find larval muscles in all stages of degeneration together with quite well-formed imaginal muscles.

The metamorphosis of the muscles in Drosophila, as described in the following account, has been studied by Robertson (1936). Further information on this subject may be found in papers by Evans (1936), Perez (1910), van Rees (1888), and Kowalevsky (1887).

The Histolysis of the Larval Muscles

Histologically the first sign of muscle degeneration is a liquefaction of the peripheral part of the muscle fiber. An advanced state of this process results in a slight swelling of the muscle bundles. As the degeneration gradually proceeds, the interior of the muscles is broken up by a separation of the fibrillae, which gives the muscle at this stage a vacuolated appearance in cross-section. Subsequently the muscles separate from their attachment and fragment. The muscle fragments known as sarcolytes are gradually consumed by phagocytes.

In the young prepupa the muscles of the head and thoracic segments already show definite signs of histolysis (partial liquefaction). However, the time of complete muscle destruction in this region varies for the different kinds of muscles. With the exception of the dilator muscles of the pharynx and three pairs of dorsal thoracic muscles, all muscles in the four anterior segments are fragmented 7 hours after puparium formation. Scattered sarcolytes of these muscles can still be found 3–4 hours later. By this time the three pairs of dorsal thoracic muscles, still showing only slight signs of degeneration, become surrounded by the myocytes of the imaginal longitudinal thoracic muscles. The number of these myocytes increases as development proceeds, while the enveloped three pairs of larval muscles break up into six-paired muscle bundles, which gradually degenerate and are dissolved 6 hours after pupation. The dilator muscles of the pharynx remain unchanged until after pupation. They apparently function in the process of the head evagination at pupation, after which they degenerate, but rather slowly. In 3-hour-old pupae they are still present but have completely disappeared 6 hours after pupation, that is, 18 hours after puparium formation.

The abdominal muscles degenerate later than most of the anterior muscles. Although slight liquefaction has already begun in these muscles at pupation, they are nevertheless functional, at this time being active in moving the abdomen of the animal in the puparium. Within 6 hours after pupation, starting in the last abdominal segment, most of the abdominal muscles have degenerated. However, one pair of muscles in each of the six abdominal segments becomes histolyzed much later in development and is still present in 42-hour-old pupae. These long-persisting larval muscles Robertson (1936) believes are helpful in establishing the segmentation of the pupal abdomen. They apparently cause each segment to telescope into the preceding one by contraction. Soon after the abdominal segmentation has become visible (at 42 hours), these muscles degenerate and are phagocytized.

The Formation of Imaginal Muscles

The longitudinal thoracic flight muscles of the imago develop from a group of presumptive muscle cells which appear in the thoracic region shortly after the formation of the puparium. These cells, termed "myocytes," envelop the persisting three pairs of larval muscles. When the larval muscles have degenerated in the 18-hour-old pupa, the myocytes have arranged themselves in the future position of the imaginal longitudinal muscles. Differentiation of these muscle primordia begins 21 hours after puparium formation and proceeds from the center of the primordial cell complex both anteriorly and posteriorly. In pupae 27 hours of age, these muscles have attained one-third, and in 42-hour-old pupae about two-thirds, of their final length. By the fiftieth hour cross-striation is first visible. Eighty-four hours after puparium formation, differentiation is almost completed, and the muscles are attached to the epidermis by short tendons.

Small spindle-shaped myocytes originating in the head region form the dilator muscles of the imaginal pharynx. The muscles of the imaginal abdomen develop from the histoblasts which form the hypoderm of the abdomen. Each histoblast in 24-hour-old pupae gives off on its inner surface small spindle-shaped cells, which in the course of development differentiate into the imaginal abdominal muscles. In 60-hour-old pupae these muscles are well formed and show cross-striation about 6 hours later.

THE ALIMENTARY CANAL AND ITS METAMORPHOSIS

Throughout the larval period the morphology of the alimentary tract is essentially the same. This generalization will not hold, however, for the mouth hooks and the chitinous pharyngeal armature, for it has been noted that changes occur in the armature during the progress of the larval stages (Fig. 4). The cells comprising the intestinal tract, including the oesophagus, pharynx, proventriculus, midintestine, and hindintestine, as well as those of associated structures, such as the Malpighian tubes and salivary glands, grow throughout the larval stages only through increases in cell size and not through cell multiplication. The cells increase markedly in volume, as do their nuclei, but mitosis among these cells is not observed. This is a continuation of the process already observed in the embryo (Sonnenblick, see Chapter 2 on embryology). As soon as the component parts of the alimentary tract and of the associated parts are laid down, mitotic activity diminishes and then is no longer noticeable. However, during this period of mitotic quiescence and of differentiation in the later embryonic stages and throughout the larval instars, endomitosis and subsequent formation of large nuclei containing polytene chromosomal complexes occur. Polytene chromosomes have been reported for the nuclei of the cells of the alimentary tract of Drosophila by Makino (1938) and Cooper (1938). In the prepupa and pupa the cells of the alimentary canal and the related salivary glands undergo histolysis and are replaced by the proliferation of certain imaginal rings of tissue, to which reference will be made below. The Malpighian tubes are an exception, for despite their relationship with the hindintestine at its anterior end (or the caudal end of the midintestine) and the polytene nature of the chromosomes in the cells of the tubes, the tubules are not destroyed at metamorphosis but persist, with slight changes, from embryo to imago.

Figure 29A represents semidiagrammatically the arrangement and general appearance of the intestinal tract in the mature larva. The photographs in Figs. 29B–I are sections through various parts of the organ system and also of whole mounts of certain structures. In the drawing, the foreintestine has been left clear, the long, coiled midintestine has been stippled, and the hindintestine has been darkened. Although the proventriculus has been left unmarked, it may be stated that the layers of that structure belong to both fore- and midintestine.

The mouth of the larva opens into the muscular pharynx which proceeds posteriorly as the narrow oesophagus. In and about the walls of the pharynx an elaborate skeleton and musculature are developed. Much, but not all, of the chitinous pharyngeal skeleton in which is inserted the sets of muscles which protract and dilate the pharynx and support the mouth hooks can be seen through the body wall of the anterior segments. A section through the powerful muscles of the dorsal pharyngeal wall is

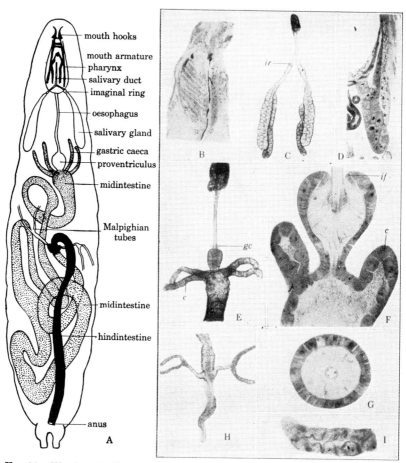

Fig. 29. The larval alimentary canal. A: The arrangement and general appearance of the intestinal tract of the mature larva, semidiagrammatically shown. B: Section through pharyngeal musculature of a mature larva. C: Whole mount of the larval salivary glands. D: Section through a larval salivary gland. E: Whole mount of anterior region of alimentary canal, showing oesophagus, proventriculus, and stomach, with gastric caeca. F: longitudinal section through proventriculus and anterior stomach in the region of the gastric caeca. G: Cross-section through proventriculus. H: Whole mount of intestine at point of junction between mid- and hindgut, showing Malpighian tubes opening into gut. I: Longitudinal section through larval Malpighian tube. c, gastric caeca; gc, wreath cells; if, imaginal ring of foregut; ir, imaginal ring of salivary gland.

shown in Fig. 29B. The nuclei are, as shown in Chapter 2 by Sonnenblick, scattered throughout the muscle bars, which thus have the appearance of a syncytium. Transverse sections through the pharynx, showing the dorsal muscles and the riblike folds of the ventral wall, are seen in Fig. 3A. Detailed studies of the larval pharyngeal skeleton structure and musculature have been made in Drosophila by M. Strasburger (1932) and in the apple maggot, Rhagoletis, by Snodgrass (1924).

The common duct of the paired tubular salivary glands, which lie on either side of the oesophagus, enters the floor of the pharynx. A whole mount of the paired glands removed from a mature larva, together with some of the neighboring fat body, is presented in Fig. 29C. The fat body cells are normally in fairly close association with the glands, often actually in contact with and bordering them. The union of the ducts to form the common duct and the entrance of the latter into the muscular pharynx are clearly indicated. Figure 29D represents a section through a salivary gland. The smaller cells of the duct and the stellate nuclei of the vacuolated fat body cells which are apposed to the gland are seen in this photograph. A more highly magnified view of a section of fat body tissue can be seen in Fig. 2.

The pharynx continues as the narrow oesophagus, a simple straight tube which passes between the brain hemispheres into the proventriculus. In front of the proventriculus is located a ring of small cells which encircles the oesophagus (Fig. 29E, gc). These cells were first described by Weissman (1864) for Musca and were named "Guirlandenzellen" (wreath cells). Their function is not known, but they have apparently no connection with the nervous system.

Figure 29E is a whole mount showing the oesophagus as a straight passageway from the hind portion of the pharynx to the bulblike proventriculus. In a longitudinal section through this region (Fig. 29F) the oesophagus is seen to be prolonged into the midintestine as an inner tube which is then reflected upon itself to join the midintestine. The proventriculus has a characteristic appearance and in transverse section (Fig. 29G), can be observed to possess an inner layer with its chitinous intima and then a region of long, clear cells which stain very lightly with haematoxylin and whose cell borders are poorly defined, followed by an outer layer of large cells which absorb the stain very well and which are identical in appearance with the cells of the "stomach" (ventriculus) and of the gastric caeca. The two inner layers can be considered part of the foreintestine, while the outer layer belongs to and is an extension of the midintestine. Where the inner and outer walls meet is a ring of imaginal cells which will later, together with the labial buds, give rise to the foregut epithelium of the adult (Fig. 29F, *if*). The imaginal ring of the foregut, as well as other parts of the tract, was described many years ago in the muscids by Kowalevsky (1887) and more recently in Drosophila by M. Strasburger (1932) and Robertson (1936).

The enlarged anterior portion of the midintestine which immediately follows the proventriculus is the so-called "stomach." From the anterior end of this region four diverticula develop; these outgrowths are blind sacs or gastric caeca (Fig. 29E). Two of the caeca are disposed dorsally, while the other two caeca push out ventrally, but all are oriented cephally. In morphology and staining capacity, the cells of the caeca are identical with those of the stomach (Fig. 29F). As indicated in the drawing, the caudal end of the stomach is looped and is then continuous with the long midintestine, which is laid in a series of loops and coils through most, and probably all, of the abdominal segments. The midgut is the longest single portion of the alimentary tract and typically has four loops (Fig. 29A) before junction is made with the hindintestine The histological picture varies in different regions of the midintestine, as indicated by M. Strasburger (1932) and E. H. Strasburger (1935). According to the former, the epithelial cells in different regions of the midgut perform the functions of secretion and of reabsorbtion. The secretory function is assigned to the cells of the stomach and central part of the midgut, whereas the reabsorbtive function is carried out by the cells of the caudal portion of the midintestine.

The Malpighian tubules, which are the principal excretory organs, are slender, lengthy, blind tubes which open at their proximal ends near the point of junction of the mid- and hindintestine (Fig. 29A). Shortly after they arise, the two basal outgrowths bifurcate, giving rise to four tubules, one pair extending anteriorly to about the region of the proventriculus and the other passing caudally to the last abdominal segment. Figure 29H is a whole mount of a portion of the gut, showing the site of origin of the tubules and the branching of the basal outgrowths. In the forms which we have studied, *D. melanogaster* and *D. virilis*, both pairs of tubules seem to be free at their distal extremities. In *D. funebris*, however, Eastham (1925) reports that the posterior pair of tubes form a closed loop, with the lumen of the tubes continuous. In wild-type flies the tubules appear bright yellow, whereas those in certain mutant strains of *D. melanogaster* are pale yellow or colorless (Beadle, 1937; Brehme and Demerec, 1942). The cells of the tubes are large, and the lumina are quite narrow and wavy in appearance (Fig. 29I). The inner wall of the epithelial cells bordering the lumen of the tubule, like the inner wall of the midgut epithelium, has a striated border, called a "brush border," composed of numerous delicate filaments. A good discussion of the histophysiology of the Malpighian tubes can be found in the text of Wigglesworth (1939).

The Malpighian tubes arise in the general region where the mid- and hindintestine meet (Fig. 29A), but it is still a question whether the tubes are outgrowths from one or the other portions of the tract. As Wigglesworth (1939) has indicated, the tubes open in some insects directly into the midintestine, whereas in others they clearly discharge into the hindintestine. In other forms, however, they open into a "neutral zone" at the site where the mid- and hindgut unite. In the Drosophila embryo (Poulson, 1937) they arise initially from the blind or inner end of the proctodaeum. However, M. Strasburger (1932) and Robertson (1936) cite some evidence for their belief that the tubules belong to the midintestine. From whole mounts alone it is obvious that this question can be decided only with difficulty, and no prolonged debate on the matter will be attempted. The point can best be decided by ascertaining the embryonic origin of the tubes, but even here there is a peculiarity, for although the tubules arise originally from the proctodaeum they seem, after the formation of the midgut, to open into the latter part of the midintestine (Sonnenblick, Chapter 2). It may very well be that the cells at the ends of the proctodaeum and of the stomadaeal ribbons which meet and form the junction of mid- and hindgut become definitive midgut cells. This would mean that the tubes could arise early as diverticula from the proctodaeum near its blind end, but that subsequent migration and differentiation result in this area later developing into the caudal part of the midintestine; this is the area, according to M. Strasburger (1932) and Robertson (1936), into which the tubes enter. The position taken here serves at any rate to bring together the observations of the investigators mentioned above. In the drawing (Fig. 29A) it has been indicated that the tubes enter the anterior region of the hindgut. With the above points in mind, it may be that the drawing is inaccurate in this respect.

From approximately the place where the Malpighian tubes enter the alimentary canal, the hindgut begins and continues as a more or less straight tube to the anal orifice in the last segment. The width of the hindintestine is less than that of the midintestine (Fig. 29A), although its muscular layer, particularly in the posterior region, is well developed.

During metamorphosis the larval alimentary tract and also the salivary glands, but not the Malpighian tubules, degenerate, and the adult structures develop from specific formative centers, either imaginal rings or islets of imaginal cells. The foreintestine is reconstituted by a ring of imaginal cells at the region where the fore- and midintestine meet (Fig. 29F), as well as by proliferation of the labial buds (Fig. 6), from which the pharyngeal epithelium is derived. The processes of histolysis occur predominantly in the prepupal stage, while imaginal bud differentiation is especially prominent during the pupal period. However, the degenerative processes as well as those of reconstruction occur simultaneously as a continuous and balanced series of events; these have been described by Robertson (1936). A new structure, the crop, appears in the early pupa as a dilation from the caudal end of the oesophagus. In the midpupal stage the crop is seen to be connected by a long narrow stalk with the oesophagus immediately in front of the imaginal proventriculus, as evident in Fig. 30C, which is a photograph of a part of a late pupal alimentary tract. The crop is thin-walled and functions primarily as a food reservoir. The larval salivary glands degenerate, and the adult glands are reformed anew from imaginal rings located at the anterior ends of the glands. This is described on pp. 348–350.

The midintestine of the mature larva has an inner layer of large epithelial cells bordering on an external basement membrane. Between the epithelial cells just under the basal membrane are scattered islets of small imaginal cells which give rise to the midgut epithelium of the adult fly. In a larva ready for pupation these islets or replacement cells have already begun to proliferate. Figure 30A is a section of the midintestine of a 1-hour-old prepupa. The large larval epithelial cells bordering the lumen of the canal, the basement membrane, and the darkly staining compact groups of replacement cells are seen. As the prepupal period progresses, the replacement cells proliferate and

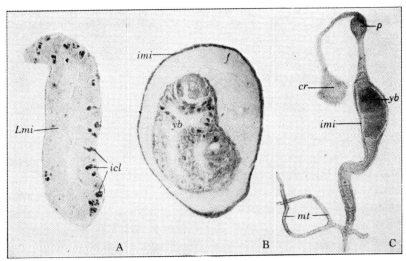

Fɪɢ. 30.   A: Longitudinal section through midintestine of young prepupa, showing replacement cells. B: Cross-section through midintestine of older pupa. The replacement cells have already formed the imaginal gut, while the larval gut forms the yellow body. C: whole mount of almost completely metamorphosed intestinal tract. *cr*, crop; *icl*, imaginal replacement cells; *imi*, imaginal midintestine; *mt*, Malpighian tubes; *yb*, yellow body; *Lmi*, larval midintestine; *p*, proventriculus.

extend beneath the membrane until they unite to form a continuous new cellular layer, the epithelium of the imaginal midintestine. Concurrently, the epithelium of the larval midgut and of the four gastric caecae (which do not reappear in the adult) is sloughed off and cast into the enteron, where it forms a distinct mass greenish yellow in color. This shed mass is usually referred to as the yellow body. The reconstituted epithelium, derived from the replacement cells, and the yellow body are seen in section of an organism in the midpupal period in Fig. 30B. The shed mass persists throughout pupal development (see the whole mount, Fig. 30C) and is discharged from the alimentary tract as the meconium soon after emergence of the imago.

A ring of imaginal tissue just caudal to the entrance of the Malpighian tubes renews the anterior part of the histolyzed larval hindgut epithelium. Kowalevsky (1887) has also described for the muscids, as has Robertson for Drosophila, a posterior abdominal imaginal disc around the anal opening, which through proliferation of its cells gives rise to the posterior part of the hindintestine. Proliferation and eventual union of the cells of the forward and back imaginal rings renew the degenerating larval epithelium.

THE DEVELOPMENT OF THE SALIVARY GLANDS

The salivary glands are accessory organs of digestion. They secrete saliva, which is mixed with the food as it is taken in, and apparently aid in moistening and digesting it. The salivary glands are paired organs. They develop as invaginations on either side of the anterior ectoderm (Sonnenblick, 1940). When the larva hatches from the egg, the salivary glands are already well developed and seemingly functional. They consist at this early stage of a pair of elongated masses of cells (Fig. 31, stage 1), located on each side of the body just below the muscles of the body wall, and run from a position slightly anterior to the brain to the third thoracic segment. Parts of the anterior larval fat body are closely connected with the gland. Each gland has a long strip of fat body attached to its outer border (Fig. 29C). This attachment is maintained throughout larval life. It seems, however, that in older larval stages the fat body is somewhat more loosely attached to the gland than in younger stages. According to Makino (1938), the number of cells for each gland in *D. virilis* at the time of hatching is about 115 and remains unchanged during larval life. The growth of the larval salivary gland is thus caused by an increase in cell size rather than by cell multiplication. The cell number of the *D. melanogaster* salivary gland has not been counted but is probably about the same. The individual cells of the gland are of the same size, but their shape is somewhat irregular. The nuclei are round, and their detailed structure is difficult to observe in young first-instar larvae. In cross-section it can be seen that about six to eight cells surround a rather small lumen which passes through the entire gland. Each gland opens anteriorly into a duct. These ducts unite medially into a common duct which leads into the floor of the pharynx. The cells which make up the walls of the ducts are much smaller than are those of the gland. The ducts are lined with a chitinous intima which forms a chitinous spiral thread or taenidium. The intima is shed at each molt.

By the end of the first instar the salivary glands have grown considerably (Figs. 31, stage 2, and 32, stage 2′). The nuclei have also increased in size. The chromatin can be seen forming a fine network around the periphery of a clear area in the nucleus, while the chromocenter stains deeply and can easily be recognized in all cells. The lumen of the gland has become somewhat wider.

During the second instar the glands continue to grow in length and width (Figs. 31, stages 3 and 4, and 32, stages 3′ and 4′). They now extend into the first abdominal segment. The cells are still uniform in size throughout the gland; except for their increase in size, the nuclei have not changed noticeably. In larvae about 30 hours of age a number of small cells appear at the border between the glands and the ducts. These cells are exceedingly small, stain deeply, and represent the Anlage cells of the imaginal salivary gland (Fig. 31, stage 4, broken arrow).

At about the second molt the cells and their nuclei in the posterior portion of the gland have become somewhat larger than those in the anterior gland region. Because of this the diameter of the gland is narrower anteriorly than posteriorly, which gives the gland its characteristic shape. In the nuclei, especially in those of the posterior gland portion, individual chromosome strands can be distinguished which show alternating deeply and lightly stained bands. The difference in the cell and nuclear size between the anterior and posterior gland portions becomes more and more pronounced as development proceeds. By 55 hours the chromosome strands are relatively wide, and the chromosome bands are more distinct. The deeply staining chromocenter is at this time still visible in all cells of the gland. The imaginal Anlage of the adult gland has increased its cell number and forms a characteristic ringlike structure around the salivary duct. In about 75-hour-old glands the chromosome strands in the nuclei of the posterior cells are so broad that it is difficult to distinguish them from the chromocenter in total mounts. As a consequence the chromocenter has become invisible in these large cells but is

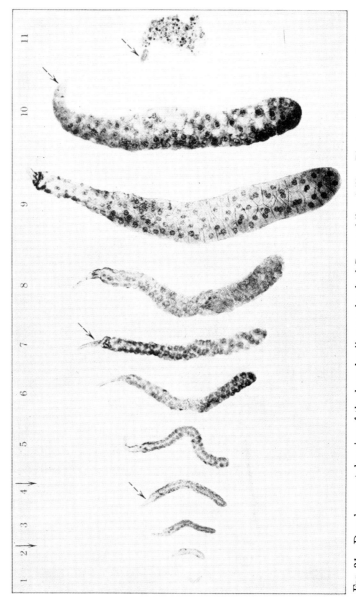

FIG. 31. Developmental series of the larval salivary gland of *Drosophila virilis*. Photomicrographs of eleven successive stages. Full arrows indicate the time of molting; broken arrows indicate position of imaginal salivary gland Anlage. After Bodenstein (1943).

still visible in the anterior half of the gland. A few hours later (about 80 hours) cell growth in the more anterior gland regions has increased so much that the chromocenter is now visible only

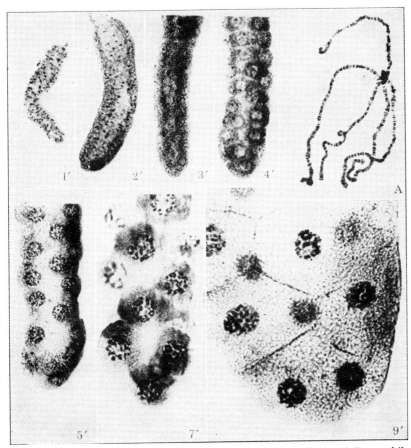

FIG. 32. Developmental series of the larval salivary gland of *Drosophila virilis*. The distal portion of the same glands as shown in Fig. 31, at a higher magnification. A: Smear preparation of a *virilis* salivary gland nucleus, stained with orcein. Note characteristic banding of chromosome strands and the nucleolus in center of picture, which stains but slightly with orcein. After Bodenstein (1943).

in the anterior third of the gland. The gradual disappearance of the chromocenter in a posterior-anterior direction is thus a good indication of the age of the gland.

Although up to this time cell growth took place by an increase in size of the nucleus and the cytoplasm simultaneously throughout the gland, a change now occurs in the cells in the posterior gland portion. The nuclei in these large distal cells either cease to grow or grow very little, whereas the cytoplasmic portion of the cell increases immensely as development proceeds. A comparison of Fig. 32, stage 7', with Fig. 32, stage 9', illustrates this point very clearly. The significance of this fact is not known. Ross (1939) observes that at about this stage secretion granules appear in the gland, which increase in number as the larva approaches the non-feeding stage before pupation. He believes that the posterior cells represent the secretory region of the gland, while the anterior cells function only as a conductive region. How valid is this distinction between the two gland portions is rather difficult to decide. Though it is certainly true that from the middle of the third instar onward the cells in the anterior region appear different from the cells in the posterior gland portion, yet it is equally true that the anterior cells grow more slowly than the posterior cells and might thus be considered younger. As it is, one finds indeed that the posterior cells of younger stages resemble the anterior cells of older stages. Moreover, the cells in the anterior gland portion of young prepupae very much resemble the posterior secretory cells of the mature larval stage at the time when the cells in the posterior portion already show clear signs of metamorphosis. Furthermore, it seems rather unlikely that the salivary gland should secrete so much more actively at the end of larval life when the larvae begin to feed less. A more plausible explanation seems to be that the apparent changes in the proximal cells are the first signs of metamorphosis. This assumption is also supported by experimental evidence (Bodenstein, 1942). Now there is one further point; as far as the cell size is concerned, there is never any sharp separation of anterior and posterior cells. The cells increase gradually in size in an anterior-posterior direction. There is a definite anterior-posterior growth gradient, as shown by the growth rate of the cells in the different parts of the gland.

In mature larvae 96 hours of age the salivary glands have reached their maximal size and extend as far backward in the body as the second abdominal segment. A fine basement membrane envelops the gland. In living condition the cytoplasm of the cells has become slightly opaque, while in previous stages it was transparent (Bodenstein, 1943a). This opaque condition is the first clear morphological sign of the beginning of metamorphosis of the gland. The cells in the posterior part of the gland become opaque usually somewhat earlier than the anterior cells; when the whole gland has become opaque, this condition is slightly more pronounced in the proximal cells. The cytoplasm stains less heavily with orcein than in previous stages, while the nuclei take this stain exceedingly well. It is characteristic of orcein-stained whole mounts of this stage that the nuclei stand out clearly against a lightly stained background (Fig. 31, stage 9; Fig. 32, stage 9'). The chromocenter is visible only in the anterior one-fifth of the gland, where the cells are still the smallest. The chromosome strands have reached their maximal width and show their characteristic banding perfectly. The chromosome strands stretch very well in smear preparations (Fig. 32A), in contrast to earlier or later stages. For this reason this stage is best suited for studies of the arrangement of the bands in the chromosome strands.

About 5 hours after puparium formation, clear signs of histolysis may be observed in the glands. In various regions of the gland, usually first in the distal part, the cells begin to vacuolate, and the cell walls rupture. The cytoplasm stains very poorly with orcein in these regions of degeneration (Fig. 31, stage 10). In living condition the gland appears opaque, while milky white zones indicate the regions of advanced histolysis. The nuclei are still intact, but the chromosome strands are clumped in the center of the nucleus and are surrounded by a clear spherical area (Fig. 31, stage 10). Only a few cells in the extreme anterior end of the gland show the chromocenter. This indicates that the small anterior cells still grow in the early prepupa, at the time when the more proximal cells have already begun to histolyze. The ring of imaginal cells at this stage is very pronounced (Fig. 31, stage 10, broken arrow), while the salivary ducts have become much shortened. The two glands may be found to be at slightly different stages of histolysis at any given time. Ten hours after puparium formation, the histolysis of the gland is far advanced. The degenerating regions within the gland have extended and have become more numerous. The nuclei are pycnotic, and the nuclear membranes break down. The basement membrane surrounding the gland ruptures, and finally, probably with the help of phagocytosis, the larval gland is dissolved about 24 hours after puparium formation. The anterior part is the last to disappear (Fig. 31, stage 11).

The time data and developmental events given in this discussion refer to *D. melanogaster,* although they are illustrated in Figs. 31 and 32 by *D. virilis* preparations.

The differentiation of the adult salivary gland from the imaginal Anlagen begins about 10 hours after puparium formation. At this time the Anlagen begin to grow out anteriorly and posteriorly. The anterior outgrowths form the ducts of the adult salivary gland. The ducts of the paired glands meet in the neck region, where they fuse and continue as a common duct leading into the pharynx. In the meantime the posterior outgrowths which form the adult salivary glands grow as a straight tube backward through the thorax, until they finally reach the first abdominal segment where each tube ends in a single coil.

SOME TECHNICAL DIRECTIONS

The method of transplantation has in recent years become a most useful tool for the Drosophila experimentalist. This transplantation technique was worked out by Ephrussi and Beadle (1936) and consists of injecting tissues into the body cavity of larvae by means of a micropipette. For such experiments it is necessary to dissect and handle the larval and pupal tissues to be implanted in a suitable physiological saline solution. The following modified Ringer solution has been used successfully (Ephrussi and Beadle, 1936): H_2O, 1000 cc; NaCl, 7.5 gr; KCl, 0.35 gr; $CaCl_2$, 0.21 gr. Larval organs remain alive in this solution for 4 hours and even longer, apparently without any harmful effects on their developmental capacities.

For transplantation experiments the following equipment is needed: a binocular dissecting microscope and injecting apparatus, a narcotizing chamber, two pairs of watchmaker forceps, a pair of iridectomy scissors, two sharp steel dissecting needles, a camel's-hair brush, a small spatula for picking the larvae out of the food, a mouth pipette, two standard pipettes, several plain and depression slides, filter paper, micropipettes (injection needles) of various bores, an Arkansas grinding stone, a microburner, Drosophila Ringer solution and ether.

If larval organs are to be transplanted, the donor larvae must be carefully washed in Ringer solution before they are dissected. For the dissection of pupal organs, the older pupae have to be freed of their pupal cases. This is done by carefully slicing off the dorsal surface of the pupal case with a sharp razor blade. After the dorsal part of the puparium is thus removed, the pupa can be lifted easily with forceps from the rest of the case. The animals are dissected under a binocular dissecting microscope in a drop of saline placed on a slide or in a depression slide filled with Ringer solution. Some investigators prefer transmitted,

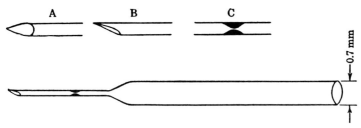

FIG. 33. Microinjection pipette. A and B: Two views of properly sharpened points. C: Part of pipette shaft with correct constriction.

others reflected, light for this manipulation. The dissection is made with sharp steel needles. A pair of fine iridectomy scissors has been found to be very helpful for the dissections. After the desired organs have been dissected out, they are transferred, preferably by a mouth pipette, into fresh Ringer solution, where they remain until transplanted.

For the transplantation, part of a Chambers microinjection apparatus is used, which consists of the following four main parts: a micropipette holder; a hypodermic syringe fitted into a holding clamp; a long piece of fine capillary brass tubing, which connects the syringe with the pipette holder; and a micropipette (Fig. 33) fastened to the pipette holder. The making of the micropipette is tricky and requires some patience. One starts with a fine glass or Pyrex capillary 0.7 mm in diameter and 0.1 mm in wall thickness. One end of this capillary is drawn out in a microburner in such a way that the newly drawn out shaft gives a smooth fine tube of equal diameter throughout its length. This can be achieved by heating a rather wide portion of the capillary and drawing it out after removal from the flame. The width of the shaft depends upon the size of the tissue to be transplanted.

The fine capillary shaft is then broken off so that its remaining length is about 1.5 cm. A fine constriction is now made near the proximal end of the shaft (Fig. 33) by holding that part of the capillary in a horizontal microflame. In so doing, one must be careful not to overheat this point, thus sealing the capillary. The constriction is very important; it serves to regulate the flow of liquid through the pipette and also blocks the piece of tissue sucked into the capillary, preventing it from being carried by the stream of liquid into the wider part of the pipette. The distance between the tip of the capillary and the constriction should not be too long, since the liquid distal from the constriction will be injected before the tissue. If too much liquid is present in this distal region, it is likely to fill the larva before the tissue is ejected, thereby making the operation unnecessarily difficult. A further important feature of a good working pipette is its tip. The end of the pipette should slope off gradually into a fine point, as shown in Figs. 32A–B. A rough point of this sort can be made by breaking off small bits of the capillary end with a steel needle, until the capillary breaks at the right sloping angle. This roughly made point can be perfected by grinding it carefully into the right shape on a hard Arkansas stone. Much time is saved in grinding the point by the use of a circular Arkansas stone mounted on a small motor. If liquid is to be injected, instead of pieces of tissue, the micropipette is drawn out to a very fine point and needs no constriction.

The host larvae have to be narcotized before they can be injected. For this, larvae which have been washed are put on filter paper and dried by rolling them gently with a fine hair brush. The larvae are now laid on a glass slide, which is placed into the narcotizing chamber. This chamber is a simple glass vial with some cotton at the bottom containing the ether and a cork stopper to close the chamber. The vial is conveniently mounted on a piece of wood or rubber padding. The larvae remain in the chamber until they become motionless, at which time they are sufficiently anaesthetized. It will frequently be found that the narcotized larvae have contracted the muscles of their body walls. If, however, a small drop of Ringer solution is put on such individuals, they usually relax immediately. Larvae which are contracted are very difficult to inject and should be discarded or narcotized anew if they are needed.

For the injection the pipette (not the pipette holder or the syringe) is filled about three-fourths with Ringer solution, and the tissue to be transplanted drawn into the pipette. The filled pipette is now laid aside, and the slide with the larva put under the microscope and adjusted so that the longitudinal axis of the larva is placed horizontally in the field. The pipette holder is now picked up with the right hand. The larva is held down by the blunt tips of a pair of forceps held in the left hand, while the pipette is inserted at any point in the abdominal segments of the larva. The hold on the larva is now released; the forceps are put aside while the left hand presses the syringe, injecting the tissue; and then the injecting needle is withdrawn. In the effort to withdraw the needle after injection, it is often held in the skin, with the result that the whole larva is lifted up. In such a case, the needle may be easily withdrawn by moistening the needle and larval skin at the point of insertion and by holding the larva gently down. In order to avoid lifting up the larva and to increase the ease of the operation, it is advisable to fasten the larva to the glass slide during the operation. This is done as follows. After the etherized larvae have been arranged on the slide in the manner desired for the operation, a drop of Ringer solution is put on each larva. The liquid, however, is sucked off again with a dry hair brush, save a small amount, which remains at the region where the larval body touches the slide. This remaining liquid is permitted to evaporate, and the salt crystals thus formed hold the larva to the slide. After the operation the crystals are dissolved by moistening the larvae, which may then be removed from the slide.

The body cavity of adult flies can also be used as a culture medium for larval and pupal tissues (Bodenstein, 1943a and b). The flies are narcotized for the operation, and the injection procedure is essentially the same as for the larval hosts.

The handling of the animals after operation depends upon their state of development. Mature larvae are kept in glass vials on several layers of moist filter paper, where they soon pupate. Younger larval stages are kept on yeasted agar slants in glass vials. Adult flies are kept in glass vials on yeasted agar slants one-half of which is covered by a small strip of filter paper. The mortality of the injected larvae may vary widely and depends to a great extent on the skill of the operator.

The fixation methods employed for the histological study of the various developmental stages obviously depends upon the special purpose of the investigation. For general histological examination, however, good results are obtained after fixation in Carnoy's fluid (alcohol absolute, 60 cc; chloroform, 30 cc; acetic acid, 10 cc) or in alcohol Bouin (picric acid, saturated in 80 per cent alcohol, 75 cc; formalin, 20 cc; acetic acid, 5 cc). While Carnoy's fluid is especially suitable for nuclear structures, it is found that alcohol Bouin preserves the cytoplasm somewhat better. The fixing fluids should be heated almost to boiling before use. It is difficult for any fixing fluid to penetrate the larval and pupal cuticle. Therefore, after a short time of preliminary fixation, the hypoderm of the animal should be punctured with a fine steel needle at several points to allow better penetration of the fixing fluid. Older pupal stages should be removed from the puparium, which is easily done by forceps after preliminary fixation. However, even the exposed pupae should be punctured. Should it be necessary to leave the puparium, some larger cuts must be made in it, or at least the operculum cut off. If only certain regions of the larva or pupa are needed for the histological examination, it is advisable to cut parts of the animal off, thus allowing more complete penetration of the fixing fluid and the paraffin.

Fixation in boiling water is often useful for special needs. Larvae thrown into boiling water usually stretch out well and can be measured accurately. The larval muscles will also show clearly after the larvae have been fixed in this way. The removal of the prepupal and the pupal stages from the hard puparium is relatively easy after the pupae have been fixed in boiling water. Pupal wings, which are quite laborious to prepare, especially in their earlier developmental stages, are easier to dissect after the pupae have been boiled.

It is rather difficult for the beginner to determine the exact location of and to identify the various larval organs within the larval body. If the larvae are fixed, coagulation prevents a clear view; if narcotized larvae are cut open, their muscles contract to such an extent that all organs seem misplaced; ether-killed larvae also are usually so much contracted that the same difficulty is encountered as with narcotized larvae. However, larvae killed by drowning in Ringer solution make a very good object for a dissection preparation. Such larvae are stretched out completely. If they are cut open along their dorsal midline with a pair of fine iridectomy scissors, the body wall can be laid aside, and all the organs observed in their normal position. With the aid of the diagrams given in other chapters of this book, the various organs should be readily identified.

For embedding, the specimens are dehydrated in the alcohol series. From absolute alcohol they are transferred into methylbenzoate celloidin solution (methylbenzoate plus 1 per cent celloidin), where they are changed once, or better, twice. When placed in methylbenzoate, the tissues float on the surface of the solution but will gradually sink down; only after this has happened can they be placed into the next change. The specimens can remain in methylbenzoate for a long time without harm. Leaving them in the third change overnight is, however, sufficient. After methylbenzoate the tissues are transferred to benzol or chloroform, where they remain for about one-half hour. They are then infiltrated with paraffin and embedded. Sections 8–10 μ in thickness are adequate for most histological studies. Interposing the methylbenzoate step between absolute alcohol and benzol has been found very useful, for this procedure reduces shrinkage, softens the tissues, and thus prevents breakage. Even the hard chitinous pupal case is easily cut after methylbenzoate treatment.

The sections are commonly stained with Heidenhain's iron haematoxylin or Delafield's haematoxylin and counterstained with orange G or eosin.

A very convenient method of making whole mounts of larval and pupal organs, especially of larval organ discs, is the following. After dissection the organs are fixed in alcohol, stained in a drop of orcein, washed in alcohol (95 per cent), and mounted in diaphane. They may also be stained directly in orcein after dissection and then fixed in alcohol and mounted. Borax carmine and Delafield's haematoxylin are also adequate stains for this purpose; the latter is especially suitable for young pupal wings. If mounted in balsam, the organs must be dehydrated in absolute alcohol after staining and cleared in xylol. Whole mounts of total larvae, in particular the younger stages, are very satisfactory when stained with Feulgen's after fixation in Carnoy's. Whole mounts of chitinous parts, such as the larval mouth armature, the stigmata, and the genital apparatus, are prepared by placing these parts in a 15 per cent potassium or sodium hydroxide solution until the non-chitinous tissues have been eroded. They are then washed in water and alcohol and mounted in diaphane or balsam.

CHRONOLOGICAL TABLE OF EVENTS, AT 25°C

A. General postembryonic stages

HOURS

0	Hatching from the egg.
25	First molt.
48	Second molt.
96	Puparium formation; puparium white.
97	Puparium yellowish.
98	Puparium fully colored.
100	Prepupal molt.
108	Pupation; cepahlic complex everted.
145	Pigmentation of eye begins.
165	Bristle pigmentation begins.
192	Pupa ready to emerge.

B. Eyes

1	The antennae-eye complex is a small saclike structure, apparently having a lumen.
24	The antennae-eye complex has increased in size; the eye stalk seems to be present and is attached to the brain.
27.	The antennae-eye complex has increased in size and is oval-shaped; the optic stalk is clearly present and attached to the brain. The complex is several cell layers thick.
32	The anterior portion of the complex is somewhat elongated.
40	The complex has increased in size. The anterior portion has widened, and the posterior portion has become broader. The differentiation of the complex into the antennae and the eye disc proper begins.
50	A distinct anterior portion and posterior portion, i.e., the antennae and eye discs, respectively, can be distinguished. A fine membrane makes up the outer wall of the complex, while the inner wall is formed by the antennae and eye disc.
56	The antennae and eye discs are now well formed; both have increased in size. They are connected by a thin layer of cells. The antennae and eye disc cells are still very similar.
62	The discs have increased in size. The antennae disc becomes folded. Some organization begins to take place in the eye disc.

HOURS

72   Definite cell clusters consisting of four cells each are present in the eye disc, which represent the precursors of the imaginal ommatidia.

82   The discs have grown greatly. The number of cell clusters has increased. The segmentation of the antennae disc has progressed.

96   The discs have reached their maximal size. The cell clusters are now arranged in regular rows. Their number approximates the number of the final ommatidia. The segmentation of the antennae disc is completed.

Puparium formation

108   Eye and antennae discs are everted.

110   The eye is a relatively thick-walled shallow cup; the ommatidia are wide and short.

120   The eye cup increases in area; the single ommatidia have become thinner.

132   The eye cup area has become still larger. The ommatidia are short and ball-like.

144   The formation of the cornea lenses begins. Hairs develop from special hair-forming cells.

156   The ommatidia begin to elongate; they are now in a typical mushroom stage. Cornea lenses and hairs are clearly visible.

157–192   The ommatidia grow rapidly in length. Differentiation is completed.

C. Wings

1   The discs are apparently already invaginated into the body cavity, but they have no lumen.

15   The disc has somewhat elongated, possessing a good lumen.

24   The Anlage grows and differentiates into the peripodial sac and disc proper.

30   The inner wall of the disc thickens.

40   The disc walls have become very long.

48   The peripodial membrane is very thin, but the disc is not yet folded. A ridge separating the anterior thoracic-forming portion from the wing-forming portion has appeared.

57   The disc has grown considerably. The second crossridge has appeared and a concentric fold begins to show.

96   Folding is completed. A shallow impression marks the point of later evagination of the wing pouch.

Puparium formation

97   Evagination starts and forms a wing pouch.

98–100   The evagination ridge is most pronounced. The wing pouch stretches posteriorly.

HOURS

102	The wing is almost entirely evaginated and is relatively thin. The remnant of the wing cavity forms the prepupal veins.
104	The wing becomes inflated.
108	The wing is a large bloated sac.
112	The wing becomes contracted.
116	Veins are present in the tip and base of the wing; they are very broad.
119	The wing has become a thin blade. The posterior crosswing is clearly visible.
120–136	The veins become gradually narrower; the wing expands somewhat in area. Venation is completed at about 136 hours.
138	The wing is fully formed; hair development has begun.
140	The wing begins to fold.
192	Ready to emerge.

D. Legs

1	Leg discs are present in the form of a thickened region in the ventral hypoderm.
2–24	Difference between leg Anlage cells and cells of the hypoderm becomes clearer.
40–48	The invagination of the discs into the body cavity, accompanied by the formation of a lumen in the disc, begins. The first differentiation of the Anlage into the disc proper and the peripodial sac takes place. A stalk connecting the disc with the hypoderm is formed.
65	The neural connections shorten; the leg discs approach each other and move nearer the brain. The invagination of the disc into the peripodial cavity begins.
81	Invagination of the discs into the peripodial cavity is far advanced, resulting in the formation of a lumen in the leg, into which mesenchyme has migrated. The formation of the leg segments begins, which is indicated by concentric folds.
96	The formation of the leg segments is completed.

Puparium formation

98	The peripodial sacs have widened. The leg segments are drawn out telescopically.
100	The peripodial sacs have opened; the legs are ready for eversion; telescoping of the leg segments is more pronounced.
102	The legs have everted.
106	Secretion of the fine pupal cuticle takes place; the legs elongate.
110	The legs have attained about their normal length and are inflated.

HOURS

115	The legs begin to deflate, diminishing the large lumen in them. The legs are now separated clearly from the pupal sheath; leg segmentation becomes visible.
120	Myocytes appear in the legs.
156	The leg muscles are well formed.
160	The chitin of the leg becomes colored; hair and bristles begin to darken.
192	Ready to emerge.

E. *The male reproductive system*

1	The testes are present. They are larger than the ovaries; each contains 36–38 germ cells.
25	The testes contain only spermatozoa.
48	The genital disc is formed.
50	The first spermatocytes are found.
60	The genital disc can be found in dissection. The testis is much larger than the ovary.
60–96	The genital disc and testis grow in size.
96	Spermatozoa are found only in the extreme anterior tip of the testes, while the rest is filled with spermatocytes. The primordia of the vasa deferentia and accessory structures are visible histologically.

Puparium formation

101	The genital disc is somewhat swollen and begins to elongate slightly. The testis is oval-shaped.
107	The anterior and posterior portions of the genital disc become distinct. The elongation of the testis begins.
112	The anterior and posterior disc portions are clearly defined.
120	The outgrowing vasa efferentia are visible as two small blunt protuberances, as are the paragonia. Nearly mature spermatozoa are already found.
126	Testes attach themselves to the vasa efferentia.
138	All parts of the internal genital ducts are relatively well formed. The oviducts and the paragonia especially have grown considerably in length. The coiling of the testis begins.
160	Pigmentation of the testis begins.
192	The imaginal reproductive system is fully established; the pupa is ready to emerge.

F. *The female reproductive system*

1	The ovaries are present, each containing 8–12 germ cells.
25	The genital disc is formed.

HOURS

48   The small genital disc can be found in dissection. The ovary is very small.

60–96   The genital disc and the ovaries grow in size.

Puparium formation

98   The genital disc is relatively large; its internal structure seems to be simpler than that of the male disc. The ovaries contain only oögonia. The ovarioles are quite well differentiated.

107   The genital disc is somewhat swollen and its anterior end narrower, with two small protuberances.

112   The anterior and posterior portions of the disc are clearly defined. From the anterior disc portion project a number of short processes; the ovaries have grown only slightly.

118   The projections on the anterior disc portion, five in number, have grown in length. They can now be identified as the single receptacle, two spermathecae, and two parovaria. A sixth outgrowth, which will give rise to the oviducts, begins to form.

123   The two oviducts can be distinguished; the tips of the parovaria and spermathecae are slightly swollen. In the ovaries the ovarioles can be faintly recognized.

130   The ovaries become attached to the oviducts. The ovaries have grown in size and are pear-shaped. The ovarioles have become more distinct but contain only oögonia.

130–180   The growth and differentiation of the various parts are completed; the first oöcytes appear.

192   The pupa is ready to emerge. The ovary contains no ripe eggs as yet.

G. The imaginal hypodermis of thorax and abdomen

Puparium formation

98   Imaginal hypodermal cells originating from the anterior region of the thoracic discs grow into the larval hypoderm of the thorax. The imaginal hypodermal cells of the genital disc grow into the hypoderm of the last abdominal segment.

100   The abdominal histoblasts appear.

101   The larval hypoderm of the thorax is largely replaced.

103   The imaginal hypoderm of the thorax is completed.

112   The bristle-forming cells appear.

126   Bristle formation begins.

130   The imaginal hypoderm of the abdomen is completed.

132   The bristles have reached one-half their full length.

136   The bristle-forming cells have reached their maximal size.

144   Bristle and bristle socket formation are completed.

HOURS

146   The secretion of the imaginal cuticle begins. The bristles begin to become hollowed out.

165   The bristle-forming cells are now very small. The pigmentation of the bristles begins.

166   Bristle pigmentation begins dorsally on head and thorax.

174   Pigmented bristles are visible on head, thorax, wings, and legs. Bristle pigmentation begins in first and second abdominal segment.

178   The abdominal bristles are fully pigmented. The pigmentation of the abdominal body wall begins.

192   Pigmentation of bristles and body wall is completed. The pupa is ready to emerge.

H. The tracheal system

1-25   First instar. The anterior spiracles are functionless and do not open to the outside. The posterior spiracles have two openings each.

25   Molting.

25-48   Second instar. The anterior spiracles are functionless; they do not open to the outside. The posterior spiracles have three openings each.

48   Second molt.

48-96   Third instar. The anterior spiracles have become functional and open to the outside. They each carry at their tips seven to nine fingerlike processes. Each of the posterior spiracles has three openings.

Puparium formation

96   Larvae become immobile; spiracles are everted.

97   The imaginal lateral spiracles originating as outgrowth from imaginal cells on the great lateral trunks have connected with the hypoderm.

98   The imaginal lateral spiracles form a tube filled with newly secreted chitin.

99   Larva contracts away from the anterior end of the puparium, thus partly withdrawing the anterior tracheal linings.

100   Prepupal molt. A lumen appears in the imaginal abdominal tracheal tubes, which connects the tracheal tracts with a pore in the pupal membrane. The stigmata are probably not functional.

108   Pupation. The anterior and posterior tracheal linings are shed. The newly formed tracheal tube linings of the prepupal stage are also withdrawn. The newly formed tracheal system in the head is everted. The newly formed imaginal prothoracic spiracles connect with the lateral larval trunks.

HOURS

110	The abdominal tracheae are still larval.
114	The pupal intima of the abdominal longitudinal trunks is broken off near the posterior end.
120–156	The posterior tracheal trunks are histolyzed. The chitinous intima shrinks into the anterior tracheal trunks, which form the large thoracic air sacs.
158	The closing device of the imaginal stigmata is established.
192	Emergence. The chitinous intima is shed through the prothoracic spiracles.

I. Muscles

Puparium formation

100	The muscles of the anterior segments begin to histolyze.
102	The myocytes of the wing muscles appear.
104	The muscles of the anterior segments are almost dissolved, except for the three pairs of dorsal thoracic muscles and some of the pharyngeal dilator muscles.
110	Most of the abdominal muscles are dissolved.
115	The dorsal thoracic muscles and the dilator muscles of the pharynx are dissolved. The wing muscles begin to differentiate.
123	The wing muscles are about one-third formed.
138	The dorsal abdominal muscles still persist. The wing muscles are about two-thirds formed.
144	All the abdominal muscles are dissolved.
146	The wing muscles become striated.
156	The imaginal muscles of the abdomen are well formed.
162	The imaginal muscles of the abdomen become striated.
180	The wing muscles are completed.
192	All imaginal muscles are formed.

J. Alimentary canal

0–95	Larval stage: All parts of the alimentary canal, with the exception of the anterior and posterior imaginal rings, grow by increase in cell size.
96	Replacement cells appear in midintestine.

Puparium formation

97	Mouth and anus are closed off.
98	Replacement cells in midintestine proliferate. The epithelial cells of the larval midintestine loosen.
99	The larval epithelium separates from the basement membrane.
100	Malpighian tubes separate from the midintestine. The midintestine closes at its ends.
101	Phagocytes appear in the muscular layer of the midintestine.

HOURS

102   The gastric caecae, the proventriculus, and the larval midintestine shed into the new imaginal midintestine, forming the "yellow body."

106   The epithelium of the foregut is replaced.

110   The hindgut is much shortened; the crop begins to form.

117   The phagocytes have disappeared from the midintestine. Phagocytes appear in the hindintestine. The proventriculus is formed.

123   The crop has differentiated into a posterior saclike structure and a narrow stalk.

132   Phagocytes have disappeared from the hindintestine.

150   The crop has become a large flattened sac.

156   The crop is folded.

180   The midintestine is still closed off.

192   The intestine has become a continuous tube.

K. Salivary glands

1   The gland is very small. All the cells are uniform in size.

20   The cells have grown and are uniform throughout the gland.

35   The cells have grown and are still uniform throughout the gland. A small group of tiny cells at the proximal tip of the gland representing the imaginal Anlage of the adult salivary gland becomes visible.

55   The cells at the distal gland portion are larger than the more proximal gland cells. The imaginal Anlage cells have increased in number and form a ringlike structure around the salivary duct. The chromosome strands show alternating deeply and lightly stained areas.

65   The cell size has greatly increased. The difference in cell size between proximal and distal gland portions is very noticeable. The imaginal ring cells have increased in number.

75   The chromosome strands have become broader, and the chromosome bands very distinct. The chromocenter is visible in the distal cells of the gland.

82   The nuclei of the salivary gland cells, especially those of the distal end of the gland, have about reached their maximal size. The imaginal ring cells have increased in number.

96   The nuclei in the distal cells are about the same size as they were at 82 hours, but the cells themselves have grown immensely. The individual strands show their characteristic banding most perfectly.

Puparium formation

100   The cells become vacuolated.

106   Histolysis of the gland is advanced; the imaginal ring begins to differentiate.

HOURS

120   The larval gland is dissolved. The imaginal Anlage has continued to differentiate.
121   The imaginal salivary gland is completed.
192   Emergence.

LITERATURE CITED

ALPATOV, W. W. 1929. Growth and variation of the larvae of *Drosophila melanogaster*. *J. Exptl. Zoöl.* **52**:407–437.

AUERBACH, CHARLOTTE. 1936. The development of the legs, wings, and halteres in wild type and some mutant strains of *Drosophila melanogaster*. *Trans. Roy. Soc. Edinburgh* **58**:787–815.

BEADLE, G. 1937. The inheritance of the color of Malpighian tube in *Drosophila melanogaster*. *Am. Naturalist* **71**:277–279.

BODENSTEIN, D. 1938. Untersuchungen zum Metamorphoseproblem I. Kombinierte Schnürungs- und Transplantationsexperimente an Drosophila. *Arch. Entwicklungsmech. Organ.* **137**:474–505.

BODENSTEIN, D. 1939. Investigations on the problem of metamorphosis. VI. Further studies on the pupal differentiation center. *J. Exptl. Zoöl.* **82**:329–356.

BODENSTEIN, D. 1942. Hormone-controlled processes in insect development. *Cold Spring Harbor Symposia Quant. Biol.* **10**:17–26.

BODENSTEIN, D. 1943a. Factors influencing growth and metamorphosis of the salivary gland in Drosophila. *Biol. Bull.* **84**:13–33.

BODENSTEIN, D. 1943b. Hormones and tissue competence in the development of Drosophila. *Biol. Bull.* **84**:34–58.

BODENSTEIN, D. 1944. The induction of larval molts in Drosophila. *Biol. Bull.* **86**:113–124.

BODENSTEIN, D. 1947. Investigations on the reproductive system of Drosophila. *J. Exptl. Zoöl.* **104**:101–152.

BREHME, K. S., and M. DEMEREC. 1942. A survey of Malpighian tube color in the eye color mutants of *Drosophila melanogaster*. *Growth* **6**, 3:351–355.

CHEN, T. Y. 1929. On the development of the imaginal buds in normal and mutant *Drosophila melanogaster*. *J. Morphol.* **47**:135–199.

CHEVAIS, S. 1937. Sur la structure des yeux Implantés de *Drosophila melanogaster*. *Arch. anat. Mikr.* Bd. 33:107–112.

COOPER, K. 1938. Concerning the origin of the polytene chromosomes of Diptera. *Proc. Natl. Acad. Sci.* **24**:452–458.

DAY, M. F. 1943. The function of the corpus allatum in Muscoid Diptera. *Biol. Bull.* **84**:127–140.

DERRICK, G. E. 1938. The development of the eye and optic tract in *Drosophila melanogaster* and its "eyeless" mutant. *Proc. Okla. Acad. Sci.* **8**:100–105.

DOBZHANSKY, TH. 1930. Studies on the intersexes and supersexes in *Drosophila melanogaster*. *Bull. Bur. Genet. Leningrad* **8**:91–158.

EASTHAM, L. 1925. Peristalsis in the Malpighian tubes of Diptera, preliminary account: with a note on the elimination of calcium carbonate from the Malpighian tubes of Drosophila funebris. Quart. J. Microscop. Sci. 69:385–398.

ENZMANN, E. V., and C. P. HASKINS. 1938. The development of the imaginal eye in the larvae of Drosophila melanogaster. J. Morphol. 63:63–72.

EPHRUSSI, B., and G. W. BEADLE. 1936. A technique of transplantation for Drosophila. Am. Naturalist 70:218–225.

EVANS, A. C. 1936. Histolysis of muscle in the pupa of the blow-fly Lucilia sericata Meig. Proc. Roy. Soc. Entomol. London, A. 11:52–54.

FRAENKEL, G., and K. M. RUDALL. 1940. A study of the physical and chemical properties of the insect cuticle. Proc. Roy. Soc. London, series B, 129, 854:1–35.

GLEICHAUF, R. 1936. Anatomie und Variabilität des Geschlechtsapparates von Drosophila melanogaster. Z. wiss. Zoöl. 148:1–66.

GOLDSCHMIDT, R. 1935. Gen und Aussencharakter, III. Biol. Zentr. 55:535–554.

HADORN, E., and J. NEEL. 1938. Der hormonale Einfluss der Ringdrüse (corpus allatum) auf die Pupariumbildung bei Fliegen. Arch. Entwicklungsmech. Organ. 138:281–304.

HASKINS, C. P., and E. V. ENZMANN. 1937. Studies on the anatomy of the respiratory system of Drosophila melanogaster. J. Morphol. 60:445–458.

HERTWECK, H. 1931. Anatomie und Variabilität des Nervensystems und der Sinnesorgane von Drosophila melanogaster (Meigen). Z. wiss. Zoöl. 139:559–663.

JOHANSSEN, O. A. 1924. Eye structure in normal and eye mutant Drosophilas. J. Morphol. 39:337–348.

KALISS, N. 1939. The effect on development of a lethal deficiency in Drosophila melanogaster: with a description of the normal embryo at the time of hatching. Genetics 24:244–270.

KEILIN, D. 1927. The chordotonal organs of the antenno-maxillary complex in the larvae of cyclorrhaphous Dipteres. Ann. Nat. Hist. 9.

KERKIS, J. 1931. The growth of the gonads in Drosophila melanogaster. Genetics 16:212–224.

KERKIS, J. 1933. Development of gonads in hybrids between Drosophila melanogaster and Drosophila simulans. J. Exptl. Zoöl. 66:477–509.

KOCH, J. 1945. Die Oenocyten von Drosophila melanogaster. Rev. suisse zool. 52:415–420.

KORSCHELT, E., and K. HEIDER. 1899. Textbook of the Embryology of Invertebrates. Vol. III. Swan Sonnenschein and Co., Ltd., London.

KOWALEVSKY, A. Z. 1887. Beiträge zur Kenntnis der nachembryonalen Entwicklung der Musciden (Calliphora). Diss. Zool. 45:542–594.

KRAFKA, J. 1924. Development of the compound eye of Drosophila melanogaster and its bar-eyed mutant. Biol. Bull. 47:143–148.

LEES, A. D., and C. H. WADDINGTON. 1942. The development of the bristles in normal and some mutant types of Drosophila melanogaster. Proc. Roy. Soc. B. 131:87–110.

MAKINO, S. 1938. A morphological study of the nuclei in various kinds of somatic cells of *Drosophila virilis*. *Cytologia* 9:272–282.

MEDVEDEV, N. N. 1935. Genes and development of characters I. The study of the growth of the imaginal discs of eyes of the wild-type larvae and the three mutants lobe, glass, and "eyeless" in *Drosophila melanogaster*. *Z. induk. Abst. Vererb. lehre*. Bd. 70:55–72.

NEWBY, W. W. 1942. A study of intersexes produced by a dominant mutation in *Drosophila virilis*, Blanco stock. *Univ. Texas Pub.* 4228:113–145.

PEREZ, C. 1910. Recherches histologiques sur la métamorphose des Muscides (Calliphora). *Arch. Zool*, esp. Ser. 5T 4:274.

PILKINGTON, R. W. 1942. Facet mutants of Drosophila. *Proc. Zool. Soc.*, London, Ser. A, 111:199–222.

POULSON, D. 1937. The embryonic development of *Drosophila melanogaster*. *Actualités sci. et ind.* 498:51. Hermann et Cie., Paris.

POWER, M. E. 1943. The effect of reduction in numbers of ommatidia upon the brain of *Drosophila melanogaster*. *J. Exptl. Zoöl.* 94:33–71.

REES, J. VAN. 1888. Beiträge zur Kenntnis der inneren Metamorphose von *Musca vomitoria*. *Zool. Jahrb. Anat.* 3:1–134.

RICHARDS, M. H., and E. Y. FURROW. 1925. The eye and optic tract in normal and "eyeless" Drosophila. *Biol. Bull.* 48:243–258.

ROBERTSON, C. W. 1936. Metamorphosis of *Drosophila melanoagster*, including an accurately timed account of the principal morphological changes. *J. Morphol.* 59:351–399.

ROSS, F. B. 1939. The postembryonic development of the salivary gland of *Drosophila melanogaster*. *J. Morphol.* 65:471–495.

RÜHLE, H. 1932. Das larvale Tracheensystem von *Drosophila melanogaster* Meigen und seine Variabilität. *Z. wiss Zoöl.* 141:159–245.

SAVELIEV, V. 1928. On the manifold effect of the gene vestigial in *Drosophila melanogaster*. *Trav. soc. naturalistes Leningrad* 58:65–88.

SNODGRASS, R. E. 1924. Anatomy and metamorphosis of the apple maggot *Rhagoletis pomonella Walsh*. *J. Agr. Research* 28:1–36.

SONNENBLICK, B. P. 1940. The salivary glands in the embryo of *Drosophila melanogaster*. *Genetics* 25:137.

SONNENBLICK, B. P. 1941. Germ cell movements and sex differentiation of the gonads in the Drosophila embryo. *Proc. Natl. Acad. Sci.* 27:484–489.

STARK, M. B., and A. K. MARSHALL. 1930. The blood-forming organ of the larva of *Drosophila melanogaster*. *J. Am. Inst. Homeopathy* 23:1204–1206.

STEINBERG, A. G. 1941. A reconsideration of the mode of development of the bar eye of *Drosophila melanogaster*. *Genetics* 26:326–346.

STEINBERG, A. G. 1943. The development of the wild type and bar eyes of *Drosophila melanogaster*. *Can. J. Research* 21:277–283.

STERN, C. 1938. Innervation of setae in Drosophila. *Genetics* 23:172–173.

STERN, C. 1941a. The growth of testes in Drosophila. I. The relation between vas deferens and testis within various species. *J. Exptl. Zoöl.* 87:113–158.

STERN, C. 1941b. The growth of testes in Drosophila. II. The nature of interspecific differences. *J. Exptl. Zoöl.* **87**:159–180.

STERN, C., and E. HADORN. 1939. The relation between the color of testes and vasa efferentia in Drosophila. *Genetics* **24**:162–179.

STRASBURGER, E. H. 1935. *Drosophila melanogaster Meig. Eine Einführung in den Bau und die Entwicklung.* Julius Springer, Berlin.

STRASBURGER, M. 1932. Bau, Funktion, und Variabilität des Darmtraktus von *Drosophila melanogaster* Meigen. *Z. wiss. Zoöl.* **140**:539–649.

VOGT, M. 1942a. Ein Drittes Organ in der larvalen Ringdrüse von Drosophila. *Naturwissenschaften* **30**:66–67.

VOGT, M. 1942b. Weiteres zur Frage der Artspezifität gonadotroper Hormone. Untersuchungen an Drosophila Arten. *Arch. Entwicklungsmech. Organ.* **141**:424–454.

VOGT, M. 1942c. Induction von Metamorphoseprozessen durch implantierte Ringdrüsen bei Drosophila. *Arch. Entwicklungsmech. Organ.* **142**:131–182.

VOGT, M. 1943a. Zur Kenntniss des larvalen und pupalen corpus allatum in Calliphora. *Biol. Zentr.* **63**:56–71.

VOGT, M. 1943b. Zur Production gonadotropen Hormones durch Ringdrüsen des ersten Larvenstadums bei Drosophila. *Biol. Zentr.* **63**:467–470.

VOGT, M. 1943c. Zur Production und Bedeütung metamorphosefördernder Hormone während der Larvenentwicklung von Drosophila. *Biol. Zentr.* **63**:395–446.

WADDINGTON, C. 1939. Preliminary notes on the development of the wings in normal and mutant strains of Drosophila. *Proc. Natl. Acad. Sci.* **25**:299–307.

WEISSMAN, A. 1864. Die nachembryonale Entwicklung der Musciden, nach Beobachtungen an *Musca vomitoria* und *Sarcophaga carnaria.* *Z. wiss. Zoöl.* **14**:187–333.

WIGGLESWORTH, V. 1939. *The Principles of Insect Physiology.* Methuen, London.

ZAVREL, J. 1935. Endokrine Hautdrüsen von Syndiamesa Branicki Now. (Chironomidae). *Pubs. faculté sci. univ. Masaryk Cis.* **213**:1–19.

5

External Morphology of the Adult

G. F. FERRIS

INTRODUCTORY NOTE

The reader of this chapter on the external morphology of the adult of Drosophila will find that it expresses ideas and to some degree employs a terminology different from what is to be found in most discussions of insect morphology and especially in discussions of the structure of flies. It therefore seems necessary to offer a rather extended explanation before engaging upon the purely morphological presentation.

The date of this writing is approximately 10 years since the author was asked to undertake the preparation of this section. In company with a graduate student, Dr. Bryant E. Rees, the work was begun with results which have been stated in an earlier paper (1939, p. 83) as follows:

It was soon found that no such paper on Drosophila could be prepared without preliminary studies on more generalized Diptera. And then in turn it was found that no such study on a generalized dipteran could be prepared without going still further back and investigating the morphology of the Mecoptera, for it is generally accepted, and is here regarded as definitely demonstrated, that in the morphology of the Mecoptera lie the keys to many of the highly specialized developments of the Diptera. It may be added that the morphology of the Mecoptera cannot be fully understood without going still further back into the morphology of such groups as the Neuroptera, in which lie the keys to certain developments in the Mecoptera.

The process thus initiated by the proposed study of Drosophila has continued until at the present moment work is under way on certain features of the Annelida and Crustacea, in which lie the keys to an understanding of the nervous system and, through this, an understanding of the segmentation of the head in insects.

The proposed study of the morphology of a single species of fly has thus led to the development of a program of work on the comparative morphology of the insects. In the course of this work nearly twenty papers on various aspects of the subject have been published; others are now in course of preparation, and it is proposed to continue the program indefinitely.

As a result of these studies it has become increasingly apparent that many of the currently accepted ideas concerning not only the morphology of the Insecta but also that of the Crustacea and of the Annelida are fundamentally erroneous and that not until these errors have been corrected and new ideas—and, above all, new and clearly defined principles—have become incorporated into the basic structure of our comparative morphology will it be possible to prepare a definitive treatment of any highly specialized form.

It will therefore appear that this study of the external morphology of Drosophila is still premature, since these ideas and principles are not yet fully developed and, above all, have not yet been applied to a sufficiently wide range of material to be fully utilized in such a study. Furthermore, they will undoubtedly for some time to come be bitterly disputed by adherents of the commonly accepted ideas. The reader of this chapter will therefore quite justifiably view with doubt the conclusions here presented. The author is thus placed in the rather uncomfortable position of having either to present statements which he is convinced are wrong or to present conclusions that are not yet widely accepted and which are peculiar to himself and not yet fully supported by published evidence.

The conclusion has been reached that intellectual honesty precludes the first course. A break with accepted ideas must be made some time. To perpetuate these ideas in a volume which will long be used as a standard work of reference is not merely to assist in delaying the development of this break but also to shorten the period of time in which the work will be of real value.

Such a consideration is important in connection with an animal which has contributed so much to the development of biological theory. Since Drosophila has through certain favorable characteristics led to the extension of our knowledge of genetics, it seems very inappropriate if it may not be used to extend our knowledge of the theories of comparative morphology. This is especially true, since, in the opinion of the present writer, genetical theory and morphological theory should travel hand in hand, each contributing to the progress of the other.

This is not the place to discuss these matters in great detail, but some reference to them cannot well be avoided if this chapter is to be intelligible to its readers and its implications are to be grasped.

Comparative morphology is in its very essence the search for continuity in evolution. Thus it is generally accepted that the nervous system of Arthropoda and that of Annelida have much in common. It becomes therefore the function of comparative morphology to attempt to trace that continuity, to discover how far it extends and whether or not it has been interrupted by the appearance of new developments. If the appearance of such developments leading to the disintegration and disappearance of old systems has been a common phenomenon, then it would appear that comparative morphology should reveal this fact. But if evolution has been a continuous process devoid of such revolutions, this too should appear.

Much of the thinking of morphologists in the past has apparently been conditioned by a belief in the occurrence of such revolutions, and the idea is not entirely unknown in the field of genetics. How far this idea has gone may be illustrated by the following examples from the comparative morphology of the insects.

Snodgrass, whose book entitled *The Principles of Insect Morphology* (1935) is widely known as the standard work of reference on this subject, has apparently accepted the idea of revolutionary change as applying rather generally throughout the insects. Thus he remarks in this book (p. 171): "In the pterothorax of the Diptera, for example, the more primitive sutures of the sternal as well as the pleural areas have become almost obliterated, and secondary grooves appear which divide the skeletal surface into parts that have little relation to those in more generalized orders." Again (p. 103): "The various sutures that appear in the definitive head capsule, with the exception of the subterminal post-occipital suture . . . have no relation to the original metamerism." Again, in regard to the mouth hooks of dipterous larvae he maintains (p. 314): "The mouth hooks of the cyclorrhaphous larvae are often called 'mandibles,' but since they are solid cuticular structures, shed with each molt, arising from the lips of the atrial cavity, which is evidently derived from the infolded neck membrane, it is not clear how the larval jaws can have any relation whatever to true mandibles."

On the basis of such statements as these it might be inferred that the comparative morphology of the Insecta supports the concept that the evolution of the group has involved quite frequently the occurrence of profound alterations of whole systems and their replacement by other systems.

With any such concept issue is here taken directly. The work which has been done in connection with the program that was initiated by the present study supports quite the contrary opinion that, except for evolution by losses which are at times extensive, no such profound alterations of whole systems have occurred anywhere in the Insecta. In earlier papers by the author (1940b, 1942, 1943a, 1943b) it has been shown that recourse to any such assumptions concerning the heads of insects in general and concerning the thorax of Diptera is entirely unnecessary. Work now under way, which may be published before this book appears in press (a paper by Edwin F. Cook on the heads of dipterous larvae), will show conclusively that the "mouth hooks" of the more specialized fly larvae are in fact the mandibles. Another paper now in preparation will trace the segmentation of the head from the Annelida to the Insecta on the basis of the nervous system and will show the fallacy—or group of fallacies—upon which current ideas concerning the composition of the insect head are based and will reveal the existence of a basic system which has endured with relatively little change since at least the Cambrian.

The opinion is here maintained that there is nothing in the structure of even the most specialized Diptera which cannot be homologized with the basic pattern of even the most generalized insects if the problem is approached with due regard to clearly definable principles and by way of an adequate series of explanatory forms. It is true that there are still problems in regard to the morphology of the Diptera and the morphology of insects in general, but the indications are that none of these problems are insoluble. They merely await their turn to be considered.

Actually there appear to have been but two revolutionary changes in the course of evolution of the Arthropoda from some annelid-like ancestor. One arose with the development and utilization of chitin, through which great modifications of the body were imposed upon the fundamental system of segments. The other was the development of wings in the insects, and this was probably not as revolutionary as it appears on the surface to have been.

It will be clear from this why it is impossible for the author of this section to present a paper on the morphology of *Drosophila melanogaster* simply as a detached insect unrelated to any underlying concepts of comparative morphology and of evolutionary processes.

STATEMENT OF PROCEDURE

It would appear that this study should meet two requirements.

On the one hand, there is the need merely for a treatment that will supply a terminology and a basis of reference for those who have to deal with the superficial structures and desire names by which to refer to them. On the other hand, there would seem to be a need for a genuine comparative morphology which is concerned with the homologies of structures. Thus the actual homologies of the structures called the "labella" and the question of whether or not the antennae are serially homologous with the segmental appendages would seem to be of considerable theoretical importance in view of the mutations which at times involve these parts and of the circumstance that certain ideas as to evolutionary processes have been based upon these mutations.

The reconciliation of these two requirements presents considerable difficulty.

In the first place the existing terminology, as employed for descriptive purposes, is almost entirely unmorphological and is to a considerable degree erroneous and misleading. It is also much confused, since different authors apply different terms to the same part or area or to different combinations of the same structures. Perhaps the best and most nearly complete statement of the descriptive terminology will be found in a recent paper by Crampton (1942), and from this paper it will be possible to gain some idea of the prevailing confusion. Thus it requires nearly a page of text to explain the variations in the use of the word "gena," a word that has no morphological meaning in the first place, but which merely designates an area regardless of what the morphological homologies of that area may be.

In the second place the genuinely morphological terminology is inadequate and to some extent misleading. Thus, the term "labella," which is utilized for the lobes of the labium, leaves the morphological homologies of these lobes entirely untouched. The term "sternopleurite" likewise is entirely misleading and erroneous, since there is no element of the sternum involved in the piece in question.

The procedure here adopted is that of approaching the problem from a purely morphological point of view and attempting first to develop an understanding of the actual homologies of the various structures. Then, as a purely secondary aim, there has been an attempt to reconcile the current morphological—and to some extent the non-morphological—terminology with the results thus obtained.

It will be necessary first to consider some of the general features of insect structure, concerning which ideas are here expressed that are at variance with those generally held.

THE HEAD

Some General Considerations

The head in the higher Diptera offers certain problems for which no answer is at present available, these having to do with the parts of the head capsule and with the structure called the *ptilinum*. Owing to the fact that the basic sutures of the head have disappeared because of fusion of parts, it is difficult adequately to define the morphological areas, but by close attention to known landmarks a reasonable interpretation can perhaps be achieved.

Upon the basis of work done by the author and his students in the program of morphological investigation which has been mentioned, the following general statement concerning the segmental composition of the head may be made.

The head of insects consists of six segments as follows: 1—the labrum or true first segment; 2—the clypeus or true second segment; 3—the antennal-ocular or true third segment; 4—the mandibular or true fourth segment; 5—the maxillary or true fifth segment; 6—the labial or true sixth segment. The dorsal portion of the labial segment is never sclerotized and does not enter into the composition of the sclerotized head capsule except ventrad of the occipital foramen, where it forms the submentum if a submentum is present.

The labrum is unquestionably a true segment which is strictly homologous with the first true segment, commonly called the prostomium, in the earthworm but not with the structure called the prostomium in the Polychaeta. The labrum is innervated from the so-called tritocerebral ganglion, which is actually the first ganglion of the ventral nerve chain and is the ganglion of the labral segment. The stomodaeum, or anterior portion of the alimentary canal, is an invagination from the labral segment and, with all muscles inserting upon it, is innervated by nerves having their origin in this same ganglion.

The clypeus belongs to the second segment of the body. While there are numerous muscles which originate upon the clypeus, there are none which insert upon it and which belong to the second segment. Consequently there are no dorsal nerves which can be ascribed to this segment in any insect that has yet been examined, although such nerves are present in some Crustacea and in the centipedes. In addition to the clypeus there is another structure which also belongs to the second segment.

There has been much difference of opinion concerning the morphological significance of the hypopharynx, which is normally a free lobe that forms the floor of the preoral chamber. We may not here review all the various theories as to the composition of this structure and need only remark that upon the basis of evidence derived from a study of the nervous system of the head from the Annelida to the Insecta, made by Laura Henry (1947, 1948), the hypopharynx belongs morphologically to the ventral wall of the clypeal segment. It is in fact merely a lobe of that segment. In forms in which the hypopharynx is well developed and possesses muscles, those muscles are innervated by a pair of nerves which occupy a position that is held by nerves of the second segment in all forms that have been examined from the Annelida to the Insecta. These facts concerning the hypopharynx will be referred to once more when we come to consider the morphological position of the opening of the salivary glands.

The antennae and the eyes belong to the third segment and are innervated from the dorsal portion of the brain. It is generally agreed by embryologists that this dorsal portion, composed of the so-called protocerebrum and deutocerebrum, is in its embryonic origin independent of the ventral nerve chain with which it fuses during ontogeny. The antennae seem certainly to be homologous with some one pair of the appendages of the brain capsule—the so-called "prostomium"—of the Polychaeta, which lies above the true third segment in these worms. It appears that, as the brain enlarged, this head lobe fused with and eventually took over the third segment, and thus the eyes and the antennae of insects are innervated from the dorsal portion of the brain. Coincident with this the segmental nerve originally supplying this segment has disappeared in the insects. It is clear, then, that the antennae are not serially homologous with the other body appendages and that the eyes are not serially homologous with appendages.

The "intercalary segment," which was invented in order to supply the tritocerebral ganglion with a segment proper to it, has no existence in fact, since this ganglion is actually the ganglion of the labral segment and has merely been displaced posteriorly.

The remaining segments of the head require no special discussion since there is no disagreement concerning them. There is, however, one structure that merits some consideration, this being the opening of the salivary duct.

It is generally agreed by embryologists that the salivary glands belong to the labial segment and that the opening of the common duct of these glands therefore belongs to that segment. This is supported also by the innervation of the glands through a pair of nerves that arise just posterior to the point of origin of the labial nerves. We may therefore accept the opening of the salivary glands as a definite landmark, and we seem to be quite safe in saying that wherever this opening may be it still belongs to the labial segment. However, there has been considerable misunderstanding concerning this.

Normally the opening of the salivary duct lies at the base of the pouch which is formed between the hypopharynx and the labium. At times, however, this opening may be borne at the tip of an apically free lobe, and when this occurs the tendency is to identify this lobe with the hypopharynx. This is morphologically unsound, for the opening of the duct does not belong to the hypopharynx. There are instances known where it seems

clear that this lobe is nothing more than a papilla arising around
the opening of the duct. This is true, for example, in the larvae
of Lepidoptera, and it is true in the sucking lice, where the papilla
becomes a long and slender needle.

These facts concerning the morphological position of the opening of the salivary duct have a definite bearing upon a matter
that we shall meet in the course of our discussion of Drosophila.

There has been much discussion in morphological works as to
whether various structures of the head are preoral or postoral.
This discussion, when considered in the light of what has just
been presented, becomes utterly meaningless. The evidence indicates that the mouth belongs to the labral segment, which is the
anteriormost segment of the body, and that consequently all structures are morphologically postoral. The difficulty has arisen because of the mistaken assumption that the prostomium of the
earthworm and the brain capsule which is called the prostomium
in the Polychaeta are homologous areas, which they definitely
are not.

The sutures on the heads of insects thus may and do in part
correspond to or at least parallel the true segmental lines. There
are certain sutures which have originated in other ways. Thus
the coronal suture, which occupies tho midline of the head capsule
in many insects, is definitely not segmental. The postfrontal
suture seems definitely to be intrasegmental but is probably not
merely a line along which the derm splits at the time of ecdysis,
since a fold corresponding in position to this suture occurs even
in the polychaete worms and in certain Crustacea.

The Head Capsule of Drosophila

(Figs. 1 and 2)

The evolutionary changes which have produced the head capsule of the more specialized Diptera have been concentrated especially in the ventral region of the head, where they have involved
the mouthparts and the labrum and clypeus. The clypeus, which
is an important portion of the sclerotized cranium in many insects, is separated in the Diptera from that capsule by a membranous area that apparently represents the expanded clypeofrontal suture and is now incorporated into the base of the pro-

boscis. Because of this the clypeus can best be considered in connection with the proboscis.

The cranium is composed almost entirely of the antennal-ocular segment, there being only small areas on the posterior aspect which can be ascribed to other segments. These areas are only partially defined by sutures.

Viewed from in front (Fig. 1A), the broad area (frons) between the great compound eyes bears near its center the antennae, which are set close together. The three ocelli are borne close together in a little triangle near the top (vertex) of the head. The only suture to be seen originates on each side in the narrow space (the gena) between the eyes and the ventral border of the cranium and passes upward, paralleling the border of the corresponding eye, and then across above the antennal foramina. This is the ptilinal suture, and it is the external evidence of that peculiar structure, the ptilinum. The ptilinum is a membranous sac that is everted by blood pressure at the time of eclosion of the adult and that serves to push off the end of the puparium. When it has performed this one function, it is withdrawn into the head and collapses, the outer derm becomes sclerotized, and the ptilinal suture remains as the evidence of the existence of the sac.

The origin of this ptilinal suture and of the ptilinum is at present quite obscure. By some authors the suture has been identified as the frontal suture, but this is in error for two reasons. In the first place, as shown elsewhere by Ferris (1942), the frontal suture does not exist in fact, having been compounded, with the aid of imagination, out of the postfrontal and and clypeofrontal sutures. And even if the frontal suture exists, the ptilinal suture could not be the same thing, since the frontal suture is supposed to embrace the median ocellus between its arms.

All that can at present be said concerning the ptilinum and the ptilinal suture is that they represent a development which is peculiar to the more specialized flies and that they have nothing to do with any structure to be found that is basic to the insect head. There is a possible explanation, but it has not yet been explored and will not here be considered.

Morphologically the entire front of the head capsule belongs to the antennal-ocular segment. In species in which the postfrontal suture is retained, this suture forms an inverted V, the

apex of which is on the midline behind the ocelli. Each arm of
the **V** passes through the position of a lateral ocellus, and at their
greatest extent these arms pass downward between the antennal

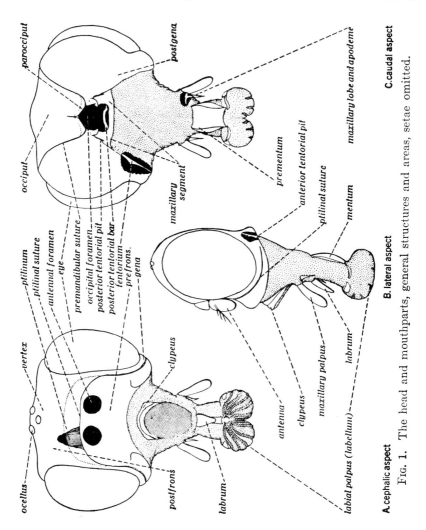

FIG. 1.   The head and mouthparts, general structures and areas, setae omitted.

foramina and the eyes, only rarely reaching the border of the
cranium. The postfrontal suture is suppressed in Drosophila,
and consequently only its presumptive course remains as an index
to the homologies of areas. On the basis of this the entire face
from the ocelli to the ventral margin of the cranium belongs to
the antennal portion of the segment. This entire area, by the

original definition of the word, constitutes the "frons"; and, although that word has become perverted by being applied to an imaginary and non-existent plate, it may still legitimately be

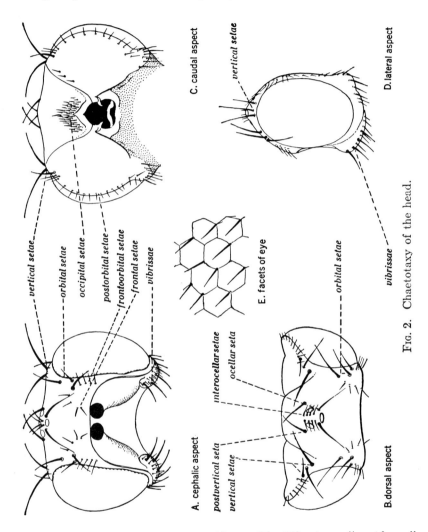

Fig. 2.   Chaetotaxy of the head.

applied to the entire area up to the ocelli. The term "postfrons" is thus applicable to the area between the ptilinal suture and the ocelli, while the term "prefrons" may be applied to the area anterior to, that is, below, the ptilinal suture.

The postfrons is marked by areas on which the surface of the derm is variously ridged or beset with minute points. The area about the ocelli remains smooth, and this difference in texture has given rise to the idea that the ocelli are set in a distinct plate (the vertical or ocellar triangle), but this area does not represent a separate sclerite.

Beneath the eyes and ventrad of the end of the ptilinal suture there is on each side an elevated ridge which has no morphological significance, but which represents an area commonly known as the "jowl," and which bears a row of setae commonly called the vibrissae.

On the posterior aspect of the head the following features may be noted (Fig. 1C).

The occipital foramen is very small and is situated at about the center of the cranium. Passing upward from this foramen there is on each side a very marked suture which extends upward and laterally until it almost meets the margin of the corresponding eye and ends. These sutures have been entirely neglected by morphologists and have been dismissed as merely secondary depressions, but they are present in many flies and require a morphological explanation. The posterior regions of the head of insects have not yet been adequately explored morphologically, but such evidence as is available indicates that these sutures represent part of the intersegmental suture separating the mandibular segment from the ocular lobes, and they are here so indicated. The area enclosed between these sutures is commonly called the occiput. A very short and rather faint median line extending forward from the foramen is all that remains of the coronal suture. The area laterad of the sutures is commonly called the postgena. This area probably is composed mostly of the mandibular segment, but since the premandibular sutures are not completely present, it is impossible to point out the limits of the mandibular segment and of the ocular lobes. However, in many forms in which the complete mandibular suture is retained it follows closely along the row of setae called the "postorbitals."

The occipital foramen is bordered anteriorly by a very narrow band which is continuous laterally with a lobe on each side and extends around the ventral side of the foramen. This border bears on each side a point of articulation with the cervical sclerites and is defined laterally by a line which is the indication of the posterior tentorial pit. This border about the foramen may be regarded as all that is left of the maxillary segment.

Extending across the ventral portion of the foramen is the small and narrow posterior tentorial bar or bridge. The tentorium is fused with the head wall for some distance and then becomes free and extends downward and laterad until it meets the anterior tentorial pit, which is here displaced far posteriorly on the ventral border of the cranium and almost at its margin. The tentorium thus assumes an almost vertical position, departing greatly from the normal longitudinal position which it has in most insects and even in many flies.

The Compound Eyes and Ocelli

As has been pointed out, the three ocelli are situated close together at the top of the head. There is nothing about them that requires any special comment.

The compound eyes are relatively very large and widely separated and are of the same size and shape in the two sexes. There is nothing of special note about them except the presence of small, stiff setae, one of which arises from each angle of meeting of the facets.

Attention may here be called to the supposed existence of an ocular sclerite which surrounds each eye. Actually no such ocular sclerite exists in any insect. The basis for the belief in such a sclerite seems to be as follows. In many insects the ocular foramen is very small, and the eye bulges out around the foramen much like the head of a mushroom on a very short stalk, usually being very closely appressed to the head. The derm immediately beneath the bulging eye is commonly deeply pigmented and forms an ocular diaphragm, but it very definitely is not a distinct sclerite. In Drosophila the ocular foramen is practically as large as the eye itself, and there is no indication of this diaphragm.

The Antennae (Fig. 3)

The antennae are of the type which is peculiar to the "brachycerous" Diptera. The first segment is very small and forms a narrow ring; the second is larger and is somewhat swollen; the third is large and somewhat bulbous; the fourth, fifth, and sixth are greatly reduced in diameter, with the fourth forming a minute ring, the fifth somewhat longer, the sixth elongate, slender, and branched. The first segment is generally known as the scape, the second as the pedicel, and all beyond the second are inclusively

called the flugellum. However, in many of the Diptera the at-
tentuated terminal segments form what is commonly called the

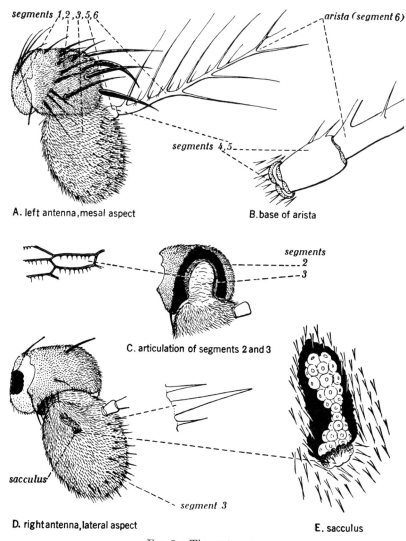

segments 1,2,3,5,6

arista (segment 6)

segments 4,5

A. left antenna, mesal aspect

B. base of arista

segments
2
3

C. articulation of segments 2 and 3

sacculus

segment 3

D. right antenna, lateral aspect

E. sacculus

FIG. 3. The antennae.

arista, and it is by no means clear that the facts of segmentation
of this portion have always been recognized.

The first three segments are beset throughout with small spines,
and in addition the first segment bears a few and the second segment numerous enlarged setae of various sizes.

The articulation of the second and third segments is of some interest. The second segment is invaginated at its apex to form a deep cup, the walls of which are marked by a reticulum of sclerotized ridges. The third segment fits into this cup and has a narrow, short basal stalk by which it articulates to the second segment.

On its outer face the third segment bears a minute invaginated pouch, presumably of a sensory nature and called by earlier authors the "sacculus."

It may be noted that in generalized insects the muscles of the antennae originate on the tentorium. Here they originate on the wall of the head capsule just above their bases.

THE MOUTHPARTS (FIGS. 4, 5, 6)

As will be found discussed in every entomological textbook, the mouthparts of the higher Diptera are much modified, there being much reduction and loss of parts, much modification of parts, and a great extension of the membranous conjunctival areas. The whole structure involves the clypeus, the labrum, the much-reduced maxillae, and the much-enlarged labium. There is no trace of mandibles. The combined structures form a "fleshy" proboscis in which any obvious resemblance to the generalized mouthparts of insects is lost. There has in the past been much dispute concerning the morphology of this proboscis, but the differences of opinion have for the most part disappeared, leaving only a few disputed questions. In general the views presented by Snodgrass (1935, p. 320) are here adopted, although there is disagreement in regard to some details.

Merely as a matter of convenience in speaking of the proboscis as a whole, it may be divided into three regions: first the basiproboscis, which includes the clypeus and the area of the submentum and which bears the maxillary palpi, the maxillary lobes, and the labrum; second the mediproboscis; and third the distiproboscis, which is made up in Drosophila of the swollen labial palpi.

From the cephalic aspect (Fig. 1) one of the most conspicuous features of the proboscis is the quite large sclerite, somewhat in the form of an inverted U, which lies islandlike in a membranous field. Snodgrass has considered this plate to be the clypeus, since the muscles operating the oral pump originate upon it. This opinion is here accepted. The connection of this plate with the pump will be considered later. The plate is divided by a transverse fold into two parts.

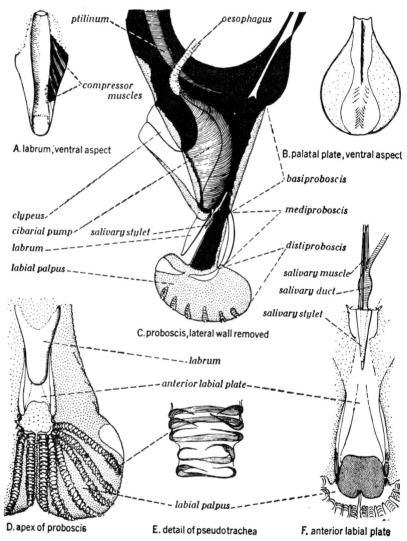

A. labrum, ventral aspect

B. palatal plate, ventral aspect

ptilinum

oesophagus

compressor muscles

basiproboscis

mediproboscis

clypeus

cibarial pump

labrum

labial palpus

salivary stylet

distiproboscis

salivary muscle

salivary duct

salivary stylet

C. proboscis, lateral wall removed

labrum

anterior labial plate

labial palpus

D. apex of proboscis   E. detail of pseudotrachea   F. anterior labial plate

FIG. 4. Details of the proboscis and mouthparts.

Distad of the clypeus and occupying the median portion of the proboscis is a small, sclerotized, apically pointed flap which is accepted by all as the labrum. Under the labrum and at its base is the very small oral opening. Just posterior to the oral opening is a very minute, slender process which bears the opening of the salivary duct at its apex.

This little lobe is usually regarded as the hypopharynx. Actually, it is not the hypopharynx at all, and the hypopharynx does not exist in this fly. This little process is, by all morphological evidence, a papilla formed about the mouth of the salivary duct. As pointed out earlier in this paper, the opening of the salivary duct belongs morphologically to the labium and has nothing to do with the hypopharynx. The mistake of calling any structure the hypopharynx if it bears the opening of the salivary duct is very commonly made, but this does not keep it from being a morphological error. This lobe is here called the salivary stylet.

The dorsal or anterior surface of the labium is occupied medially by an elongated plate of variable sclerotization, to which the term "anterior labial plate" may be applied.

The apical region of the proboscis forms the two large membranous lobes that are commonly known as the labella. The morphological origin of these lobes has been much disputed. Without entering into the various arguments it may here merely be remarked that the evidence now at hand quite definitely indicates that these lobes represent the labial palpi.

It can be shown in many flies that each palpus consists of two segments, and it is probable, on theoretical grounds, that such is the case in Drosophila, but here the fusion of parts has proceeded so far that it would be difficult to indicate where the line between the two is to be drawn. Each lobe consists merely of a sac that is entirely membranous except for small plates at the base. Basally the two sacs unite with each other. On the dorsal or anterior surface there are two minute, sclerotized plates which articulate with the elongated dorsal labial plate. From each of these two little plates there radiate six furrows in the derm. These furrows are the pseudotracheae. They extend from their basal origin over the surface of the palpus to points slightly around the outside of each lobe and there end. The furrows are kept open by narrow, somewhat variably shaped, transverse, sclerotized bands, the points of which are free from the derm.

The maxillae are very greatly reduced. Arising from the basiproboscis there is on each side the slender, one-segmented maxillary palpus. In the angle formed by the palpus and the proboscis there arises a very small pointed process which represents all that is left of the maxilla other than the palpus. This can scarcely be identified more closely than merely to call it the maxillary lobe. The musculature of this will be described later.

As viewed from the caudal aspect (Fig. 1C), the structures of the proboscis are as follows. The basiproboscis is entirely membranous. This area of membrane undoubtedly includes the portion of the labial segment which, if sclerotized, would be called the submentum, and involved in it are undoubtedly areas belonging to the maxillary segment, but these cannot be differentiated.

The mediproboscis is occupied mostly by a somewhat rectangular plate to which the labial palpi articulate apically. Some authors have thought it necessary to apply a special term to this plate, calling it the theca, but there is no reason to regard it as anything more than the area called by Snodgrass the prementum. Morphologically this is formed by the fusion of the parts which in the maxillae are called the stipes.

THE ORAL PUMP (FIG. 5)

The oral opening leads into a large, sclerotized structure which constitutes the sucking pump. This is formed in a rather peculiar manner. Its floor is a strongly sclerotized plate in the form of a deep, elongated trough. The sides of the trough are continued until they meet and fuse with the lateral margins of the external clypeal plate. The dorsal wall is formed by a somewhat pear-shaped plate, which fits, with its small end posteriorly, into the trough and which fuses with the ventral walls along the sides of the trough, the line of fusion being clearly evident as a dark line. The little dorsal plate bears a high median ridge to which muscles attach as will be shown later, and on its oral surface it bears a number of small setae.

The trough narrows posteriorly and opens into the oesophagus.

MUSCULATURE OF THE PROBOSCIS (FIGS. 5 AND 6)

It is not the intention here to enter into a detailed study of the musculature of all the parts in Drosophila, since such a study is beyond the scope of this section. However, the muscles of the proboscis are comparatively few and are concerned in the morphological interpretation of some of the skeletal parts to such an extent that they should be considered.

Snodgrass (1935, p. 323) has expressed this opinion: "Most of the muscles, however, appear to be special adaptations to the functions of the fly proboscis, and they cannot be satisfactorily

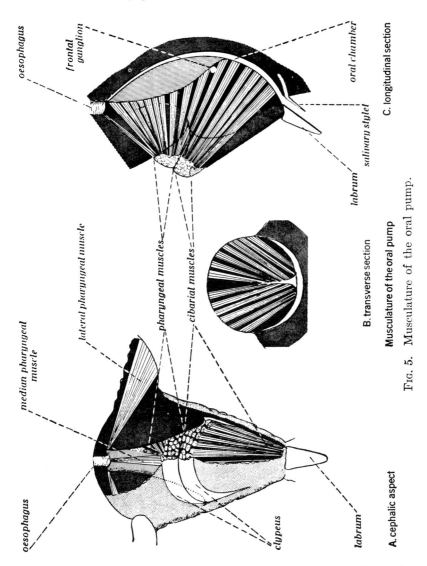

Fig. 5.   Musculature of the oral pump.

homologized with the muscles of the mouthparts in biting insects." Here we have once more the expression of an implied belief in evolutionary discontinuity, a belief which is here definitely rejected. However, only a very extended study of a long series of forms ranging from the generalized type through the modified mouthparts of the Mecoptera, the more generalized flies,

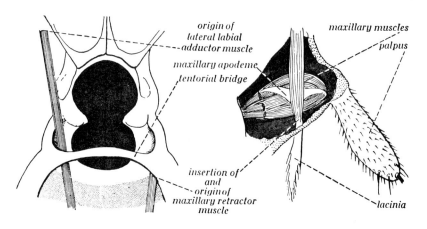

A. internal structures near occipital foramen

B. muscles of maxilla

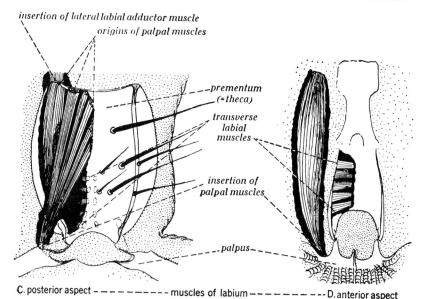

C. posterior aspect – – – – – – – – – muscles of labium – – – – – – – – – D. anterior aspect

FIG. 6.   Musculature of labium and maxilla.

and finally into the highly specialized forms will make it possible to trace homologies definitely. There exist certain excellent studies of the mouthparts of the higher flies, but none of these studies is of a comparative nature, being for the most part merely descriptive accounts of what is to be seen in a single fly species. They therefore are of little or no aid in the present study, and to a certain degree only suggestions can here be made as to homologies of some of the muscles that are present in Drosophila.

The Muscles of the Oral Pump (Fig. 6). The sclerotized portions of this apparatus have already been described. Snodgrass (1944, p. 70) has stated that the pump in the cyclorrhaphous Diptera is composed of the cibarial portion alone, while the pharyngeal portion is not developed. This opinion is not supported by the evidence of the musculature in Drosophila, the evidence indicating that the pump is composed either of the pharyngeal portion alone or of the fused cibarial and pharyngeal portions.

At the posterior end of the pump, where the oesophagus begins, there inserts on each side a large muscle which extends laterally and originates on the cranial wall just below and anterior to the eye. This seems to correspond to the lateral pharyngeal muscle of other insects and is here so designated.

The external clypeal plate, as has been noted, is divided transversely by a fold. On the area posterior to this fold there originate numerous muscle bundles that insert upon the median ridge of the dorsal plate of the pump and also on the plate itself. On the area anterior to the fold there originate numerous such bundles which insert upon this plate just behind the oral opening. The two groups of muscles thus somewhat oppose each other. Between the two groups lies the frontal ganglion, and the connectives from this ganglion to the brain seem to embrace the anteriormost muscles, thus suggesting that all the muscles are actually pharyngeal. However, the anterior group is here considered to be composed of cibarial, and the posterior group of pharyngeal, muscles.

The Labrum (Fig. 4A). The writer, after repeated dissections, has been unable to find any muscles that move the labrum as a whole, but within the labrum are small muscle bundles that may be regarded as the labral compressors.

The Maxillae (Fig. 6B). The little external maxillary lobe here called the lacinia is continuous with a very small, sclerotized, T-shaped apodeme that extends into the head cavity like the handle of a dagger. This apodeme seems quite definitely to be homologous with the very large and conspicuous "maxillary rods" that appear in many other flies and that have been the source of much argument among morphologists and much quite unnecessary mystification. Since this apodeme is very definitely maxillary in those forms in which maxillary vestiges other than the palpus remain, its homologies are to be sought in some persistent structure of the maxilla. The most likely candidate for our choice is the apodeme of the muscle called by Snodgrass the cranial flexor of the lacinia or, later and perhaps much more appropriately, the retractor of the maxilla. This is an extremely persistent muscle which in many insects develops a very large apodeme and which inserts upon the base of the maxillary lobe called the lacinia. Ordinarily the muscle attached to it originates on the dorsal, posterior wall of the head. Consequently, this apodeme is here considered to be the apodeme of the maxillary retractor muscle.

In Drosophila one arm of the crossbar of the T impinges upon the base of the maxillary palpus and seems to serve as the fulcrum upon which the palpus is hinged, although the two structures are not attached to each other. From the upright of the T a long, slender muscle passes upward and originates on the cranial remnant of the maxillary segment, just below and laterad of the occipital foramen. From the crossbar of the T several small muscle bundles extend to the clypeal phragma just posterior to the base of the labrum. The exact homology of these muscles is not clear and cannot be indicated until studies on other flies have been completed. They are here simply called maxillary muscles.

These muscles inserting upon the T-shaped apodeme are all that serve the maxilla in Drosophila.

The Muscles of the Labium (Figs. 6A, C, D). Attention has been called already to the elongate plate on the anterior or dorsal surface of the labium and to the quadrangular plate on the posterior surface. The establishment of the homologies of this posterior plate depends upon the muscles that are connected with it. It should be noted that the plate is divided lengthwise into three portions, there being a broad central portion which is demarked on each side from a narrow lateral portion by a strongly sclerotized line. No explanation for these side pieces is at present available. Whether they represent separate parts which have fused with the median plate or are merely subdivisions of the whole plate cannot be determined at present.

The musculature is as follows.

Upon each basal angle of this median plate there inserts a long, narrow muscle which passes upward between the posterior tentorial bar and the posterior wall of the head and originates on the area just dorsad and laterad of the occipital foramen, an area that presumably belongs to the mandibular segment.

From the same basal angle a single muscle originates and passes in the opposite direction to the anterior or dorsal side of the labium and inserts upon the dorsal articulation of the palpus. From the median, basal half of the plate a bundle of muscles passes to the ventral articulation of the palpus, where it inserts. All these therefore are the palpal muscles, and in number and position they seem to be quite in accord with what is to be seen in generalized insects. From the median, distal portion of the plate two rows of little muscles pass transversely through the labium and attach to the dorsal or anterior labial plate, distad of the apparent opening of the salivary duct. The homologies of these little muscles are not clear.

The identity of the other muscles and consequently the identity of the median plate seem to be quite clear. This plate represents the "prementum" of generalized insects as described by Snodgrass (1935, p. 146, Fig. 84). The muscles inserting upon its basal angles correspond exactly with the lateral labial adductors, except that their origins are not upon the tentorium but on the head wall. The identity of the papal muscles is clear.

The Salivary Duct (Fig. 6D). The common duct of the salivary or labial glands swells into a slight bulb just before entering the salivary stylet which has previously been described. Attached to the base of the stylet dorsally is a pair of small muscles which originate on the ventral side of the oral chamber and are without much doubt the muscles called by Snodgrass the dorsal salivary muscles. In more generalized forms these originate on the hypopharynx.

CHAETOTAXY OF THE HEAD (FIG. 2)

The head capsule bears a number of setae, some of which are relatively very large and all of which occupy quite definitely fixed positions. These setae are designated by terms derived from their positions. The terminology is indicated in the accompanying figures as nearly in accord with the commonly accepted system as the writer has been able to determine it.

THE THORAX

General Considerations

A concept of the organization of the thorax is here adopted which is in many respects different from that generally accepted. The divergences have especially to do with the plates generally called sternites. The views here presented have their beginning in embryological work done many years ago by Heymons, who concluded that the lateral plates of the thorax, commonly called pleurites, are actually parts of a subcoxal segment of the legs. This view was adopted by Snodgrass, but the latter author attempted to combine this concept with the conventional ideas which include the presence of ventral plates or sternites as a part of the basic system of the thoracic plates. This subject has been discussed by the present writer (1940b), and the opinion has been adopted—and will here be maintained—that such sternal plates are not a part of the basic pattern in winged insects. That true sternites, that is, plates developed from genuinely sternal centers of sclerotization, are present in many insects is undeniable. But such plates appear only subsequent to certain other developments, and in perhaps the majority of all insects they are not present at all or may be considered to involve only the limited areas about the openings of the sternal apophyses. The attempt to force these plates into the structure of all winged insects has led to much confusion and to the virtual impossibility of attaining sound interpretations of the homologies of these parts as long as such an attempt is continued. It seems to be this which has led to the remark of Snodgrass, already quoted, to the effect that the sutures on the thorax of the higher flies have little to do with those of the more generalized insects.

The conventional concept of the insect thorax assumes that in cross-section the walls of each segment involve a dorsal plate or tergite, a lateral plate or pleurite on each side between the wings and the leg bases, and a ventral plate or sternite which lies between the leg bases.

The concept here adopted assumes that these walls involve merely a dorsal plate or tergite and a pair of limb bases. These limb bases are in the form of an irregular, truncated cone, with the apex downward. The bases of each pair of these cones meet and are at times closely pressed together along the midline of the venter, enclosing between them certain structures which lie on the midline and which alone can be regarded as truly sternal. These structures are the sternal apophyses, the paired apophyses of the furcae, and the unpaired apophyses of the spinae. By certain processes, which can be definitely traced, these truly sternal elements may expand in some insects and come finally to appear as definite sternal plates.

The Generalized Pterothorax

(Figs. 7, 8, 9)

The two wing-bearing segments of insects are usually quite closely united to each other and are commonly referred to collectively as the "pterothorax." These wing-bearing segments in all insects have followed a course of evolution which departs more or less widely from that followed by the prothorax. Usually these segments are much more complicated than is the prothorax. Because of these complications their structures have been the subject of very considerable disagreement among morphologists, and some preliminary explanations seem necessary. In order to make the whole subject more readily understandable there are here included three illustrations from an earlier paper (Ferris and Pennebaker, 1939). These illustrations of the thoracic structures of a neuropteran, *Agulla adnixa* (Hagen), are based upon a species in which the pterothorax is probably as closely similar to the pterothorax of the primitive winged insects as anything that now exists.

We need look no farther than this insect in order to see clearly a substantiation of the idea that the body wall of each of these segments is composed of a series of dorsal plates and of the walls of the limb bases or subcoxae. In fact, if we were to remove these limb bases, there would be but little of the thorax left. Note how each limb base appears as an irregular frustum of a cone of which the ventral or median walls are much reduced but are still clearly present and how the cones of each pair press against each other along the midline of the venter, forming an apparent median

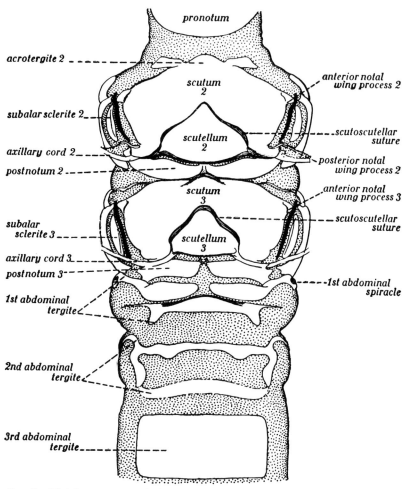

FIG. 7. Notal structures of the pterothorax of a generalized winged insect, *Agulla adnixa* (Hagen), of the Order Neuroptera.

surface line which in these illustrations is labeled the "line of invagination of pleura and sternum." It was later realized that there is no sternum involved other than the small area contained in the sternal apophyses and that the word is misleading. Consequently the term "discriminal line" was later applied to this median fold.

Note how the structures appearing on the lateral aspect are continuous with those which are revealed when the members of

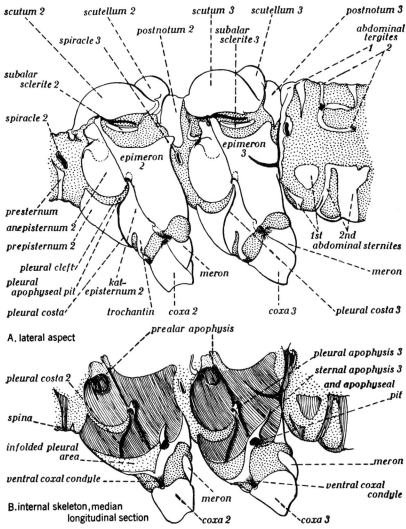

Fig. 8. Subcoxal (= pleural) structures of the pterothorax of a generalized winged insect, *Agulla adnixa* (Hagen), of the Order Neuroptera. Lateral aspect.

each pair are drawn apart. Note in the figure of the lateral aspect (Fig. 8A) the line or fold extending from the coxal condyle to the corresponding wing base and bearing the pleural apophysis. This is the "pleural fold or suture" and is a primary landmark.

Note the dark line (Fig. 9) indicated as the pleural costa which extends from near the opening of the pleural apophysis around to the ventral side of the body and divides the episternum into two areas, the anterior of which is called the pre-episternum.

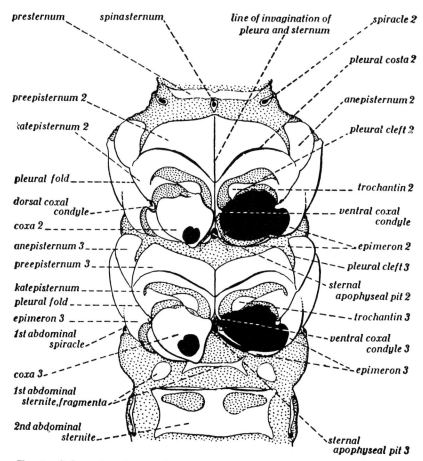

FIG. 9. Subcoxal and sternal structures of the pterothorax of a generalized winged insect, *Agulla adnixa* (Hagen). Ventral aspect.

This also is a primary landmark, and the changes in proportion of the areas anterior to and posterior to this line account for much of the difference in appearance which is to be seen in different groups of insects.

A study of these illustrations will show the basic structures from which the thorax of a fly such as Drosophila have been derived through processes involving little more than fusion of parts, changes of proportions, or loss of parts. The thoracic structures of every pterygote insect that the writer has seen in the course of several years of work (with the possible exception of the dragonflies) can be derived from this pattern without invoking any processes other than these.

One difficulty is involved in the application of this concept. Its use makes it difficult to formulate a satisfactory terminology for the parts involved, and because of this occasional compromises between a strictly morphological terminology and a purely topographical terminology, with some resulting discrepancies, are at present inescapable.

The Thorax of Drosophila

(Figs. 10, 11, 12, 13)

With this background we may now proceed to an examination of the thorax of Drosophila.

THE CERVICAL REGION AND THE PROTHORAX (FIGS. 10 AND 11)

Lying in the neck membrane there is on each side a pair of plates called the cervical sclerites, which serve as a link between the head and the prothorax. In this species the anterior plate of each pair is extremely small and seems to be fused with the posterior plate to such an extent that the two are practically one. Anteriorly this compound plate articulates with the head at a point on the side of the occipital foramen and posteriorly with the episternum of the prothorax. There has been much argument concerning these plates, and they have been considered by some to be vestiges of a neck segment, but certain recent work on the nervous system supports the generally accepted conclusion that they belong to the prothorax.

The prothorax itself is greatly reduced, serving merely as a support for the prothoracic legs. Its dorsal portion, or notum, is merely a narrow collar extending across the thorax and almost concealed beneath the bulging anterior margin of the mesonotum. It is closely fused posteriorly with the mesonotum and laterally with the subcoxal (= pleural) elements of its own segment, being separated from the latter, however, by a fold.

The subcoxal region consists of a single large piece from which the leg is swung by a dorsal condyle. From this condyle there

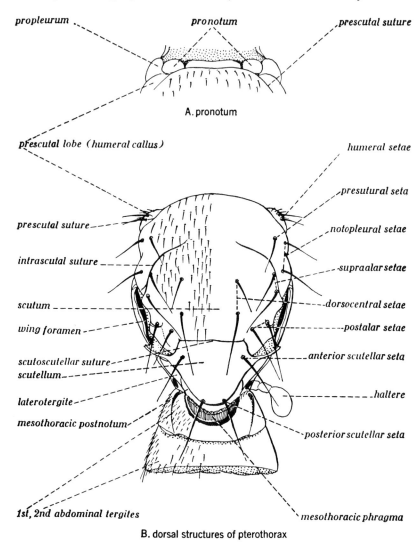

propleurum   pronotum   prescutal suture

A. pronotum

prescutal lobe (humeral callus)   humeral setae

                    presutural seta

prescutal suture   notopleural setae

intrascutal suture   supraalar setae

scutum   dorsocentral setae

wing foramen   postalar setae

scutoscutellar suture   anterior scutellar seta
scutellum

laterotergite   haltere

mesothoracic postnotum   posterior scutellar seta

1st, 2nd abdominal tergites   mesothoracic phragma

B. dorsal structures of pterothorax

FIG. 10.   Notal structures of the thorax of Drosophila.

extends upward a fold that divides the pleural area into a large anterior portion or episternum and a smaller posterior area, the epimeron. This fold is the pleural fold or "suture," and from it arises the pleural apophysis which extends into the body and fuses with the sternal apophysis.

The episternum is continued by a very narrow, sclerotized isthmus around to the ventral side of the body, and there expands into a large plate which meets its companion from the opposite

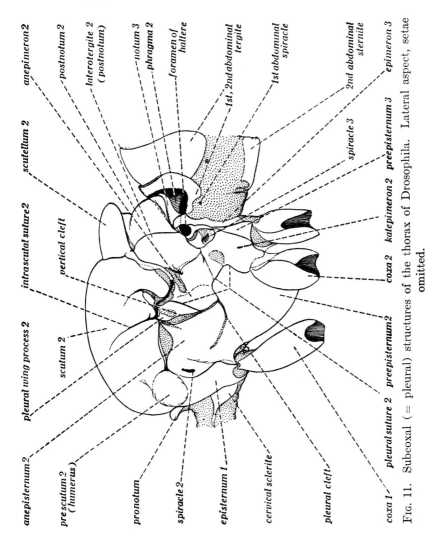

Fig. 11. Subcoxal (= pleural) structures of the thorax of Drosophila. Lateral aspect, setae omitted.

side of the body along a clearly defined median line, the two plates together forming an area that separates the coxae. Ordinarily this ventral piece is considered to be the "sternite," but morphologically it is clear that it is merely a part of the ventral arc of the subcoxa, and it is here included in the episternum.

Just posterior to this median area is a median plate which surrounds the openings into the sternal apophyses. This area may be regarded as truly sternal and is here identified as the furcasternum.

No trochantin is present, and there is no ventral articulation of the coxae.

THE PTEROTHORAX (FIGS. 10, 11, 13)

Having in mind the general discussion which has been presented and the structures of a typical pterothorax as shown in the illustrations of *Agulla adnixa*, we may turn to the pterothorax of Drosophila.

First of all it will be noted that all the segments of the thorax have fused into an almost solid box, with the intersclerital membranes almost obliterated. Then it will be noted that there has been a tremendous enlargement of the mesothorax and a reduction of the metathorax, these changes being associated with the fact that the mesothoracic wings carry the entire flight function, while the metathoracic wings have been reduced to the halteres, which have become organs of equilibrium.

It will be necessary to keep a close account of the various landmarks in order to follow the various modifications in detail which have occurred in addition to the more conspicuous changes.

The Notum (Fig. 10). Tipping the anterior end of the thorax up, it is possible to see the pronotum. It is closely fused with the mesonotum, and along the line of fusion a phragma is formed which probably represents in part the acrotergite or prescutum of the mesonotum. The prescutum is continued as a narrow band along the mesonotum and swells out on each side to produce a prescutal lobe that to systematists is known as the "humeral callus."

The next plate, which occupies the greater part of the mesonotum, is the scutum of the mesothorax. About midway of its length a depressed line extends from near each wing base to about a third of the way across the scutum. These lines are the "intrascutal sutures."

Posteriorly, just behind the wing bases, the scutum is separated by a deeply impressed line from the small, shield-shaped scutellum. The latter is produced laterally into narrow arms which lead into the axillary cord and thus to the posterior border of the wings.

Behind the scutellum and almost completely concealed by it is a large, transverse plate which dips sharply downward, and the lateral margins of which are carried around to the sides of the body just behind the wings. This plate dips far downward into the body and forms a deep phragma. Its lateral portions are separated from the median part by a fold and have the appearance of constituting distinct plates, although they are probably not so. There has been much confusion concerning this plate. By some it has been regarded as the notum of the metathorax, while others have considered the lateral parts to belong to the pleural areas of that segment. According to the views of Snodgrass, which are here accepted, it is the postnotum of the mesothorax. The lateral pieces are commonly called laterotergites.

The metonotum is very greatly reduced. It is possible that the postnotum of the mesothorax actually belongs to the metathorax, but if it does not, all that is present dorsally of the latter segment is contained in the posterior wall of the deep phragma.

The Subcoxal or Pleural Areas (Figs. 11 and 13). From the lateral aspect also most of the structures that are to be seen belong to the mesothorax. A consideration of these parts may very well begin with the dorsal condyle of the middle leg.

Beginning at this point there is to be seen a deeply impressed line which extends by a sinuous course from the leg articulation upward to the base of the wing. This is the subcoxal fold or pleural suture, which has already been indicated as one of the most persistent landmarks of the thorax. The sinuous course which it here pursues is characteristic of the more specialized Diptera, for in the more generalized Diptera and in most other insects it forms an almost straight line from the articulation to the wing base. It is the surface indication of an infolding, which in some insects forms a deep phragma, that functions as a support both for the wing and for the articulation of the leg. The phragma is to some degree developed in Drosophila and follows the sinuous course of the furrow.

The subcoxal (= pleural) elements of the mesothorax are thus divided into a posterior series and an anterior series, the former constituting what is commonly called the epimeron and the latter the episternum.

We may consider the episternum first. One peculiarity of Drosophila, to which reference has previously been made, is the complete secondary sclerotization of the membrane between the pro-

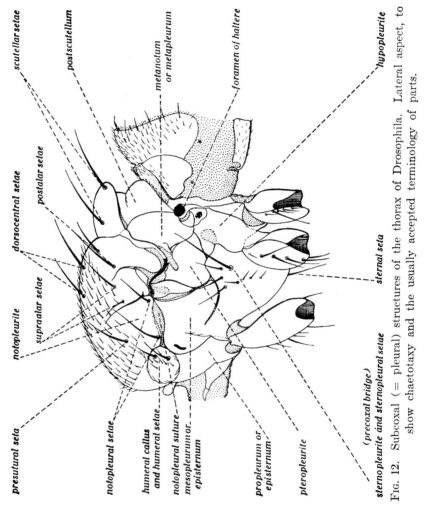

FIG. 12. Subcoxal (= pleural) structures of the thorax of Drosophila. Lateral aspect, to show chaetotaxy and the usually accepted terminology of parts.

thorax and mesothorax. Ordinarily the mesothoracic spiracle lies in this membrane, but here it is embedded in a sclerotized plate that is continuous with the other sclerotized elements of both segments. This fact makes it impossible to define the exact limits of all the parts.

Extending forward from the pleural fold almost to the anterior border of the episternal region is a distinct line or suture which

Snodgrass (1935, p. 185) has called the "episternoprecoxal suture." A reference to the accompanying figure of the thorax of Agulla (Fig. 8A) will show at once that this corresponds in all respects to the line there called the pleural cleft. This cleft divides the episternal region into an upper "anepisternum" and a lower "katepisternum."

The anepisternum is divided vertically by a cleft, which for part of its length coincides with the pleural folds and which leaves an isolated piece of the episternum just below the wing base. This cleft is here called the vertical cleft. It is distinctive of some of the higher Diptera.

The katepisternum passes around in front of the coxa to the ventral side of the body, its continuity unbroken. It meets the corresponding piece from the other leg, and along the midline this point of meeting forms the "discriminal line." Conventionally this ventral area would be assumed to contain some element of the sternum, and also the area in front of the coxa has been called the precoxal bridge. From the point of view here adopted it is clear that no sternal element enters into the picture at all, and it is also clear that the term "precoxal bridge" has no morphological meaning.

If one refers again to the accompanying illustration of Agulla (Fig. 9) there will be seen the line called the pleural costa, which divides the katepisternum into two parts, the anterior part being called the pre-episternum. It will be noted that in Drosophila this line does not appear. Evidence from other flies indicates that in the Diptera the line tends to move posteriorly until it nearly or even quite corresponds with the posterior margin of the episternum, thus leaving the pre-episternum as the sole possessor of the entire area. It is for this reason that the name "pre-episternum" is used for the entire ventral area in Drosophila.

Along the midline between the coxae the subcoxa is continued as a very narrow band, which carries the ventral articulation of the coxa and passes around the openings of the small mesosternal apophyses to connect with the portion of the area behind the coxa.

The trochantin is entirely lacking, and there is consequently no trochantinal articulation.

Returning to the lateral aspect of the thorax, we see posterior to the pleural fold a large and very irregular plate. This is the

epimeron. It is divided into a dorsal plate or anepimeron and a
ventral plate or katepimeron by a furrow which corresponds to

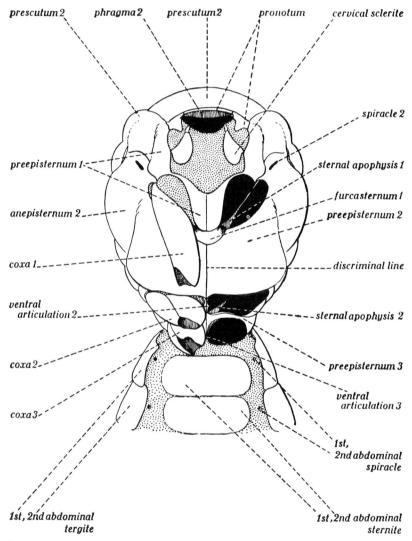

FIG. 13. Subcoxal and sternal structures of the thorax of Drosophila. Ventral aspect.

and is probably a continuation of the pleural cleft of the episternum. The katepimeron passes on down around the posterior side
of the coxa to meet the ventral prolongation of the episternum.

The Metathorax. The subcoxal elements of the metathorax are greatly reduced. The pleural fold is clearly recognizable, extending upward to the base of the haltere. The episternum passes down and around the coxa and carries the very inconspicuous ventral articulation of the coxa, but there terminates. The epimeron is not continued downward around the posterior side of the coxa. The metathoracic spiracle is borne on a small plate just below the base of the haltere.

The Thoracic Spiracles (Fig. 15). The position of the spiracles has already been pointed out. Both spiracles, as is usual in most insects, are noticeably large. They seem to be the same in structure. In both the opening is merely a slit which is bordered by a fringe of fimbriate processes forming a filter.

CHAETOTAXY OF THE THORAX (FIGS. 10 AND 12)

The large hairs of the thorax occupy quite definitely fixed positions for which systematists have devised an elaborate terminology. These setae, with the names commonly employed, are indicated in Fig. 15, in which also the non-morphological terminology, as far as it differs from the morphological terminology, is indicated. As seen from the dorsal aspect, these setae are indicated also in Fig. 10. It seems unnecessary to enter into any detailed description of them.

It should be noted that this fly is relatively hairless and that a complete terminology of the setae in the more hairy flies is much more elaborate than that here employed.

THE LEGS (FIG. 14)

The legs require no special consideration, the normal parts being present.

In the case of the foreleg there is a sexual dimorphism due to the presence in the male of a longitudinal comb or ctenidium of enlarged setae on the metatarsus or basal tarsal segment. The accompanying figures (Fig. 14) should obviate the necessity for any extended discussion.

THE WINGS AND WING BASES (FIGS. 15A, 15B, 16)

The wing bases show the axillary sclerites in a pattern that clings quite closely to the typical form. There is one small sclerite which seems to be additional to the usual number, and the medial plates are somewhat reduced, but otherwise there is nothing at all peculiar. The terminology of the various parts

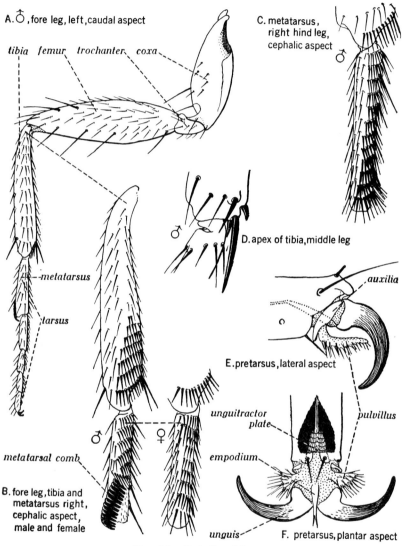

FIG. 14. Details of legs.

may be obtained by reference to the figures. These are labeled in accord with the views of Snodgrass (1935, pp. 219, 228).

The only special complication in regard to the wings rises from the circumstance that there are varying terminologies for the parts and for the veins. There is a morphological terminology, and there are various non-morphological terminologies which have been employed by systematists.

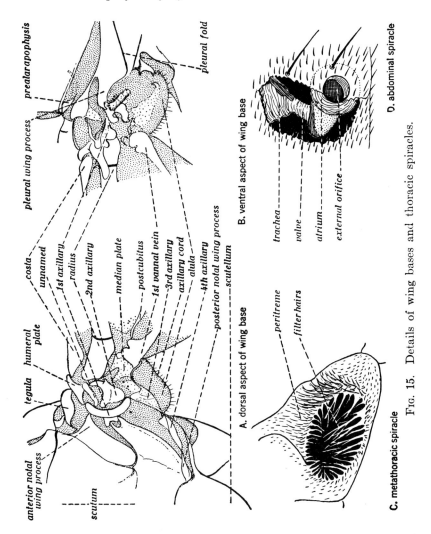

FIG. 15. Details of wing bases and thoracic spiracles.

The morphological terminology (Fig. 16A) has been presented, as nearly as we can determine it, in accord with the views indicated by Snodgrass (1935, pp. 221 et seq.).

It should be noted that the little basal lobe on the posterior border of the wing, which is called by some systematists the alula (Fig. 16B), would not be so regarded by Snodgrass, the term "alula" being used by him for the membranous area between this lobe and the body.

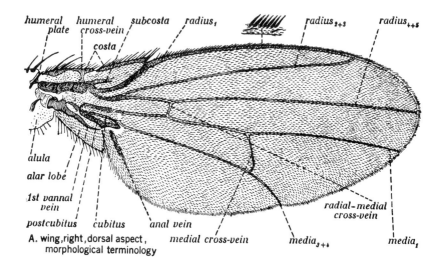

A. wing, right, dorsal aspect, morphological terminology

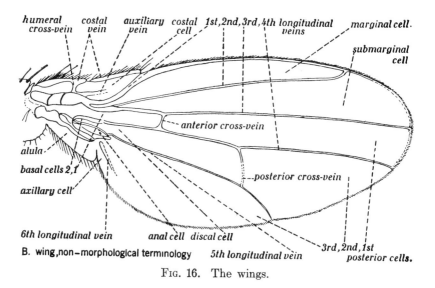

B. wing, non-morphological terminology

FIG. 16. The wings.

The non-morphological terminology most commonly employed is indicated in Fig. 16B. In this the cells are named. In Fig. 16A each cell would bear the name of the vein which forms its anterior border, and these names are therefore not indicated for the cells.

THE HALTERES (FIG. 17)

Without any question the halteres are the morphological equivalents of the metathoracic wings, although certainly they have come far from their morphological antecedents. It may be noted that there are a few flies in which the halteres are definitely winglike.

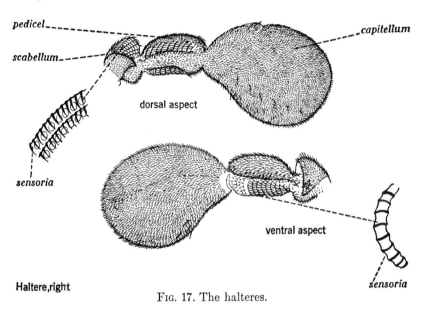

FIG. 17. The halteres.

The terminology here employed (Fig. 17) is derived largely from Lowne's study of the blowfly. The haltere is divided into three parts: a basal portion or "scabellum," a median portion or "pedicel," and the swollen apical portion or "capitellum." On the dorsal side of the scabellum and pedicel are peculiar structures which are the surface indications of sensory organs.

THE ABDOMEN

Segmentation

(Fig. 18)

The segmentation in both sexes has been disturbed by secondary developments so that the primitive number of segments is not found. Furthermore, but seven pairs of spiracles are present.

The female seems to be more generalized and may be considered first. On the basis of studies previously made on the gener-

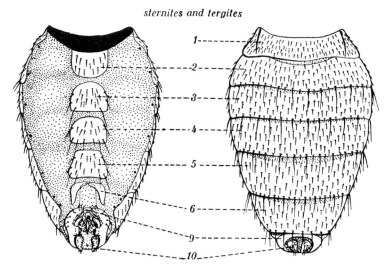

A. abdomen of male, ventral aspect   B. abdomen of male, dorsal aspect

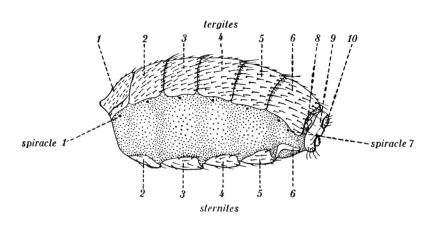

C. abdomen of male, lateral aspect

Fig. 18. Segmentation of the abdomen of the male.

alized family Tipulidae it appears that the genital opening of the female in the Diptera, as in most insects, is between the eighth and ninth abdominal segments. If this is assumed to be the case here, the spiracles must be assigned to the first seven segments. The eighth segment is therefore without a spiracle, as in the Tipulidae. After the eighth segment there is recognizable only a small segment, forming what might be called the proctiger and bearing the anus, this segment having a dorsal and a ventral plate (Fig. 19). The identity of this segment is open to doubt, but it is here considered to be the ninth, which would mean that the tenth and eleventh have become entirely membranous and are represented by nothing more than the membrane about the anus.

In the male (Fig. 18) the situation is still more complicated. It is well known that in the males of the more specialized Diptera segments may disappear entirely. Here, apparently, the seventh segment has disappeared or at least has lost all sclerotization, while its spiracle has moved forward until it is closely associated with the sixth tergite. The eighth segment seems to be represented, as far as any sclerotization is concerned, merely by a small plate on each side (Fig. 20), which represents a vestige of the tergite. The ninth is represented by a very large tergite and a much smaller sternite, while the tenth is represented by a pair of sclerotized plates lying alongside the anus. The eleventh segment must be regarded as entirely membranous.

In the unspecialized segments there is a very large tergal plate (Fig. 18) and a very small sternal plate. The sternite of the first segment seems to be lacking in both sexes.

Terminalia of the Female
(Figs. 19, 20)

Modification begins with the eighth segment, which has no spiracle and no definite sternite. Ventrally this segment bears two elongate plates the bases of which are connected by a narrow, heavily sclerotized bar. In the light of what is found in the Tipulidae these plates are here considered to represent the gonopods of the eighth segment, being all that remains of the ovipositor. The vulva lies between the apices of these gonopods.

From the wall of the uterus there arise three irregular tubes which extend into the body. Two of these terminate internally, each in a mushroom-shaped, sclerotized body which is a spermatheca. The third terminates as a simple tube.

Terminalia of the Male

(*Figs. 20, 21, 22*)

In probably almost all insects the genital opening of the male is between the ninth and tenth segments. This fact affords a

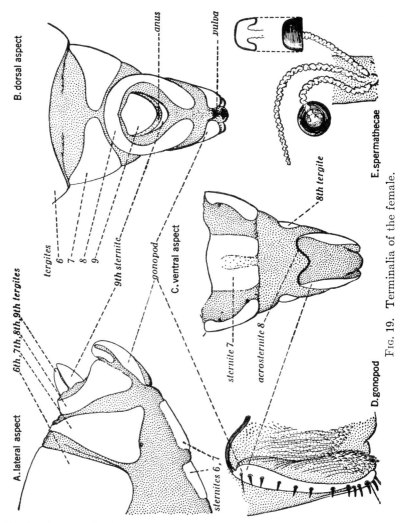

FIG. 19. Terminalia of the female.

basis for conclusions regarding the numbering of the segments, and the conclusions here presented are so based.

Modification begins with the sixth segment. The spiracle of the seventh segment has moved forward and become associated

with the posterior margin of the sixth tergite. The sixth sternite
is divided medially into two slightly elongate plates which bear
no setae. The seventh segment is apparently entirely membran-

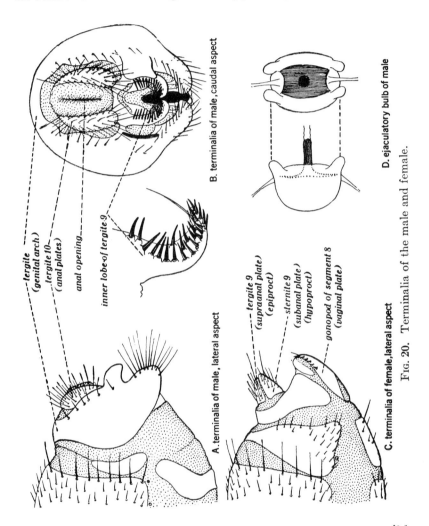

B. terminalia of male, caudal aspect

D. ejaculatory bulb of male

tergite
(genital arch)
tergite 10
(anal plates)
anal opening
inner lobe of tergite 9

A. terminalia of male, lateral aspect

tergite 9
(supraanal plate)
(epiproct)
sternite 9
(subanal plate)
(hypoproct)
gonopod of segment 8
(vaginal plate)

C. terminalia of female, lateral aspect

Fig. 20. Terminalia of the male and female.

ous and is represented only by its spiracle. It is of course possible
that this segment has merely fused with the sixth, but there is no
evidence to support such a conclusion.

As far as sclerotization is concerned, the eighth segment is rep-
rented merely by a small, presumably tergal, plate on each
side of the body in the membrane between the sixth and ninth

tergites, although it is probable that at least a part of this membranous area belongs to the eighth segment.

The ninth segment is very highly modified. Its most conspicuous feature is a very large dorsal plate which is produced down-

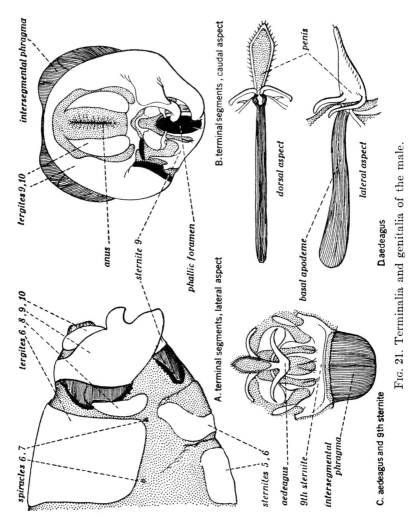

Fig. 21. Terminalia and genitalia of the male.

ward on each side past the ventral side of the body. In part this plate is certainly the tergite, but there is a suggestion that other elements are involved in it, as will be shown later. Viewed from the side, this plate shows a prominent lobe on its posterior border, and below this it is continued into a prominent point. Anteriorly the dorsal portion of the plate is continued beneath the interseg-mental membrane and into the body as a large phragma. From the posterior aspect it can be seen that the plate, beneath the pro-jecting posterior lobe, is continued as a narrow band across be-neath the anus. From this band there arises a pair of curved, fingerlike lobes that are free except at their bases and that have their mesal margins set with stout, black setae.

The ninth sternite also is an elaborate structure. Its base is continued into the body as a large, platelike phragma. Its pos-terior margin is produced into several fingerlike processes, which are not free. The mesal pair of these processes encloses the foramen of the aedeagus.

The homologies of these parts will be considered later.

The tenth segment shows merely a pair of sclerotized plates flanking the anus. The eleventh segment is entirely membranous.

THE AEDEAGUS (FIGS. 21, 22)

The external portion of the aedeagus consists of a median, sclerotized, pointed lobe, through which the ejaculatory duct dis-charges. At the base of this lobe and immovably attached to it are four slender, recurved processes.

From the base of the aedeagus a long, slender apodeme extends into the body.

The ejaculatory duct leading into the body from the penis enters the swollen ejaculatory bulb. This is here illustrated, but any discussion is left to the section on internal anatomy.

MUSCULATURE AND HOMOLOGIES OF THE AEDEAGUS AND ASSOCIATED PARTS (FIG. 22)

It should be noted that at the present time there exists a very high degree of confusion concerning the genital structures of male insects because of the great variety of form which they present. So great is this diversity that the feeling has developed that these structures cannot be reduced to any consistent system. The writer, however, is convinced that such a system can be shown to exist. Certain work now being done by one of his students is pointing toward that much-to-be-desired result but is at present incomplete and will not here be anticipated except in regard to one or two essential matters.

The fundamental difficulty has to do with two pairs of movable structures associated with the genitalia, these being the styles or claspers of the ninth segment with their coxites and the parameres.

The styles belong to the original segmental appendages of the ninth segment, and in their most distinctive form consist of the coxite and a terminal lobe which is apparently identical with the

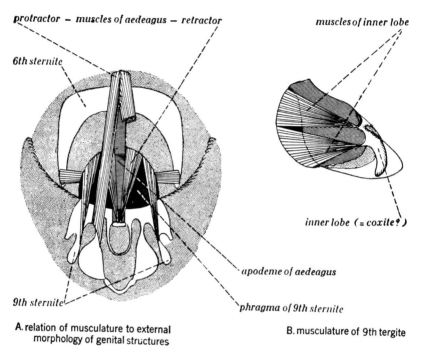

A. relation of musculature to external
    morphology of genital structures

B. musculature of 9th tergite

Fig. 22. Musculature of terminalia and genitalia of male.

coxal style of such insects as the Thysanura. The style does not represent the telopodite of the segmental appendage. The coxite in at least some insects tends to articulate with the tergite of the ninth segment. The style is moved by two opposing muscles which have their origins on the coxite.

The parameres are usually regarded as secondary developments arising from the base of the aedeagus, but unpublished evidence now at hand from the work of a student (Joel Gustafson) indicates that they are modifications of and derived phylogenetically from the eversible sacs of the ninth coxite of such forms as the Thysanura. In the Thysanura the muscles of the parameres originate on this coxite.

The confusion concerning these structures has arisen in part because their phylogenetic origin has not yet been traced through a series of forms beginning with the Thysanura, in part from the fact that they present many vagaries of form, and in part from the fact that both parameres and style occur simultaneously in but few insects that have been carefully studied. Thus Crampton (1942, p. 85) has been led into the idea that the styles and the parameres are actually the same thing, which they definitely are not.

The whole subject is still unclear, but on the basis of such evidence as is now available we may proceed to examine the possible homologies of these genitalic parts in Drosophila. This requires a study of the musculature.

The aedeagus itself is actuated by two pairs of muscle bundles. Two very large bundles insert upon the inner end of the basal plate, and one of them originates on each side on one of the fingerlike lobes of the ninth sternite. It is evidently these muscles which by contracting cause the aedeagus to be protruded, and they may therefore be designated as the protractors of the aedeagus.

The second pair of muscles originates upon the inner face of the basal phragma of the ninth sternite and insert upon the base of the aedeagus itself. These evidently function to retract that structure and may therefore be designated as the retractor muscles of the aedeagus.

One other pair of muscles originates upon this phragma and inserts upon another pair of the lobes of the ninth sternite. The identity of these muscles is not clear, and they must for the present be left unnamed.

The fingerlike lobes arising from the inner side of the projecting terminal lobe of the ninth tergite and bearing a row of black setae are movable, even though connected to the ninth tergite by a narrow, sclerotized isthmus. They are actuated by two opposing muscles which have their origin on the basal phragma of this tergite. Because of the arrangement of these muscles the opinion is here adopted that these lobes in all probability represent the coxopodites of the segmental appendages of the ninth segment, the styles having been lost. Since this opinion can be confirmed only by extended studies, they are here labeled merely as the inner lobes of the ninth tergite.

It would thus appear that there are no parameres in this fly. The little processes at the base of the aedeagus certainly cannot be regarded as such, since they are immovable processes of the aedeagus.

There is in the external morphology of this species no indication of the 360° rotation of the terminalia which has been called upon by other authors in order to explain certain conditions connected with the internal structures. The writer holds to the opinion that some explanation, other than this supposed rotation, should be sought.

Abdominal Spiracles

There are, as has been noted, seven pairs of abdominal spiracles. All these lie in the pleural membrane close to the ventral margins of the tergites. All are of a single type. The external orifice has no filter apparatus and leads directly into the more or less spherical atrium, which in turn opens into a somewhat conical pouch from which a slender tube connects with the trachea.

LITERATURE CITED

CRAMPTON, G. C. 1942. The external morphology of the Diptera. In The Diptera or True Flies of Connecticut, State Geological and Natural History Survey of Connecticut. *State of Connecticut Public Document* 47:10–174; Figs. 1–17.

FERRIS, G. F. 1940a. The morphology of *Plega signata* (Hagen) (Neuroptera: Mantispidae). *Microentomology* **5**, part 2:33–56; Figs. 6–20.

FERRIS, G. F. 1940b. The myth of the thoracic sternites of insects. *Microentomology* **5**, part 3:87–90.

FERRIS, G. F., 1942. Some observations on the head of insects. *Microentomology* **7**, part 2:25–62; Figs. 10–27.

FERRIS, G. F. 1943a. Some fundamental concepts in insect morphology. *Microentomology* **8**, part 1:2–7.

FERRIS, G. F. 1943b. The basic materials of the insect cranium. *Microentomology* **8**, part 1:8–24; Figs. 1–6.

FERRIS, G. F., and PHYLLIS PENNEBAKER. 1939. The morphology of *Agulla Adnixa* (Hagen) (Neuroptera: Raphidiidae). *Microentomology* **4**, part 5:121–142; Figs. 59–71.

FERRIS, G. F., and BRYANT E. REES. 1939. The morphology of *Panorpa nuptialis* Gerstaecker (Mecoptera: Panorpidae). *Microentomology* **4**, part 3:79–108; Figs. 36–51.

HENRY, LAURA M. 1947–1948. The nervous system and the segmentation of the head in the Annulata. *Microentomology* **12**:59–110; Figs. 23–45. *Microentomology* **13**:1–48; Figs. 1–16.

PETERSON, ALVAH. 1916. The head-capsule and mouth-parts of Diptera. *Illinois Biological Monographs* **3**, 2:1–112; plates 1–25.

REES, BRYANT E., and C. F. FERRIS. 1939. The morphology of *Tipula reesi* Alexander (Diptera: Tipulidae). *Microentomology* **4**, part 6:143–178; Figs. 72–91.

SNODGRASS, R. E. 1935. *The Principles of Insect Morphology*. McGraw-Hill, New York.

SNODGRASS, R. E. 1944. The feeding apparatus of biting and sucking insects affecting man and animals. *Smithsonian Misc. Coll.* **104**, 7:1–113; Figs. 1–39.

6

The Internal Anatomy and Histology of the Imago of Drosophila Melanogaster*

ALBERT MILLER †

INTRODUCTION

Despite their great diversity in anatomical details, all insects have an internal organization and histological structure that are fundamentally uniform throughout the class. The general body plan of the adult pomace fly, *Drosophila melanogaster* Meigen, is thus typically insectan in character. The skeletal framework of the body consists of a segmented, chitinous external cuticle secreted by a single underlying layer of epidermal cells. The hardened or sclerotized areas of the body wall and appendages are joined by flexible membranes, which permit movement and, in some places, a certain amount of distention by expansion. Sense organs are situated on the surface of this exoskeleton, and muscles are attached to its inner side. A tubular alimentary canal extends through the central axis of the body from the mouth at the anterior pole to the anus at the posterior pole. Above the alimentary canal there is a mediodorsal blood vessel through which the blood passes forward to the head, where it is poured into the haemocele, or definitive body cavity, to flow backward and bathe all the organs. The central nervous system consists

---

*This study was done in 1940–1941 at the Department of Genetics, Carnegie Institution of Washington, Cold Spring Harbor, Long Island, N. Y.

† Department of Tropical Medicine and Public Health, School of Medicine, the Tulane University of Louisiana.

of a dorsal brain in the head and a medioventral ganglionated nerve cord below the alimentary canal. Respiration is effected through a system of branching tubes that open to the outside through lateral breathing pores and permeate to all the tissues of the body. The reproductive organs in both sexes are situated posteriorly and have an external opening below and morphologically anterior to the anus. Masses of fat tissue lie within the haemocele. Excretory and incretory (endocrinal) functions are subserved by special organs closely associated with the digestive and circulatory systems. The predominant cellular elements of the body comprise epithelia, muscle, fat, and nerve tissue. In addition there are blood cells, special secretory elements, and sex cells. Except for the adipose tissue, connective tissue of the type found in vertebrates and mesoglial or parenchymatous tissue are entirely absent.

Drosophila is, however, a highly specialized insect. Morphological specialization is evident in all the organ systems and is of the type found also in related muscoidean flies. Prominent among the internal modifications which contrast with the structure of many generalized insects are the coiling of the alimentary canal, the absence of a true proventriculus, and the presence of a pedunculate crop; modified head musculature correlated with the highly specialized ("sponging") mouthparts; enormous development of the mesothoracic flight muscles and the reduction of the metathoracic musculature; condensation and cephalization of the central nervous system and modification of the stomodaeal nervous system; extensive development of air sacs in the respiratory system; foreshortening of the heart, expansion of its anterior chamber, and reduction of the dorsal diaphragm; asymmetric disposition and reciprocal transverse displacement of the two testes; complete rotation of the male genital segments and presence of an ejaculatory "pump"; and development of a "uterus" in the female in which an egg may be temporarily retained. Some of these features are characteristic of the Diptera in general, but many are entirely absent or merely incipient in more primitive flies and are found only in the highly evolved Muscoidea. On the other hand, some of the specializations evident in Drosophila are developed to a still more advanced degree in calyptrate muscids like Musca and Calliphora.

The voluminous literature on Drosophila contains no comprehensive account of the internal anatomy of the imago, but several of the organ systems have been the subject of individual and detailed study. The digestive system has been described by Marie Strasburger (1932), the nervous system and sense organs by Hertweck (1931),[1] and the reproductive organs by Nonidez (1920) and Gleichauf (1936). A synoptic account of these systems has also been presented by E. Strasburger (1935). The variability and development of the reproductive organs have received especial attention (Dobzhansky, Stern, and others), and gametogenesis has been extensively studied (Guyénot and Naville, 1929, 1933; Huettner, 1930). Additional anatomical observations are to be found in a number of other papers (e.g., Robertson, 1936) and these will be cited where pertinent. The circulatory, respiratory, and muscular systems have heretofore been almost entirely neglected.[1] The anatomy of larger muscoidean flies has been studied intensively, and the works of Lowne (1890–1895), Ritter (1911), Graham-Smith (1930, 1934, 1938), and others on the blowfly Calliphora, and of Hewitt (1914) on the housefly *Musca domestica*, contain much information of value in morphological and physiological studies of Drosophila. General textbooks particularly useful are those of Berlese (1909), Schröder (1928), Weber (1933), Imms (1934), Snodgrass (1935), and Wigglesworth (1939).

The following account attempts to present a detailed but concise description of the "normal" anatomy and histology of the imago of *Drosophila melanogaster* Meigen, based on original studies of wild-type Swedish-*b* flies. Each organ system was investigated in manual dissections of living and fixed specimens and in serial sections. The descriptions of systems studied by earlier authors were subjected to confirmation and supplemented by additional detail and reinterpretation in the light of general insect morphology. In preparing the illustrations, particular care was taken to represent the gross anatomy clearly without sacrificing topographical accuracy; and it is hoped that the photographs, despite their inherent shortcomings, will afford a

---

[1] Since this writing (1941) papers on the brain by Power (1943) and flight muscles (*D. repleta*) by Williams and Williams (1943) have appeared.

more objective record of histological details than would diagrammatic drawings. All illustrations are original, except that for the sense organs it proved expedient to redraw some of Hertweck's figures. The standard terminology of insect morphology used by Snodgrass (1935) has been adhered to wherever possible, and his system of abbreviation adopted for labeling the illustrations.

Living flies, etherized and dipped in 95 per cent alcohol to facilitate wetting, were dissected in saline solution or dilute (25 per cent) sea water by means of sharpened needles in matchstick handles. Fixed material was dissected in 70 per cent alcohol and examined directly, with or without staining, in alcohol or after clearing in glycerin or cedar oil. Specimens bisected while embedded in paraffin, freed of paraffin by means of xylol, stained, and then mounted in cedar oil or glycerin also assisted in studies of gross anatomy. Hot formol-acetic-alcohol (Kahle's or Dietrich's formula: D.) proved a satisfactory general fixative for both gross and sectioned material, though other fixatives were also employed—namely, Schaudinn's (S.), Gilson's (G.), alcoholic Bouin's (B.), Bouin-Allen's (B.A.), and Carnoy's (C.) fluids. Piercing the body or cutting off appendages enables better fixation. In sectioning, cellosolve and butyl alcohol were used for dehydration and rubber-paraffin for embedding. Material was cut in sagittal, transverse, and frontal planes at thicknesses of 5, 8, 10, and 25 μ. The stains used included Delafield's haematoxylin and eosin (H.E.), iron haematoxylin (I.H.), Mallory's connective tissue stain (M.), Feulgen's and light green (F.L.G.), and chlorozol black (C.B.: Darrow, 1940); for whole mounts borax carmine, methylene blue, light green, picrofuchsin, and Mallory B. Mallory's and chlorozol black are particularly useful for staining chitinous structures in sections, as well as for certain cytological details (Mallory's for muscle striations, chlorozol black for fiber tracts within nerve ganglia).

Although a combination of methods, including whole mounts, gross dissections, and microtome sections, were used for the study of each organ system, specific mention may be made of the following. For the muscular system fixed material was bisected in 70 per cent alcohol, stained with Mallory B or picrofuchsin, and dissected in glycerin; exoskeletons cleared by 10 per cent potas-

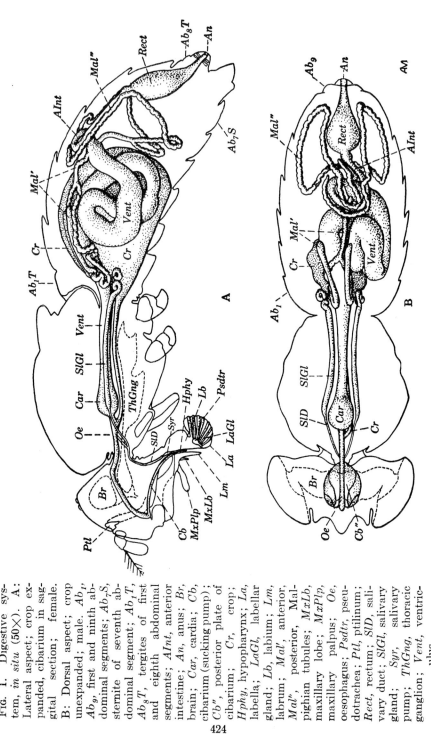

FIG. 1. Digestive system, *in situ* (50×). A: Lateral aspect; crop expanded, cibarium in sagittal section; female. B: Dorsal aspect; crop unexpanded; male. Ab_1, Ab_9, first and ninth abdominal segments; Ab_7S, sternite of seventh abdominal segment; Ab_1T, Ab_8T, tergites of first and eighth abdominal segments; $AInt$, anterior intestine; An, anus; Br, brain; Car, cardia; Cb, cibarium (sucking pump); Cb'', posterior plate of cibarium; Cr, crop; $Hphy$, hypopharynx; La, labella; $LaGl$, labellar gland; Lb, labium; Lm, labrum; Mal', anterior, Mal'', posterior, Malpighian tubules; $MxLb$, maxillary lobe; $MxPlp$, maxillary palpus; Oe, oesophagus; $Psdtr$, pseudotrachea; Ptl, ptilinum; $Rect$, rectum; SlD, salivary duct; $SlGl$, salivary gland; Syr, salivary pump; $ThGng$, thoracic ganglion; $Vent$, ventriculus.

424

sium hydroxide were useful adjuncts. For the heart, the dorsal wall of the abdomen was stained with light green, Mallory's, or picrofuchsin and examined in glycerin. For the respiratory system, fresh material was cleared in glycerin and the tracheae were studied while still filled with air, appearing silvery in reflected light and black in transmitted light; injection *in vacuo* with trypan blue (Hagmann, 1940) proved inadvantageous and erratic. The peripheral nerves are difficult to trace in dissections, but staining with picrofuchsin, light green, or silver nitrate is helpful.

The drawings are based on *camera lucida* tracings of dissections and sections. Where the body is drawn in a sagittal plane, the outline of the female is shown; the frontal plane represents the male. The photomicrographs were taken on Agfa Isopan film, using Bausch and Lomb optics and appropriate filters; they were not retouched.[2] Magnification on the printed page and the fixative and stains employed, designated by the abbreviations noted above, are given for each figure.

THE DIGESTIVE SYSTEM

The digestive system comprises the alimentary canal and its associated organs: namely, the mouthparts, labellar glands, salivary glands, and Malpighian tubules (Figs. 1, 3). The last-named are the chief excretory organs but are morphologically diverticula of the alimentary canal.

The *mouthparts* form a protrusible sucking proboscis for the ingestion of liquid food (see External Morphology and Fig. 1). The soft, expansible distal lobes, or labella (La), bear a series of open collecting channels, the pseudotracheae ($Psdtr;$ Figs. 1A, 11), 10–12 μ in diameter, that converge medially and conduct the food liquors to the tip of the labrum (Lm). This organ, the inner or posterior wall of which is the epipharynx ($Ephy;$ Fig. 2A), lies in a groove along the front of the labium (Lb) and forms a tubular canal (FC), 14–21 μ in diameter, up which the food is drawn by the action of the sucking pump or cibarium (Cb) in the base of the proboscis.

[2] The author gratefully acknowledges the skillful assistance of Miss A. M. Hellmer, artist, who inked the drawings, and Miss R. Stewartson, photographer, Department of Genetics, Carnegie Institution of Washington.

The cibarium ("pharynx" of M. Strasburger, 1932) consists of an arched posterior plate (*Cb"*) fixed in position by its lateral connections with the clypeus, and a similarly arched anterior plate (*Cb'*) which fits against the concave side of the fixed plate (Figs. 1, 3). These sclerotized structures, including the external clypeus, together constitute the stirrup-shaped "fulcrum" of the basiproboscis or rostrum (Hewitt, 1914; Graham-Smith, 1930; *et al.*). The two plates of the cibarium are joined together along their edges (Fig. 2B) and enclose a flat lumen between them that communicates with the labral food channel at the lower end and with the oesophagus at the upper end. The anterior plate is spatulate, bears a median apodemal carina (*Ap*), and has attached to it the large dilator muscles of the cibarium, which have their origin on the clypeus (Figs. 2B, 25A; muscles *11* and *12*). Their contraction draws the anterior plate forward, enlarges the cibarial lumen, and thus produces suction. Elasticity of the plate probably causes it to spring back into position when the muscles are relaxed. However, the possibility also exists that differential action of the muscles attached respectively to the carina at the upper portion of the plate (muscle *11*) and to the plate's expanded ventral portion (*12*) may provide a measure of antagonistic action, permitting dilatation in one end of the cibarium while the other end remains compressed, and thus controlling the direction of flow of liquids being ingested or regurgitated. The labral-cibarial and cibarial-oesophageal junctions are probably valvular in function, but circular muscles occur only around the oesophagus, those near the cibarium being thicker than elsewhere. The anterior plate bears two rows of upward-projecting sense hairs (Figs. 3, 29), while in transection the posterior plate shows fine vertical striae and has special pores with which sense cells are associated (*CSO;* Figs. 2B, 35D). These and the sense organs on the labella and labrum are considered under Sense Organs. The cibarium represents morphologically the external preoral cavity; the true mouth opening and beginning of the alimentary canal proper are at the junction of cibarium and oesophagus.

Between the labrum and the floor of the labial groove lies the short, slender, and pointed hypopharynx (*Hphy;* Figs. 1A, 2A), extending downward about one-third the length of the labrum and enclosing the salivary canal (*SC*), which is connected proximally with the salivary pump (see below). The muscles of the proboscis are considered under Muscular System. For a full account of the structure and mechanism of the proboscis of muscoidean flies, the student should refer to the detailed description of that of Calliphora by Graham-Smith (1930) ; for its comparative morphology, see Chapter 5 and Snodgrass (1944).

The *alimentary canal* is a modified tube extending from mouth to anus (Figs. 1, 3). Its course is direct in the head and thorax

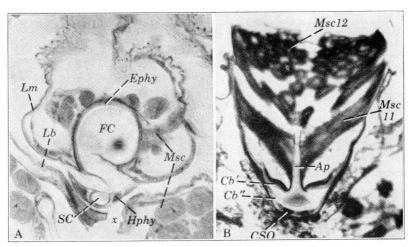

Fig. 2. Mouthparts. A: Portion of transection of proboscis at base of labrum (D., F.L.G.; 660×). B: Transection of cibarium or sucking pump (G., C.B.; 280×). *Ap*, apodeme; *Cb'*, anterior, *Cb"*, posterior, plate of cibarium; *CSO*, cibarial sense organ; *Ephy*, epipharynx; *FC*, food canal; *Hphy*, hypopharynx; *Lb*, labium; *Lm*, labrum; *Msc*, muscles; *Msc11, 12*, dilator muscles of cibarium; *SC*, salivary canal; *x*, artificial break.

Note. In this and other legends abbreviations in parentheses indicate the fixative and stains employed, as listed on page 423.

but considerably convoluted in the abdomen, terminating in the last abdominal segment. With a total length of nearly 7 mm, it is three to three and one-half times as long as the body. On histological and developmental grounds, three principal divisions are distinguishable: the foregut, the midgut, and the hindgut, corresponding respectively to the stomodaeum, mesenteron, and proctodaeum of the embryo. Foregut and hindgut are characterized by the presence of a true chitinous intima covering the epithelium; this is lacking in the midgut. The foregut includes the oesophagus (in head and prothorax), the crop (a thoracic diverticulum extending into the abdomen), and the stomodaeal valve within the

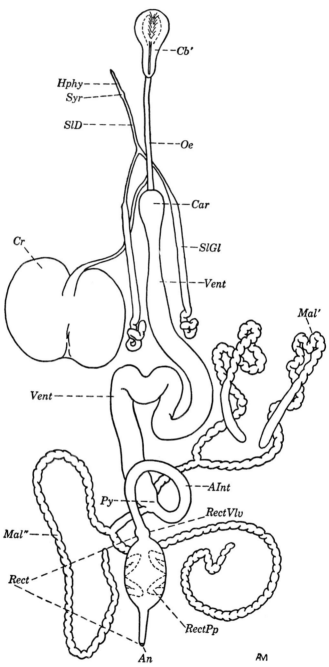

FIG. 3. Alimentary canal and salivary apparatus, dorsal aspect; crop expanded (50×). *Py*, pyloric region; *RectPp*, rectal papilla; *RectVlv*, rectal valve; other abbreviations as in Fig. 1.

cardia (in mesothorax). The midgut or mesenteron consists of the ventriculus (mesothorax to fourth abdominal segment), its anterior end forming the cardia. The hindgut includes the anterior intestine (fourth and fifth abdominal segments) and the posterior intestine or rectum (fifth to tenth abdominal segments), terminating in the anus. Specialized valves occur at the junction of oesophagus and ventriculus and between the anterior intestine and rectum; in addition, circular muscles at other junctions (e.g., where the oesophagus joins the cibarium and where the stalk of the crop joins the oesophagus) probably act as sphincters. There are two pairs of Malpighian tubules, each pair arising from a single stalk that springs from the pyloric region of the ventriculus. The parts of the alimentary canal are without color, except for that imparted by contained food or waste material (see sections on crop and Malpighian tubules, below). The structure of the alimentary canal of Drosophila bears a close resemblance to that of Calliphora. The latter has been the subject of an elaborate description by Graham-Smith (1934), which the student will find an excellent guide to more detailed study than is here presented.

The *oesophagus* (*Oe*) is a slender tube, 550 μ long and 20–36 μ in diameter, extending from the inner end of the cibarial pump through the cephalic ganglionic mass and the narrow cervical region into the anterior part of the thorax, where it enters the cardia. Its epithelium is 1.5 μ thick, has indistinct cell boundaries, and is thrown into longitudinal folds. An intima 0.8 μ thick is present, and a single layer of circular muscles forming bands 4–6 μ wide and 3 μ thick surrounds the tube. (The muscles near the cibarium are 6–7.5 μ thick.) No longitudinal muscles are evident (Fig. 4A).

The *crop* (*Cr*) is a pedunculate sac evaginated from the oesophagus a short distance before the cardia. Its stalk is slender (830 μ by 18–27 μ) in the thorax, where it extends beneath the mesenteron, and widens so as to be narrowly funnelform (175–250 μ long and 60–85 μ in diameter) where it joins the saccular portion. The latter is bilobed and lies in the anterior part of the abdomen. The sac is capable of great distention by ingested liquid food, its size being dependent upon the age and nourishment of the fly. Its capacity, calculated on a basis of size, is probably about 0.05 cu mm. In flies that have recently emerged, the sac is unexpanded, 200–270 μ long and 300–360 μ wide, with its walls compressed into compact folds; it lies in the first and second abdominal segments (Fig. 1B). After feeding, the crop becomes greatly distended, to as much as eight or ten times its original volume (e.g., 520 by 690 by 350 μ), so that it may extend back to the caudal part of the

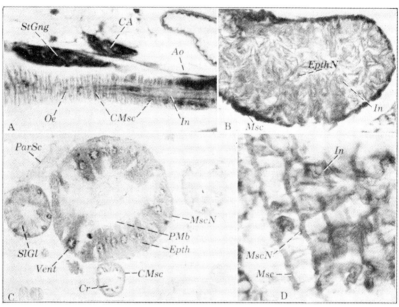

FIG. 4. Oesophagus, crop, and thoracic ventriculus. A: Portion of oblique sagittal section of oesophagus anterior to junction with crop (*junction at left*) (B., H.E.; 490×). B: Sagittal section of unexpanded crop sac (G., C.B.; 370×). C: Transections of thoracic ventriculus, crop stalk, and salivary glands (C., F.L.G.; 350×). D: Tangential section of crop sac, showing muscle plexus on surface (D., C.B.; 490×). *Ao,* aorta; *CA,* corpus allatum; *CMsc,* circular muscle; *Cr,* crop; *Epth,* epithelium; *EpthN,* epithelial nucleus; *In,* intima; *Msc,* muscle; *MscN,* muscle nucleus; *Oe,* oesophagus; *ParSc,* parenteric tracheal sac; *PMb,* peritrophic membrane; *SlGl,* salivary gland; *StGng,* stomodaeal ganglion; *Vent,* ventriculus.

fourth abdominal segment (Fig. 1A), while lack of food and water for 24–48 hours results in a marked decrease in volume. Enlargement of the sac causes distention of the entire abdomen and crowds the other organs against the body wall and backward; in such individuals the crop contents are discernible as a grayish translucent area in the abdomen during life. The distended crop is displaced so that it lies on the right side of the abdomen, as a result of the sinistral coiling of the neighboring mesenteron (Fig. 1A). When dissected out in saline solution, the crop is pale amber in color, owing to its contents, and may show muscular contractions in the dilated portion of the stalk and in the sac. The crop is lined throughout with an intima, which in the funnelform region bears many slender spines, directed posteriorly or erect. In the stalk the epithelium forms narrow longitudinal folds projecting into the lumen, the cell nuclei occurring either in these folds or basally between them (Fig. 4C). The musculature seems to consist solely of wide bands of circular muscles, 1.5 μ thick and 3–7 μ wide, in both the slender and dilated parts of the stalk (Fig. 6A). In the sac thick muscle fibers, 3–5 μ in diameter, extend around the base of each lobe, converging posteriorly, and the distended portions of the wall are covered with a plexus of branched and interlacing fibers 1–3 μ thick (Fig. 4D). Epithelium is barely evident in the sac, but epithelial nuclei (1.5–3 μ) are discernible within the crowded and compact folds of intima in teneral flies (Fig. 4B); when the crop is expanded these nuclei (1 μ) are widely separated and difficult to detect. The intima of the sac appears about 1 μ thick in teneral flies but becomes considerably thinner when stretched by expansion of the crop. The crop serves to receive and store liquid food, passing it on to the ventriculus for digestion (cf. Graham-Smith, 1914).

The *cardia* (*Car*) appears as an ovoid swelling, 100–140 μ in diameter and 140–210 μ long, at the anterior end of the ventriculus and lies in the mesothorax. It is a saccular modification of the ventricular wall which covers the stomodaeal valve formed by the invaginated end of the oesophagus. The cardia in Drosophila has usually been called the "proventriculus," but this term is properly applied only to the part of the foregut, variously modified in other insects, that lies anterior to the stomodaeal valve. Like most sucking insects, Drosophila has no differentiated proventriculus.

In longitudinal section (Fig. 5) the cardiac region is seen to contain three layers of epithelium, the inner two forming the stomodaeal valve and the outer constituting the wall of the cardia. The fusiform stomodaeal valve consists of a deep fold of the foregut epithelium (*a, b*) projecting into the cardia, the space between the two layers of the fold filled with longitudinal muscle fibers. The stomodaeal intima continues over both layers, forms irregular transverse ridges within the apical portion of the valve, and terminates on the outer side of the valve near its base; its termination marks the junction (*j*) of foregut and midgut. The greater part of the valvular epithelium consists of vacuolate cells, cuboidal (*a*) in the inner layer but gradually becoming columnar (*b*) in the outer layer at the thickest part of the valve. Where the vacuolate cells end, there is a small cluster of apical vacuoles directly beneath the intima, representing a narrow band girdling the valve

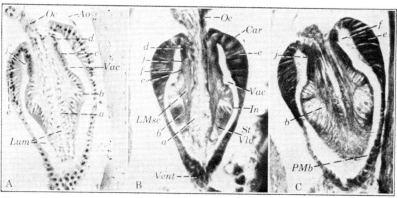

Fig. 5. Cardia. Sagittal sections (210×). A: Stained with Feulgen's and light green (Schaudinn's fixative). B: Stained with Mallory B (Dietrich's fixative). C: Stained with iron haematoxylin (Carnoy's fixative). *a*, inner epithelium of valve; *Ao*, aorta; *b*, vacuolate outer epithelium of valve; *c*, non-vacuolate foregut epithelium; *Car*, cardia; *d*, midgut epithelium at base of valve; *e*, tall cells of cardia; *f*, secretion; *In*, intima; *j*, junction of foregut and midgut; *LMsc*, longitudinal muscles; *Lum*, lumen; *Oe*, oesophagus; *PMb*, peritrophic membrane; *StVlv*, stomodaeal valve; *Vac*, group of vacuoles; *Vent*, ventriculus.

at this point. In the short zone between these vacuoles and the termination of the intima, the epithelial cells contain no vacuoles (*c*). Beyond the intima the midgut epithelium of the cardia (*d*) begins, with its more deeply staining and granular cytoplasm. The wall of the cardia surrounding the base of the stomodaeal valve is composed of tall, narrow cells (*e*), 20–28 μ high; posteriorly it becomes thinner, with low (7 μ) cells, and finally merges with the digestive epithelium of the ventriculus.

On the inner surface of the epithelium in the anterior part of the cardia, sections show either small globules (Fig. 5B, *f*) or a more or less continuous layer of a refringent substance (Fig. 5C, *f*). This is presumably a formative stage of the peritrophic membrane, secreted by the cells in this region and drawn back into the ventriculus as a thin (0.3 μ) cylindrical sheath enclosing the food contents (cf. Snodgrass, 1935, p. 368). It is definitely membranous immediately behind the tall cells of the cardia and extends from there all the way back to the rectum.

The cardia is covered externally by thin, scattered muscle fibers like those elsewhere on the ventriculus.

The cardia and stomodaeal valve may be regarded as having the dual function of a sphincter to prevent regurgitation of the ventricular contents and an organ to produce and mold the peritrophic membrane.

The *ventriculus* (mesenteron or stomach; *Vent*), the longest part of the alimentary canal (5 mm), is a thick-walled tube that has a straight section ("Magen" of M. Strasburger, 1932) in the thorax and a coiled portion lying in the first four abdominal segments. The straight section is narrow (80 μ) behind the cardia, increases in diameter posteriorly (to 130 μ), and merges directly into the abdominal portion. Here the diameter is fairly uniform, but local constrictions and dilatations (90–225 μ) may be present, perhaps peristaltic in origin or the result of fixation. [In freshly emerged flies, part of the abdominal ventriculus is temporarily greatly distended by the swallowing of air. This forces blood into the wings and causes them to expand. Later the air disappears from the gut, apparently diffusing into the body (Eidmann, 1924; cf. Calliphora, Fraenkel, 1935).] In the abdomen the ventriculus courses dorsally and toward the right and then describes two vertical coils that lie side by side and progressively farther to the left (Figs. 1, 3). The direction of coiling in the first loop is downward, forward, upward, and backward (clockwise when viewed from the left); the direction then reverses abruptly and in the second loop is forward, downward, and backward (counterclockwise, paralleling the first loop). Finally, the posterior portion of the ventriculus courses upward and mesad between the gonads into the fourth abdominal segment, where it terminates in an abruptly narrowed pyloric region that joins the hindgut and receives the Malpighian tubules. These coils lie above and behind the undistended crop and may be more or less oblique to the median plane of the body. When the crop expands, it crowds the coils to the left and upward against the body wall. M. Strasburger (1932) found no significant variability in the loops of the mesenteron in 132 wild-type flies, or in several mutants investigated.

Histologically the ventriculus consists of a thick epithelium surrounded by a latticework of fine muscle fibers and loosely lined by the peritrophic membrane. The epithelium is a single layer of large digestive cells between the bases of which there is an occasional small regenerative cell. In the thoracic ventriculus (aside from the thin epithelium near the cardia), the digestive cells are more or less columnar in form, 25–30 μ high, with a clear apical border, spongy cytoplasm which may show vertical striations basally, and a large, centrally located, vesicular nucleus; in transverse sections they project irregularly into the lumen (Figs. 4C, 6A). The digestive cells gradually decrease in height in the abdominal ventriculus and acquire a distinct striated or "brush" border about 2 μ thick. In the coils this border may appear as a tuft covering the projecting end of each cell (Figs. 6B, d, and 7A, B) ; where the cells are more cuboidal or flatter (as in the posterior part of the ventriculus), it forms a continous layer over the epithelium (Fig. 6B). Occasionally the epithelium consists of alternate tall and low cells, the tall cells pedunculate in form, with a light apical border, and the recess above each low cell filled with a mass of brush border, giving a false appearance suggesting goblet cells (Fig. 6C). The posterior cells may be as low as 5 μ, with flattened nuclei, and sometimes have a brush border 7 μ high merging distally with a layer of coagulum of similar thichness outside the peritrophic membrane. In teneral flies the epithelium of the thoracic ventriculus is sometimes of the brush-bordered cuboidal type of cells instead of the clear-bordered columnar type mentioned above. The regenerative cells are infrequent, are sometimes wedge-shaped with a tapering apex, and have denser cytoplasm than the active cells; their nuclei are also denser, smaller, and situated near the base (Fig. 6D). The digestive epithelium ends abruptly at the point where the Malpighian tubules arise and is succeeded by a thin and simple squamous epithelium in the short pyloric region (Fig. 7A, Py).

The varied histological appearance of the digestive epithelium is probably due, at least in part, to different stages of physiological activity when fixed, but the significance or sequence of the changes has not been determined. The true nature of the striated border is also obscure, but it is not a layer of cilia (cf. Zilch, 1936).

The peritrophic membrane extends throughout the ventriculus, lying more or less free of the epithelium and separating it from the contained food materials (Fig. 6B).

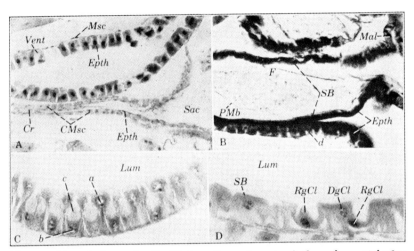

Fig. 6. Ventriculus. A: Sagittal section of part of thoracic ventriculus and crop (S., F.L.G.; 210×). B: longitudinal sections of abdominal ventriculus, showing flat posterior epithelium and taller, more anterior epithelium below (D., M.B; 210×). C: Longitudinal section of wall of abdominal ventriculus, showing another appearance presented by the epithelium (B., I.H.; 475×). D: Transection of wall of abdominal ventriculus, showing regenerative cells of epithelium (D., F.L.G.; 490×). *a*, tall epithelial cell; *b*, low epithelial cell; *c*, space filled with "brush border"; *CMsc*, circular muscle; *Cr*, crop stalk; *d*, projecting end of cell capped by striated border; *DgCl*, digestive cell; *Epth*, epithelium; *F*, food material; *Lum*, lumen; *Mal*, Malpighian tubule; *Msc*, muscle; *PMb*, peritrophic membrane; *RgCl*, regenerative cell; *Sac*, crop sac; *SB*, striated border; *Vent*, ventriculus.

The musculature of the ventriculus is weakly developed and consists of inner circular and outer longitudinal fibers, 2 μ in diameter. Neighboring fibers lie some distance apart, the longitudinal muscles being most widely separated. In section the minute inner fibers lie beneath or between the bases of the epithelial cells and are outwardly covered by a thin membrane that seems to extend between the longitudinal fibers.

The ventriculus is the principal seat of digestion.

The *anterior intestine* (*AInt*) is a short and slender tube 700–900 μ long and 50–70 μ in diameter, extending from the pyloric region of the ventriculus to the rectal valve. It describes a hori-

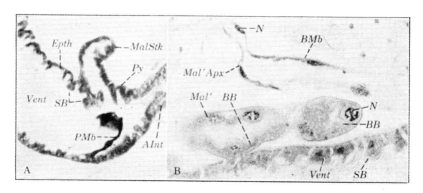

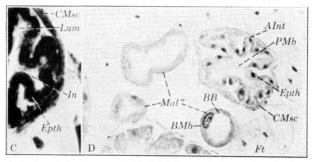

FIG. 7. Anterior intestine and Malpighian tubules. A: Sagittal section through posterior ventriculus, showing pyloric junction and basal stalk of one pair of Malpighian tubules (B., H.E.; 210✕). B: Sagittal and oblique sections of anterior Malpighian tubule, showing thin-walled apical region and unmodified region (D., F.L.G.; 375✕). C: Portion of transection of anterior intestine (D., C.B.; 490✕). D: Transections of anterior intestine and posterior Malpighian tubule (D., F.L.G.; 375✕). *AInt*, anterior intestine; *BB*, brush border; *BMb*, basement membrane; *CMsc*, circular muscle; *Epth*, epithelium; *Ft*, fat cells; *In*, intima; *Lum*, lumen; *Mal'*, anterior, *Mal''*, posterior, Malpighian tubules; *Mal'Apx*, apical portion of anterior Malpighian tubule; *MalStk*, stalk of Malpighian tubules; *N*, nucleus; *PMb*, peritrophic membrane; *Py*, pyloric region; *SB*, striated border; *Vent*, ventriculus.

zontal counterclockwise loop in the dorsal part of the fourth abdominal segment. This loop is rarely incomplete or absent and normally projects forward beyond and above the end of the ventriculus (Figs. 1, 3). The rectal valve is visible externally as a slight thickening of the intestine, followed by a constriction where it joins the rectum. The intestinal epithelium is moderately thick (3 μ), lined by a true cuticular intima (1–1.5 μ), and forms four or five loose longitudinal folds which project into the lumen (Figs. 7C, D). The muscle coat consists of a single layer of well-developed and closely placed circular fibers 2–4 μ thick and 3–5 μ wide. In section a very thin interconnecting membrane is apparent between adjacent fibers. Longitudinal fibers seem to be absent. The epithelium of the rectal valve is modi-

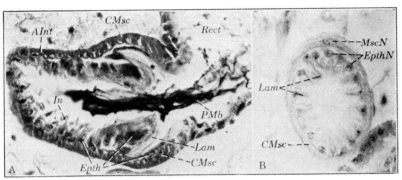

FIG. 8. Rectal valve. A: Sagittal section (D., H.E.; 350$\times$). B: Transection (B.A., F.L.G.; 490$\times$). *AInt*, anterior intestine; *CMsc*, circular muscle; *Epth*, epithelium; *EpthN*, epithelial nucleus; *In*, intima; *Lam*, lamina; *MscN*, muscle nucleus; *PMb*, peritrophic membrane; *Rect*, rectum.

fied to form about twenty thin laminae that extend lengthwise into the lumen (Fig. 8). These plates are covered with a thickened intima and bear minute spines on their inner edges. The circular muscle fibers in this region are thicker than elsewhere, forming a strong sphincter, and are distinctly "tubular" in nature, with a central core of sarcoplasm in which the nuclei are located (see Muscular System). The lumen of the anterior intestine contains remnants of food material enclosed in the peritrophic membrane. The rectal valve, by means of its plates and spines, is probably instrumental in drawing the peritrophic membrane backward and, when closed, in holding it in position.

The *posterior intestine* or *rectum* (*Rect*) lies along the mediodorsal line from abdominal segment 5 or 6 to the anal segment (10). It consists of a broadly expanded, thin-walled rectal sac (250 μ long, 250 μ wide, and 160 μ high) that tapers posteriorly to the narrow rectum proper (250 μ long and 70–95 μ in diameter), terminating at the anus (Figs. 1, 3). Projecting from the side walls into the lumen of the rectal sac are two pairs of large conical rectal papillae (*RectPp*), one pair behind the other (Fig. 3). The epithelium of the posterior intestine (Figs. 9B, C) is, for the most part, very thin (1–3 μ), with scattered and indistinct nuclei and a thin intima (0.5 μ). A series of thickened epithelial folds radiates a short distance from the rectal valve, and here the intima is minutely spined. Each rectal papilla (Figs. 9A, B) is a conical thick-walled invagination of the epithelium 100–150 μ high and 65–85 μ in diameter basally, containing a narrow central

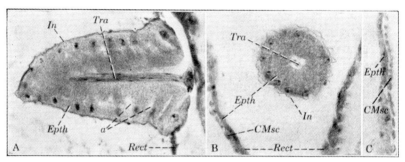

Fig. 9. Rectal papillae and rectal sac. A: Sagittal section of rectal papilla (D., H.E.; 350×). B: Transection of rectal papilla and sagittal section of wall of rectal sac (S., F.L.G.; 350×). C: Sagittal section of wall of rectal sac (S., F.L.G.; 350×). *a,* intercellular spaces; *CMsc,* circular muscle; *Epth,* epithelium; *In,* intima; *Rect,* wall of rectal sac; *Tra,* tracheae.

cavity. The latter is occupied by a bundle of five tracheae, which arise from the visceral branch of the sixth dorsal segmental trachea of the abdomen (see Respiratory System) and give off tracheoles which ramify between the cells of the epithelium. The intima covering the papilla bears a few weak spinose processes at the tip but does not differ in thickness from that elsewhere in the rectum. The large epithelial cells, 25–30 μ high, radiate from the central cavity and are broadest near the intima, where the nuclei are located. Their cytoplasm is granular and somewhat denser in the end toward the tracheae. Restricted spaces (*a*) are evident between the cells, representing the intercellular sinus described by Graham-Smith (1934) in *Calliphora*, within which tracheoles occur. The muscular coat of the posterior intestine consists of a single outer layer of thick, closely set circular fibers (Figs. 9B, C, 40B), underlaid by inconspicuous, widely spaced inner longitudinal fibers. The circular fibers are 3 μ thick and

10 μ wide in the sac, and 3–5 μ thick in the posterior rectum; the longitudinal fibers are 3–5 μ wide and 1 μ thick. At the bases of the papillae muscles are absent, but a thin membrane, penetrated by the tracheae, extends from the surrounding fibers and covers these areas. The peritrophic membrane is indistinct and fragmentary in the rectal sac, which in sections is filled with a granular coagulum of fecal wastes. Wigglesworth (1939) regards the rectal papillae of insects as water-absorbing organs, and it has been suggested that in the Diptera their movement and spinous tips may serve to rupture the peritrophic membrane (Engel, 1924). In the narrow posterior end of the rectum the epithelium forms numerous longitudinal folds, the circular muscles act as an anal sphincter, and dilator muscles extend to the surrounding body wall.

The *Malpighian tubules* (*Mal*) are two pairs of long slender tubes, about 2 mm long and 30 μ in diameter, arising on single stalks from opposite sides of the pyloric ventriculus and following a fixed course among the other viscera (Figs. 1, 3). The stalk of the anterior pair (*Mal'*) extends forward, and the two tubules are closely applied to the lateral walls of the first section of the abdominal ventriculus. Each tubule winds forward in close convolutions to the mesophragma, then turns abruptly back and extends straight along the ventriculus above the convoluted portions, ending in the third abdominal segment. The stalk of the posterior pair of tubules (*Mal''*) proceeds downward from its point of origin before forking. The tubules diverge to right and left, each looping under a lateral oviduct or a testicular coil to course upward to the side of the rectal sac, then turning forward to flank the anterior intestine, and ending independently in close relation to the intestinal loop. Tracheal branches bind the tubules to adjacent organs.

Both the stalks and the tubules are about 30 μ in diameter during life and have an undulating or twisted outline owing to the bulging epithelial cells that constitute their walls. While the posterior tubules are unmodified and simply closed at the distal ends, the straight terminal portions of the anterior pair are thin-walled and bladderlike and may be distended as much as four times their normal diameter by white or greenish white contents (visible through the dorsal abdominal wall). These contents are probably calcium salts (cf. Eastham, 1925; Wigglesworth, 1939). The unmodified portions of all tubules are color-less, amber, or pale yellow in life, although they become opaque white in desiccated individuals owing to the accumulation of solid waste products.

Histologically the tubules consist of large epithelial cells sur-rounded outwardly by a distinct basement membrane and enclos-ing a narrow lumen. Actually these cells are lozenge-shaped, each curled to form the entire circumference of the tubule, the cells interlocking side by side in a single row so that adjacent nuclei lie on different sides of the zigzag lumen. Thus in cross-section a single cell may constitute the wall of the tube, the large nucleus (8 by 9 to 9 by 13 μ) lying to one side of the lumen (Fig. 7D). The cytoplasm is spongy and in life contains pale yellow granules up to 1 μ in diameter. The lumen is lined by a filamentous or "brush" border (Fig. 7B), more deeply staining than the cyto-plasm and with the structural details often obscured by fixation. In the thin-walled distal portion of the anterior tubules the cells are much flattened, the nuclei bulge inwardly, and the brush border is thin and vague (Fig. 7B). In the stalks the epithelial cells are smaller, and their brush border forms a lining directly continuous with that of the ventriculus (Fig. 7A).

A few superficial circular and longitudinal muscle fibers are present on the stalks, and these parts sometimes undergo peristal-tic contractions in saline solution.

The Malpighian tubules are excretory organs, their urinary waste products passing out of the body by way of the hindgut.

The *salivary glands* (*SlGl*) are two slender, unbranched epithe-lial tubes, 900 μ long and with a uniform diameter of about 35 μ, extending straight through the thorax on either side of the ven-triculus and ending blindly in a compact coil in the base of the abdomen (Figs. 1, 3). Their anterior ends taper abruptly into slender salivary ducts, 7 μ in diameter, which unite under the nerve cord in the cervical region to form an unpaired duct (*SlD*); this runs down along the posterior wall of the proboscis and is continuous with the salivary canal (Fig. 2A), which runs through the hypopharynx and opens at its tip into the labial groove. At the base of the hypopharynx the duct is dilated to form a small chamber, the salivary pump or valve (*Syr*). A slender dilatory muscle (*13*, Figs. 25B, 26A) arising on the posterior wall of the cibarium is attached to the flexible anterior wall of the chamber;

its action may eject the salivary secretion as in other Diptera (e.g., mosquito, blowfly) or merely regulate the flow by opening the valve. No other muscles are associated with the salivary glands, except a slender suspensory fiber attached near the anterior end of each near the cardia. The salivary ducts are annulated chitinous tubes with minute epithelial nuclei at wide intervals on the outside of the walls. The epithelial cytoplasm is imperceptible, and the chitinous intima alone, stiffened by trans-

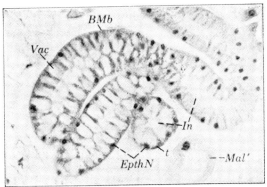

Fig. 10. Salivary gland. Section through coiled portion, showing longitudinal and transverse planes (D., F.L.G.; 400×). *BMb*, basement membrane; *EpthN*, epithelial nucleus; *In*, intima; *Mal'*, anterior Malpighian tubule; *t*, transection of gland; *Vac*, vacuole.

verse ridges (taenidia), constitutes the wall. The duct is continued into the gland as a very thin intima lining the lumen constituting the intra-acinar duct, 6–9 μ in diameter. The glandular epithelium consists of vacuolate cuboidal cells 10–15 μ high and 9–12 μ wide having small basal nuclei (3 μ) and bounded outwardly by a thin basement membrane (Figs. 4C, 10). Five to ten cells surround the lumen in cross-section. Most of the volume of each cell is occupied by a single large vacuole, with the cytoplasm limited to the base and sides. The salivary secretion which accumulates in the vacuoles presumably diffuses through the intimal lining directly to enter the intra-acinar duct; no intracellular collecting tubules like those found in other insects (Berlese, 1909; Weber, 1933) have been detected. The exact function of the saliva has not been demonstrated. It may possibly serve to dissolve solid foodstuffs before ingestion; in Calliphora the saliva contains amylase.

The *labellar glands* (*LaGl*) are two pyriform groups of cells, 45 μ by 30 μ, one in each lobe of the labella (Fig. 1A). Each gland consists of about seven large epithelial cells surrounding an irregular lumen that opens medially, through a short small-celled duct, in close proximity to the tip of the labrum (Fig. 11). A thin basement membrane surrounds the gland externally, and a thin intima lines the lumen.

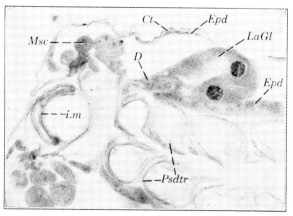

FIG. 11. Labellar gland. Portion of frontal section through labella, showing gland and tip of labrum (D., F.L.G.; 550×). *Ct,* cuticula; *D,* duct of labellar gland; *Epd,* epidermis; *LaGl,* labellar gland; *Lm,* labrum; *Msc,* muscle; *Psdtr,* pseudotrachea.

THE CIRCULATORY SYSTEM AND ASSOCIATED TISSUES

The circulatory system is of the "open" type characteristic of insects, consisting of a dorsal vessel, the haemocele or definitive body cavity, accessory pulsatile organs and septa, and the blood or haemolymph which bathes all the internal organs. Associated with and performing their functions directly through the medium of the circulatory system are the nephrocytes, oenocytes, corpus allatum, and fat body.

The *dorsal vessel* is differentiated into a posterior pulsatile heart, lying in the abdomen, and an anterior aorta extending through the thorax and into the head (Fig. 12).

The *heart* (Ht_{1-4}) is a four-chambered tube, about 1.1 mm long, extending medially from the mesophragma to the sixth abdominal segment close to the dorsal body wall, to which it is attached by suspensory fibrils (*SFb*), a rudimentary dorsal diaphragm (*DDph*), and several pairs of "alary muscles" (*AlMsc*). The caudal end of the heart is closed and is attached to both the sixth tergite and the dorsal wall of the rectal sac. The chambers of the heart are dilatations of the tube about 40 μ wide, separated by narrower portions (best defined in lateral view) where paired cell masses (*HtCl*) project into the lumen. The large first chamber (*Ht$_1$*) is funnelform, widening to 120 μ anteriorly at its junction with the expanded end of the aorta behind the mesophragma, and occupies abdominal segments 1 and 2. The second and third chambers are subcylindrical and occupy segments 3 and 4 respectively, while the fourth chamber (*Ht$_4$*) tapers posteriorly through segment 5 to the middle of segment 6 (Fig. 13A). Here the slender caudal portion is abruptly expanded, and the closed end is stretched between its points of attachment on the rectum and dorsal body wall. Behind the middle of each chamber there is a pair of oblique lateral slits (*Ost$_{1-4}$*). At each slit the heart wall is deeply inflected to form a pair of inwardly projecting flaps that act as a valve with the true ostial opening at its tip, for the intake of blood.

The wall of the heart consists of a single muscular layer, 1.5–3 μ thick, containing circular striated fibers and scattered lateral nuclei (Fig. 13). On the ventral side an outer layer of longitudinal muscles (Figs. 13B, F) extends the entire length of the heart, except for the slender caudal end. In whole mounts the wall of the large first chamber appears to be made up of a succession of wide bands (Fig. 13D, *b*), each band containing a number of striated fibrillae (e.g., eight to ten) and one ovoid nucleus in an island of cytoplasm on each side. Each lateral half of such a unit is probably derived from a single cardioblast. In sections the large nuclei (7 by 4 μ) are seen to lie within the heart wall, and the surrounding nonstriate cytoplasm may bulge inward (Figs. 13C, G). Dorsally the fine suspensory fibrils extend as processes from the heart wall (Figs. 13B, G). The larger fibrils sometimes show indications of striations, and a minute nucleus is occasionally detected. Between the chambers of the heart one or two large spongy cells project inward from the wall on each side (Figs. 13A, E, *HtCl*). The ostiolar valves consist of striated muscular flaps, each with a single nucleus (Figs. 13A, D, E). The striate ventral longitudinal fibers are fifteen or more in

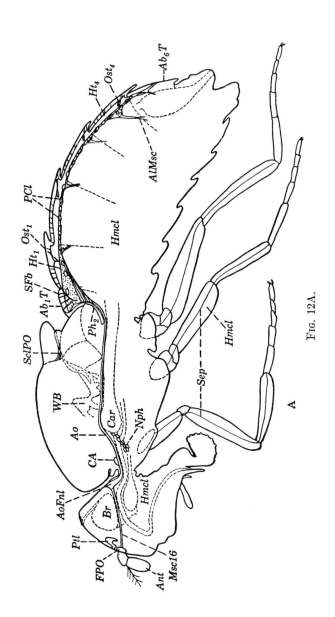

Fig. 12A.

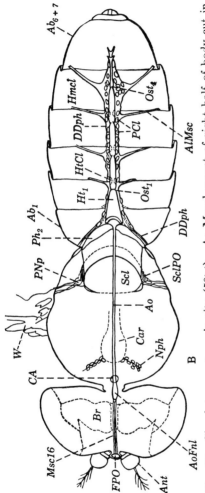

FIG. 12. Circulatory system, *in situ* (50×). A: Mesal aspect of right half of body cut in parasagittal plane to left of heart; anterior dorsal diaphragm indicated by stippling. B: Ventral aspect of dorsal half of body. Ab_1, first, Ab_6T, sixth, Ab_{6+7}, fused sixth and seventh, abdominal segments; Ab_1T, first, Ab_6T, sixth, Ab_{6+7}, fused sixth and seventh, abdominal tergites; $AlMsc$, alary muscle of heart; *Ant*, antenna; *Ao*, aorta; *AoFnl*, aortic funnel; *Br*, brain; *CA*, corpus allatum; *Car*, cardia; *DDph*, dorsal diaphragm; *FPO*, frontal pulsatile organ; *Hmcl*, haemocele; Ht_1, first, Ht_4, fourth, chamber of heart; *HtCl*, heart cells delimiting chambers; *Mscl6*, muscles of frontal pulsatile organ; *Nph*, thoracic nephrocytes; Ost_1, Ost_4, ostia; *PCl*, pericardial cells (abdominal nephrocytes); Ph_2, mesophragma; *PNp*, posterior notal wing process; *Ptl*, ptilinum; *Scl*, internal cavity of scutellum; *SclPO*, scutellar pulsatile organ; *Sep*, longitudinal septum of leg; *SFb*, suspensory fibril; *W*, wing; *WB*, wing base.

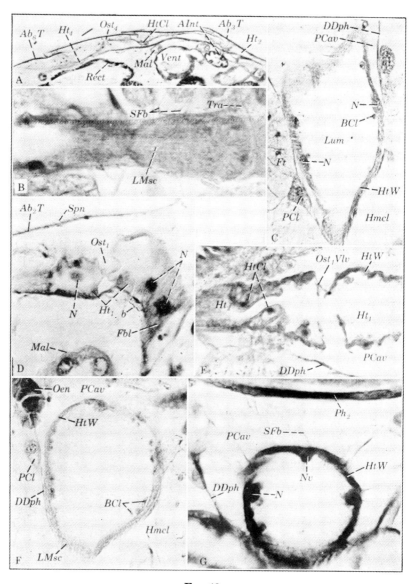

Fig. 13.

number, arranged side by side in a single layer, and have individual minute nuclei scattered along their lengths (Fig. 13B). Sections also show a nerve fiber running along the mid-dorsal line projecting into the lumen of the heart, and blood cells lying free in the lumen or applied to the inner wall (Figs. 13F, G).

The *aorta* (*Ao*) is a very slender thin-walled tube extending from the heart to the occipital region of the head, about 1 mm long and 3–10 μ in diameter (Fig. 12). At its junction with the heart, the aorta is widened to form a flattened triangular chamber against the posterior side of the mesophragma. From here it runs forward directly above the alimentary canal and beneath the great median dorsal muscles of the thorax, and finally opens into the haemocele behind the brain. The open end is a delicate membranous funnel (*AoFnl*) attached posteriorly to the head wall along the internal median occipital ridge. A pair of very slender muscles (*Msc16*) extends from its anterior edge through the oesophageal canal of the brain to the frontal accessory pulsatile organ between the antennae. It is undetermined whether the relation of these muscles to the aorta serves merely to keep the end open or actually assists in propulsion of the blood. The wall of the aorta is thin and membranous, shows no muscle striations, and has a few widely spaced elliptical nuclei, 12 by 2 μ, along its length (Figs. 4A, 5A, 18A). The aorta.tends to be circular in cross-section, except where it is depressed by neighboring structures—for example, between the overlying corpus allatum and the oesophagus or the stomodaeal ganglion. No direct con-

---

FIG. 13. Heart. A: Sagittal section of dorsal portion of abdomen, showing second, third, and fourth chambers of heart (B.A., I.H.; 100$\times$). B: Frontal section, showing ventral wall of second chamber of heart (D., C.B.; 350$\times$). C–G: First chamber of heart; C, oblique transection (D., F.L.G.; 350$\times$); D, tangential parasagittal section (B.A., I.H.; 350$\times$); E, frontal section (D., C.B.; 350$\times$); F, transection (D., F.L.G.; 350$\times$); G, transection (D., C.B.; 350$\times$). Ab_2T, second, Ab_3T, third, Ab_6T, sixth, abdominal tergites; *AInt*, anterior intestine; *b*, muscle band; *BCl*, blood cell; *DDph*, dorsal diaphragm; *Fbl*, fibrillae; *Ft*, fat cells; *Hmcl*, haemocele; Ht_1, first, Ht_2, second, Ht_4, fourth, chamber of heart; *HtCl*, heart cells delimiting chambers; *HtW*, wall of heart; *LMsc*, longitudinal muscle; *Lum*, lumen; *Mal*, Malpighian tubule; *N*, nucleus; *Nv*, nerve; *Oen*, oenocytes; Ost_1, first, Ost_4, fourth, ostium; Ost_1Vlv, valve of first ostium; *PCav*, pericardial cavity; *PCl*, pericardial cell; Ph_2, mesophragma; *Rect*, rectum; *SFb*, suspensory fibers; *Spn*, spinule (microchaeta); *Tra*, trachea; *Vent*, ventriculus.

nection has been detected between the aorta and the thoracic nephrocytes described below.

The *dorsal diaphragm* (*DDph*), which in a typical insect separates a definite dorsal sinus (pericardial cavity) from the perivisceral sinus below, is not well developed in Drosophila (Fig. 12). For the most part, it is reduced to a free-edged, narrow fibrous mesh along each side of the heart, bearing the pericardial cells and attaining its greatest width in the fifth abdominal segment. Attached to it are four pairs of delicate and inconspicuous muscle fibers (*AlMsc*) that extend laterad to the front edges of tergites 3–6. The last pair of these is the thickest and strongest. (Another pair may be present between segments 1 and 2, but this is uncertain.) Along the first chamber of the heart the diaphragm extends upward to the body wall and contains closely spaced longitudinal muscles that stretch from the ventral wall of the heart to the junction of metathorax and abdomen. Thus, in sections, a membranous diaphragm delimiting a distinct pericardial cavity is evident only in the first two abdominal segments (Figs. 13C, E, F, G); elsewhere fibrils extending from the heart to the pericardial cells may be detected, but the dorsal sinus consists only of an ill-defined and inconstant space between the fat body and the heart. (Tracheae pass above the heart, and in segment 2 the true pericardial sinus contains a mass of fat cells.) The paired muscles and the anterior longitudinal muscles represent the alary muscles of the heart of other insects. The paired muscles are single cordlike striated fibers, 2 μ in diameter, with elongate isolated nuclei. Toward the heart they branch and seem to be continuous with the fine reticular fibers of the diaphragm. The end attached to the tergite is also splayed. In the thin membranous diaphragm of the first two abdominal segments occasional minute nuclei may be detected, which belong to the longitudinal muscle fibers there. No membrane can be detected in the meshes of the reticular diaphragm, and the minute fibers appear as fine homogeneous strands.

The *haemocele* (*Hmcl*) is the general body cavity and as such surrounds all the internal organs. Since the body is compactly organized, the blood-filled perivisceral spaces consist for the most part of narrow and irregular intercommunicating passages between the organs. The subdivision of the body cavity into major dorsal, perivisceral, and ventral sinuses that occurs in many insects (see Snodgrass, 1935) is not evident in Drosophila, where the dorsal diaphragm is poorly developed and a ventral diaphragm is entirely absent. There is, however, in each leg a distinct longitudinal septum dividing the cavity into a dorsal and ventral sinus and consisting of a very delicate membrane that extends from the epidermis of one side of the leg to the other side and bears the leg trachea along its center (Fig. 14).

Accessory pulsatile organs are located in the lateral angles of the scutellum (Fig. 12, *SclPO*) and in the front of the head between the bases of the antennae *(FPO)*. Movement of only the scutellar organs has been observed in the living fly, but the pulsatile nature of the frontal organ is indicated by its anatomical relations and the presence of corresponding structures in other insects. Each of the two *scutellar organs* consists of a small band of muscle fibers spanning the angle where the scutellum joins a ridge that runs down to the base of the wing. The muscle fibers are arranged in a flat

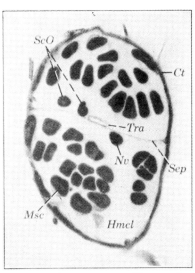

Fig. 14. Leg septum. Transection of femur of third leg (S., H., 400×). *Ct*, cuticula; *Hmcl*, haemocele; *Msc*, muscle; *Nv*, nerve; *ScO*, scolopophorous organs; *Sep*, longitudinal septum; *Tra*, trachea.

sheet, attached at both ends and by fine lateral fibrils to the body wall. A small group of fat cells lies mesad of the muscle band. These scutellar organs resemble those described for muscoidean flies by Thomsen (1938) and presumably function in a similar way, drawing blood into the haemocele of the scutellum from the region at the base of the wing. Actual circulation of blood within the wings has not been detected in Drosophila, though it occurs in the larger species studied by Thomsen, which also possess alar pulsating organs. The *frontal organ* is a thin-walled ovoid ampulla 23 by 10 μ, located medially against the front wall of the head, directly beneath the ptilinal fold and between the antennae. Blood cells are found within it in sections, and the pair of muscles (*16*, Figs. 25, 26A, 27A) that traverse the brain and are posteriorly attached to the aorta are inserted on its rear wall. The structural details require further study, and the manner of communication with the haemocele has not been determined. Accessory pulsatile organs in a similar position have been described for the bee and certain other insects (Freudenstein, 1928; Snodgrass, 1935; Wigglesworth, 1939), and it seems probable that the frontal ampulla in Drosophila, activated by the attached muscles, may assist circulation in the antennae. No pulsatile organs in the legs and wings have been recognized in Drosophila, but they may be present.

The *blood* or *haemolymph* consists of a colorless liquid plasma and free blood cells or haemocytes (*BCl*). The latter are not very numerous and are most evident in sections within the lumen of the heart, where they appear as small, more or less spherical cells 4.5–6 μ in diameter (Figs. 13C, F). Most of these cells are uninucleate and have homogeneous cytoplasm, but some show several clumps of chromatin or one or more large vacuoles. Adequate study would require use of a smear technique and due regard for physiological changes, especially those reflecting metamorphic processes.

The *course of the blood* has not been traced, but it presumably resembles that which occurs in larger Diptera and in the honeybee (cf. Freudenstein, 1928; Thomsen, 1938). Blood from the surrounding haemocele is drawn into the heart through the lateral ostial valves, propelled forward by contractions of the muscular chambers (pulse beat about 140 per minute in physiological saline solution), poured into the head through the open end of the aorta to circulate about the brain and into the antennae (aided by the frontal pulsatile organ) and proboscis, thence backward to the thorax and abdomen. A secondary stream courses down each leg on one side of the septum to the tarsus and then returns on the other side of the septum to the thorax, while the scutellar organs aid the flow through the lateral and dorsal parts of the thorax. Extrusion of the ptilinum to rupture the puparium, expansion of the wings after emergence, dilation of the labella, and perhaps, to some extent, protrusion of the proboscis when feeding are all dependent on general or localized changes in internal blood pressure.

The *nephrocytes* (*Nph, PCl*) are discrete, large, ovoid, colorless, uninucleate or binucleate cells attached to the dorsal diaphragm in the abdomen and near the anterior end of the ventriculus in the thorax (Fig. 12). In the abdomen they are serially arranged in a row of twenty to twenty-five on each side of the heart, from the first to the sixth segment, and are the so-called "pericardial cells." In the thorax there are about ten on each side anterolaterad of the cardia, associated with the lower wall

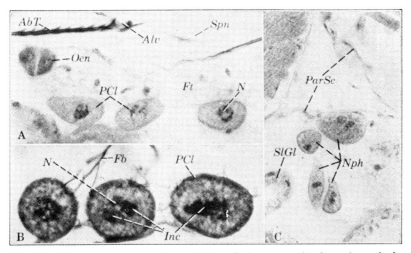

Fig. 15. Nephrocytes. A: Pericardial cells in parasagittal section of abdomen (S., F.L.G.; 540×). B: Pericardial cells of a recently emerged fly (D., C.B.; 560×). C: Thoracic nephrocytes in frontal section of thorax (D., F.L.G.; 400×). *AbT,* abdominal tergite; *Alv,* setal alveolus; *Fb,* fibril of dorsal diaphragm; *Ft,* fat cells; *Inc,* inclusions; *N,* nucleus; *Nph,* thoracic nephrocytes; *Oen,* oenocytes; *ParSc,* parenteric tracheal sac; *PCl,* pericardial cells; *SlGl,* salivary gland; *Spn,* spinule (microchaeta).

of the trachea leading to the first spiracle and with the muscle fiber that extends from the cardia to the anterior end of the underlying salivary gland. The thoracic cells (*Nph*), 28–35 μ by 10–17 μ, have finely reticulate or spongy cytoplasm, a definite birefringent peripheral membrane, and two subcentral spherical nuclei each about 4 μ in diameter and lying close together (Fig. 15C). The abdominal cells (*PCl*) are larger than the thoracic cells, 35–70 μ long and 15–30 μ wide (usually about 40 by 25 μ), and contain only one nucleus, 8–10 μ in diameter (Figs. 15A, B). In recently emerged flies they appear more turgid and contain more conspicuous inclusions (Fig. 15B) than in older individuals, where the cytoplasm may be less dense, lack inclusions (Figs. 15A, C), and have a clear area around the nucleus. Basophilic inclusions occur in both the thoracic and abdominal cells in the form of large masses or smaller granules. In young flies large masses are found close to the poles of the nucleus in the abdominal cells (Fig. 15B), while smaller granules surround the nuclei of the thoracic cells. Older flies may show a few small inclusions near or apart from the nuclei in either type of cell. The abdominal cells are attached to the lower side of the membranous dorsal diaphragm in the basal segments (Figs. 13C, F); further back they are connected to each other and to the heart by the diaphragm fibrils (Fig. 15B), the connection to the heart often appearing double in cross-sections. The loosely grouped thoracic cells seem to be incorporated in a delicate membrane that holds them in place. The function of the nephrocytes is not definitely known: they may serve an intermediary function in metabolism or excretion. It has been found in other insects that they take up colloidal particles from the blood (Wigglesworth, 1939).

The *oenocytes* (*Oen*) are large, acidophilic, colorless, uninucleate cells occurring in metameric masses in abdominal segments 1–7 (Fig. 16). Typically, there is a dorsal strand on each side of the segment, extending from the spiracle to the heart beneath the posterior edge of the tergite, and also an isolated short transverse band (sometimes divided medially into two patches) lying over the posterior border of the sternite. The sternal mass is absent in segment 1, where the sternite is vestigial, and in segment 7 of both sexes; it is also absent in segment 6 of the male. There is only one dorsal strand on each side in segments 6 and 7 of the male, near its posterior border—the separate bands from the two spiracles converging laterally (Figs. 16B, C). In the female, segments 6 and 7 have individual dorsal strands (Fig. 16A). The dorsal strands of segment 1 in both sexes extend forward along the sides of the heart outside the dorsal diaphragm and then laterad behind the mesophragma; in other segments they end at the heart. The oenocytes at the lower end of the dorsal strand lie against the spiracular trachea; from there, however, they do not follow the dorsal segmental trachea but extend backward and upward on the inner side of the peripheral fat body, under the posterior edge of the tergite and outside the dorsal longitudinal muscles. The sternal masses of oenocytes likewise lie beneath, and here tend to envelop, the ventral longitudinal

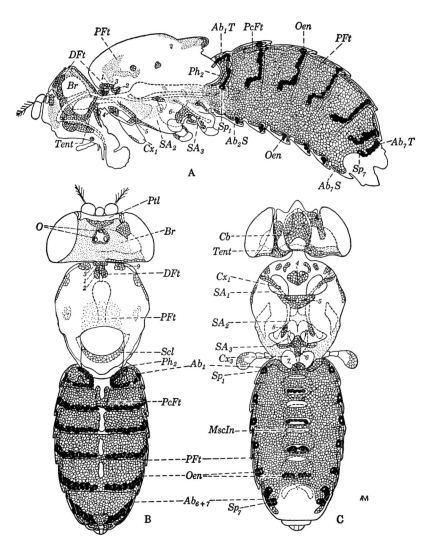

FIG. 16. Distribution of adipose tissue and oenocytes; deep fat of abdomen omitted (36×). Thin layers of peripheral fat indicated by stippling, thicker masses by reticulations, strands of oenocytes in black. A: Mesal aspect of right half of body, female. B: Ventral aspect of dorsal half of body, male; in head the dorsal fat is shown at left, deeper fat at right. C: Dorsal aspect of ventral half of body, male; in head the fat below brain is shown at left, fat at level of brain at right. Ab_1, first, Ab_{6+7}, sixth and seventh, abdominal segments; Ab_2S, second, Ab_7S, seventh, abdominal sternites; Ab_1T, first, Ab_7T, seventh, abdominal tergites; Br, brain; Cb, cibarium; Cx_1, first, Cx_3, third, coxa; DFt, deep fat; $MscIn$, muscle insertion; O, ocelli; Oen, oenocytes; $PcFt$, pericardial fat; PFt, peripheral fat; Ph_2, mesophragma; Ptl, ptilinum; SA_1, prothoracic, SA_2, mesothoracic, SA_3, metathoracic, sternal apophyses; Scl, scutellum; Sp_1, first, Sp_7, seventh, abdominal spiracles; $Tent$, tentorium; 1–9, specific masses of fat in thorax.

453

muscle fibers, so that in longitudinal sections both dorsal and ventral groups of oenocytes lie in the pockets formed by the intersegmental folds. Histologically the oenocytes are strongly eosinophilic cells, 19 by 5 μ, with dense microvacuolate cytoplasm, a vesicular nucleus (3.5 μ), and a columnar, cuneate, or polygonal form, depending upon the plane of section (Fig. 17). They are compactly grouped in unilayered bands, a dorsal strand showing about four to eight cells in cross-section. The oenocytes of adult Diptera have been termed "synoenocytes" by Zavrel to dis-

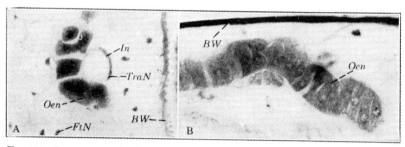

Fig. 17. Oenocytes. A: Transection of strand of oenocytes and of spiracular trachea, second abdominal segment (B.A., I.H.; 490$\times$). B: Longitudinal section of dorsal strand of oenocytes in sixth abdominal segment (D., F.L.G.; 490$\times$). *BW*, body wall; *FtN*, fat-cell nucleus; *In*, intima of trachea; *Oen*, oenocytes; *TraN*, nucleus of tracheal epithelium.

tinguish them from larval oenocytes, which they replace (Hanström, 1939). Their function is still undetermined, though in some insects they have been regarded as organs of intermediary metabolism and of internal secretion (Snodgrass, 1935; Hanström, 1939; Wigglesworth, 1939).

The *corpus allatum* (*CA*) is a single rounded mass of tissue, 20–30 μ long by 20–30 μ wide by 10–13 μ thick, lying upon the dorsal wall of the aorta in the prothorax, directly above and in front of the stomodaeal ganglion, which is beneath the aorta (Figs. 12, 18A, 4A). It is overlaid by a small mass of fat, through which run the dorsal transverse tracheal commissure of the prothorax and the cervical muscles that arise from the medial pronotal apodeme. The innervation of the corpus has not been determined, but there are probably nervous connections with the adjacent stomodaeal ganglion. The corpus contains about fourteen vesicular nuclei, situated near the periphery. The dense, moderately acidophilic cytoplasm shows no cell boundaries and either is homogeneous throughout or encloses a few large vacuoles in the dorsal region (Fig. 18). This organ, apparently not previously described, corresponds to part of the "ring gland" of the larva (Scharrer and Hadorn, 1938; Scharrer, 1941b; Vogt, 1941c, 1942).[3] The adult organ probably serves an endocrinal function, perhaps associated with the development of eggs as in the grasshopper (Pfeiffer, 1939, 1940), the roach (Scharrer, 1943), and the reduviid bug Rhodnius (Wigglesworth, 1939); cf. Vogt, 1940a, b, 1941a, b, and Thomsen, 1940, 1941. Special cells in

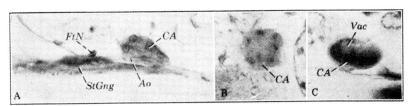

Fig. 18. Corpus allatum. A: Corpus allatum and stomodaeal ganglion in sagittal section (B., H.F.; 490×). B: Frontal section of corpus allatum (D., F.L.G.; 490×). C: Sagittal section of corpus allatum, showing vacuoles (B.A., I.H.; 490×). Ao, aorta; CA, corpus allatum; FtN, fat-cell nucleus; StGng, stomodaeal ganglion; Vac, vacuole.

the stomodaeal ganglion, or corpus cardiacum, and in the central nervous system may likewise be incretory in function (see Nervous System).

The adipose tissue (Ft) consists of masses or sheets of fat cells and is collectively termed the "fat body." It occurs beneath the integument and deeper in the haemocele, filling most of the space within the body that is not occupied by other organs. The distribution of the adipose tissue is shown in Fig. 16. In the head and thorax both the peripheral (PFt) and deep fat masses (DFt) are indicated; in the abdomen, which contains the largest amount, most of the deep fat that lies among the viscera has been omitted. In general, peripheral fat occurs in sheets immediately below the integument in areas where no muscle attachments are present. In the head and thorax the peripheral fat is very thin (one cell layer, 5–15 μ thick), but in the abdomen it may attain consider-

[3] Day (1943b) has subsequently described the metamorphosis of the ring gland of Lucilia, homologizing its adult derivatives as corpus allatum, corpora cardiaca, and hypocerebral ganglion, and has reported experiments on its functions (1943a).

able bulk (18–25 μ thick). The peripheral fat of the abdomen is divided into overlapping segmental masses under the tergites, but ventrally it forms a continuous sheet interrupted only by the sternal groups of oenocytes. The genital segments contain little or no fat. The deep fat occurs in circumscribed masses that occupy definite positions in the head and thorax; in the abdomen it invests the viscera, its abundance varying with the physiological condition of the fly, but here also circumscribed masses are discernible—for example, independent masses enveloping each spermatheca and accessory gland in the female. No fat is found in the appendages, except a little in the coxae, halteres, and proboscis.

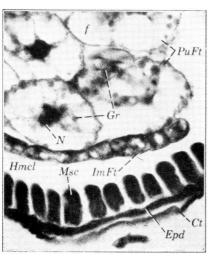

Fig. 19. Adipose tissue. Pupal and imaginal fat cells in frontal section of abdomen of young fly (D., C.B.; 550×). *Ct*, cuticula; *Epd*, epidermis; *f*, fat vacuole (negative image of oil globule); *Gr*, granular inclusions in cytoplasm; *Hmcl*, haemocele; *ImFt*, imaginal fat cells; *Msc*, tergosternal muscle; *N*, nucleus; *PuFt*, pupal fat cells.

The adipose tissue consists of uninucleate cells more or less completely filled with oil globules, so that in sections prepared by ordinary histological methods the cytoplasm forms a meshwork of narrow strands radiating from the nucleus and enclosing large vacuoles.

Recently emerged flies contain two types of fat cells (Fig. 19). Most of the deep fat consists of large, separate, spherical cells (diameter 30–40 μ) with a central densely granular nucleus (7 by 9 μ), large fat vacuoles, and granular inclusions up to 6 μ in diameter (waste products?) in the cytoplasm (*PuFt*). These cells float freely out of the body when dissected in saline solution and are presumably carried over from the pupal period into imaginal life. The other type of fat, mainly peripheral, occurs as sheets of small rectangular cells (4–13 μ) closely adherent in single layers, filled with small vacuoles, and having small, dense nuclei (2–3 μ) but no cytoplasmic inclusions (*ImFt*). This is young imaginal tissue.

In older individuals all the fat in the body appears histologically similar, regardless of the position or shape of the mass. The cells are crowded, highly vacuolate, large, and polygonal in form (25 by 25 to 50 by 20 μ), show no inclusions, and have a central ovoid or distorted nucleus (8 by 8 to 3.5 by 12 μ) containing dense, finely divided chromatin and a deeply staining nucleolus (Figs. 7D, 15A, 17A, etc.).

It has not been determined whether the "pupal fat" of young flies is replaced by or converted into the type of tissue found in older individuals; the former, with complete disintegration of the waste-containing pupal cells, is possible as a delayed process of metamorphosis. The waste-containing cells probably correspond to the "urate cells" described in other insects, but it is questionable whether they should be considered truly distinct from the fat-forming and -storing cells or trophocytes. The distinction probably depends on different physiological conditions of similar cells (cf. Snodgrass, 1935; Wigglesworth, 1939). Evans (1935) describes chemical reactions indicating that the viscid granular inclusions in the pupal fat of adult Lucilia are proteid in nature; they disappear before the cells are attacked by leucocytes, differing from Calliphora (Pérez, 1910) in this regard.

THE RESPIRATORY SYSTEM

The complex respiratory system of insects consists of branching tubes, the tracheae, that open to the exterior through spiracles, ramify throughout the body, and end in minute tubules, the tracheoles, where gaseous exchange between air and tissues occurs. In Drosophila, as in other higher Diptera and many other insects, certain portions of the principal tracheae may be dilated to form thin-walled air sacs, which probably enable more efficient aeration of the tissues. By reflected light, the air-filled tubes and sacs appear silvery through the body wall during life and when freshly dissected in aqueous media or glycerin.

The *tracheae* are ectodermal epithelial tubes lined by a transversely ridged chitinous intima, which varies in thickness from

0.7 μ in the larger tracheae to 0.3 μ in the smaller ones [4] (Figs. 17A, 20). The ridges, or taenidia, serve to stiffen the walls of the tubes and prevent their collapse. The tracheal epithelium is very thin and almost imperceptible, with small, widely spaced, and flattened nuclei 4.5 by 1.5 μ (Figs. 9A, 17A). The tracheae branch freely and range in diameter from about 14 μ in the major undilated trunks to 0.8 μ in the finer subterminal and terminal

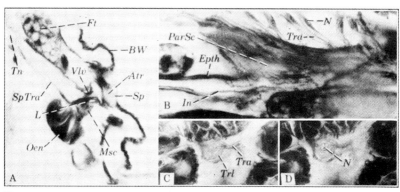

FIG. 20. Tracheal structures. A: Frontal section through first abdominal spiracle (D., C.B.; 400×). B: Longitudinal section of posterior portion of parenteric air sac (S., M.; 480×). C, D: Developing tracheole coiled in mother cell in ovary of recently emerged fly; C and D are two optical levels of the same cell (D., M.; 480×). *Atr*, atrium; *BW*, body wall; *Epth*, epithelium; *Ft*, fat cells; *In*, intima; *L*, lever of closing apparatus; *Msc*, muscle; *N*, nucleus; *Oen*, oenocytes; *ParSc*, parenteric air sac; *Sp*, spiracle; *SpTra*, spiracular trachea; *Tn*, taenidia; *Tra*, trachea; *Trl*, tracheole; *Vlv*, valve.

branches. In the *air sacs* both intima and epithelium are attenuated and the nuclei sparsely distributed; taenidia are present in the narrow tubular sacs but absent in the more dilated sacs (Figs. 15C, 20B). Their delicate walls are consequently very flexible and may collapse or expand with changes in pressure in and around them. (Rhythmic inflation and deflation of the dorsal

[4] These dimensions are the apparent thicknesses of the refractile lining of the tracheae in sections, including the taenidia, and are undoubtedly exaggerated by reflection and refraction. Richards and Anderson (1942a, b) have found that the electron microscope reveals the intertaenidial membrane of the larger tracheae in several insects to be 0.01–0.02 μ thick after drying (probably less than 0.02 μ in life), while in tracheoles the membrane may be only 0.005 μ thick. The same is probably true in Drosophila.

sacs of the head are sometimes observed through the transparent body wall and may be due to changes in cephalic blood pressure.) The tracheae give off from their walls the minute *tracheoles*, which spread over and penetrate between the adjacent tissue cells. The tracheoles are smooth-walled,[5] branched or unbranched chitinous tubules, 0.3 μ in diameter, which develop coiled within single cells and have one end communicating with the tracheal lumen. The rounded uninucleate mother cells are found in recently emerged flies (Figs. 20C, D). In older tracheoles the cytoplasm and nucleus of the mother cell are not evident in sections.

The *spiracles*, or external tracheal orifices, in Drosophila number nine pairs: an anterior mesothoracic pair between propleura and mesopleura, a posterior metathoracic pair between the bases of the halteres and the mesomera (hypopleura), and seven abdominal pairs in the pleural membrane beneath tergites 1–7 (see External Morphology and Fig. 21). The closing mechanism of each thoracic spiracle consists of a two-lipped external rim or peritreme provided internally with a short ventral muscle (Fig. 26B, *76, 97*). This is attached to the neighboring body wall, and its contraction draws the elastic lips together (cf. Snodgrass, 1935, Fig. 231D). The abdominal spiracles have a rigid circular peritreme 19 μ in diameter, and the opening leads into a short tubular vestibule or atrium lined by inward-projecting circular lamellae (exaggerated taenidial ridges). Where the inner end of the atrium joins the trachea, the passage is constricted by a deep fold of the anterior wall (Fig. 20A). Attached to this fold is a sclerotized lever so arranged that the muscle extending from its tip to a rigid portion of the atrial wall acts as an occlusor, bringing the fold against the sclerotized opposite wall of the passage (cf. Snodgrass, Fig. 232H). No dilator muscles are apparent, so that the elasticity of the valvular apparatus probably opens the passage.

The arrangement of the tracheae in Drosophila is shown in Figs. 21 and 22. The basic pattern of the tracheal system is metameric in origin, resulting from anastomosis of similar pri-

---

[5] Although no taenidia are visible under the oil immersion lens, the electron microscope has shown them in the tracheoles of other insects. This invalidates a distinction between tracheae and tracheoles on the basis of presence or absence of taenidia (Richards and Anderson, 1942a, b).

mary tracheae in successive segments. However, the degree of modification in this specialized insect is so high that the general plan is obvious only in the central abdominal segments. Here the tracheae are slender throughout and are nowhere dilated to form air sacs. In the anterior abdominal segments considerable departure from the basic plan is evident, but the principal features are still readily recognizable, and the development of air sacs is represented only by partial dilatation of one pair of trunks. In the thorax the unequal development of the segments (mesothorax enlarged at the expense of pro- and metathorax) obscures the basic plan, and the picture is further complicated by the extensive development of air sacs in this region. The head likewise has a highly specialized tracheal system, richly provided with air sacs. In the posterior part of the abdomen the tracheal system has undergone reduction in keeping with the modification of the terminal segments; the eighth pair of abdominal spiracles found in generalized insects is absent in Drosophila. Since the least specialization occurs in the abdomen, the tracheation of this region will be described first.

Tracheation of the Abdomen

Each of the first seven segments of the abdomen is provided with a pair of spiracles (Sp_{1-7}), situated laterally in the pleural membrane close to the edge of the tergites. In the female (Fig. 21A) all spiracles lie near the center of their respective segments; in the male (Figs. 21B, C) the first five pairs are similarly placed, but the sixth and seventh pairs both lie under the edge of the large sixth tergal plate (fused sixth and seventh tergites?). Each spiracle communicates with the interconnecting system of tracheae by means of a short spiracular trachea ($SpTra$), separated from the atrium (Atr) by the closing apparatus described above.

The general plan of the abdominal tracheation consists of (1) a pair of longitudinal lateral trunks ($LTra$), connecting the successive spiracular tracheae on each side; (2) a pair of zigzag dorsal trunks ($DTra$), extending through segments 1–8; (3) an ascending dorsal segmental trachea ($DSgTra$) in each of the first seven segments, connecting the lateral and dorsal trunks on each side; (4) six transverse dorsal commissures ($DCom$), uniting the dorsal trunks in segments 1–6; (5) visceral tracheae (*VscTra*), arising mainly from the lateral trunks and dorsal segmental tracheae; and (6) superficial dorsal (*DSTra*) and ventral (*VSTra*) tracheae, arising from or near the spiracular tracheae. From the main vessels, at more or less definite points, spring smaller tracheae that supply the adjacent tissues. The principal tracheal trunks lie in the haemocele between the somatic muscle layer and the viscera, hence near the body wall; the dorsal commissures pass above the heart.

The netlike interconnections and segmental nature of the main tracheae are most apparent in segments 4–7. In a typical segment, such as 5, three main tracheae arise from the spiracular trachea, one going cephalad, one caudad, and one dorsad. The cephalic and caudal vessels extend to the spiracular tracheae in the neighboring segments and thus form a section of the lateral trunk. The dorsal vessel is the dorsal segmental trachea; it courses upward and forks into an anterior and a posterior branch, which in turn anastomose respectively with the posterior branch of the preceding segment and the anterior branch of the succeeding segment. These branches thus serially united make up the zigzag dorsal longitudinal trunk. Right and left dorsal trunks are connected by transverse commissures that arise from the angles where successive posterior and anterior segmental branches unite. Consequently the commissures tend to be intersegmental in position, as between segments 3–4, 4–5, and 5–6. The last one, however, is more centrally located in segment 6. The vessels described thus form, through their anastomoses, a large-meshed tracheal net under the sides and top of the segments concerned. A segment also has, on each side, a dorsal and a ventral superficial trachea that usually arise from the walls of the spiracular trachea, and a variable number of visceral tracheae originating on the dorsal segmental trachea and the lateral trunk (see below).

Although the same fundamental plan holds for the first three segments, modifications make it less apparent. The lateral trunks are the same as in the other segments, but the dorsal trunks are distended to form a pair of slender, tapering air sacs (*AbSc*) in segments 1 and 2. These sacs lie deeper in the body than do the more posterior, unmodified dorsal trunks. Each is joined to the first three spiracular tracheae of the same side by connectives

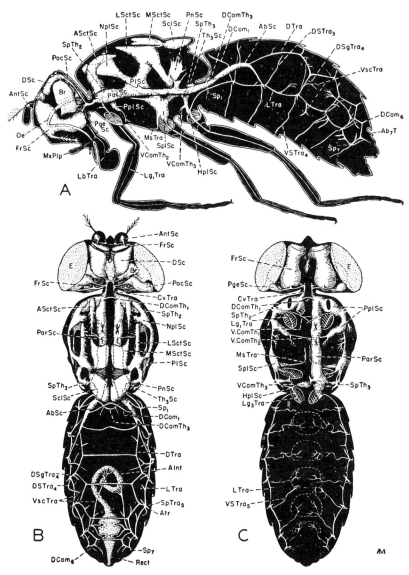

FIG. 21.

FIG. 21. Respiratory system (36×). A: Mesal aspect of right half of body, showing air sacs and principal tracheae; female. B: Dorsal aspect of body with dorsal wall removed, showing dorsal system of air sacs and tracheae, and their relation to brain and hindgut (*stippled*); male. C: Dorsal aspect of ventral half of body, showing ventral air sacs and tracheae; male; left parenteric sac omitted; openings (*black*) in propleural sacs and in lateral tracheae in front of metathoracic spiracles communicate with the dorsal system of air sacs; coxal cavities cross-hatched, abdominal sternites in broken outline. *AbSc*, abdominal air sac; *Ab$_7$T*, seventh abdominal tergite; *AInt*, anterior intestine; *AntSc*, antennal sac; *ASctSc*, anteroscutal sac; *Atr*, atrium of spiracle; *Br*, brain; *CvTra*, cervical trachea; *DCom$_1$*, first, *DCom$_6$*, sixth, abdominal dorsal commissures; *DComTh$_1$*, prothoracic, *DComTh$_3$*, metathoracic, dorsal commissures; *DSc*, dorsal sac of head; *DSgTra$_4$*, dorsal segmental trachea of fourth abdominal segment; *DSTra$_3$*, *DSTra$_4$*, dorsal superficial tracheae of third and fourth abdominal segments; *DTra*, dorsal tracheal trunk; *E*, compound eye; *FrSc*, frontal sac; *HplSc*, hypopleural sac; *LbTra*, labial trachea; *Lg$_1$Tra*, *Lg$_3$Tra*, leg tracheae; *LSctSc*, lateroscutal sac; *LTra*, lateral tracheal trunk; *MSctSc*, medioscutal sac; *MsTra*, mediosternal trachea; *MxPlp*, maxillary palpus; *NplSc*, notopleural sac; *Oe*, oesophagus; *ParSc*, parenteric sac; *PgeSc*, postgenal sac; *PlSc*, pleural sac; *PnSc*, postnotal sac; *PocSc*, postocular sac; *PplSc*, propleural sac; *Rect*, rectum; *SclSc*, scutellar sac; *Sp$_1$*, first, *Sp$_7$*, seventh, abdominal spiracle; *SplSc*, sternopleural sac; *SpTh$_2$*, anterior (meso-), *SpTh$_3$*, posterior (meta-), thoracic spiracle; *SpTra$_5$*, spiracular trachea of fifth abdominal segment; *Th$_3$Sc*, metathoracic sac; *VComTh$_{1,2,3}$*, pro-, meso-, and metathoracic ventral commissures; *VscTra*, visceral trachea; *VSTra$_{4,5}$*, ventral superficial tracheae of fourth and fifth abdominal segments.

that are really dorsal segmental tracheae which have become shortened and extend mesad rather than dorsad, the third forming the tapering end of the sac. Two dorsal commissures extend between the air sacs in segments 1 and 2; a third commissure ($DComTh_3$), anterior to the others and located beneath the mesophragma, is regarded as a metathoracic trachea. This modification involves a cephalic concentration of the dorsal vessels so that the portion of the dorsal trunk corresponding to the posterior branch of dorsal segmental trachea 3 is considerably lengthened and extends from the air sac in segment 2 to the dorsal commissure between segments 3 and 4. The abdominal air sacs, which are always tubular and never baglike as in higher Diptera (Musca, Hewitt, 1914; Calliphora, Lowne, 1890–95; Fraenkel, 1935), are continuations of the parenteric thoracic sacs, while the lateral trunks connect with the distended bases of the tracheae that enter the metathoracic legs.

Caudally each lateral trunk ends at the seventh spiracle, and the dorsal trunk reaches into the eighth segment as the posterior branch of dorsal segmental trachea 7.

The principal tracheae supplying the viscera ($VscTra$) arise from the lateral trunks between the spiracles (from spiracle 2 back) and from the fourth to seventh dorsal segmental tracheae. In general they seem to supply the immediately adjacent organs (e.g., the ovaries are profusely supplied by branches between spiracles 3, 4, and 5, and from dorsal segmental trachea 4). The visceral branch from dorsal segmental trachea 6, however, not only supplies the neighboring rectal papillae but also sends a long branch forward into segments 4 and 5 to the loop of the anterior intestine (Fig. 21B). Although the number and position of the visceral tracheae are not strictly constant, they usually arise as follows: from the lateral trunks, one between spiracles 2 and 3; one or two between spiracles 3 and 4; one between spiracles 4 and 5; two between spiracles 5 and 6; two or three between spiracles 6 and 7. None, one, or two arise from the fourth dorsal segmental trachea, and one each from the fifth, sixth, and seventh dorsal segmental tracheae. Smaller branches arise from the dorsal trunks, and the air sacs may also give off a visceral branch between their two dorsal commissures.

The superficial dorsal ($DSTra$) and ventral ($VSTra$) tracheae may arise directly from the spiracular trachea or from the lateral trunks near the spiracular junction. There is generally one dorsal and one ventral superficial trachea in each spiracle-bearing segment. The ventral trachea extends to the midline of the sternum, its branches supplying the ventral wall of the segment (Fig. 21C), while the dorsal trachea is limited largely to the dorsolateral area beneath the dorsal trunk. The mediodorsal wall is supplied mainly by fine branches from the dorsal trunks.

Tracheation of the Thorax

The thorax is highly specialized for locomotor functions, and its tracheation is correspondingly modified (Fig. 21). The primary morphological pattern of the tracheal system is obscured by relative displacement of its elements and by the extensive development of air sacs. Both these types of modification seem to be correlated principally with the enormous development of the dorsal muscles of the mesothorax as indirect flight muscles, their bulk inducing relocation of neighboring structures and their intense physiological activity necessitating a highly efficient respiratory mechanism, probably afforded by the air sacs. There are superficial and deep-lying air sacs, adjacent to or between the muscle masses and along the alimentary canal. The relationship of the thoracic to the abdominal tracheation indicates that the superficial sacs are derivatives of the lateral tracheae, while the deeper ones are modified dorsal tracheae displaced inwardly by the overlying dorsal muscles.

The lateral tracheal system on each side is modified into a series of interconnecting air sacs lying near the body wall and connecting directly with the spiracles. Three sacs—propleural ($PplSc$), sternopleural ($SplSc$), and hypopleural ($HplSc$)—lie above the bases of the legs, connect together and to the first abdominal spiracle by lateral tracheae, and give off each a single trachea ($Lg_{1-3}Tra$) to the corresponding leg (cf. Fig. 14). The anterior thoracic spiracle ($SpTh_2$) opens into the propleural sac, the posterior spiracle ($SpTh_3$) into the hypopleural sac. Left and right prothoracic leg tracheae, propleural sacs, and hypopleural sacs are connected by three transverse ventral commissures ($VComTh_{1-3}$), the posterior one dilated and saccular medially, and each sternopleural sac gives off a trachea ($MsTra$) to the mediosternal region. From the lateral trachea between the sternopleural and hypopleural sacs on each side arises a branching series of air sacs that underlie the dorsolateral and dorsal area of the mesothorax: namely, postnotal ($PnSc$), lateroscutal ($LSctSc$), medioscutal ($MSctSc$), scutellar ($SclSc$), pleural ($PlSc$), and notopleural ($NplSc$) sacs, all joined by broad connectives as shown in the figures. A small diverticulum, the metathoracic sac (Th_3Sc), arises just above the posterior spiracle.

The deeper tracheal system consists of a pair of elongate parenteric air sacs ($ParSc$) and associated diverticula. The parenteric sacs are direct continuations of the dorsal trunks of the abdomen; they flank the digestive tube in the thorax and become the cervical tracheae ($CvTra$) in the neck region. Each parenteric sac anastomoses anteriorly with the lateral system of the same side through a wide medial extension of the propleural sac and posteriorly through a short connective near the posterior spiracle. From the sacculate anterior connective arises a sickle-shaped dorsal diverticulum, the anteroscutal sac ($ASctSc$), which extends upward and backward between the dorsal muscles. Short dorsal diverticula also arise from the parenteric sacs about a third of the way back and again near the level of the posterior spiracles, supplying the lower dorsal muscles (Fig. 20B). The two parenteric sacs are conected by a short, straight dorsal commissure in the prothorax ($DComTh_1$) and by a longer curved commissure under the mesophragma ($DComTh_3$), both passing above the alimentary canal.

The paired nature of the thoracic air sacs is readily apparent in dorsal view (Fig. 21B). The left and right systems are connected only by the two dorsal and three ventral commissures mentioned above. An interesting correlation exists between the position of the air sacs and the external color pattern. When the fuscous trident pattern of the scutum is present, it is coextensive with the underlying medioscutal air sacs, the edges of the fuscous area (except the median tine) and of the air sacs coinciding. The dusky scutellum is entirely occupied by the scutellar sacs.

The air sacs give off from their walls a profusion of tracheal branches direct to the great flight muscles. The connecting tracheae also give off branches to supply adjacent muscles and viscera.

Tracheation of the Head

The cephalic tracheal system (Figs. 21, 22) is also modified to form a series of large paired air sacs, which occupy most of the space between the other organs in the head capsule. A single

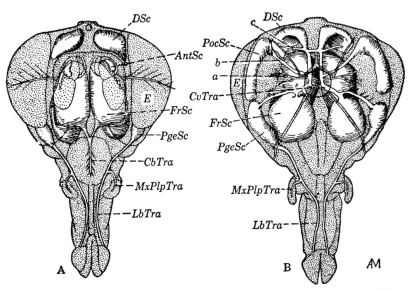

FIG. 22. Tracheation of head. Head represented as a transparent object with the tracheal system in white (57×). A: Anterior aspect. B: Posterior aspect. *a, b, c,* lateral branches of ascending trachea (*a* and *c* to brain); *AntSc,* antennal air sac; *CbTra,* trachea to cibarial muscles; *CvTra,* cervical trachea; *DSc,* dorsal sac; *E,* compound eye; *FrSc,* frontal sac; *LbTra,* labial trachea; *MxPlpTra,* trachea of maxillary palpus; *PgeSc,* postgenal sac; *PocSc,* postocular sac.

pair of tracheal trunks, the cervical tracheae (*CVTra*) arising from the parenteric sacs of the thorax, enters the head through the foramen magnum. Each trunk immediately forks into a dorsal and a ventral branch. The dorsal branch ascends to the vertex, closely parallel to its mate of the opposite side, to which it is usually joined by a short commissure in the occipital region. Three lateral branches are given off, the first (*a*) and third (*c*) supplying the brain directly, the second (*b*) leading to the elongate postocular air sacs (*PocSc*). At the top of the head the ascending branch expands to form a large dorsal air sac (*DSc*), which extends forward to the ptilinal suture. The ventral branch of each cervical trunk courses laterad and connects first with the posterior end of the frontal air sac (*FrSc*) and then with the postgenal air sac (*PgeSc*) that lies behind the eye along the lower posterior border of the head. The frontal sacs are the largest in the head. They flank the cibarium and extend forward and upward around the suboesophageal ganglion and brain to the level of the antennae and ptilinal suture. Dorsally each frontal sac gives off a fine branch to the antenna, where a small air sac (*AntSc*) is found in the second segment and three minute sacs in the third segment. Anteroventrally each frontal sac gives off a branch which joins that of the opposite side, forming a short median trachea (*CbTra*), which supplies small branches to the pumping muscles of the cibarium. Each postgenal air sac tapers ventrally into a trachea (*LbTra*) that extends down the proboscis, gives off a branch to the palpus (*MxPlpTra*), and supplies the labial muscles. From the tracheal trunks and air sacs arise very fine lateral branches to neighboring parts of the head, the brain being especially richly supplied by numerous branches from the caudal tracheal complex, and the eyes from the postocular air sacs. Adjacent air sacs may be in contact when normally expanded but are interconnected only as described above.

THE MUSCULAR SYSTEM

The muscular system, comprising skeletal and visceral muscles, consists of a multitude of contractile fibers arranged in groups or layers. The skeletal or somatic muscles are attached to the inner surface of the body wall, extend between articulated or flexibly joined areas, and serve to move the parts of the exoskeleton; in some places they extend from the body wall to certain viscera. The visceral or splanchnic muscles are constituents of the walls of the digestive, circulatory, and reproductive organs (q.v.) and usually have a peristaltic or compressing function. The skeletal muscles consist of groups of elongate parallel or converging fibers, or sometimes of single fibers, and differ widely in size and strength according to the number and size of these fibers. Individual fibers range in length from 20 μ (occlusor muscle of anterior thoracic spiracle) to 260 μ (cibarial dilator muscles), and in diameter from 3 μ (dorsal abdominal muscles) to 18 μ (cibarial muscles), while the fiberlike bundles constituting the indirect flight muscles reach 800 μ in length and 120 by 60 μ in cross-section. The visceral muscle fibers are usually thinner (1.5 to 5 by 9 μ in diameter) and occur either as discrete parallel strands forming transverse or longitudinal layers around tubular organs (Figs. 4A, 8) or as branching and interlacing fibrils forming a plexus over the surface of the crop and ovaries (Figs. 4D, 44D). Living muscles are colorless and of a gelatinous consistency.

Both skeletal and visceral muscles consist of striated fibers, but the striations may be obscure in the finest visceral fibers. In general, a fiber is an elongate, multinucleate bundle of minute parallel fibrillae (sarcostyles or myofibrils), accompanied by varying amounts of undifferentiated fluid cytoplasm (sarcoplasm), and surrounded by a thin surface membrane (sarcolemma). Three types of fibers may be distinguished in adult Drosophila: namely, thin fibers with superficial nuclei (fine visceral muscles; Fig. 4D), "tubular" fibers with axial nuclei (most skeletal and many visceral muscles; Figs. 23A, 8A), and the "fibrillar muscle" type with large fibrillae and interstitial nuclei (indirect flight muscles; Fig. 24). The plexiform peritoneal muscles of the ovary (Fig. 44D) and endothelioid muscles of the heart wall (Fig. 13D) are specialized types described in conjunction with the organs concerned. The alary muscles of the heart and the muscles of the mesenteron consist of very slender fibers, 1.5–2 μ in diameter, perhaps including only one sarcostyle, with an occasional bulging nucleus that seems to be located directly under the sarcolemma. In the "tubular" fibers, which constitute most of the muscles and probably grade into the first type along the digestive tract, the nuclei (1.5 by 0.7 to 4.5 by 1.5 μ; 12 by 3 μ in muscle *16*) are located in a central core of granular sarcoplasm surrounded on all sides by numerous slender sarcostyles (0.5 μ in diameter), and the entire fiber is covered by a close-fitting sarcolemma (Fig. 23A). In the skeletal muscles of this type the nuclei are very numerous and lie close together in a single row (two widely separated rows in the broad flat fibers of the mesothoracic tergal depressor of the trochanter), but in visceral muscles only an occasional isolated nucleus is seen (Fig. 8B). The "fibrillar" indirect flight muscles (nos. *45* to *48*), the largest muscles of the body, consist of a great many parallel fibrillae, 0.7 μ in diameter, arranged in bundles that are polygonal in cross-section and 14–21 μ

in diameter (Fig. 24). The fibrillae constituting a bundle seem to be bound together by encircling tracheae, and all the bundles of a given muscle are collectively enclosed in a single membranous covering (sarcolemma?). The nuclei (1 by 3 to 1.5 by 4.5 μ) are scattered beneath this membrane and between the bundles of fibrillae. Although the indirect flight muscles of other insects

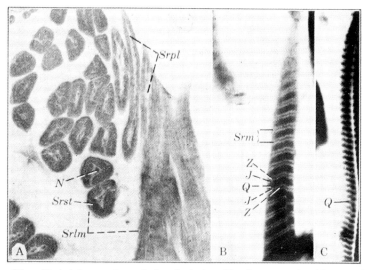

Fig. 23.   Skeletal muscles of the "tubular fiber" type. A: Transections and longisections of cibarial dilator muscles (C., F.L.G.; 530×). B, C: Parasagittal sections of cephalic and coxal muscle fibers, showing striations (S., M.; 530×). *J*, light disc; *N*, nucleus; *Q*, dark disc; *Srlm*, sarcolemma; *Srm*, sarcomere; *Srpl*, sarcoplasm; *Srst*, sarcostyles (fibrils); *Z*, telophragma.

are yellowish, in Drosophila they are colorless like the other muscles of the body. The granular "sarcosomes" that occur between the fibrillae in some insects are perhaps represented by minute interfibrillar granules 0.6–1 μ in diameter, apparent in some sections.

The striations in all types of fibers consist of alternating light and dark bands extending across the entire width of the fiber, and are made up of aligned isotropic and anisotropic segments of the individual fibrillae. In fixed and stained tissue, fine transverse membranes (telophragmata, Krause's or Z membranes), attached to the sarcolemma, divide the fiber into sarcomeres (Figs. 23B, C). Each sarcomere contains a central dark Q disc and a narrow light *J* disc at each end. The dark disc is stained blue and the light discs red by Mallory's connective-tissue stain. In some fibers, perhaps in a different stage of contraction, each dark disc is crossed by a central pale band (median disc *H* or Henson's line). In still other fibers the telophragma occurs in the center of a dark "contraction disc" formed by the approximated halves of divided *Q* discs of adjoining sarcomeres. A deli-

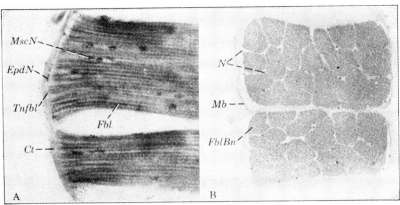

Fig. 24. Skeletal muscles of the "fibrillar" type. A: Longisections of anterior ends of dorsal median indirect wing muscles (B., H.E.; 490×). B: Transections of same (C., F.L.G.; 350×). *Ct,* cuticula; *EpdN,* epidermal nucleus; *Fbl,* muscle fibrilla; *FblBn,* bundle of fibrillae; *Mb,* membrane surrounding entire muscle; *MscN, N,* muscle nuclei; *Tnfbl,* tonofibrilla.

cate membrane, the mesophragma *M,* traversing the median disc, and also narrow granular accessory discs, *N,* have been described in insect muscle (Wigglesworth, 1939) but have not been detected in the Drosophila material at hand. In the fibrillar muscles Mallory's stain may show single red bands alternating with two blue bands with intervening lavender zones, but this may be due to the division of the blue *Q* disc that is said to accompany strong contraction.

The ends of the skeletal muscles are attached to the cuticle or to the epidermis, usually through intermediary "tonofibrillae," which may be very short as in the indirect flight muscles (Fig. 24A). Some muscles are attached to long chitinous tendons which extend from the cuticle—for example, the trochanteral depressors (*34, 66,* Fig. 26B) and axillary muscles (*50, 53–56,* Fig. 26C).

The arrangement of the skeletal muscles is shown in Figs. 25–28. The simplest plan is found in the unmodified abdominal segments, where only dorsal longitudinal, ventral longitudinal, and lateral vertical muscles are present (Figs. 26A, 27). The terminal segments contain the more intricate muscles of the genitalia. In the thorax also the muscles may be classified into dorsal, ventral, and lateral categories, but their arrangement is highly complex to provide for the movement of the wings, legs, and head (Figs. 26, 27). The head (Fig. 25) contains a specialized musculature concerned mainly with feeding (movement of the proboscis and activation of the sucking pump). There are about 150 pairs of functional skeletal muscle groups in the adult female, which may be estimated to include about 4000 fibers.

The musculature has much in common with that of other insects and corresponds most closely to that of other higher Diptera, like Calliphora. Intersegmental and interspecific homologies of individual muscles are evident in varying degree, but a detailed and complete interpretation from this point of view would require phylogenetic considerations beyond the scope of the present study. The muscles have therefore been arbitrarily numbered to facilitate reference, and their evident or suggested homologies are simply indicated in the list that follows. Except for the specialized head muscles, the homologies are based mainly upon the general plan of insect musculature outlined by Snodgrass (1935). For a detailed comparative myology of insects, the reader is referred to Berlese (1909).

Recently emerged flies have a number of muscles in the head and abdomen that later degenerate and disappear completely. Some of these temporary muscles are pupal and others perhaps larval in origin. They include (a) muscles of the frons, vertex, genae, and ptilinum, consisting of many fibers that extend to the tentorium and cibarium for retraction of the ptilinum after emergence; (b) a pair of large dorsal oblique fibers and (c) a pair of intersegmental lateral fibers for each of the unmodified abdominal segments; and (d) several smaller muscles in the basal abdominal segments. Of these, only (b) and (c) are included in the figures (Fig. 26A, muscles in abdomen distinguished by broken outlines).

In the following list, the muscles are designated in accordance with their points of insertion (rather than origin), thus in the case of extrinsic muscles indicating the part moved. The numbers are those given in the figures, followed in parentheses by homologous designations of other authors. All except a few median muscles (*10, 13, 18b, 151*) are bilaterally paired.

Muscles of the Head

The names in parentheses are those used by Graham-Smith (1930) for apparently homologous muscles in Calliphora.[6]

Muscles of proboscis: *1* (retractors of rostrum), *2* (accessory retractors of rostrum), *3* (flexors of labrum). *1* and *2* retract proboscis, *3* flex proboscis. (Figs. 25, 26A, 27)

Muscles of maxillary apodeme: *4* (adductor of apodeme). Probably assist extension of haustellum. (Figs. 25, 26A)

Muscles in labrum: *5* (dilators of labrum-epipharynx). Probably compress labrum and dilate food canal. (Figs. 25A, 26A)

Muscles in haustellum: *6* (retractors of paraphyses; extend labella), *7* (retractors of furca; spread labella), and *8* (transverse muscles of haustellum; assist extension of labella). (Figs. 25, 26A)

Muscles of cibarium: *9* (retractors of fulcrum), *10* (with *9*, probably rotate cibarium upward into head), *11* and *12* (dilators of pharynx; operate cibarial sucking pump: see Digestive System). (Figs. 25A, 26A, 27B)

Muscles of salivary pump: *13* (gracilis). Operate salivary pump or valve: see Digestive System. (Figs. 25B, 26A)

Muscles of second antennal segment: *14, 15*. (Figs. 25A, 26A, 27A)

Muscles of frontal pulsatile organ: *16* (see Circulatory System). (Figs. 25, 26A, 27A)

Temporary muscles: *17*, from vertex to cibarium; *18a*, from ptilinum to cibarium (may include homologues of Graham-Smith's retractors of the oesophagus); *18b*, a large transverse fiber from one side of ptilinum to the other, above level of antennae; *19a*, from tentorium to frons; *19b*, from postgena to anterior portion of gena. Many fibers, apparently for retraction of ptilinum and front of head after emergence from puparium; not known to be present at time of pupation (cf. Robertson, 1936). All soon degenerate and are not found in older flies. (Fig. 31A)

[6] For the musculature of the mouthparts, see also Chapter 5 on External Morphology. The terms used there correspond to the numbers in the list given here as follows: *1*, lateral labial adductor muscle; *2*, maxillary retractor muscle; *4*, maxillary muscles; *5*, labral compressor muscles; *6* and *7*, palpal muscles; *8*, transverse labial muscles; *9*, lateral pharyngeal muscles; *10*, median pharyngeal muscle; *11*, pharyngeal muscles; *12*, cibarial muscles; *13*, dorsal salivary muscles.

For a detailed account of the musculature and action of the muscoidean proboscis, see Graham-Smith (1930). Calliphora has several muscles not detected in Drosophila: namely, Graham-

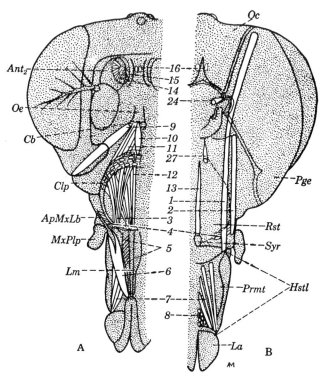

FIG. 25. Skeletal muscles of the head. Right half of head represented as a transparent object with the muscles in white (temporary muscles omitted; proboscis extended; 75×). A: Anterior aspect. B: Posterior aspect. Numbers refer to muscles listed in text; *Ant₂*, second segment of antenna; *ApMxLb*, apodeme of maxillary lobe; *Cb*, cibarium; *Clp*, clypeus; *Hstl*, haustellum or midiproboscis; *La*, labella; *Lm*, labrum; *MxPlp*, maxillary palpus; *Oc*, occiput; *Oe*, oesophagus; *Pge*, postgena; *Prmt*, prementum; *Rst*, rostrum or basiproboscis; *Syr*, salivary pump (syringe).

Smith's flexors of the haustellum, retractors of the oesophagus, extensors of the haustellum, and epipharyngeal muscles.

Muscles of the Thorax

In this list, the letters in parentheses indicating probable homologies are those used by Snodgrass (1935) to designate the thoracic muscles of pterygote insects in general.

PROTHORAX

Cervical muscles, moving the head: Dorsal *20, 21, 22, 23;* lateral *24;*
and ventral *25, 26, 27* (Snodgrass, H). (Figs. 25B, 26, 27B)

Muscles of first coxa: tergopleural promotor *28* (I?); pleural remotor and abductor *29* (J?); pleural promotor *30* (M?); sternal
anterior rotator *31* (K); sternal posterior rotator *32* (L); sternal
adductor *33* (N). (Figs. 26A, C, 27B)

Muscles of trochanter: extracoxal depressor *34* (P); intracoxal depressor *35* (Q); intracoxal lateral and mesal levators *36, 37* (O).
(Figs. 26B, 27B, 28)

Muscles of femur: reductor *38* (R). (Fig. 28)

Muscles of tibia: levator *39* (S) and depressors *40, 41* (T). (Fig.
28)

Muscles of tarsus: levator *42* (U) and depressor *43* (V) of the metatarsus. (Other tarsal segments have no muscles; Fig. 28.)

Muscles of the pretarsus: depressor attached to long tendon *44* (X)
of unguitractor plate; traced as far as femorotibial joint, but
muscle not identified. (Fig. 28)

MESOTHORAX [7]

Indirect wing muscles (fibrillar muscles): six pairs of dorsal median
muscles *45a–f* (mA); two lateral oblique dorsal muscles *46a, b*
(lA); three tergosternal muscles *47a–c* (C); two tergal remotors of
coxa *48a, b* (J) attached to meron 2, which has become part of the
thoracic wall (Figs. 26, 27). These sets of muscles probably provide the main propelling force in flight. The bases of the wings
are so related to the thoracic wall that the dorsal median muscles
depress the wings by longitudinal arching of the scutum, and the
vertical muscles elevate them by drawing down the scutum. The
tergal remotor of the coxa, though phylogenetically a leg-muscle,
has been transformed into a flight muscle having the same fibrillar
type of structure as the other indirect wing muscles (Fig. 24).
The large extracoxal depressor of the trochanter *66*, on the other
hand, is composed of "tubular" fibers like the other leg muscles;
whether it takes an active part in flight is unknown. The pleurosternal muscle *59* may possibly bear some relation to flight
movements.

Direct wing muscles *49–58* (tubular fibers), attached to the roots of
the wings and serving to extend and flex the wings and to produce
torsion of the wing surfaces during flight (see Ritter, 1911, on Calliphora, and Snodgrass, 1935, for mechanics of flight). The insertions of some of these muscles are difficult to ascertain, but they
may be tentatively classified as follows (Fig. 26C): muscles of
prealar apophysis *49* (cf. Fig. 27B), *50* (B?; *49* corresponds to
Ritter's abductor, which draws wing horizontally forward; *50* is
his m. gracilis of undetermined function); basalar muscles *51, 52*
(E'; *51* is Ritter's pronator, which depresses anterior border of wing,

[7] Cf. Williams and Williams (1943) on *Drosophila repleta.*

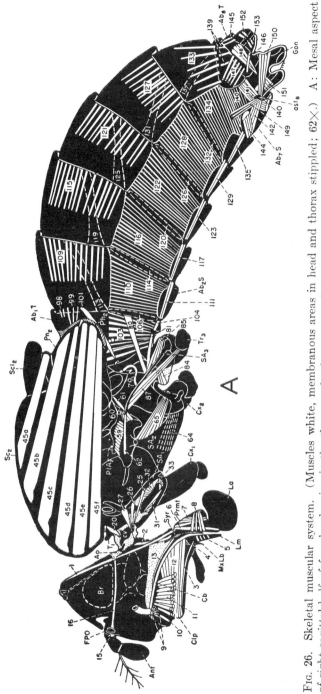

Fig. 26. Skeletal muscular system. (Muscles white, membranous areas in head and thorax stippled; 62×.) A: Mesal aspect of right sagittal half of female, showing in the thorax only the more medial muscles; temporary abdominal muscles in broken outline.

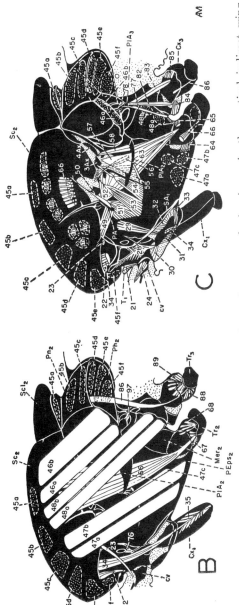

B: Mesal aspect of right half of thorax, with medial muscles removed, showing only the vertical indirect wing muscles and trochanteral muscles; sites of insertion of dorsal median muscles indicated in broken outline. C: Mesal aspect of right half of thorax with medial and indirect wing muscles removed to show the direct wing muscles and other muscles which lie against the body wall; sites of insertion of indirect wing muscles indicated in broken outline. Numbers refer to muscles listed in text; Ab_2S, and Ab_7S, second and seventh abdominal sternites; Ab_1T, Ab_8T, first and eighth abdominal tergites; Ant, antenna; Ao, aortic funnel; ast_8, acrosternite of eighth abdominal segment; $3Ax$, $4Ax$, third and fourth axillaries; Br, brain; Cb, cibarium; Clp, clypeus; cv, cervical sclerite; Cx_{1-3}, coxae; FPO, frontal pulsatile organ; Gon, gonopod (ovipositor); La, labella; Lm, labrum; Mer_2, meron of mesothorax (hypopleurite); $MxLb$, maxillary lobe; PA, prealar apophysis; $PEps_2$, preëpisternum (sternopleurite); Ph_2, mesophragma; PlA_2, mesopleural, PlA_3, metapleural, apodeme; Pn_2, postnotum of mesothorax; $Prmt$, prementum; SA_{1-3}, sternal apophyses; Sc_2, scutum; Scl_2, scutellum; Syr, salivary pump; T_1, pronotum; Tr_2, Tr_3, second and third trochanters.

52 is his m. anonymous of undetermined function); muscles of first
axillary *53, 56* (cf. Ritter's levators, which raise wing and draw it
backward); muscles of third axillary *54, 55* (D; either flexors, or
correspond to Ritter's first supinator, which depresses anal portion
of wing); subalar muscle *57* (E″; corresponds to Ritter's adductor,
which draws wing backward toward body); muscles of fourth axillary or axillary cord (?) *58* (correspond to Ritter's second supinator, which depresses anal portion of wing) and *58a, b* (two "external muscles of the fourth axillary" of Williams and Williams,
1943; not illustrated: laterad of *57* and *58*).

Pleurosternal muscles *59, 60* (G; m. latus of Ritter). (Figs. 26A, 27B)

Lateral intersegmental muscle *61* (F). (Figs. 26A, 27B)

Ventral longitudinal muscle *62* (H). (Figs. 26A, 27B)

Muscles of second coxa: pleural remotor *63* (3E″?); sternal promotor
64 (K); sternal remotor *65* (L). (Figs. 26A, C, 27B)

Muscles of trochanter: extracoxal depressor *66* (P); intracoxal depressor *67* (Q); levator *68* (O). (Figs. 26B, 27)

Muscles of other leg segments: *69–75*, as in first leg. (Not illustrated.)

Occlusor muscle of anterior spiracle *76*. (Figs. 26B, 27B)

METATHORAX

Muscles of haltere: *77, 78*. (Figs. 26C, 27B)

Pleurosternal muscle *79* (G); lateral intersegmental muscle *80* (F);
ventral longitudinal muscle *81* (H). (Figs. 26A, 27B)

Muscles of third coxa: pleural promotor *82* (M); pleural remotor *83*
(3E″?); sternal promotor *84* (K); sternal remotor *85* (L). (Figs.
26A, C, 27B)

Muscles of trochanter: tergal and sternal extracoxal depressors *86, 87*
(P); intracoxal depressor *88* (Q); levator *89* (O). (Figs. 26A, B,
27B)

Muscles of other leg segments: *90–96*, as in first leg. (Not illustrated.)

Occlusor muscle of posterior spiracle *97*. (Fig. 26B)

Muscles of the Abdomen

SEGMENT 1

Dorsal longitudinal muscles: median *98*, lateral *99, 100* (probably support abdomen); oblique lateral *101, 102*. (Figs. 26A, 27)

Lateral tergosternal muscles *103*. (Figs. 26A, 27B)

Ventral longitudinal muscles *80, 81, 104*. (Figs. 26A, 27B)

Spiracular occlusor muscles *105*. (Fig. 20A)

Temporary muscles: internal lateral muscle *106* along intersegmental
fold between segments 1 and 2 (Fig. 26A); oblique sternopleurals
107, 108 to thoracoabdominal junction (not illustrated).

SEGMENTS 2–6

Dorsal muscles *109, 115, 121, 127, 133:* fifteen to twenty-five longitudinal fibers on each side, single or in groups of two or three; re-

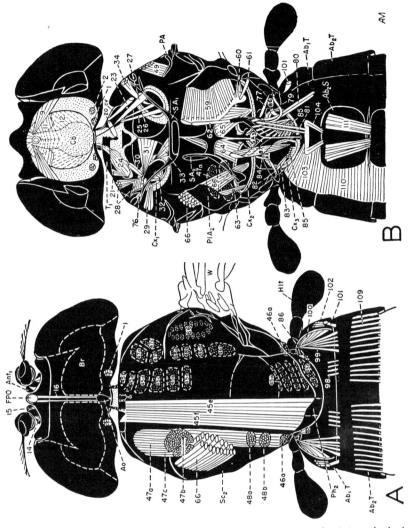

FIG. 27. Skeletal muscular system. (Muscles white; sites of muscle insertions in broken outline; membranous areas in head and thorax stippled; temporary muscles omitted; 64×.) A: Ventral aspect of dorsal half of head, thorax, and base of abdomen, showing dorsal muscles; thoracic muscles of left side removed and insertions indicated. B: Dorsal aspect of ventral half of head, thorax, and base of abdomen, showing lateral and ventral muscles; direct wing muscles and vertical indirect wing muscles omitted; only the right or left muscle of each pair is shown in the thorax. Numbers refer to muscles listed in text; Ab_2S, second abdominal sternite; Ab_1T, Ab_2T, first and second abdominal tergites; Ant_1, first segment of antenna; Ao, aortic funnel; Br, brain; Cb, cibarium; cv, cervical sclerite; Cx_{1-3}, coxae; FPO, frontal pulsatile organ; Hlt, haltere; PA, prealar apophysis; Ph_2, mesophragma; PlA_2, mesopleural apodeme; SA_{1-3}, sternal apophyses (SA_2, mesofurca, SA_3, metafurca); Sc_2, scutum; T_1, pronotum; W, wing.

tractors of tergites (musculi dorsalis externi mediales and laterales of Snodgrass). (Figs. 26A, 27A)

Lateral tergosternal muscles *110, 116, 122, 128, 134:* about twenty fibers in each segment; compressors, throwing pleural membrane into longitudinal folds (musculi lateralis externi). (Figs. 26A, 27B, 19)

Ventral muscles *111, 117, 123, 129, 135:* about six fibers on each side; retractors of sternites (musculi ventralis interni mediales). (Figs. 26A, 27B)

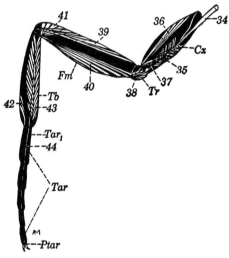

FIG. 28. Muscles of leg. Posterior aspect of left foreleg, represented as a transparent object with muscles in white (66×). Numbers refer to muscles listed in text; *Cx,* coxa; *Fm,* femur; *Ptar,* pretarsus; *Tar,* tarsus; *Tar₁,* metatarsus; *Tb,* tibia; *Tr,* trochanter.

Spiracular muscles *112, 118, 124, 130, 136.* (Not illustrated.)

Temporary oblique dorsal muscles *113, 119, 125, 131, 137:* a single large fiber on each side; retractors of tergites (m. dorsalis externi lateralis). (Fig. 26A)

Temporary internal lateral muscles *114, 120, 126, 132:* a single large vertical fiber overlying the intersegmental region on each side; compressors (m. laterales interni). (Fig. 26A)

(Transverse abdominal muscles are represented only by the alar muscles of the heart (q.v.). There are no protractors or dilators of the abdominal segments, expansion being due to internal pressure.)

SEGMENT 7 (FEMALE)

Dorsal *139,* lateral *140, 141,* ventral *142,* and spiracular muscles *143;* sternal muscles *144* to uterus. (Fig. 26A)

SEGMENT 8 (FEMALE)
    Dorsal muscles *145* to supra-anal plate; *146* to subanal plate; *147* to
    uterus; *148* to gonopod. (Fig. 26A)
    Ventral muscles *149* to uterus; lateral *150* and transverse ventral
    muscles *151* of gonopod. (Fig. 26A)

ANAL SEGMENT (FEMALE)
    Dorsal and ventral rectal muscles *152, 153.* (Fig. 26A)
    (For the musculature of the male terminalia, see Chapter 5 on Ex-
    ternal Morphology.)

THE NERVOUS SYSTEM

The nervous system consists of nerve centers, or ganglia, and
peripheral nerves which convey sensory stimuli from the sense
organs and motor stimuli to the muscles and glands. The nerve
centers constitute a large somatic central nervous system, comprising the brain and ventral cord, and a smaller visceral or
stomodaeal system, associated with the dorsal wall of the
oesophagus. The nerves of the peripheral system are bundles of
fibers which arise from ganglionic cells and from sensory cells
of the body wall and viscera.

The principal cellular elements of the nervous system are the
nerve cells, or neurones. Each consists of a cell body, the neurocyte, from which fibrous processes arise that may be differentiated
as a principal branch, the axon (often provided with a collateral
branch), and smaller branches, the dendrites. The processes end
in terminal arborizations of fine branching fibrils. Neurones are
unipolar, bipolar, or multipolar when they have, respectively,
one, two, or more than two processes arising directly from the
neurocyte, and are of three functional types: (1) sensory cells
(bipolar and multipolar) with the neurocytes in or near the body
wall, sense organs, or viscera, and their axons forming afferent
fibers going to the ganglia; (2) association neurones (usually
unipolar) with the neurocytes and processes contained entirely
within the ganglia and their connectives; and (3) motor neurones
(usually unipolar) with their neurocytes and collateral branches
within the ganglia and their axons forming efferent fibers to the
muscles and glands. Sensory and motor neurones may communicate directly through close contact (synapse) of their terminal
branches, or through synapses with one or more association neu-

rones. Besides the neurones, the nervous system includes supporting or glia cells within the ganglia and a nucleated neurilemma, or sheath, covering the surface of the ganglia and peripheral nerves.

In the ganglia the neurocytes are crowded in the peripheral regions, forming the cortex, and their massed fibrous processes constitute the internal neuropile or "Punctsubstanz." In sections of Drosophila stained with ordinary histological dyes, several types of cells are recognizable (Fig. 32C). The neurocytes vary in size and staining properties. *Small ganglion cells* (*SGC*) have a round nucleus 2–3 μ in diameter with an irregular central clump of chromatin, and the cytoplasm forms an inconspicuous pale-staining rim. They are probably association cells. *Large ganglion cells* (*LGC*) have round or oval nuclei 4–5 μ in diameter, the chromatin irregularly clumped into one or more masses, and the cytoplasm, though pale-staining, may be more copious (5–9 μ across) than in the small cells and occasionally contains one or two vacuoles. They presumbly include both motor and association neurones. A third type of ganglion cell, apparently the *giant cells* of Hertweck (*GtCl*), is also large, with round or ovoid nuclei from 4 by 5 to 5 by 7 μ containing one or two masses and scattered granules of chromatin, and abundant cytoplasm (6 by 9 to 12 by 17 μ) that is dense, dark-staining, and more or less basophilic. Although no granules have been detected in the cytoplasm of these cells (which, however, may sometimes be microvacuolate), their general appearance suggests a glandular function, and they may correspond to the neurosecretory cells described in other insects (Hanström, 1939; Day, 1940; Scharrer, 1941a, b). They occur singly or in groups among the other ganglion cells in definite positions as noted below. The nuclei of the *neuroglia cells* (*NglCl*) are deeply staining and elliptical or flattened, 0.7–1.5 μ thick and 1.5–3 μ long. The cytoplasm of these cells is obscure, though sometimes appearing as a thin membrane extending from the nuclei; it probably is in the form of slender fibrous processes (glia fibers; Scharrer, 1939). Neuroglia cells are found mainly between the cortex and neuropile, around some of the fibrous bodies of the brain, and scattered among the fibers of the larger nerve trunks. The nuclei of the *cells* (*NlmCl*) *of the neurilemma* (perilemma; Scharrer, 1939) are likewise densely chromatinic and flattened, 0.7 μ thick and 3–4.5 μ long.

They occur singly at wide intervals beneath the thin (0.3 μ) membrane (neural lamella) that forms a continuous covering over the surface of all ganglia and nerves.

The peripheral nerves which arise from the ganglia consist of closely packed, non-medullated nerve fibers, the axons of both sensory and motor neurones (cf. Fig. 33, $CvCon$). Smaller groups of fibers are given off as branches from the larger nerves, the terminal branches which supply scattered sense organs and motor endings consisting of single fibers (Fig. 34A). The smallest branches as well as the main trunks are covered by the neurilemma, and the larger nerves also contain scattered interstitial neuroglia nuclei. (For studies on the finer structure of insect nerves, see Richards, 1944.)

The sensory and motor nerve endings in the epidermis, muscles, and viscera have not been studied, except those associated with special sense organs (q.v.). An integumentary nervous system has not been detected with the unspecialized methods used but should be sought by means of neurological techniques.

The Central Nervous System

The central nervous system of insects typically consists of a supraoesophageal nerve mass, or brain, connected with a series of segmentally arranged median ganglia, which, with their paired connectives, constitute a ventral nerve cord. In Drosophila, however, as in other muscoidean flies, the ganglia have been consolidated into two composite masses of nervous tissue located in the head and thorax, and no ganglia occur in the abdomen (Figs. 29, 30). The cephalic mass consists of the supraoesophageal (Br) and suboesophageal ($SoeGng$) ganglia, which are compactly merged together around the oesophagus. The cephalic mass is connected by a thick stalk, the cervical connective ($CvCon$), to the thoracic mass. The latter is a fusion product of the thoracic ($ThGng$) and abdominal ganglia ($AbGng$), the greater part made up of the three thoracic ganglia and only a small portion representing the abdominal ganglia. The only external evidences of metamerism in the thoracic mass are the ventral constrictions between the thoracic nerve centers and the segmental distribution of the peripheral nerve trunks that arise from it.

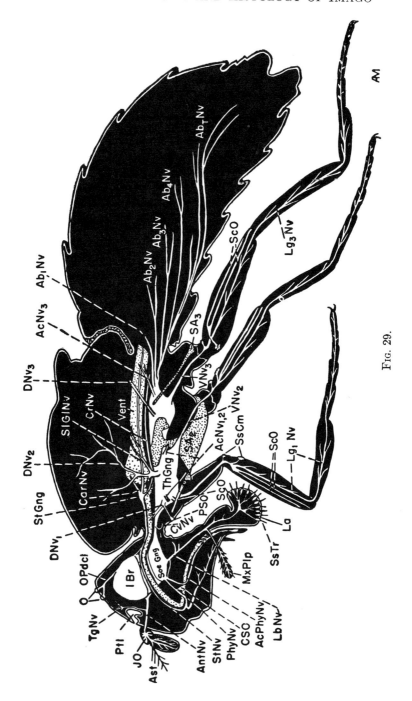

FIG. 29.

Fig. 29. Nervous system, *in situ*. Mesal aspect of right sugittal half of female, showing the central and stomodaeal nervous systems and their principal nerves. Body represented as cut to left of midline, the paired nerves shown being those of the left side except for the antennal nerve; brain in section. (Nerve tissue white, alimentary canal and apophyses stippled; 60×.) $Ab_{1-4}Nv$, nerves of first to fourth abdominal segments; Ab_TNv, terminal abdominal nerves; $AcNv_{1,2,3}$, first, second, and third accessory nerves; $AcPhyNv$, accessory pharyngeal nerve; $AntNv$, antennal nerve; Ast, arista; $1Br$, proto-cerebrum; $CarNv$, nerve to cardia; $CrNv$, nerve to crop; CSO, cibarial sense organ; $CvNv$, cervical nerve; $DNv_{1,2,3}$, first, second, and third dorsal nerves; JO, Johnston's organ; La, labella; $LbNv$, labial nerve; Lg_1Nv, Lg_3Nv, nerves of first and third legs; $MxPlp$, maxillary palpus; O, ocelli; $OPdcl$, ocellar pedicel; $PhyNv$, pharyngeal nerve; PSO, prosternal sense organ; Ptl, ptilinum; SA_2, SA_3, sternal apophyses (mesofurca and metafurca); ScO, scolopophorous organ; $SlGlNv$, nerve to salivary gland; $SoeGng$, suboesophageal ganglion; $SsCm$, sensilla campaniformia; $SsTr$, sensilla trichodea; $StGng$, sto-modaeal ganglion; $StNv$, stomodaeal nerve (recurrent nerve, combined behind brain with nervus ganglii occipitalis); $TgNv$, tegumentary nerve; $ThGng$, thoracic ganglionic mass; $Vent$, ventriculus; VNv_2, VNv_3, second and third ventral nerves.

Hertweck (1931) has described the nervous system of Drosophila and should be consulted for additional details and illustrations. The present account incorporates the results of further study and reinterprets some of the structural details in the light of the general discussion of the insectan nervous system given by Snodgrass (1935).

The Cephalic Ganglionic Center and Its Nerves

The cephalic ganglionic mass (450 μ by 260 μ by 160 μ) lies transversely in the head capsule between the compound eyes (Fig. 30). It is pierced by a narrow median canal through which pass the oesophagus, the anterior stomodaeal nerve ($StNv$, Fig. 29), and a pair of muscle fibers (muscles of the frontal pulsatile organ, Figs. 12, 26A, 27A, 31A): The supraoesophageal portion, or brain proper, includes the median protocerebrum ($1Br$) and its lateral optic lobes (OpL); the deutocerebrum ($2Br$), marked externally by a pair of low anterior swellings from which the antennal nerves ($AntNv$) arise; and the vestigial tritocerebrum, which is not differentiated externally. (The generally more anterior regions are here dorsal and dorsocaudal in position, so that the brain is "bent backward" and the deutocerebrum lies farther forward than the protocerebrum.) A single ocellar nerve ($OPdcl$) arises medially from the posterodorsal region of the protocerebrum and forks into three short branches under the ocelli (O), each branch terminating in a ganglionic thickening (Fig. 31A). The optic lobes appear as broad lateral extensions of the protocerebrum connecting directly with the compound eyes (E), a deep constriction near the junction with the eye marking the location of the outer chiasma (see below). Each of the thick antennal nerves, 17 μ in diameter and arising from the deutocerebrum, gives off a branch to the dorsal wall of the head ($TgNv$, nervus tegumentalis of Hertweck). Posteriorly, at the oesophageal canal, two minute and inconspicuous nerves of the stomodaeal system (q.v.) enter the protocerebrum. Laterad of the oesophageal canal the complex includes the fibers of the circumoesophageal connectives, joining the supraoesophageal region with the suboesophageal ganglion ($SoeGng$). The latter constitutes the ventral portion of the cephalic nerve mass, gives off three pairs of ventral nerves to the proboscis, and is joined posteriorly to the unpaired cervical connective ($CvCon$, cephalo-

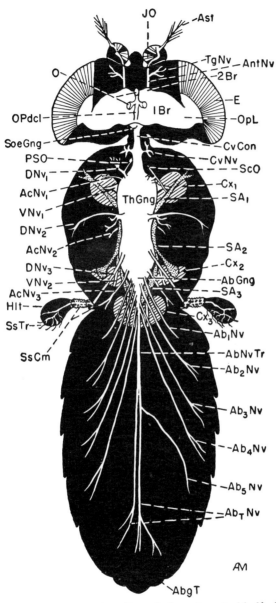

FIG. 30. Nervous system, *in situ*. Dorsal view of ventral half of male, showing the central nervous system. Compound eyes represented as sectioned
horizontally; prothoracic dorsal and accessory nerves of right side and
scolopophorous organ of left side omitted. (Nerve tissue white, apophyses
stippled, coxal cavities cross-hatched; 60×.) *AbGng*, abdominal ganglionic
center; *Ab₅Nv*, nerve of fifth abdominal segment; *Ab₉T*, ninth abdominal
tergite; *2Br*, deutocerebrum; *CvCon*, cervical connective; *Cx₁₋₃*, coxal
cavities; *E*, compound eye; *Hlt*, haltere; *OpL*, optic lobe; *SA₁*, prosternal
apophysis; *VNv₁*, first ventral nerve; other abbreviations as in Fig. 29.

thoracic nerve strand of Hertweck, formed by fusion of paired connectives from the thoracic nerve center).

Of the ventral nerves arising from the suboesophageal ganglion, the first and second pairs are, respectively, the pharyngeal nerves (*PhyNv*) and accessory pharyngeal nerves (*AcPhyNv*) of Hertweck. They leave the ganglion anteriorly, near the oesophageal canal. The two nerves on each side always arise close together, but they may appear either separate or united at the extreme base. In both conditions, however, the neurofibrillar cores of these nerves coalesce, the coalescence occurring within the periphery of the ganglion when the bases of the nerves are separate or, when united, in the fused base (25 μ or less in length) outside the ganglion. The pharyngeal nerve, 7 μ thick, passes forward and, after giving off a branch to the cibarial sense organs (*CSO*), loops beneath muscle *9* (Fig. 25A) near the cibarial-oesophageal junction to anastomose medially with the stomodaeal nerve and with the pharyngeal nerve of the other side. From the region of these anastomoses a single median nerve extends ventrad and gives off a number of branches to the dilator muscles (*11, 12*) of the cibarial pump.[8] The accessory pharyngeal nerves, 5 μ thick, also course forward, closely parallel to the pharyngeal nerves, but penetrate at a lower level into the space between the anterior cibarial plate and the dilator muscles. Here the nerves pass downward on each side, innervate the sense hairs of the anterior cibarial plate, and enter the labrum to connect with its muscles and sense organs. The third pair of suboesophageal nerves, the labial nerves of Hertweck (*LbNv*), 10 μ thick, arises from the ventral side of the ganglion and gives off branches to the muscles of the basiproboscis and haustellum, the maxillary palpi, and the labellar glands and sense organs (*SsTr*). A very fine nerve extends from the base of the accessory pharyngeal nerve to the labial nerve of the same side, close to the suboesophageal ganglion.

[8] Ferris and his students (personal correspondence) homologize the pharyngeal nerves with the frontal ganglion connectives of more primitive insects and have termed the median nerve of the cibarial muscles "the procurrent nerve." Although no definite frontal ganglion has been detected in Drosophila by the present author, the anastomosis of these nerves with the stomodaeal (= recurrent) nerve supports this homology. Hertweck's accessory pharyngeal nerves are regarded logically as the labral nerves, and his labial nerves as compound maxillary-labial nerves. (Cf. Henry, 1947–1948.)

Histologically the cephalic ganglionic center consists of a dense mass of nerve fibers (*Npl*, neuropile or medullary tissue), surrounded by an irregular cortical layer of ganglion cells (neurocytes, *GngCl*) and enclosing denser fibrous bodies, compact masses of nerve endings (glomeruli), and special fibrous tracts (Fig. 31). These structures constitute the intricately connected association centers of the brain, elucidation of which would require more intensive study by neurological methods than has yet been accorded them in Drosophila (cf. Cajal and Sánchez, 1915, Hanström, 1928, and Snodgrass, 1935, for other insects, including Calliphora).[9] The surface of the brain is covered by a very thin neurilemma with widely scattered nuclei, and tracheae penetrate into the nervous tissue.

In the *protocerebrum* there are recognizable the corpora pedunculata, pons cerebralis, corpus centrale, corpora ventralia, the optic tract, the commissural tract connecting the corpora ventralia, and other fibrous masses and tracts (Fig. 31; cf. Hertweck, 1931, Figs. 15 and 16). The various corpora and the pons are dense clusters of fibers and glomeruli arising principally from association neurones, the cell bodies of which are located in the cortex. (Globuli cells—special groups of differentiated association cells—are not apparent as such in Drosophila.) The corpora pedunculata (mushroom bodies) are paired bodies lying one on each side of the median part (pars intercerebralis) of the protocerebrum. Each has an expanded end, a compound "calyx" (*CpdClx*, Fig. 31F) formed of pear-shaped fiber masses, in the posterodorsal cortical region, and a long stalk (*CpdStk*, Figs. 31D, E) which extends ventrocephalad through the neuropile and forks into a median root ("Balken" of Hertweck), ending in front of the corpus centrale, and a dorsal root ("Rückläufiges Stiel"), ending in a small prominence directly under the neurilemma in the anterodorsal part of the brain. The pons cerebralis (*Pncr*, Figs. 31E, F) is a curved fibrous mass lying in the dorsomedian region between the calyces of the corpora pedunculata. The corpus centrale (*Cc*), principal association center of the brain, is a conspicuous ovoid body lying medially above the oesophageal canal (Figs. 31A, B, D, E). It consists of four glomerulous masses: a large,

[9] Since this was written, a detailed description of the brain of Drosophila has been presented by Power (1943).

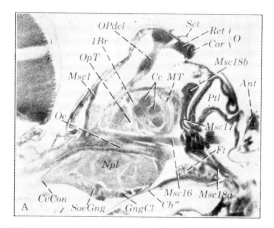

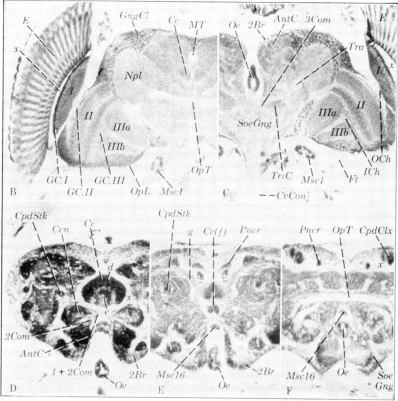

Fig. 31.

FIG. 31. Cephalic ganglionic center (160×). A: Sagittal section of upper portion of head, through brain and suboesophageal ganglion (D., C.B.). B: Left half of frontal section through protocerebrum, optic lobe, and compound eye, above oesophagus (D., C.B.). C: Right half of frontal section through deutocerebrum, suboesophageal ganglion, and optic lobe at a lower level than B (*same specimen*). D–F: Portions of alternate vertical transections, at successively more posterior levels, through brain and suboesophageal ganglion; fibrous bodies separated and their boundaries accentuated by faulty fixation (G., C.B.). *Ant*, antenna; *AntC*, antennal center; *1Br*, protocerebrum; *2Br*, deutocerebrum; *Cb″*, posterior plate of cibarium; *Cc*, corpus centrale; *Cc(f)*, tubercles of corpus centrale; *1 + 2Com*, combined protocerebral and deutocerebral commissural tracts; *2Com*, deutocerebral commissural tract; *3Com*, tritocerebral commissural tract; *Cor*, corneal lens; *CpdClx*, calyx of corpus pedunculatum; *CpdStk*, stalk of corpus pedunculatum; *CvCon*, cervical connective; *Cvn*, corpus ventrale; *d*, dorsal mass of corpus centrale; *E*, compound eye; *e*, oval mass of corpus centrale; *Ft*, fat cells; *g*, fiber bundle to deutocerebrum (olfactoria globularis tract); *GC.I*, *GC.II*, *GC.III*, ganglion cells of periopticon, epiopticon, and opticon; *GngCl*, ganglion cells (neurocytes); *I*, periopticon; *II*, epiopticon; *IIIa*, anterior portion of opticon; *IIIb*, posterior portion of opticon; *ICh*, inner chiasma; *Mscl*, *16*, *17*, *18a*, *18b*, head muscles; *MT*, median fiber bundle; *Npl*, neuropile; *O*, ocellus; *OCh*, outer chiasma; *Oe*, oesophagus; *OPdcl*, ocellar pedicel; *OpL*, optic lobe; *OpT*, optic tract; *Pncr*, pons cerebralis; *Ptl*, ptilinum; *Ret*, retinal cells; *Set*, seta (ocellar bristle); *SoeGng*, suboesophageal ganglion; *Tra*, trachea; *TriC*, tritocerebral center; *x*, artificial break or shrinkage space.

bean-shaped dorsal mass (*d*, Fig. 31D) overlying an anterior oval mass (*e*, "Ellipsoidkörper"), and two small ventral masses (tubercles of the central body; *f*, Fig. 31E). The corpora ventralia (*Cvn*, Fig. 31D) are a pair of glomeruli lying ventrolaterally, above the antennal glomeruli of the deutocerebrum; these corpora are connected by a fibrous commissural tract (*1Com*), which passes beneath the corpus centrale along with the deutocerebral commissural tract. The optic tract (*OpT*, Figs. 31A, B, F; "Hirnbrücke" of Hertweck; central commissure of Power) is a conspicuous transverse bundle of fibers connecting the optic centers of the two compound eyes; it passes behind the corpus centrale. Other fiber tracts run in various directions in tne protocerebrum, including a median bundle of fibers that extends vertically in the anterior region (*MT*, Figs. 31A, B), and tracts going to the ocellar nerve and deutocerebrum (*g*, Fig. 31E; olfactoria-globularis tracts of Hanström, 1928, and Power, 1943).

Each *optic lobe* (Figs. 31B, C) contains three optic centers; namely, a distal periopticon (lamina ganglionaris or ganglion opticum externum, *I*), a median epiopticon (medulla externa or ganglion opticum medium, *II*), and a proximal opticon (medulla interna or ganglion opticum internum, *IIIa, b*). These are highly differentiated synaptic regions in which the terminal branches of retinal and optic lobe neurones are associated (see Cajal and Sánchez, 1915, on the blowfly). The periopticon (*I*) receives optic nerve fibers (forming a postretinal fiber layer) from the compound eye and consists of an outer granular layer of ganglion cells (*GC.I*) and an inner medullary layer of parallel nerve fibers arranged perpendicular to the outer layer. The periopticon is connected to the epiopticon by the outer chiasma (*OCh*), where nerve fibers cross in a horizontal plane. The epiopticon (*II*) is a large, laminated mass of terminals bordered externally by a narrow layer of ganglion cells, which lie within the inner angle of the chiasma and form the "cuneiform ganglion" (*GC.II*). The epiopticon is connected to the opticon by the inner chiasma (*ICh*), where interspersed ganglion cells (*GC.III*) are also present. The fibers of the inner chiasma cross in a vertical plane and lie between the two masses of nerve endings that constitute the laminated anterior (*IIIa*) and posterior (*IIIb*) "capsules" of the opticon. The larger anterior capsule is connected to the protocerebral bodies and to the opticon of the opposite side by fibers in the optic tract. The optic tract also sends direct fibers to the epiopticon anteriorly.

The *deutocerebral lobes* (*2Br*, Figs. 31C, D, E) contain peripheral ganglion cells (including motor neurones of the antennae) and numerous glomeruli (*AntC*) in the neuropile which connect with the antennal nerves, protocerebral fiber bundles (*g*, olfactoria-globularis tracts), the circumoesophageal fiber tracts to the suboesophageal ganglion, and the deutocerebral commissural tract (*2Com*). The last-named passes above the oesophagus, connecting the right and left sides beneath the corpus centrale, and is confluent with the commissural tract of the corpora ventralia (Fig. 31D).

The *tritocerebrum* is obscure and vestigial. Theoretically its right and left components should lie beneath the deutocerebral lobes in association with the circumoesophageal fiber tracts and should be connected by a commissure beneath the oesophagus (not above as indicated by Hertweck). Transverse fibers found directly beneath the neurilemma in the anterior part of the suboesophageal ganglionic mass may represent this tritocerebral commissure (*3Com*, Fig. 31C) connecting inconspicuous centers (*TriC*) lying behind the antennal centers.

The *suboesophageal ganglion* (*SoeGng*) consists of neuropile tissue and cortical ganglion cells, with most of the latter ventral in position (Figs. 31A, C). It theoretically incorporates structures corresponding to the mandibular, maxillary, and labial nerve centers that are apparent in more primitive insects. However, the internal structure and fiber tracts have not been studied in sufficient detail to reveal this tripartite nature or to establish neurologically the homologies of the three pairs of nerves that arise from this region. If the anterior nerves represent frontal connectives and labral nerves which originate from the tritocerebrum in lower insects (see footnote 8, p. 488), their position in Drosophila suggests that tritocerebral elements are also included in the suboesophageal ganglionic mass. Two longitudinal dorsal fiber tracts run through the neuropile and extend through the cervical connective (*CvCon*) to the thoracic nerve mass, indicating that the connective is a fused pair of nerve strands. The cervical connective has a vertical thickness of 17 μ and a width of 28 μ.

The cephalic ganglionic center contains all the types of cells already described. The cortical cells of the optic lobes (Figs. 31B, C) and of the dorsal part of the protocerebrum, around the calyces of the corpora pedunculata, are of the small ganglionic cell type. Those in the region of the corpora pedunculata correspond to the globuli cells of other insects but are not set apart as distinct groups. Their grouped nature is evidently more marked in *Musca domestica* (Hanström, 1928, Fig. 608). Both small and large ganglion cells are found elsewhere in the cortex. Giant cells occur in several places; namely, a group of twenty to thirty cells in the intercerebral furrow at the upper end of the median fiber bundle, a smaller group at its lower end, and a single cell or small clusters of cells in the posterior cortex laterad of the corpora pedunculata and in the cortex between the protocerebrum and optic lobes. The nuclei of the periopticon are somewhat larger than those of the other optic lobe neurocytes. Neuroglia is most evident around the corpus centrale and at the inner boundary of the cortex.

THE THORACIC GANGLIONIC CENTER AND ITS NERVES

The thoracic nerve center $(ThGng + AbGng)$ is an elongate, compact mass of ganglionic tissue, 450 μ long, 160 μ wide, and 110 μ high, with rounded ends and three paired ventral bulges where the nerve trunks to the legs arise (Figs. 29, 30). It extends longitudinally in the ventral part of the thorax below the alimentary canal and is embraced behind the middle by the arms of the mesofurca (SA_2) and posteriorly by the metafurca (SA_3), which support and hold it in place. It is joined anteriorly by the cervical connective to the suboesophageal ganglion of the head and gives off posteriorly a median nerve trunk to the abdomen $(AbNvTr)$. A pair of cervical nerves $(CvNv)$ to the muscles of the neck springs from the sides of the cervical connective. From each of the three thoracic segmental centers, which constitute the greater part of the ganglionic mass, there arise three pairs of nerves: namely, dorsal (DNv_{1-3}), accessory $(AcNv_{1-3})$, and ventral nerves (VNv_{1-3}). In addition to these the prothoracic center has a pair of very short nerves connecting directly to two scolopophorous sense organs (ScO) that lie close to the anterior end of the ganglion. Of the dorsal nerves, the first pair (DNv_1)

innervates the prothoracic muscles and the prosternal sense organ (*PSO,* through a basal branch on each side that is illustrated as an independent nerve by Hertweck); the second pair (*DNv$_2$;* for variability, see Hertweck) innervates the indirect and direct wing muscles and the sense organs of the wings; and the third and thickest pair (*DNv$_3$*) goes to the halteres (*Hlt*). The accessory nerves (*AcNv$_{1-3}$*) also innervate muscles, the first extending toward the anterior spiracles and the third toward the posterior spiracles. The three pairs of ventral nerves (*VNv$_{1-3}$;* 17 μ in diameter) innervate the muscles of the coxae and extend to the tips of the legs (*Lg$_{1-3}$Nv*), giving off branches to the muscles and sense organs therein. In the wings, nerves occur in the costal, auxiliary, first, and third longitudinal veins (Hertweck, 1931, Pl. 4, Fig. 4).

The small caudal portion of the ganglionic mass, behind the roots of the metathoracic ventral nerves, is the abdominal nerve center (*AbGng*). From it arise paired nerves (*Ab$_1$Nv, Ab$_2$Nv*) to the first and second abdominal segments and a thick median trunk (*AbNvTr*), 10–14 μ in diameter, that gives off nerves (*Ab$_{3-5}$Nv*), 3–4 μ thick, to the more posterior segments, most of them arising near the base of the trunk. Some distance further back the trunk forks, usually more posteriorly in the male than in the female, and the terminal branches (*Ab$_T$Nv*) innervate the genital segments and internal reproductive organs. The mode of branching and origin of the fork seem to be variable and not in full agreement with Hertweck's figures (the nerves are difficult to trace with certainty). In the male the trunk or its fork passes above the testes, and in the female either one branch of the fork or nerves from both branches pass above the lateral oviducts.

Internally the thoracic ganglionic mass is largely neuropile tissue, with the ganglion cells limited principally to the lateral and ventral periphery (Fig. 32). Fiber tracts traverse the neuropile in various directions, some arising from the roots of the peripheral nerves, others forming longitudinal connectives and transverse commissures (Fig. 32A). The fiber tracts of the cervical connective enter the neuropile anteriorly (Fig. 32B). The cortex is a mixture of large and small ganglion cells and contains groups of giant cells (Fig. 32C). About a dozen giant cells are found next to the neuropile at the outer side of each of the second and third ventral nerve roots, and about twenty medially beneath the neuropile of the abdominal ganglionic center. Neuroglia cells are scattered in and around the neuropile, and neurilemma cells occur at the periphery of the ganglion.

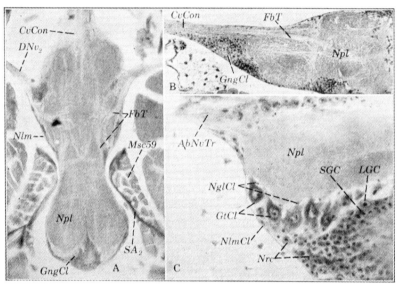

FIG. 32. Thoracic ganglionic center. A: Frontal section of thoracic ganglion (D., C.B.; 140×). B: Sagittal section of anterior end of thoracic ganglion (S., F.L.G.; 210×). C: Sagittal section of abdominal ganglionic center at posterior end of thoracic ganglionic mass (B., H.E.; 480×). *AbNvTr*, abdominal nerve trunk; *CvCon*, cervical connective; *DNv₂*, second dorsal nerve; *FbT*, fiber tract; *GngCl*, ganglion cells (neurocytes); *GtCl*, giant cell; *LGC*, large ganglion cell; *Msc59*, pleurosternal muscle; *NglCl*, neuroglia cell; *Nlm*, neurilemma; *NlmCl*, neurilemma cell; *Npl*, neuropile; *Nrc*, neurocytes; *SA₂*, sternal apophysis (mesofurca); *SGC*, small ganglion cell.

The Stomodaeal Nervous System

The stomodaeal nervous system (also designated as the visceral, stomatogastric, or sympathetic system) consists of a single elongate stomodaeal ganglion (*StGng*), lying between the aorta and oesophagus near the cardia, and a number of visceral nerves (Fig. 29). The stomodaeal ganglion (pharyngeal, oesophageal, hypocerebral, or ventricular ganglion) is probably homologous with the corpora cardiaca and occipital (hypocerebral) ganglion of other insects (see below).

Two nerves arise anteriorly from the stomodaeal ganglion and extend forward along the oesophagus in the cervical region to the aortic funnel at the rear of the brain. Here the left nerve enters the back of the protocerebrum to the left of the oesophagus. The right nerve, which is thicker (6 μ) than the left (3 μ), divides into two nerves. One of these enters the protocerebrum to the right of the oesophagus. The other continues forward as a single nerve (StNv) along the oesophagus and over the cibarial junction, innervating the oesophagus and anastomosing anteriorly with the pharyngeal nerves. (See description of the nerves of the cephalic ganglionic center.) The single nerve along the anterior part of the oesophagus corresponds to the recurrent nerve of other insects, although no frontal ganglion is present, as such, at its anterior end. The nerves entering the back of the brain are apparently the nervi ganglii occipitalis of Snodgrass (1935), the left one extending back to the stomodaeal ganglion independently, the right one coalescing with the recurrent nerve to form the thicker nerve that runs back to the ganglion along the upper right side of the oesophagus. In Lucilia, Day (1943b) describes a single nerve corresponding to these in the cervical region as compounded of the recurrent nerve and two nervi corporis cardiaci I (Hanström's terminology).

From the posterior end of the stomodaeal ganglion there arise a pair of nerves to the cardia (CarNv), a nerve that follows the stalk of the crop (CrNv), and two nerves that run back along the salivary glands (SlGlNv). The heart may be innervated from the stomodaeal ganglion; no nerve connections have been traced, but what appears to be a dorsal median nerve is present in the wall of the first chamber of the heart (Fig. 13G). Connections with the corpus allatum have not been detected but should be sought by appropriate neurological methods.[10] The corpus allatum lies above the anterior end of the stomodaeal ganglion, separated from it by the aorta (Figs. 4A, 18A).

The stomodaeal ganglion (70 by 15 μ) contains a small amount of neuropile axially but consists principally of two types of cells (Fig. 33; also Figs. 4A, 18A). Those in the anterior portion (EoCl) measure about 5 by 10 μ, have dense, microvacuolate,

[10] Day (1943a, b) has demonstrated profuse innervation of the ring gland by nerves from the "hypocerebral ganglion" in Lucilia and Sarcophaga.

strongly eosinophilic cytoplasm, and ovoid vesicular nuclei 4–4.5 μ in diameter with a central clump of chromatin. The posterior cells (*Nrc*) are similar in size, but the microvacuolate cytoplasm stains lightly and is less conspicuous, while the nuclei (4 by 6 μ) contain several masses of chromatin. Neurilemma nuclei are evident around the ganglion and its nerves. The anterior cells are probably endocrinal in function.

The stomodaeal ganglion corresponds, at least in part, to the corpora cardiaca in hemipterous insects described by Pflugfelder,

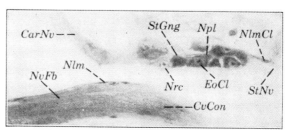

Fig. 33. Stomodaeal ganglion and cervical connective. Sagittal section (B., H.E.; 535×). *CarNv*, nerve to cardia; *CvCon*, cervical connective; *EoCl*, eosinophilic cell; *Nlm*, neurilemma; *NlmCl*, neurilemma cell; *Npl*, neuropile; *Nrc*, neurocyte; *NvFb*, nerve fibers; *StGng*, stomodaeal ganglion; *StNv*, stomodaeal nerve (recurrent nerve).

and cited by Hanström (1939) with the suggestion that the ring gland of muscoidean larvae may "represent a stage in the ontogenetic development of the corpora cardiaca." This has since been borne out by the studies of Vogt (1941c, 1942) on Drosophila and of Day (1943b) on Lucilia, published since the foregoing description was written in 1941. Vogt (1942) regards the deeply staining ("fuchsin") cells in adult Drosophila as corpus cardiacum and the lightly staining cells as hypocerebral ganglion; the former and the corpus allatum, but not the hypocerebral cells, are derived from the larval ring gland.

THE SENSE ORGANS

The sense organs are specialized epidermal structures for the reception of external stimuli (contact, vibrations, tastes, odors, light). Only a brief account of the sense organs of Drosophila can be given here, based largely upon the findings of Hertweck (1931), amended by additional observations and by reclassification of the organs into the categories defined by Snodgrass (1935). For histological details of the various types of organs, see the latter author; also Debaisieux (1938) and Hsü (1938).

The cuticula (Ct) of the body wall, which is secreted by the underlying single-layered epidermis, varies in thickness from 1–1.5 μ in the proboscis (Fig. 11) and abdominal terga (Fig. 13D) to 3 μ in the head capsule and thoracic scutum (Fig. 24A). There is no correlation between thickness and sclerotization, so that the flexible membranous areas are as thick as, and in the contracted ptilinum (5–6 μ) much thicker than, some of the hardened areas or sclerites. In unsclerotized areas the cuticula consists mainly of endocuticula, which is colorless in unstained sections and has an affinity for aniline blue. A thin external layer and the spinules, pale yellow in color and stained by orange G and acid fuchsin, probably include epicuticula and a small amount of exocuticula. Sclerotized cuticula consists of a refractile outer epicuticula, 0.7 μ thick, and a homogeneous exocuticula which is yellowish in color and is stained by orange G and acid fuchsin; no endocuticula is evident. (For detailed studies of insect cuticle by means of the electron microscope, see Richards and Anderson, 1942c.) The cuticula bears two types of sclerotized hairlike processes. The macrochaetae (Set), or bristles, are true setae, 3–10 μ thick and 60–400 μ long; they have an internal cavity, are articulated at the base within a circular socket or setal alveolus, and are scattered widely over the sclerotized areas of the body and appendages in definite positions (Figs. 15A, 31A). The microchaetae (Spn) are simple, solid cuticular outgrowths or spinules, 0.7 μ thick and 6–12 μ long (much smaller on the proboscis), which cover the entire surface of the body and appendages, arranged close together on both sclerotized and exposed unsclerotized areas (Figs. 13D, 15A, 17A).

Fundamentally, each type of sensillum, or unitary sense organ, includes a modified portion of the cuticula, one or more sense cells (sensory neurones or sensory epidermal cells) connected to a nerve, and associated epidermal cells. These epidermal cells are obscure in many of the sensilla in adult Drosophila, perhaps as a corollary of the degenerate condition of the general epidermis (hypodermis), which in most parts of the body wall is reduced to a thin (0.7 μ) protoplasmic layer showing only an occasional nucleus (Fig. 13D, Ab_2T). The thickest epidermis (5–7 μ) is in the labella (Figs. 11, 34C). In the pleural areas of the abdomen of recently emerged flies it is about 3 μ thick (Fig. 19).

Sensilla trichodea (*SsTr*) and *sensilla chaetica*, or setiform sense organs, are sense hairs and bristles found on all parts of the body and the appendages. Each consists of a seta that is flexibly attached to the cuticula and receives the distal process of a single bipolar sense cell (Fig. 34A), or the fibers of a compact group of such cells, located near its base. When the setae are thick-walled and have one sense cell, they are presumably tangoreceptors; when thinner-walled and provided with a group of sense cells, they may be chemoreceptors. Such setae occur on the general body wall, legs, wings, halteres (Figs. 30, 35C), antennae (Fig. 35A; setae 23 μ long, 0.7 μ wide at base), palpi, labella (Figs. 29, 34B, C; 45 by 2 μ), anal plates (Fig. 34A; 70 by 2 μ), and internally on the anterior cibarial plate (Fig. 29; supplied by accessory pharyngeal nerve; see also External Morphology). The macrochaetae of the body are presumably sensory but have no conspicuous sense cells. The prosternal organ (*PSO*, Figs. 29, 30) is a medioventral pocket in the cervical membrane containing two groups of sensilla trichodea; these and patches of minute sensilla (9 by 1.5 μ) at the base of each trochanter may be stimulated by changes in the position of the head and legs relative to neighboring parts. The marginal hairs of the labella, the hairs in the cibarium, and hairs on the tarsi having multiple sense cells are probably all taste organs (evidence of a gustatory sense in the tarsi is presented by Hertweck). The antennal arista (*Ast*, Figs. 29, 30) is provided with a nerve and may be functionally similar to the sense hairs, although morphologically it is the reduced terminal antennal segments.

Sensilla basiconica (*SsBc*), or sensory pegs and cones, differ from sensilla trichodea in that the external process is peglike or conical. The "taste papillae" of the labella (Figs. 34B, C; two to seven small pegs between each two pseudotracheae, one larger peg at the end of each pseudotrachea), sensilla on the epipharynx, or inner side of the labrum (nine on each side of median line, including an irregular linear group of seven near the middle and two near the tip; pegs 1 μ high and cones 2–5 μ long arising from circular bases 1.5–4.5 μ wide), and the "surface cones" of the third antennal segment (Fig. 35A; 12 μ by 1.5 μ) and the palpi are of this type. Cones occur also on the trochanters and at the base of the coxae. The external process is thin-walled, and a group of sense cells is present, so that they are probably chemoreceptors for taste and smell. Experimental results obtained by Begg and Hogben (1946) indicate that pegs and cones on the antennae detect odor and humidity differences.

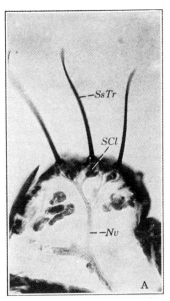
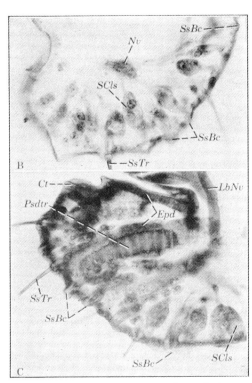

FIG. 34. Setiform sense organs. A: Parasagittal section through anal plate of male, showing sensilla trichodea (B.A., I.H.; 480×). B: Portion of parasagittal section through labella, showing sensilla trichodea and basiconica (B., H.E.; 480×). C: Another section similar to B (D., C.B.; 480×). *Ct*, cuticula; *Epd*, epidermis; *LbNv*, labial nerve; *Nv*, nerve; *Psdtr*, pseudotrachea; *SCl, SCls*, sense cell(s); *SsBc*, sensillum basiconicum; *Sstr*, sensillum trichodeum.

Sensilla coeloconica, or peg organs sunken below the surface of the cuticula, are found in the sensory pit (sacculus, *Sacl*, Fig. 35A) of the third antennal segment ("pit cones": thirty to forty in each pit, 5 μ by 1 μ, with single sense cells; chemoreceptors).

Sensilla campaniformia (SsCm), or dome organs, consist each of a small, thin-walled convexity in the cuticula, attached to a

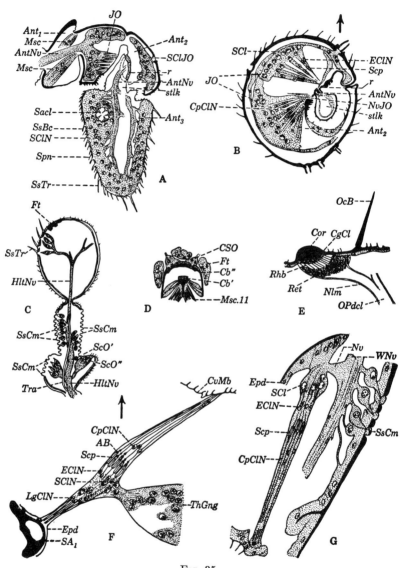

Fig. 35.

single sense cell (Fig. 35G). Their function is uncertain: chemo-reception and stress detection have both been suggested. They occur on the wings, halteres, and legs. Those of the wings are located on the veins as follows: four groups on the base of radius, one sensillum near the auxiliary or subcostal vein, two on the costal vein, four isolated sensilla along the third longitudinal or radius$_{4+5}$ vein, and one on the anterior cross vein. The stalks of the halteres bear many of these sensilla (Figs. 35C, 30, and External Morophology): nine rows on the dorsal side and seven rows on the ventral side of the pedicel, five rows and a small sunken group on the scabellum. On the legs the sensilla, with domes 5 μ in diameter and 1 μ high, are clustered in small groups on each trochanter and at the base of each femur (Fig. 29).

The *cibarial sense organ* (*CSO;* fulcral organ of Hertweck) is a pair of sense-cell clusters overlying minute pits in the posterior cibarial plate near its upper end (Figs. 35D, 29, 2B). They are innervated by the pharyngeal nerves. The structure of the sensilla has not been sufficiently elucidated to permit their classification.

FIG. 35. Sense organs. Redrawn and adapted from Hertweck (1931). A: Longitudinal section of antenna. B: Transection through second antennal segment, showing Johnston's organ. C: Longitudinal section of haltere. D: Transection through upper end of cibarium, showing cibarial sense organ. E: Vertical section through ocellus. F: prothoracic scolopophorous organ. G: Section through base of radius vein of wing, showing alar scolopophorous organ and sensilla campaniformia. *AB*, apical body of scolops; *Ant*$_{1,2,3}$, first, second, and third antennal segments; *AntNv*, antennal nerve; *Cb'*, anterior, *Cb"*, posterior, plate of cibarium; *CgCl*, corneagenous cells; *Cor*, corneal lens; *CpClN*, cap cell nucleus; *CSO*, cibarial sense organ; *CvMb*, cervical membrane; *EClN*, enveloping cell nucleus; *Epd*, epidermis; *Ft*, fat cells; *HltNv*, nerve of haltere; *JO*, Johnston's organ; *LgClN*, ligament cell nucleus; *Msc*, muscle; *Msc.11*, dilator muscle of cibarium; *Nlm*, neurilemma; *Nv*, nerve; *NvJo*, nerve to Johnston's organ; *OcB*, ocellar bristle; *OPdcl*, ocellar pedicel; *r*, sclerotized ring of articular membrane between second and third antennal segments; *Ret*, retinal cells; *Rhb*, rhabdoms; *SA*$_1$, sternal apophysis of prothorax; *Sacl*, sacculus (sensory pit); *SCl*, sense cell; *SClJo*, sense cells of Johnston's organ; *SClN*, sense cell nucleus; *ScO'*, small scolopophorous organ; *ScO"*, large scolopophorous organ; *Scp*, scolops; *Spn*, spinule (microchaeta); *SsBc*, sensillum basiconicum; *SsCm*, sensillum campaniformium; *SsTr*, sensillum trichodeum; *stlk*, stalk of third antennal segment; *ThGng*, thoracic ganglion; *Tra*, trachea; *WNv*, wing nerve; *arrows* point cephalad.

Sensilla scolopophora (*ScO*), the scolopophorous or "chordo-tonal" organs, are compound sense organs characterized by the presence of a minute capsulate sense rod or scolops (3–8 μ long) at the apex of each sense cell. Each sensillum consists of a distal cap cell attached to the cuticula, an enveloping cell enclosing the scolops, and a sense cell connected to a nerve and attached directly (Fig. 35G) or by means of a proximal ligament cell (Fig. 35F) to another part of the body wall. An organ made up of a group of such sensilla thus extends between two separate points on the body wall. The scolopophorous organs are presumably detectors of vibratory stimuli and are found in the prothorax (one on each side between the cervical membrane and the pro-sternal apophysis, with about 20 scolopes; Figs. 29, 30, 35F); in the femora (two in each, a large one with about 20 aligned scolopes and a small one with 8 scattered scolopes, extending from the base to the tip of the femur; Figs. 14, 29); in the wings (one extending obliquely within the base of radius proximad of the auxiliary vein, with about 20 scolopes; Fig. 35G); in the halteres (two at right angles to each other in the scabellum, one with 18, the other with 8–9 scolopes; Fig. 35C); and in the second antennal segment (a large number of crowded sensilla filling almost the entire interior and extending between the outer wall and a sclerotic ring in the distal articular membrane, constituting "Johnston's organ," which may have a static function; *JO*, Figs. 29, 30, 35A, B).

The *ocelli* (*O*), or simple eyes, three in number, are arranged triangularly on a small prominence on the vertex of the head, with the median ocellus foremost (Figs. 29, 30, and External Morphology). They are of an amber color in the wild-type fly and oval in form (30 by 40 μ). Each consists of a single thick planoconvex lens (*Cor*, Figs. 31A, 35E), an underlying layer of transparent cuboidal corneagenous cells (*CgCl*), and a second layer of retinal cells (*Ret*, optic sense cells) which form a swelling (ocellar nerve center) at the end of the attached branch of the ocellar pedicel (*OPdcl*). The upper ends of the retinal cells form a hyaline zone where their sensory borders are grouped together to form rhabdoms or receptive rods (*Rhb*). Their lower ends contain the nuclei and minute brownish pigment granules (absent in mutant white). The ocellar nerve fibers from the retinal cells are probably short and associate within the ocellar

center with the terminals of brain neurones whose fibers constitute the ocellar pedicel.

The *compound eyes* (*E*) are two large, convex groups of highly
specialized photoreceptors occupying the sides of the head (Figs.
16, 22, 30). They are bright red in the wild-type fly, 0.42 mm
high, 0.33 mm wide, and 0.15 mm deep. Each eye consists

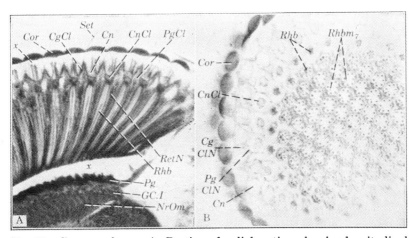

Fig. 36. Compound eye. A: Portion of radial section, showing longitudinal
sections of ommatidia (D., C.B.; 350×). B: Portion of horizontal section,
showing transections of ommatidia at successively deeper levels toward
right center (S., F.L.G.; 350×). *CgCl*, corneagenous cell; *CgClN*, corneagenous cell nucleus; *Cn*, crystalline cone; *CnCl*, cone cells; *Cor*, corneal
lens; *GC.I*, ganglion cells of periopticon; *NrOm*, "neurommatidium" (optic
nerve terminals in periopticon of optic lobe); *Pg*, pigment; *PgCl*, secondary
pigment cell; *PgClN*, secondary pigment cell nucleus; *RetN*, retinular
cell nucleus; *Rhb*, rhabdom; *Rhbm_7*, seventh rhabdomere; *Set*, seta; *x*,
artificial break or shrinkage space.

of 680–700 contiguous cylindrical units, the ommatidia, which
radiate from the optic lobe of the brain and end at the surface
as a honeycomb of hexagonal lenses, or facets, in the cuticula
(Figs. 36, 31B). The ommatidia are 70–125 μ long, and the
facets 17 μ in diameter. At each intersection of the boundaries
of three adjacent facets a seta (sensillum trichodeum?) projects
from the surface of the eye (*Set*, Fig. 36A).

Each ommatidium (Figs. 36, 37) consists of a distal corneal
lens (*Cor*), two corneagenous pigment-bearing cells (*CgCl*) surrounding a crystalline cone (*Cn;* pseudocone), four cone cells
(*CnCl*) which secrete the cone and lie beneath it, a retinula (*Ret*)

of eight parallel sense cells around an axial rhabdom (Rhb), and a sheath of about twelve secondary pigment cells ($PgCl$) which are shared by adjacent ommatidia. A fenestrated basement

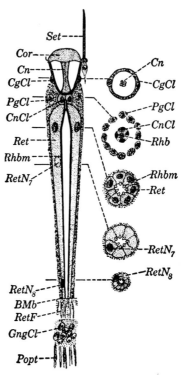

Fig. 37. Ommatidium of compound eye. Diagram redrawn and adapted from Hertweck (1931). Longitudinal section and transections at levels indicated by horizontal lines. *BMb*, basement membrane; *CgCl*, corneagenous cell; *Cn*, crystalline cone; *CnCl*, cone cell; *Cor*, corneal lens; *GngCl*, ganglion cell; *PgCl*, secondary pigment cell; *Popt*, periopticon; *Ret*, retinular cell; *RetF*, optic nerve fiber; $RetN_{7,\,8}$, nuclei of seventh and eighth retinular cells; *Rhb*, apex of rhabdom; *Rhbm*, rhabdomere; *Set*, seta.

membrane (BMb) underlies all the ommatidia and is pierced by the optic fibers ($RetF$) from the retinular cells and by tracheae.

The corneal lens is outwardly convex and inwardly slightly concave with a central convexity. The corneagenous (primary) pigment cells form a conical cup filled in life with a hyaline substance which constitutes the pseudocone. When fixed, the pseudocone may appear as four contiguous axial strands extending distally from the cone cells, which lie at the pointed lower end of the cup and surround the tip of the rhabdom.

The rhabdom, or sense rod, consists of a bundle of seven rhabdomeres (spread apart except at their distal ends in fixed material) that extend along the inner edges of the elongate retinular cells, these cells forming a rosette in cross-section. The eighth retinular cell, restricted to the base of the ommatidium, is much reduced and has neither a rhabdomere nor a detectable optic fiber; its nucleus ($RetN_8$) is near the basement membrane. Of the other seven, six have their nuclei near the upper end, and the optic fibers which arise from their bases terminate in groups (the "neurommatidia") in the periopticon of the optic lobe (Fig. 36A), while the seventh has its nucleus ($RetN_7$) lower and its fiber differentiated as a separate "long fiber" in the periopticon. The rhabdomere of the seventh cell is thinner than the others and in transection lies next to a projection extending inward from the ommatidial boundary (Fig. 36B).

The secondary pigment cells have their nuclei at the level of the cone cells and contain an abundance of coarse pigment granules which extend between the ommatidia to the basement membrane. Pigment granules occur also in the corneagenous cells and between the optic nerve fibers under the basement membrane. According to Johannsen (1924), two kinds of pigment are present, a wine-purplish red and an ochre yellow; differences in their relative abundance and distribution determine the mutant variations in eye color. The retinulae are yellow-tinged, but no granules are present. (For other comparative histological studies of normal and mutant eyes, see Krafka, 1924; Richards and Furrow, 1925; Cochrane, 1937; and Waddington and Pilkington, 1943.)

THE MALE REPRODUCTIVE SYSTEM

The reproductive system of the male (Fig. 38) consists of paired testes (Tes), paired vasa deferentia (Vd; vasa efferentia of other authors) dilated in part to form seminal vesicles (Vsm), paired accessory glands ($AcGl$), an unpaired ejaculatory duct (Dej) with an appended ejaculatory bulb (Bej; "sperm pump"), and the external genitalia (see External Morphology). The ejaculatory duct includes two portions: a dilated, thick-walled anterior duct (Dej'; vas deferens of other authors), extending

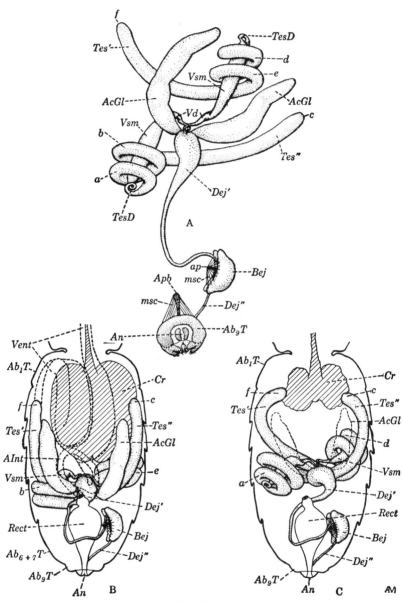

Fig. 38.

from the vasa deferentia to the ejaculatory bulb, and a slender, thin-walled posterior duct (*Dej"*; ejaculatory duct of other authors) extending from the bulb to the aedeagus. (See Miller, 1941, regarding terminology.) Morphologically the short "testicular duct" (*TesD*) connecting the testis and seminal vesicle may be regarded as either a vas efferens, corresponding to the vas efferens at the base of each sperm tube in the compound testes of other insects, or as a remnant of the mesodermal vas deferens of more generalized insects. The relative positions of the organs are shown in Fig. 38A, where they are spread apart for clarity; ordinarily the testicular coils are more compact and the twists of the vasa deferentia more loosely disposed. All the organs are colorless or nearly so in recently emerged individuals, but testes, seminal vesicles, and vasa deferentia gradually become bright butter-yellow; the color fades in fixed material.

Each of the two *testes* (*Tes*) is a single long sperm tube, about 1.9 mm long and 90–110 μ in diameter, with the basal half helically coiled and the distal half forming a freely extended arc. The coiled portion is spirally wound twice about a center occupied by the testicular duct and some portion of the seminal vesicle, the two gyres (*a*, *b*, and *d*, *e*) lying side by side. The testicular duct (*TesD*) at the base of the testis describes a more

FIG. 38. Male reproductive system (47✕). A: Dorsal aspect of reproductive tract of an old individual with well-distended seminal vesicles; organs spread apart but arranged to show their relative positions, with gyres of testicular coils separated and testicular duct drawn out of its normally central position within the first gyre. B: Dorsal aspect of abdomen with distended crop, showing reproductive system in place and typical transverse position of testicular coils; crop cross-hatched, overlying parts of alimentary canal indicated by dotted lines. C: Dorsal aspect of abdomen with undistended crop, showing oblique position sometimes assumed by uncrowded testes, especially in young individuals; crop cross-hatched, outline of overlying accessory glands dotted. *a*, first (basal) gyre of coil of right testis; Ab_1T, $Ab_{6\,+\,7}T$, Ab_9T, tergites of first, combined sixth and seventh, and ninth, abdominal segments; *AcGl*, accessory gland; *AInt*, anterior intestine; *An*, anus; *ap*, apodeme of ejaculatory bulb; *Apb*, basal apodeme of aedeagus; *b*, second gyre of coil of right testis; *Bej*, ejaculatory bulb; *c*, tip of right testis; *Cr*, crop; *d*, first gyre of coil of left testis; *Dej'*, anterior ejaculatory duct; *Dej"*, posterior ejaculatory duct; *e*, second gyre of coil of left testis; *f*, tip of left testis; *msc*, muscle fibers; *Rect*, rectum; *Tes'*, left testis; *Tes"*, right testis; *TesD*, testicular duct; *Vd*, undistended portion of vas deferens; *Vent*, ventriculus; *Vsm*, seminal vesicle.

or less complete third gyre. The testes of recently emerged teneral flies are turgid and thick and have proportionately less of their length extending freely from the coil. The normal position of the testes within the abdomen (Figs. 38B, C) is asymmetrically bilateral, each with the coiled portion lying ventrolaterally in a vertical plane on one side of the body and the free end extending to the opposite side. The free end (c, f) corresponds in position to the entire pupal testis, the coiled portion secondarily attaining its position on the opposite side as it develops by posterior growth of the ellipsoidal pupal testis. (The vasa deferentia are also reversed in position and become twisted by the spiral growth of the attached testis: Stern, 1941.) The left testis (Tes'; that with its tip on the left) lies more anterior, and its free part crosses the body in front of the right testis (Tes"). In situ the junction of the testis and seminal vesicle is directed cephalad in the anterior testis and caudad in the posterior testis. The former is dextrorotatory, spiraling clockwise and posteriad, while the latter is levorotatory, spiraling counterclockwise and anteriad (Fig. 38A). The testes are not mirror images of each other, since when viewed from the base, the direction of coiling from base to tip in both is counterclockwise and toward the ejaculatory duct. The free ends of the testes cross the ventral side of the body and then extend upward and forward laterad of the accessory glands. The segmental position of the testes varies with the age of the fly and the degree of expansion of the crop (q.v.), so that the coiled portions may lie loosely in abdominal segments 3–5 (Fig. 38C) or be compressed and crowded back into 4 and 5, or even partly into 6 (Fig. 38B); the tips extend into segments 2 and 3 (Miller, 1941). Gleichauf (1936) found that of 135 wild-type flies, 4.44 per cent had both testicular coils on the right side and 6.67 per cent had both on the left side of the body; in one instance, the cephalocaudal relations of left and right testis were the reverse of normal.

The wall of the testis (Fig. 39A, C) consists of two very thin layers containing widely spaced nuclei and having a combined thickness of 1.5–2 μ: an inner sheath (Sh'; tunica interna of Keuchenius, 1913, and inner membrane of Stern, 1941), and an outer sheath or pigment cell layer (Sh"; tunica externa, pigment cells of Stern). The latter is composed of flattened polygonal cells that contain minute yellow pigment granules and during development are "amoeboid," spreading from the testis over the attached vas deferens (Stern and Hadorn, 1939; Stern, 1941). In sections, the large nuclei of the pigment cells are most evident in recently emerged flies, and there measure about 13 μ long, 7 μ wide, and 1.5 μ thick. The nuclei of the inner sheath are smaller and flatter (4 by 0.5 μ).

The entire sperm tube is filled with germ cells in various stages of maturation (Figs. 39A–E), but a distinct zonation into successive developmental regions is not evident, except that spermatogonia are limited to the extreme tip of the tube. Unlike the condition in certain other insects, neither a special apical cell of the testis nor cyst cells enclosing groups of germ cells are apparent.

The spermatogonia (Spg) form a compact mass about 50 μ long at the apex of the sperm tube; they are 5 μ in diameter, with a small amount of homogeneous cytoplasm and densely chromatinic nuclei 2.5 μ in diameter (Figs. 39A, C). Elsewhere in the sperm tube groups of spermatocytes, spermatids, and developing spermatozoa are closely intermingled, forming a cellular layer 6–22 μ thick that usually occupies only about one-half of the circumference of the tube (Figs. 39B, D).

Spermatocytes (Spc) are most conspicuous in the distal free half of the testis, extend into the adjoining gyre of the coil, but are absent in the basal gyre. The early primary spermatocytes (15 μ) contain large nuclei, 8–10 μ in diameter, with diffuse chromatin and a conspicuous nonchromatinic plasmosome or nucleolus, stained by haematoxylin but not by the Feulgen method (Figs. 39A, C). In later spermatocytes the nucleolus is absent, disappearing shortly before the leptotene stage (Huettner, 1930). Meiotic division figures and secondary spermatocytes (nuclei with a faint chromatin reticulum) are very infrequent.

The early spermatids (Spd) are 20 by 10 μ, with a round vesicular nucleus 5 μ in diameter and a dark-staining nebenkern (a spherical mitochondrial mass) of similar size. Later stages show the chromatinic material condensing into a slender sperm head, 5–7 μ long, attached to a long filamentous tail, which becomes about 900 μ in length and 0.2 μ or less in thickness.

The sperm occur in compact bundles (SpBnd), 6 μ in diameter, with the heads clustered within a "nutritive cell" (Guyénot and Naville, 1929). These bundles each contain about 50–100 sperm

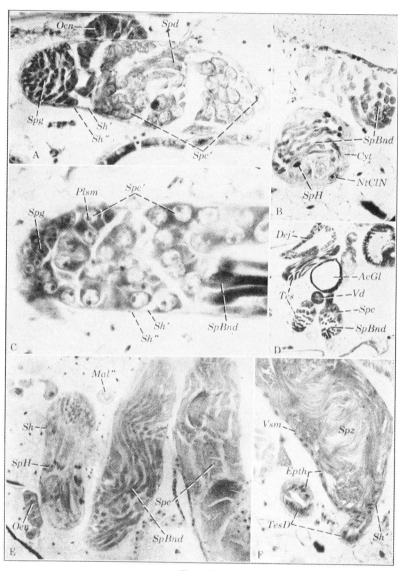

Fig. 39.

and are found throughout the testis, but only those in the coiled portion show deeply staining mature heads (Fig. 39E). In the more distal parts of the coil the clusters of sperm heads are surrounded by diffuse and seemingly degenerate masses of cytoplasm (Fig. 39B); more basally they are merely associated with the testicular wall. Peripheral vesicular nuclei (7 μ) that probably belong to the nutritive cells are also present in the coil (Fig. 39B). The sperm bundles are longitudinally arranged and less crowded in the free end of the testis, densely massed and much convoluted in the coil. Mature spermatozoa are already present when the fly emerges from the puparium. They show no movement either within the testes or in the seminal vesicles. (Those in the sperm-storing organs of the female are frequently vibratile when the parts are examined in saline solution.)

The *testicular duct* (*TesD*), connecting the testis and vas deferens, is 80 μ long and 23 μ in diameter, and makes a single spiral turn (Fig. 38A). Its narrow lumen is surrounded by a thick epithelium (6–12 μ) containing small, closely spaced nuclei (Fig. 39F). An inconspicuous inner sheath and a thin outer pigment sheath, continuous with those of the testis and totaling 1 μ in thickness, extend around the duct as in the vas deferens. The testicular ducts are said to be of mesodermal origin and hence may represent true mesodermal vasa deferentia or vasa efferentia, although they appear like unexpanded portions of the ectodermal and definitive vasa deferentia.

Fig. 39. Testes and vasa deferentia. A: Longisection of tip of testis (S., F.L.G.; 375×). B: Transverse section through first gyre (*below*) and oblique section through second gyre of coiled portion of testis (S., F.L.G.; 375×). C: Longisection of tip of testis (B.A., I.H.; 490×). D: Portion of sagittal section of abdomen, showing transections of testes and associated ducts (B.A., I.H.; 100×). E: Tangential sections of first (*left*) and second gyres of one testicular coil and of ascending free portion of other testis (*right*) (S., F.L.G.; 210×). F: Longisection of proximal end of distended seminal vesicle and transection of testicular duct (S., F.L.G.; 375×). *AcGl*, accessory gland; *Cyt*, degenerate cytoplasm; *Dej'*, anterior ejaculatory duct; *Epth*, epithelium; *Mal"*, posterior Malpighian tubule; *NtClN*, nutritive cell nucleus; *Oen*, oenocytes; *Plsm*, plasmosome; *Sh*, sheath; *Sh'*, inner sheath; *Sh"*, outer sheath (pigment cell layer); *SpBnd*, bundle of spermatozoa; *Spc*, spermatocytes; *Spc'*, primary spermatocytes; *Spd*, spermatid; *Spg*, spermatogonium; *SpH*, sperm heads; *Spz*, sperm; *Tes*, testis; *TesD*, testicular duct; *Vd*, vas deferens; *Vsm*, seminal vesicle.

The *vas deferens* (*Vd*), 350 μ long and 12–24 μ in minimum diameter, has its distal portion more or less enlarged to form the seminal vesicle (*Vsm*), 160–220 μ long and 50–100 μ in diameter. The latter, when undistended, lies more or less obliquely within the second gyre of the testicular coil (Fig. 38C); when it is extremely distended (to two or three times the diameter of the sperm tube), the gyres are pushed down toward the testicular duct. The seminal vesicles taper into the twisted narrow portions of the vasa deferentia, and these unite just before the junction with the ejaculatory duct (Fig. 38A). The unexpanded wall of the vas deferens in recently emerged flies consists of three distinct layers: namely, a thick inner epithelium (10 μ) of columnar cells containing small nuclei; a thin inner sheath (2–3 μ; outer epithelium of Stern and Hadorn) with inconspicuous, widely spaced, and flattened nuclei (3 by 1 μ); and an outer sheath of finely reticular cytoplasm 3–8 μ thick, containing an occasional large nucleus (6 by 10 μ). The latter layer is formed by pigment cells which migrate from the testis during pupal development (Stern and Hadorn, 1939; Stern, 1941) and is yellow during life. The middle layer, or inner sheath, is continuous with that of the testicular duct and the inner sheath of the testis. In older flies the layers are difficult to distinguish in sections, the inner epithelium being most evident, with the cells flattened to 3 μ in the dilated vesicle (6 μ in the narrow ducts), while the nuclei of the middle layer are spread wide apart and the pigment sheath is hardly discernible (Fig. 39F). The ectodermal origin of the vasa deferentia and vesicles (Dobzhansky, 1930; Gleichauf, 1936) implies that a chitinous intima may be present, but only suggestions of a very thin lining are detectable in sections. (The term "vasa deferentia" is used here in a functional sense and does not connote true homology with the similarly disposed and paired mesodermal vasa of more generalized insects.) The seminal vesicles contain no sperm in recently emerged teneral flies, but later each is expanded and filled by a dense mass of mature spermatozoa, the heads still grouped in bundles and the tails much convoluted (Fig. 39F). The sperm extend into the narrow part of the vas deferens up to its juncture with the ejaculatory duct. The united vasa open into the latter at the tip of a conical cell mass which projects into its lumen.

The *anterior ejaculatory duct* (*Dej'*) is about 1.2 mm long, with an anterior dilated portion 120 μ in diameter tapering to a narrow tube 20 μ in diameter, which enters the ejaculatory bulb (Fig. 38A). In the abdomen the duct extends dorsally, then curves downward and to the left, and finally loops over the rectum to the right side of segment 5 or 6 (Figs. 38B, C). This relation to the rectum is the result of a complete clockwise rotation of the terminal abdominal segments during pupal development (Gleichauf, 1936). The duct wall consists of a single thick layer of epithe-

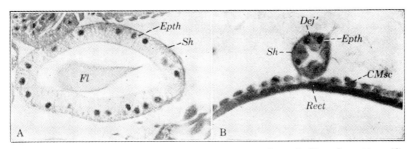

Fig. 40. Anterior ejaculatory duct. A: Transection of dilated portion (S., F.L.G.; 375×). B: Transection of narrow portion and longisection of rectal wall (S., F.L.G.; 490×). *CMsc*, circular muscle; *Dej'*, anterior ejaculatory duct; *Epth*, epithelium; *Fl*, coagulated fluid; *Rect*, rectum; *Sh*, sheath.

lial cells lined with a thin chitinous intima and surrounded by an inconspicuous outer sheath (Fig. 40). The lumen contains a fluid that in sections appears as a granular coagulum which, like the contents of the accessory glands, stains blue with Mallory's connective-tissue stain. The coagulum is, however, denser than that in the accessory glands. No sperm are stored in the duct. The duct undergoes strong peristaltic contractions in saline solution, but no muscle fibers or striations are evident. The contractility must be attributed to the outer sheath or, less probably, to the epithelium itself. The latter in the dilated portion of the duct consists of large polygonal cells, more or less cuboidal and 14 μ high in section, each with a densely chromatinic nucleus and spongy cytoplasm which is sometimes vacuolate, suggesting a glandular nature (Fig. 40A). At the tip the epithelium is invaginated to form a prominence through which the vasa deferentia and accessory glands open. In the narrow portion of the duct four to six subconical epithelial cells are present in cross-section (Fig. 40B). The outer sheath is of uniform thickness (0.6 μ) throughout the length of the duct and contains minute scattered nuclei.

The *ejaculatory bulb* (*Bej*) is a thick-walled, bean-shaped capsule, 200–300 μ long and 100–150 μ in diameter, with two blunt projections at each end, a sclerotized rodlike apodeme arising

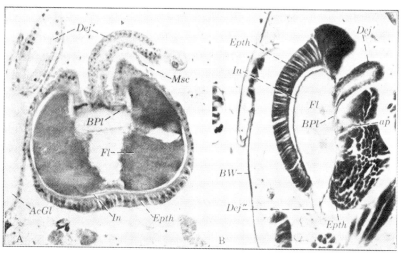

Fig. 41. Ejaculatory bulb. A: Transection of bulb at level of junction with anterior ejaculatory duct (S., F.L.G.; 210×). B: Oblique sagittal section of bulb, with dorsal epithelium of one lobe in tangential section at right (S., H.; 190×). *AcGl*, accessory gland; *ap*, apodeme; *BPl*, basal plate of apodeme; *BW*, body wall; *Dej'*, anterior ejaculatory duct; *Dej"*, posterior ejaculatory duct; *Epth*, epithelium; *Fl*, coagulated fluid; *In*, intima; *Msc*, muscle.

from its upper side, and muscle fibers radiating from the end of the rod to the side walls (Fig. 38A). The anterior ejaculatory duct enters the bulb on the upper side near one end, and the posterior duct arises between the paired projections of the opposite end. The ejaculatory bulb usually lies to the right of the rectal sac in segment 5 or 6 (Figs. 38B, C), but its location and orientation are variable (see Gleichauf, 1936). The walls of the bulb consist of a thick (11–24 μ) columnar epithelium lined with a well-developed intima (1 μ), enclosing a chamber filled with a clear viscous fluid (Fig. 41). The flat roof of the chamber bears the upright apodemal rod, whose attached muscles presumably serve to compress the chamber during ejaculation; expansion of the bulb probably results from elasticity of its walls. The fluid in the bulb (absent in recently emerged individuals) appears dense and homogeneous in sections and extends into the posterior ejaculatory duct. It stains red with Mallory's stain (orange when fuchsin is omitted), but near the mouth of the anterior ejaculatory duct an admixture of blue stain is evident, indicating the presence of fluid from the duct in this region. This supports Gleichauf's contention that the sperm must pass through the cavity of the bulb and not by-pass it through a separate passage within the basal plate of the apodeme, as Nonidez (1920) believed. (Rosenblad, 1941, describes several types of diverticula that are present on the ejaculatory bulb of other species of Drosophilinae.)

The *posterior ejaculatory duct (Dej")* is a very thin-walled tube, 300 μ long and 11–18 μ in diameter, continuous with the intima lining the ejaculatory bulb (Figs. 38A, 41B). It passes below the rectum and opens between the ninth and tenth abdominal segments at the tip of the intromittent organ or aedeagus (Figs. 38B, C). Its wall is chitinous and 0.5 μ thick, the only indication of a covering epithelium being small scattered nuclei on its outer surface. The lumen is filled with fluid having staining properties like that in the ejaculatory bulb.

The *accessory glands (AcGl; paragonia)* are a pair of elongate sacs, 425 by 130 μ, that open separately into the dilated portion of the anterior ejaculatory duct just below the vasa deferentia (Fig. 38A). They lie beneath the dorsal wall of abdominal segments 3 and 4, mesad of the ends of the testes, and curve downward and forward to join the ejaculatory duct ventrally (Figs. 38B, C). The wall of the glands (Fig. 42) consists of large binucleate cells, squamous or cuboidal in section (2–7 μ thick), covered externally by a very thin nucleated sheath (0.5 μ). A refractile inner border is sometimes evident that may be an intima, since the glands are of ectodermal origin (Gleichauf, 1936). Most of the binucleate epithelial cells are polygonal in surface view (Fig. 42B, a) and have homogeneous cytoplasm, but in the apical portion of the gland there are interspersed swollen, ovoid cells which contain large vacuoles and project into the lumen (Fig. 42A). The vacuolate cells are evidently in an active secretory phase. The interior of the gland is filled with a colorless, cloudy fluid that forms a granular or spongy coagulum in sections and stains blue with Mallory's connective-tissue stain but is not as dense as the contents of the anterior ejaculatory duct. The outer sheath contains occasional inconspicuous and minute nuclei. No muscle striations are evident, but the sheath may be the seat of the very active contractions exhibited by living glands in saline solution. The glands open into the ejaculatory duct through the invaginated prominence which surrounds the mouth of the vasa deferentia. The secretion of the accessory glands is probably necessary for effective fertilization of the female, since

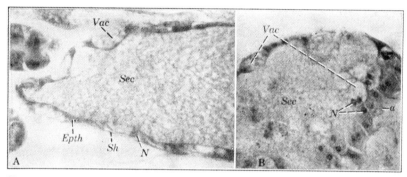

Fig. 42. Accessory gland of male. A: Longisection of tip of gland (S., F.L.G.; 490×). B: Portion of tangential section, showing binucleate epithelial cells in surface view (S., F.L.G.; 480×). *a*, non-secreting cell; *Epth*, epithelium; *N*, nuclei; *Sec*, secretion; *Sh*, sheath; *Vac*, vacuole.

sperm taken from the seminal vesicles and injected into the uterus produces offspring much more rarely than does normal ejaculate that is artificially transferred (Gottschewski, 1937).

The structure of the external genitalia is described under External Morphology and by Gleichauf (1936), who also considers the musculature (cf. Graham-Smith, 1938, Morrison, 1941, and Crampton, 1942).

THE FEMALE REPRODUCTIVE SYSTEM

The female reproductive system, occupying the posterior two-thirds of the abdomen, consists of two ovaries, an efferent duct system, three sperm-storing organs, and two accessory glands (Fig. 43). The ovaries (*Ov*) are joined by two short lateral oviducts (*Odl*) to a common median oviduct (*Odc*), and this in turn enters the pouchlike genital chamber or vagina (*GC*). The latter opens to the exterior through the vulva (*Vul*), situated between the gonopods (*Gon*), which form a retractile ovipositor (see External Morphology). Opening into the genital chamber are a ventral seminal receptacle (*SmRcp*), a pair of spermathecae (*Spt*), and a pair of accessory glands (*AcGl*).

The *ovaries* (*Ov*) are compact and pyriform and lie bilaterally in the third, fourth, and fifth abdominal segments, surrounded by adipose tissue. They flank the coiled midgut and may be displaced or crowded by the expansion of the crop. Although their tapered distal ends are unattached—no terminal ligament being present—they are held in place by the numerous tracheal branches with which they are supplied. The size of the ovaries varies with the extent of the development of the eggs and with the nurture of the individual, ranging from 220 by 350 to 900 by 1100 μ. The abdomen is distended by their growth. Each ovary consists of a compact group of parallel egg tubes or ovarioles (*Ovl*) which ordinarily vary in number from 10 to 20. The ovarioles are held together by a thin connective-tissue envelope, the peritoneal sheath (*PSh*), and converge basally where they are attached to the calyx (*Clx*) of the lateral oviduct.

An ovariole consists of a moniliform series of four to six egg chambers, tapering distally and provided with a terminal filament (*TF*) and a basal stalk or pedicel (*Pdcl*). The entire ovariole is surrounded by a thin, nucleated membrane, the epithelial sheath (*ESh*), attached to the peritoneal sheath by a slender distal strand and continuous basally with the epithelium of the oviduct. There is no separate tunica propria within the epithelial sheath other than the basement membrane of the follicle cells.

The terminal filament is a short single row of transversely compressed cells and is attached to the end chamber, or germarium (*Grm*), which is 20 μ wide and 50–70 μ long (Fig. 44A). The germarium contains the oögonia apically and is divided transversely into incipient compartments that become more distinct basally, where groups of sixteen small rounded cells are separated from their neighbors by thin walls of flattened follicle cells. The oögonia have the largest nuclei (4–6 μ) in the germarium and frequently show mitoses.

The basal chamber of the germarium merges into the first of the series of egg chambers which constitute the greater part

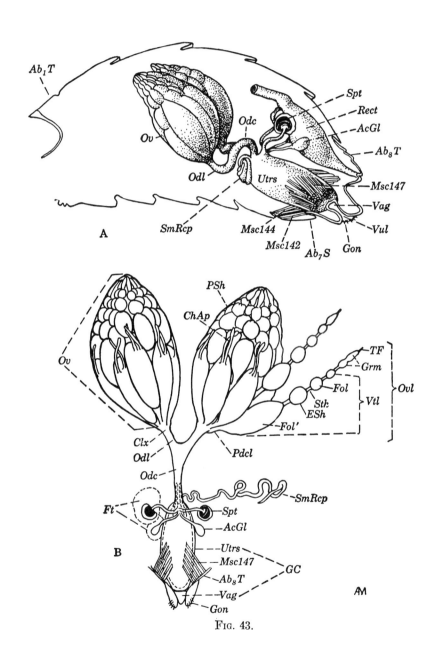

Fig. 43.

(vitellarium or zone of growth, *Vtl*) of the ovariole (Figs. 43B, 44A, B). The egg chambers or follicles (*Fol*) are successively larger toward the base of the ovariole, increasing from spheres with an initial diameter of 25 μ to ovoids 450 by 170 μ (Fig. 43B, *Fol'*), the size of the fully formed egg. Each chamber consists of a single layer of follicular epithelium enclosing a compact group of sixteen cells and is joined to the adjacent chamber at each end by a narrow stalk of small cells (*Stk*, Figs. 44A, B). Fourteen of the contained cells are nurse cells (*NrCl*). The other two undergo premeiotic phases, but only one continues development as a primary oöcyte (*Ooc*); the other degenerates and presumably acts as a trophocyte for the growing egg, like the nurse cells (Guyénot and Naville, 1933; Stolfi, 1938). Cytoplasmic bridges are said to connect the oöcyte and nurse cells. The ovarioles are thus of the meroistic polytrophic type, with an alternating succession of nurse cells and oöcytes.

In the younger follicles the nurse cells increase enormously in size, while the functional oöcyte, at the posterior pole of the follicle, remains relatively small. As yolk is deposited, the egg grows at the expense of the nurse cells, the follicle becomes greatly distended, and finally only degenerate granular remains of the nurse cells persist at the anterior end of the fully formed egg. During their growth, the originally dense and homogeneous cytoplasm of the nurse cells becomes spongy, and the deep-

---

Fig. 43. Female reproductive system. A: Lateral aspect of abdomen with left wall removed, showing reproductive organs and rectum in place; posterior vagina in sagittal section and uterus distended by an egg (60×). B: Dorsal aspect of reproductive organs with two ovarioles separated from the ovary and the seminal receptacle uncoiled; egg in uterus and fat surrounding left spermatheca and accessory gland indicated by broken outlines (45×). *Ab₇S*, seventh abdominal sternite; *Ab₁T*, *Ab₈T*, first and eighth abdominal tergites; *AcGl*, accessory gland; *ChAp*, chorionic appendage of egg; *Clx*, calyx; *ESh*, epithelial sheath; *Fol*, follicle; *Fol'*, basal follicle where formation of egg is completed; *Ft*, fat tissue; *GC*, genital chamber; *Gon*, gonopod; *Grm*, germarium; *Msc142*, ventral muscles of seventh abdominal segment; *Msc144*, sternal muscles to uterus; *Msc147*, tergal muscles to uterus; *Odc*, common oviduct; *Odl*, lateral oviduct; *Ov*, ovary; *Ovl*, ovariole; *Pdcl*, pedicel; *PSh*, peritoneal sheath; *Rect*, rectum; *SmRcp*, seminal receptacle; *Spt*, spermatheca; *Stk*, interfollicular stalk; *TF*, terminal filament; *Utrs*, uterus; *Vag*, (posterior) vagina; *Vtl*, vitellarium; *Vul*, vulva.

staining spherical nuclei increase in diameter from 3 μ to 40 μ (Figs. 44A, B, C). Painter and Reindorp (1939) give evidence

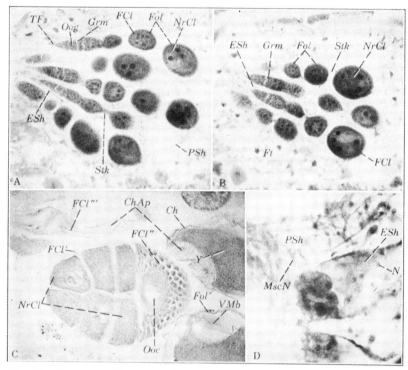

Fig. 44. Ovary. A, B: Parasagittal sections of ovary of a young fly, showing portions of several ovarioles (D., H.E.; 170×). C: Portion of oblique transection through base of ovary containing completed eggs; central follicle cut obliquely through anterior end of growing oöcyte (D., F.L.G.; 190×). D: Portion of tangential section of ovary, showing surface view of peritoneal sheath and ovariolar epithelial sheath (D., C.B.; 480×). *Ch,* chorion; *ChAp,* chorionic appendage; *ESh,* epithelial sheath; *FCl,* follicular epithelium; *FCl',* flattened follicular epithelium around nurse cells; *FCl'',* tall follicular cells around oöcyte; *FCl''',* flattened follicular epithelium around chorionic appendages; *Fol,* egg follicle; *Fol',* follicular epithelium around completed egg; *Ft,* fat cells; *Grm,* germarium; *MscNn,* muscle nucleus; *N,* nucleus; *NrCl,* nurse cell; *Ooc,* primary oöcyte; *Oog,* oögonia; *PSh,* peritoneal sheath; *Stk,* interfollicular stalk; *TF,* terminal filament; *VMb,* vitelline membrane; *Y,* yolk of egg.

that these nuclei undergo endomitosis and probably attain a 512-ploid condition.

The oöcyte has more granular cytoplasm, which becomes charged with yolk spherules as the egg develops; the nucleus is pale-staining and vesicular in prophase (28 by 16 μ; Fig. 44C) and finally becomes a compact group of chromosomes in first meiotic metaphase at the anterodorsal third of the fully formed egg (cf. Guyénot and Naville, 1929 and 1933, for cytology). A vitelline membrane appears around the egg, presumably formed by the oöcyte. The chorion and its dorsal appendages (Fig. 43B, *ChAp*) are secreted by the surrounding follicle cells in the last chamber of the ovariole.

The follicular epithelium in the smaller follicles is composed throughout of small cuboidal cells 3–6 μ high with central nuclei (*FCl*, Fig. 44A), but during growth of the oöcyte it differentiates into a thin (0.5 μ), sparsely nucleated membrane (*FCl'*) over the nurse cells and a thick columnar layer (15–45 μ), with basal nuclei, around the oöcyte (*FCl"*, Fig. 44C). As the chorion of the egg is formed, the latter follicular cells decrease in height and finally form a thin squamous covering of hexagonal cells (Fig. 45B, *Fol'*) whose outlines are impressed upon the chorion as a reticulate sculpturing. This sculpturing is also found on the dorsal chorionic appendages (*ChAp*) which during development lie beside the degenerating nurse cells and are individually ensheathed in a thin membrane of attenuated follicular cells (Fig. 44C). The micropylar cone at the anterior pole of the egg forms within a mass of small follicular cells which at this stage intervene between the egg and the degenerate nurse cells.

When the egg leaves the ovariole, the follicular cells are presumably left behind; degenerate granular masses with pycnotic nuclei are frequently present in the pedicel below the basal egg chamber and seem to include both nurse cell and follicle cell material. The basal pedicel of the ovariole is merely the constricted epithelial sheath, short when a fully formed egg is present and temporarily longer after the egg has passed into the oviduct and while the next follicle is held back by the crowding of large eggs in neighboring ovarioles.

The epithelial sheath (*ESh*, Figs. 44A, B, D) is a thin membranous layer 0.4 μ thick containing nuclei 1 μ in diameter and minute, anastomosing, transverse fibrillae which are 1 μ wide and cross-striated (Fig. 44D); it is closed basally by the massed epithelium of the calyx. The peritoneal sheath (*PSh*) that envelops the entire ovary (i.e., the ovarioles collectively) is also membranous (0.5 μ) and contains a more conspicuous network of fine, branching muscle fibers, 0.3–1.5 μ wide, faintly striated and with occasional nuclei at the junctions (Figs. 44A, D). Both the epithelial and peritoneal sheaths are apparently stretched by increase in size of the ovary, since their nuclei are closer together and more evident in the small ovaries of young individuals. The epithelial sheaths and the peritoneal sheath may be regarded as continuous respectively with the epithelial and muscular layers of the oviduct.

The *lateral oviducts* (oviducti laterales, *Odl*) are short tubes 150 μ in length, each with an ill-defined expansion, the calyx (*Clx;* most evident in small ovaries), at the ovarial junction. They unite medially to form the *common oviduct* (oviductus communis, *Odc*) which is about 450 μ long and arches dorsally to enter the anterior part of the uterus (Fig. 43A). When empty, the lateral tubes and the median tube are 50 μ in diameter, but they stretch during passage of eggs from ovary to uterus. The walls of the oviducts consist of a single-layered epithelium, lined with a chitinous intima and surrounded by circular muscles (Figs. 45A, D). The undistended epithelium is thickly massed in the calyces, thin and longitudinally folded in the lateral and common oviducts where the lumen is wide (9–45 μ; Fig. 45A), and thick and more or less cuboidal in the common oviduct where the lumen is narrow (7–15 μ) near the junction with the uterus (Fig. 45D). The cells are small, 1.5–6 μ high, with nuclei 1.5 by 3 μ in diameter. An intima is present in the common and lateral oviducts, indicating their ectodermal origin (cf. Dobzhansky, 1930); whether it is also present in the calyces has not been determined. The intima attains a thickness of 2.5 μ in the common oviduct and there contains vertical striations which are perhaps indicative of a highly elastic structure (Fig. 46). The muscles are a single layer of contiguous circular fibers with isolated nuclei (Fig. 45D); the thicker fibers (diameter 3 μ) near the uterus appear to be of the "tubular" type; those of the lateral oviducts are much thinner (1.5 μ). The common oviduct enters the uterus dorsally between the openings of the ventral seminal receptacle and the dorsal spermathecal and glandular ducts (Figs. 43A, 45D).

The *genital chamber* or *vagina* (*GC, Vag*) is an elongate muscular pouch, the larger anterior part of which has been termed the uterus (*Utrs*). Its external opening, between the gonopods (*Gon*) of the eighth (and ninth?) abdominal segment, is the

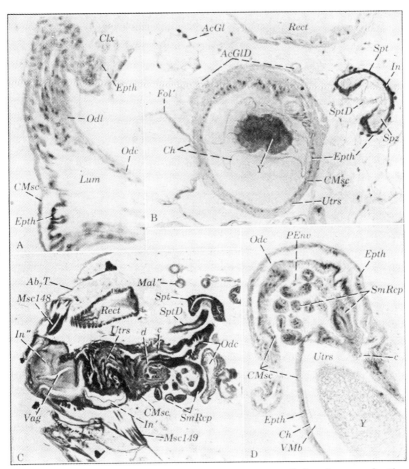

Fig. 45. Efferent ducts and accessory structures of female reproductive system. A: Longisection of lateral oviduct and adjoining common oviduct (C., F.L.G.; 400×). B: Portion of transection through fifth abdominal segment, showing radial sections of an accessory gland and a spermatheca and transection of uterus containing an egg; fertilized female (D., F.L.G.; 210×). C: Parasagittal section of posterior abdomen, showing radial section of a spermatheca and longisection of the undistended genital chamber; fertilized female (B., H.E.; 108×). D: Sagittal section of common oviduct, seminal receptacle, and anterior portion of uterus containing egg; fertilized female (D., H.E.; 160×). Ab_7T, seventh abdominal tergite; $AcGl$, accessory gland; $AcGlD$, duct of accessory gland; c, common oviduct opening into uterus; Ch, chorion; Clx, calyx; $CMsc$, circular muscle; d, seminal receptacle opening into uterus; $Epth$, epithelium; Fol', follicle around completed egg in ovary; In, intima; In', specialized vaginal intima; In'', vulvar cuticula; Lum, lumen; Mal'', posterior Malpighian tubule; $Mscl48$, muscle to gonopod; $Mscl49$, ventral muscle to uterus; Odc, common oviduct; Odl, lateral oviduct; $PEnv$, peritoneal envelope; $Rect$, rectum; $SmRcp$, seminal receptacle; Spt, spermatheca; $SptD$, spermathecal duct; Spz, sperm; $Utrs$, uterus; Vag, (posterior) vagina; VMb, vitelline membrane; Y, yolk of egg.

525

vulva (*Vul*), which serves as both a copulatory orifice and an exit for the eggs. When empty, the uterus is contracted, and its walls are thrown into folds (Fig. 45C). It can hold but one egg at a time and is then smoothly distended, 480 μ long and 200 μ in diameter (Figs. 43, 45B, D). Slender muscle fibers (*Mscl47*) extend posterolaterally from the uterus to abdominal tergite 8; others extend ventrally from the posterior vagina to the posterior border of sternite 7 (*Mscl44*) and ventrolaterally to acrosternite 8 (*Mscl49*, Figs. 26A, 45C). (These muscles are evidently the misnamed "uterine papillae" of Gleichauf, 1936.) Circular muscles, several layers (6–15 μ) deep when contracted, encircle the uterus but do not extend over the posterior vagina (Figs. 45B, D). The epithelium is of small cuboidal cells 3–12 μ high, flattened to 1.5 μ when stretched, and with the distal cytoplasm sometimes vertically striate (Fig. 45B). It is covered with a chitinous intima which in the uterus is thin (0.7 μ) and unspecialized. On the ventral wall of the posterior vagina the intima is 3.5–7 μ thick, composed of a thin granulose surface layer connected to the epithelium by thin vertical partitions that correspond more or less to the boundaries of the epithelial cells and appear as fibers in longisection (*In'*, Fig. 45C). The attachment of ventral muscles to the epithelium in this region and the absence of restraining circular muscles suggest that this peculiar structure may be due to mechanical stress during secretion of the intima. It may also enable greater expansion of the lumen when the egg is extruded. In the region of the vulva, where the vaginal wall forms five deep longitudinal folds, the intima becomes a very thick cuticula (up to 25 μ), with a dense external layer and a spongy inner zone, the former including a surface layer of exocuticula (*In"*, Fig. 45C).

The *seminal receptacle* (*SmRcp*) is a compactly coiled tube applied to the anterior end of the uterus, below the common oviduct (Fig. 43A). The tube is long (1.5–2.7 mm) and slender, with the proximal half constituting a duct 22 μ wide and the distal half consisting of a slightly dilated portion 28 μ wide and a narrow closed end 18 μ wide (Fig. 43B). It enters the uterus just ventrad of the mouth of the oviduct (Figs. 43, 45C, D). The wall of the receptacle consists of cuboidal and columnar epithelium, 3.5 μ thick in the distal portion, 7 μ proximally, lined by a thin intima and surrounded by a nucleated sheath 1–3 μ thick

(Fig. 46). The diameter of the lumen is 2.5–4.5 μ in the proximal duct (a) and 12–19 μ in the distal part (b). The coil as a whole is covered by a delicate, sparsely nucleated peritoneal envelope, 0.3 μ thick, which appears to lack muscle fibers except possibly near its attachment to the uterine muscles (Fig. 45D). Nerve fibers are abundant between the coils. In virgin females the receptacle contains a liquid, appearing as a reticular coagulum in

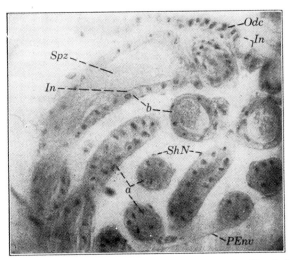

FIG. 46. Seminal receptacle. Transection of coil; fertilized female (D.; F.L.G.; 535×). a, proximal duct; b, distal portion; In, intima; Odc, common oviduct; $PEnv$, peritoneal envelope; ShN, sheath nucleus; Spz, sperm.

sections; after fertilization the lumen is filled with spermatozoa, arranged longitudinally with their heads toward the tip of the tube (Fig. 46).

The *spermathecae* (Spt) are a pair of mushroom-shaped organs, each a dark brown capsule 70 μ wide and 40 μ high, connected to the uterus by a slender duct 145 μ long and 20 μ in diameter. The capsules normally lie next to the rectum, one on each side and each surrounded by a small mass of fat (Fig. 43). In life they are sometimes visible through the dorsal body wall of the fifth or sixth abdominal segment. The ducts enter the dorsal wall of the uterus close together directly behind the mouth of the oviduct, on a low papillate elevation to which the accessory glands are also attached (Fig. 43A). Each capsule resembles an inverted double-walled bowl and is covered externally by a single layer of cuboidal epithelial cells 4–12 μ high, with large nuclei and homogeneous cytoplasm (Fig. 45B, C). The capsule is the sclerotized intima, 0.8 μ thick, secreted by these cells. In fertilized females the lumen is filled with a concentrically coiled mass of spermatozoa (Fig. 45B). The duct, which arises from the inside of the "bowl" (Fig. 45C), has a thin epithelium surrounded by a layer of about twenty-four slightly spiraled longitudinal muscle fibers; the lumen is narrow (10 μ) and is lined with a trachealike intima (0.8 μ)   The latter is ringed throughout with closely spaced ridges and near the capsule bears small downward-projecting cuticular spines.

Studies of the variability of the female organs have given special attention to the spermathecae. Sturtevant (1925–1926) records 1/330 and 14/84 individuals of two groups of *Drosophila melanogaster* as having three spermathecae instead of two, and also found partial doubling of a spermatheca in some cases. Dobzhansky (1924) gives data on differences in proportion of length to width of the spermathecal capsules, and Gleichauf (1936) notes that their position may be bilaterally symmetrical (normal, 80 per cent of observed individuals), the right or left spermatheca may be further forward than the other, or both may lie on either the right or the left side of the body. These variations in position may be conditioned by the contraction of the uterus after oviposition. The spermathecae have different forms in various species of Drosophilidae: see Sturtevant (1921, 1925–1926).

The *accessory glands* (*AcGl;* colleterial glands, parovaria) are a pair of small pyriform bodies 80 μ by 45 μ, lying behind the spermathecae, each connected to the uterus by a delicate duct 180 μ long and 11 μ in diameter (Fig. 43). Each gland is invested in adipose tissue which may adhere to that surrounding the neighboring spermatheca. The gland wall consists of a single layer of polygonal cells, each of which typically contains a large vacuole with a minute acidophilic granule toward the lumen; the nuclei are located basally between adjacent vacuoles (Figs. 45B, 47). The lumen contains a finely granular or homogeneous secretion in fixed sections and is probably lined by a very thin intima. The tapering duct consists of an extremely thin nucleated epithelium, lined with a chitinous intima irregularly ringed with sharp ridges that project into the lumen (Fig. 47A). The two ducts open into the uterus close together just behind the spermathecal ducts.

The *function* of the female reproductive system is to produce eggs and provide for their fertilization and deposition. The egg follicles formed near the distal end of the ovarioles descend within the epithelial sheath as they increase in size. At the base of the ovariole the follicle cells and nurse cells degenerate, and

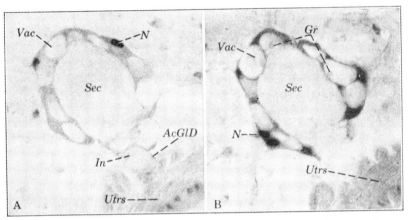

Fig. 47. Accessory gland of female (D., F.L.G.; 535×). A: Radial section, showing part of duct. B: Succeeding tangential section, showing granules in vacuoles. *AcGlD*, duct of accessory gland; *Gr*, granule; *In*, intima; *N*, nucleus of glandular epithelium; *Sec*, secretion; *Utrs*, uterus; *Vac*, vacuole.

the egg leaves the ovary, passes down the oviducts, and comes to rest in the uterus. The movement of the egg is produced by the action of the muscular network of the peritoneal sheath of the ovary, the contractile fibrillae in the epithelial sheath of the ovariole, and the muscles of the oviduct walls—all of which may show rhythmic contractions in saline solution.

The details of insemination are described by Nonidez (1920). In the uterus the egg lies with its micropyle near the opening of the seminal receptacle and its chorionic appendages still in the oviduct (Fig. 43B). During the preceding copulation, the sperm are deposited in the uterus and soon swim into the seminal receptacle, which is filled first, and into the spermathecae, their movement perhaps stimulated by secretions from the accessory glands. After an egg is in the uterus, sperm enter the space around its anterior end and penetrate through the micropyle.

There is evidence that sperm stored in the seminal receptacle are probably used first, those from the spermathecae later. No movement of the peritoneal sheath of the seminal receptacle has been detected in saline solution, but the tube itself undergoes slow, irregular transverse contractions; and the spermathecae sometime show spasmodic whipping movements, evidently due to the oblique longitudinal muscles of their ducts. Some of the contained spermatozoa are frequently observed to flicker and vibrate within the storage organs. Nonidez described the sperm as turning around and swimming out of the seminal receptacle to reach the egg. It is unknown whether their release is influenced by the containing organs or to what extent the exits of these organs are controlled by the surrounding muscles of the oviduct and uterus.

Anderson (1945), from studies of mutants in which accessory glands and spermathecae were partly or entirely absent, concludes that the spermathecae are necessary for high fertility, not only acting as sperm-storing organs but evidently being essential to the survival of sperm in the seminal receptacle, while the accessory glands have only minor or supplementary functions in this regard.

Maturation and a variable degree of embryonic development may occur while the egg is still in the uterus. Oviposition is effected by the muscles of the uterus and genital segments, which are arranged to enable peristalsis and the protrusion and retraction of the vulva (see Muscular System). A female may lay several hundred eggs.

According to Stolfi (1938), no fully formed eggs are present in the ovaries until 3 days after emergence from the puparium, although oöcytes appear before eclosion. The maximum number of completed eggs is present on the sixth to tenth days; on the twelfth to thirteenth day few or none remain, and the ovaries revert to the condition of the newly emerged fly. Indications are that thereafter the cycle is repeated, the time interval probably depending upon environmental factors. (For a physiological study and illustrations of ovarian growth in *Drosophila hydei*, see Child, 1942.)

For studies on the reproductive system of intersexes, see Dobzhansky and Bridges (1928) and Dobzhansky (1930). The reproductive organs of higher muscoidean flies are described in detail by Graham-Smith (1938).

LITERATURE CITED

ANDERSON, R. C. 1945. A study of the factors affecting fertility of lozenge females of *Drosophila melanogaster*. *Genetics* **30**:280–296.

BEGG, M., and L. HOGBEN. 1946. Chemoreceptivity of *Drosophila melanogaster*. *Proc. Roy. Soc.*, Ser. B (Biol. Sci.) **133**:1–19.

BERLESE, A. 1909. *Gli Insetti*. Vol. I. Milan.

CAJAL, S. R., and D. SÁNCHEZ. 1915. Contribución al conocimiento de los centros nerviosos de los insectos. *Trabajos Lab. Invest. Biol. Univ. Madrid* **16**:109–139.

CHILD, C. M. 1942. Differential dye reduction in ovaries of *Drosophila hydei*. *Physiol. Zoöl.* **15**:13–29.

COCHRANE, F. 1937. An histological analysis of eye pigment development in *Drosophila pseudo-obscura*. *Proc. Royal Soc. Edinburgh* **57**:385–399.

CRAMPTON, G. C. 1942. The external anatomy of the Diptera. *In* The Diptera or true flies of Connecticut (1st fasc.). *Connecticut State Geol. and Nat. Hist. Surv. Bull.* **64**:10–165.

DARROW, M. A. 1940. A simple staining method for histology and cytology. *Stain Tech.* **15**:67–68.

DAY, M. F. 1940. Neurosecretory cells in the ganglia of Lepidoptera. *Nature* **145**:264.

DAY, M. F. 1943a. The function of the corpus allatum in muscoid Diptera. *Biol. Bull.* **84**:127–140.

DAY, M. F. 1943b. The homologies of the ring gland of Diptera Brachycera. *Ann. Entomol. Soc. Am.* **36**:1–10.

DEBAISIEUX, P. 1938. Organes scolopidiaux des pattes d'insectes. II. *Cellule* **47**:79–202.

DOBZHANSKY, TH. 1924. Ueber den Bau des Geschlechtsapparats einiger Mutanten von *Drosophila melanogaster*. *Z. induktive Abst. u. Vererbungsl.* **34**:245–248.

DOBZHANSKY, TH. 1930. Studies on the intersexes and supersexes in *Drosophila melanogaster*. [Russian with English summary.] *Izv. Bur. Genet. (Leningr.)* **8**:91–158.

DOBZHANSKY, TH., and C. B. BRIDGES. 1928. The reproductive system of triploid intersexes in *Drosophila melanogaster*. *Am. Naturalist* **62**:425–434.

EASTHAM, L. 1925. Peristalsis in the Malpighian tubules of Diptera. Preliminary account: with a note on the elimination of calcium carbonate from the Malpighian tubules of *Drosophila funebris*. *Quart. J. Microscop. Sci.* **69**:385–398.

EIDMANN, H. 1924. Untersuchungen über Wachstum und Häutung der Insekten. *Z. Morphol. Ökol. Tiere* **2**:567–610.

ENGEL, E. O. 1924. Das Rektum der Dipteren in morphologischer und histologischer Hinsicht. *Z. wiss. Zoöl.* **122**:503–533.

EVANS, A. C. 1935. Some notes on the biology and physiology of the sheep blowfly, *Lucilia sericata* Meig. *Bull. Entomol. Research* **26**:115–122.

FRAENKEL, G. 1935. Observations and experiments on the blow-fly (*Calliphora erythrocephala*) during the first day after emergence. *Proc. Zoöl. Soc. London* **1935**:893–904.

FRAENKEL, G. 1939. The evagination of the head in the pupae of cyclorraphous flies. *Proc. Roy. Entomol. Soc. London*, A, **13**:137–139.

FREUDENSTEIN, K. 1928. Das Herz und das Circulationssystem der Honigbiene. *Z. wiss. Zoöl.* **132**:404–475.

GLEICHAUF, R. 1936. Anatomie und Variabilität des Geschlechtsapparates von *Drosophila melanogaster* (Meigen). *Z. wiss. Zoöl.* **148**:1–66.

GOTTSCHEWSKI, G. 1937. Künstliche Befruchtung bei Drosophila. *Naturwissenschaften* **25**:650.

GRAHAM-SMITH, G. S. 1914. *Flies in Relation to Disease.* Cambridge.

GRAHAM-SMITH, G. S. 1930. Further observations on the anatomy and functions of the proboscis of the blow-fly. *Parasitology* **22**:47–115.

GRAHAM-SMITH, G. S. 1934. The alimentary canal of *Calliphora erythrocephala* L., with special reference to its musculature, and to the proventriculus, rectal valve, and rectal papillae. *Parasitology* **26**:176–248.

GRAHAM-SMITH, G. S. 1938. The generative organs of the blow-fly *Calliphora erythrocephala* L., with special reference to their musculature and movements. *Parasitology* **30**:441–476.

GUYÉNOT, E., and A. NAVILLE. 1929. Les chromosomes et la réduction chromatique chez *Drosophila melanogaster* (cinèses somatiques, spermatogenèse, ovogenèse). *Cellule* **39**:27–84.

GUYÉNOT, E., and A. NAVILLE. 1933. Les premières phases de l'ovogenèse de *Drosophila melanogaster*. *Cellule* **42**:214–230.

HAGMANN, L. E. 1940. A method for injecting insect tracheae permanently. *Stain Tech.* **15**:115–118.

HANSTRÖM, B. 1928. *Vergleichende Anatomie des Nervensystems der wirbellosen Tiere unter Berücksichtigung seiner Funktion.* Berlin.

HANSTRÖM, B. 1939. *Hormones in Invertebrates.* Oxford.

HENRY, L. M. 1947–1948. The nervous system and the segmentation of the head in the Annulata. *Microentomology* **12**:65–110, **13**:1–48.

HERTWECK, H. 1931. Anatomie und Variabilität des Nervensystems und der Sinnesorgane von *Drosophila melanogaster* (Meigen). *Z. wiss. Zoöl.* **139**:559–663.

HEWITT, C. G. 1914. *The House-fly.* Cambridge.

HSÜ, F. 1938. Étude cytologique et comparée sur les sensilla des insectes. *Cellule* **47**:7–60.

HUETTNER, A. 1930. The spermatogenesis of *Drosophila melanogaster*. *Z. Zellforsch.* **11**:615–637.

IMMS, A. D. 1934. *A General Textbook of Entomology.* 3d ed. London and New York.

JOHANNSEN, O. A. 1924. Eye structure in normal and eye-mutant Drosophilas. *J. Morphol. and Physiol.* **39**:337–349.

KEUCHENIUS, P. E. 1913. The structure of the internal genitalia of some male Diptera. *Z. wiss. Zoöl.* **105**:501–536.

KRAFKA, J., JR. 1924. Development of the compound eye of *Drosophila melanogaster* and its bar-eyed mutants. *Biol. Bull.* **47**:143–148.

LOWNE, B. T. 1890–1895. *The Anatomy, Physiology, Morphology and Development of the Blow Fly (Calliphora erythrocephala).* 2 Vols. London.

MILLER, A. 1941. Position of adult testes in *Drosophila melanogaster* Meigen. *Proc. Natl. Acad. Sci.* **27**:35–41.

MORRISON, F. O. 1941. A study of the male genitalia in calyptrate Diptera, based on the genus Gonia Meigen (Diptera: Tachinidae). *Can. J. Research* **19**(D):1–21.

NONIDEZ, J. F. 1920. The internal phenomenon of reproduction in Drosophila. *Biol. Bull.* **39**:207–230.

PAINTER, T. S., and E. C. REINDORP. 1939. Endomitosis in the nurse cells of the ovary of *Drosophila melanogaster*. *Chromosoma* **1**:276–283.

PÉREZ, C. 1910. Recherches histologiques sur la métamorphose des Muscides. *Arch. zool. exptl. et gén.*, ser. 5, **4**:1–274.

PFEIFFER, I. W. 1939. Experimental study of the function of the corpora allata in the grasshopper, *Melanoplus differentialis*. *J. Exptl. Zoöl.* **82**:439–461.

PFEIFFER, I. W. 1940. Further studies on the function of the corpora allata in relation to the ovaries and oviducts of *Melanoplus differentialis*. *Anat. Record* **78**, suppl.: 39–40.

POWER, M. E. 1943. The brain of *Drosophila melanogaster*. *J. Morphol.* **72**:517–559.

RICHARDS, A. G., JR. 1944. The structure of living insect nerves and nerve sheaths as deduced from the optical properties. *J. N. Y. Entomol. Soc.* **52**:285–310.

RICHARDS, A. G., JR., and T. F. ANDERSON. 1942a. Electron micrographs of insect tracheae. *J. N. Y. Entomol. Soc.* **50**:147–167.

RICHARDS, A. G., JR., and T. F. ANDERSON. 1942b. Further electron microscope studies on arthropod tracheae. *J. N. Y. Entomol. Soc.* **50**:245–247.

RICHARDS, A. G., JR., and T. F. ANDERSON. 1942c. Electron microscope studies of insect cuticle, with a discussion of the application of electron optics to this problem. *J. Morphol.* **71**:135–183.

RICHARDS, M. H., and E. Y. FURROW. 1925. The eye and optic tract in normal and "eyeless" Drosophila. *Biol. Bull.* **48**:243–258.

RITTER, W. 1911. The flying apparatus of the blow-fly. *Smithsonian Misc. Coll.* **56**:1–76.

ROBERTSON, C. W. 1936. The metamorphosis of *Drosophila melanogaster*, including an accurately timed account of the principal morphological changes. *J. Morphol.* **59**:351–400.

ROSENBLAD, L. E. 1941. Description of ejaculatory sac diverticula in certain Drosophilinae. *Am. Naturalist* **75**:285–288.

SCHARRER, B. C. J. 1939. The differentiation between neuroglia and connective tissue in the cockroach (*Periplaneta americana*). *J. Comp. Neurol.* **70**:77–88.

SCHARRER, B. 1941a. Neurosecretion. II. Neurosecretory cells in the central nervous system of cockroaches. *J. Comp. Neurol.* **74**:93–108.

SCHARRER, B. 1941b. Endocrines in invertebrates. *Physiol. Rev.* **21**:383–409.

SCHARRER, B. 1943. The influence of the corpora allata on egg development in an orthopteran (*Leucophaea maderae*). *Anat. Record* **87**:471.

SCHARRER, B., and E. HADORN. 1938. The structure of the ring-gland (corpus allatum) in normal and lethal larvae of *Drosophila melanogaster*. *Proc. Natl. Acad. Sci.* **24**:236–242.

SCHRÖDER, C. 1928. *Handbuch der Entomologie*. Jena.

SNODGRASS, R. E. 1935. *Principles of Insect Morphology*. New York and London.

SNODGRASS, R. E. 1944. The feeding apparatus of biting and sucking insects affecting man and animals. *Smithsonian Misc. Coll.* **104**, 7:1–113.

STERN, C. 1941. The growth of testes in Drosophila. I. The relation between vas deferens and testis within various species. II. The nature of interspecific differences. *J. Exptl. Zoöl.* **87**:113–158, 159–180.

STERN, C., and E. HADORN. 1939. The relation between the color of testes and vasa efferentia in Drosophila. *Genetics* **24**:162–179.

STOLFI, J. E. 1938. *Ovarian Development in Drosophila melanogaster.* Thesis for degree of M.S., New York University, New York.

STRASBURGER, E. 1935. *Drosophila melanogaster Meig. Eine Einführung in den Bau und die Entwicklung.* Berlin.

STRASBURGER, M. 1932. Bau, Funkton, und Variabilität des Darmtractus von *Drosophila melanogaster* Meigen. *Z. wiss Zoöl.* **140**:539–649.

STURTEVANT, A. H. 1921. The North American species of Drosophila. *Carnegie Inst. Wash. Pub.* 301:1–150.

STURTEVANT, A. H. 1925–1926. The seminal receptacles and accessory glands of the Diptera with special reference to the Acalypterae. *J. N. Y. Entomol. Soc.* **33**:195–215; **34**:1–21.

THOMSEN, E. 1938. Ueber den Kreislauf im Flügel der Musciden, mit besonderer Berücksichtigung der akzessorische pulsierenden Organe. *Z. Morphol. Ökol. Tiere* **34**:416–438.

THOMSEN, E. 1940. Relation between corpus alatum and ovaries in adult flies (Muscidae). *Nature* **145**:28.

THOMSEN, E. 1941. Ringdrüse und Corpus allatum bei Musciden. *Naturwissenschaften* **29**:605–606.

VOGT, M. 1940a. Die Förderung der Eireifung innerhalb heteroplastisch transplantierter Ovarien von Drosophila durch die gleichzeitige Implantation der arteigenen Ringdrüse. *Biol. Zentr.* **60**:479–484.

VOGT, M. 1940b. Zur Ursache der unterschiedlichen gonadotropen Wirkung der Ringdrüse von *Drosophila funebris* und *Drosophila melanogaster.* *Arch. Entwicklungs-mech.* **140**:525–546.

VOGT, M. 1941a. Anatomie der pupalen Drosophila-Ringdrüse und ihre mutmassliche Bedeutung als imaginales Metamorphosezentrum. *Biol. Zentr.* **61**:148–158.

VOGT, M. 1941b. Weiterer Beitrag zur Ursache der unterschiedlichen gonadotropen Wirkung der Ringdrüse von *Drosophila funebris* und *Drosophila melanogaster.* *Naturwissenschaften* **29**:80–81.

VOGT, M. 1941c. Bemerkung zum Corpus allatum von Drosophila. *Naturwissenschaften* **29**:725–726.

VOGT, M. 1942. Ein drittes Organ in der larvalen Ringdrüse von Drosophila. *Naturwissenschaften* **30**:66–67.

WADDINGTON, C. H., and R. W. PILKINGTON. 1943. The structure and development of four mutant eyes in Drosophila. *J. Genetics* **45**:44–50.

WEBER, H. 1933. *Lehrbuch der Entomologie.* Jena.

WIGGLESWORTH, V. B. 1939. *The Principles of Insect Physiology.* London.

WILLIAMS, C. M., and M. V. WILLIAMS. 1943. The flight muscles of *Drosophila repleta.* *J. Morphol.* **72**:589–599.

ZILCH, A. 1936. Zur Frage des Flimmerepithels bei Arthropoden. *Z. wiss. Zoöl.* **148**:89–132.

7

Collection and Laboratory Culture

WARREN P. SPENCER*

INTRODUCTION

Although Drosophila is best known in its role as a genetic tool for the solution of problems dealing with the mechanism of heredity, it has of recent years become useful in the approach to the broader problems of organic evolution through cytogenetic studies of wild populations of various species. Furthermore, it is becoming an increasingly popular form in the attack upon problems of general physiology and animal behavior. In zoölogical teaching it is an ideal animal for the demonstration of many biological principles. It has been used for demonstrating the life cycle of a holometabolous insect, various tropisms, nutrition studies, and (with its parasite, Eucoila sp.) the relations and life cycles of insect parasitoids.

Though Drosophila will continue to be known chiefly as an organism well adapted to the study of many problems of theoretical biology, its economic role is by no means negligible. Boyce (1928), N. A. Patterson (1935), Bliss and Broadbent (1935), and Broadbent and Bliss (1936) have used *Drosophila melanogaster* as an experimental animal in the study of several poisons, especially hydrocyanic acid as insecticide. Ark and Thomas (1936) have shown that adults of *D. melanogaster* may harbor *Bacillus amylovorus* (the fire-blight organism of hops) for 8 days. When larvae were fed on medium containing this organism, it persisted through the pupal to the adult stage. Kikkawa and Peng (1938) point out, "Judging from their habits, *D. busckii* and *D. repleta* probably take an even more important role than *D. melanogaster*

* Biology Department, College of Wooster.

535

in this respect." *Drosophila funebris*, a species attracted to human and animal excrement, should be added to this list. The author (Spencer, unpublished) carried out an experiment on the possible role of *D. melanogaster* as a carrier of microörganisms. It was found that Drosophila, confined in a cage with a culture of small flagellates in one container and sterile medium in another container, soon inoculated the medium by transfer of flagellates, while a control cage showed no such transfer. As Drosophila frequently feed on fruits and other uncooked foods of man, they could and undoubtedly do from time to time play the role traditionally assigned to *Musca domestica* as disease carrier. Dove (1937) reports one case of human intestinal myiasis with *D. funebris* larvae as the infecting source. Textbooks of human parasitology list Drosophila species as agents in intestinal myiasis and as vectors of disease-producing bacteria.

Kamizawa (1936) has reported extensive damage done by *D. suzukii* to cherries in Japan. Seventy to eighty per cent of the fruits may be attacked; the flies may lay eggs on sound fruits, beginning in May. De Coursey (1925) points out that *D. melanogaster* may become a pest in groceries and homes and suggests a practical control in the form of a trap baited with fermenting banana. At the Ohio Agricultural Experiment Station *D. melanogaster* prove a pest in breeding on apple medium used in rearing the Oriental fruit moth. Sergent and Rougebief (1926) have called attention to the double role played by Drosophila in spreading yeast and keeping down mold growth in wineries. Drosophila, and particularly mutants such as vestigial, are used in feeding tropical fish. While adequate quantitative data are as yet lacking, the enormous numbers which may be trapped in very limited areas would indicate that Drosophila play a not inconsiderable role in the balance of nature. This brief résumé of their economic importance is by no means complete, but it does serve to show that the genus has more than an academic interest. However, in the final analysis the chief importance of Drosophila will continue to be its use as a tool in the investigation of theoretical problems in pure science, the sure and only foundation for advance in applied fields.

Although the successful collection, handling, and culture of Drosophila are largely dependent upon a knowledge of their ecology, physiology, and behavior, it has seemed advisable to devote a separate chapter to technical methods. Paradoxically many of the problems involved have been obscured by the ease and uniform success with which *Drosophila melanogaster* have for many years been reared in laboratories the world over. The subject of the culture of *D. melanogaster* has been approached through a considerable experience with many other species of Drosophila. In the rearing of these, inherent technical problems have from time to time been more apparent, and it has become necessary to work out their solutions. These may then be applied to advantage to *melanogaster*.

In Muller's (1939) bibliography of Drosophila, of the 2965 titles listed only one deals with collecting (Dobzhansky, 1936). Sturtevant has published a few notes on collecting methods in his taxonomic monograph (Sturtevant, 1921). In recent years a number of papers have appeared dealing with the analysis of samples of wild populations of several different species of Drosophila. Yet there is a surprising lack of information on methods used and conditions under which the samples were captured. Perhaps there is little to be learned from the haphazard catching of a few flies, but the author feels that there is no better introduction to the ecology, physiology, and evolution of insects than their intelligent collection at different seasons and under a variety of conditions. Indeed, the problems of population mechanics can best be understood through actual field contact with the organism in question. As an introduction to field work, the comprehensive monograph of Uvarov (1931) on insects and climate cannot be recommended too highly.

While literature on Drosophila collecting is almost non-existent, the author has been fortunate in having the opportunity to discuss at length methods and problems involved with A. H. Sturtevant, Th. Dobzhansky, and J. T. Patterson, all of whom have done extensive collecting of this genus. He is particularly indebted to Harrison Stalker, with whom he has worked in the field many times, and to whom much of the credit is due for data and methods here presented.

It should be understood that there is no "best" method of collecting. The technique used will depend upon the goal set, that is, maximum number of flies, maximum number of species, maximum number of one particular species, small sample of one species, adequate sample of one species from a given population in space or time or both, and upon such considerations as expense and time available for the work. In general it may be said that successful collecting consists of securing a maximum of flies with a minimum expenditure of time and materials, and in having the sample secured adequate in respect to the problem involved. This goal will not often be attained; but, if the collector is able to analyze the factors which have intervened, his work is not a failure.

BAIT

Several investigators have carried on experiments on reactions of Drosophila to various odors and baits. These studies have led rather uniformly to the conclusion that *D. melanogaster* is attracted in greater numbers to fermenting banana than to any other of a large variety of natural and synthetic substances, including fruits, oils, alcohols, and acetic and other acids. Sturtevant (1921) suggests that for trap collecting it is advisable to use many kinds of fruit. He lists banana, pineapple, tomato, and peach as especially satisfactory. However, Dobzhansky, Patterson, Sturtevant, and Stalker find fermenting banana the best all-purpose bait. DeCoursey (1925), after trying a series of volatile oils, alcohol, and organic acids, finds decaying fruits more attractive. He recommends fruits artificially fermented by mixing in yeast. His conclusion seems valid that the superior quality of banana is not so much due to its inherent flavor or attractive quality as to the fact that fermenting banana mash retains its attractive quality longer than do other fruits. As he points out, many other fruits, such as apples and grapes, ferment very quickly but rapidly lose their attractive power and become dry. The citrus fruits, on the other hand, have too large a water content to make a wholly satisfactory bait. Dobzhansky (1936) states:

Fermenting banana mash is most satisfactory as bait. Ripe bananas are mashed with the aid of a spoon or a fork; some drops of fresh yeast solution are added, and the bottle (trap containing this bait) is left standing for about 24 hours before use; if dry yeast is used this time is considerably lengthened. The bait remains good for at least 4 or 5 days after first used.

We use ripe or over-ripe bananas mashed into a smooth pulp. To this is added the skins after running them through a meat grinder.

The skins add bulk and help to serve as a base for the retention of the liquid fraction. A suspension of baker's yeast in water (about $\frac{1}{4}$ lb to 4 gal of bait) is thoroughly mixed in. The yeast may be added either at the time the traps are set out or not longer than 36 hours before, depending upon when the traps will first be visited for taking flies (see trapping procedure below). When this fermenting banana mash is too wet, cleansing tissue or cellucotton may be added at the time it is placed in the trap. The paper adds bulk and absorbs and holds excess moisture. It also provides footing for flies in the trap.

Bananas are generally available in markets throughout the territories in which collecting is likely. Because of their superior quality as an all-purpose bait the writer strongly recommends their use in preference to inferior materials which may be somewhat cheaper. However, where bananas are not available or their price is prohibitive, other media may be substituted with good results.

Fermenting cornmeal-molasses-agar has been found fairly satisfactory. Its drawing power is greatly increased if young Drosophila larvae are working through it. Flies seem to be attracted particularly by the release of fermentation products, and the presence of fly larvae as they tunnel through the medium serves the double purpose of accelerating yeast growth and increasing the surface from which attractive odors diffuse. However, the use of living larvae in bait is not generally recommended unless care is taken to recover all traps. In addition to fruits listed above, pears, plums, cantaloupe, and grapes may be suggested. A collector can usually find bait: rotting barrel cactus in the desert, ripe fungi in the woods, fresh garbage, or wind-fall fruit will serve the purpose.

It is often possible to find the food store on which local Drosophila populations are living and in some cases to utilize this material. Fungi may be mashed up and used in traps. Generally speaking, however, fungi very rapidly lose their attractive power. A variety of baits will probably result in the taking of more species in a given region. Certainly baits have a differential attraction for various Drosophila species. Sturtevant contributes this case (unpublished communication). On March 8, 1937, a trap consisting of a quart jar half full of fermenting bananas and banana skins was placed close to a heap of decaying citrus, which was yielding more than 99 per cent *D. hydei* by net and other methods of collecting. Table 1 gives the counts of flies taken

TABLE 1

FLIES TAKEN IN ONE FERMENTING BANANA TRAP FROM AN ENVIRONMENT WHERE THE POPULATION WAS OVER 99 PER CENT OF THE SPECIES
Drosophila hydei

Species	Males	Females	Total	Per Cent of Total
hydei	0	1	1	0.7
busckii	1	0	1	0.7
hydei-like *	5	6	11	7.5
immigrans	8	11	19	12.9
melanogaster	59	46	105	71.4
pseudoobscura	1	0	1	0.7
simulans	5	4	9	6.1

* Undescribed species.

from this trap on March 10. The bait was in this case at a particular stage (rather fresh) which attracted other species, particularly *melanogaster* and *simulans,* much more effectively than *hydei.*. Had an estimate of the Drosophila population been based entirely on this collection record, a very erroneous idea of species

TABLE 2

COLLECTIONS FROM BANANA AND FUNGUS TRAPS SET IN TOWN AND IN THE WOODS IN PARALLEL AND INDICATING DIFFERENTIAL EFFECTIVENESS OF BAITS (September 1940, Wooster, Ohio)

Species	Traps Set in Town		Traps Set in the Woods	
	Banana	Fungus	Banana	Fungus
affinis	0	0	3	0
algonquin	0	0	10	0
busckii	0	52	0	0
funebris	6	14	0	0
immigrans	1	0	9	0
mahican	0	0	1	0
melanica	0	0	15	0
melanica-like *	0	0	18	0
melanogaster	830	1	13	0
putrida	0	0	0	27
robusta	0	0	97	2

* *nigromelanica.*

distribution at this point would have resulted. As a matter of fact other banana traps, probably in a different stage of fermentation, set in this vicinity, gave a truer picture of the population.

Table 2 gives the results of some parallel fungus and banana traps set in Ohio in September, 1940. The data clearly indicate that species differ in their reactions to these baits.

TRAPS

Sturtevant (1921) and Dobzhansky (1936) both suggest a bottle such as an ordinary fly culture bottle for a trap. Bait is placed in this, a string is tied around the neck, and the bottle is tied to a bush or branch of a tree. This is a simple, inexpensive trap, convenient to use and fairly satisfactory. In the dry, semi-arid southwest such traps may be exposed for a few hours in the late afternoon, and the collection made during this time. In the more humid east better collections are secured when traps are exposed over a period of several days. To avoid flooding from rain Sturtevant ties the trap in a horizontal position. The author and Harrison Stalker worked three summers on improving methods of trapping. Through a considerable amount of trial and error experimentation, accompanied by a study of reactions of flies to baits, traps, and environmental factors, we worked out the method of trapping presented here. However, under special conditions and for certain purposes other types of traps are better. Some of these will be discussed later.

It is taken for granted that in most collecting the object is to cover a considerable territory and secure a large sample of flies in a limited time. The time of the collector is assumed to be more valuable than a few extra dollars put into traps and bait in the course of a week's collecting. The trap is constructed by tying a 24-in. length of string through a hole made in the lip of an ordinary paraffined paper drinking cup (a cup 2½ in. in diameter and 2½ in. deep is satisfactory). These come packed in long cartons, 100 cups per carton, and 25 cartons in a packing box. The strings may be tied, and the cups replaced in their cartons. A package of 2500 traps weighs 25 lb and occupies less than 2½ cu ft (an ample supply for an extended collecting trip). If collecting is being done in a rainy season, the hole may be made in the bottom of the cup, and the string looped through so that the cup can be suspended horizontally. (See Fig. 1.) The amount of bait to be added depends on the length of time the trap will be exposed before the collection is made and on climatic conditions. If a trap is to be left out for 3 or 4 days, it should be filled a third to half full of bait (in a dry climate the trap should not be left that long). If collection is to be made the day after setting, the bait should be yeasted a day or so before putting out; but if

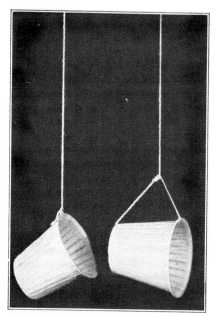

FIG. 1.   Paper-cup traps.

collection is to be made 3 or 4 days after setting the traps, the yeast may be added and thoroughly mixed into the bait shortly before the traps are set. Traps are hung on the lower branches of trees or on bushes. The trap may be a few inches to a few feet from the ground but should hang free and preferably in a place easily approached without disturbing it.

Flies are removed from the trap by means of a "collector." (See Fig. 2.) This is made from a heavy pasteboard cylinder, 5 in. or more in diameter. Such cylinders are used as a core for shipping strips of linoleum or rugs and may be secured from drygoods houses. A section of this cylinder 24 in. long is cut, and over one end a piece of cheesecloth is fastened by heavy rubber bands or shellac. The cheesecloth is of a mesh which will not allow Drosophila to pass through and is single thickness to give maximum illumination. A large cotton plug, covered with cheesecloth, is fashioned to fit snugly into the open end of the cylinder.

In collecting, the cylinder is held in the left hand at an angle of about 45 degrees, with the cheesecloth end up and the other end brought close to the point where the trap string is tied to the tree or bush. With the right hand the trap is then grasped suddenly and carried into the open end of the cylinder and as far up toward the cheesecloth end as the string will allow it to go. The trap is then tapped repeatedly against the inside of the cylinder to dislodge those flies which are still in or on it. When the flies are disturbed, they fly upward and toward the light and collect at the cheesecloth end of the cylinder. They are so strongly phototropic that, except under unusual circumstances mentioned below, they stay on or near the cheesecloth and do not attempt to traverse the dark cylinder and fly out the open end. It is not

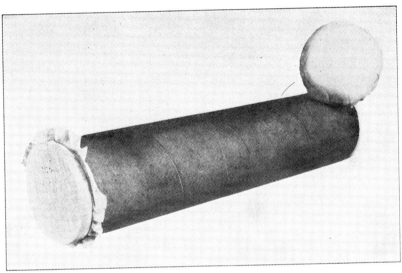

Fig. 2. "Collector" used in removing flies from traps.

necessary to close the "collector" with the cotton plug. The worker may go from trap to trap emptying them of flies. In the course of this procedure the cylinder should be held horizontal or, better, with the cheesecloth end slightly elevated. The jarring incident to walking and carrying the cylinder keeps those few flies which start to wander down toward the open end from getting far before they turn back toward the light. A few Drosophila do fly out, but when the collecting is good their number is negligible in comparison to those which are captured.

When the round of perhaps thirty to fifty traps has been made, the cotton plug is placed in the open end and slowly pushed along the tube until there is a space of an inch or so between it and the cheesecloth. The tube is then set down, cheesecloth end up; a piece of cleansing tissue with a little ether on it is placed on the cheesecloth, and the end of the cylinder is covered with a disc of cardboard. The flies are quickly etherized, the tube is inverted, the cotton plug is removed, and then etherized flies are poured directly into transporting containers, which may be vials or bottles. Collections made in this way are almost pure samples of Drosophila.

Beetles which are frequently present in the traps are negatively phototropic and soon crawl down and out the lower dark end of the tube. Large flies are much less strongly phototropic and are not held for long at the closed end of the tube. The same is true of most other large insects. Those which remain with the Drosophila can easily be separated out when the material is etherized. Where the collecting is good enough so that on an average four or five flies are found in a trap, this method is superior to any we have tried. Where the collecting is so poor that three or fewer flies are found per trap, it is inferior to some other procedures.

Repeated collections may be made from these paper cup traps and fresh bait added. However, they are ordinarily discarded after one setting. They have the advantage of being very simple in construction and inexpensive (about 300 for $1.00 in large quantities). They allow ready diffusion of attractive odors. Flies enter them more readily than glass traps, partially because the paper gives a better footing. Many times the difference in reactions of flies on lighting on paper and on glass has been observed. To test the efficiency of bottle and cup traps, collections were made in woods at six stations from 30 paper cup traps and 30 bottle traps set parallel. The cup traps yielded 774 flies and the bottle traps 316, or a ratio of better than 2:1 in favor of the cups. Flies are more easily removed from the paper traps, for the paper allows a better surface on which the fly may run or take off in flight. In cup traps there is the disadvantage of the bait drying up more rapidly than when contained in glass. Where it is necessary to leave traps for more than 4 or 5 days before collecting from them, some design other than paper cups is better.

In a sense the devices described above are not traps at all, as flies may leave them readily. Their successful use is based upon a knowledge of the feeding habits of Drosophila in relation to the diurnal temperature rhythm discussed below. It may seem preferable to use some device that would actually trap the flies, and to this end Stalker has devised a "retainer" trap. This consists of an ordinary glass culture bottle into the neck of which is fitted a cone of fine copper netting with an opening about a quarter of an inch in diameter at its apex. Such a trap may prove of use where the investigator can return to pick up his collection only at some time of the day or night when flies normally have left the traps because of heat or cold. Ordinarily, however, when collection is made at the right time of day, an open trap will on the average yield more flies than a "retainer" trap. Furthermore flies remaining in these traps in the heat of the day are likely to be in bad condition.

The author has recently done some work on a trap designed to supply food as bait and a natural environment within the trap into which the flies will retreat during the diurnal temperature peaks. Either a tin can or a pasteboard carton about 5 in. in diameter is loosely filled with a mixture of dead leaves and humus and banana yeast bait. This is set on the ground in the shade and may be covered with a coarse wire netting to keep out small mammals. Such a trap, if left for several days, collects large numbers of flies and has the advantage of retaining them long beyond the critical high or low temperatures at which flies leave cup or bottle traps. The flies can easily be removed at any time of day by placing the "collector" over the mouth of the trap and shaking it repeatedly. On the Mesa Verde, Colorado, the author found this type of trap superior to cup traps, as the Drosophila population was not abundant and too few flies were visiting cup traps to give a profitable return on time spent. As these traps are larger, they last longer without rebaiting them, and they are effective even after considerable rain.

J. T. Patterson and his colleagues at the University of Texas have done very extensive Drosophila collecting. Dr. Patterson has kindly furnished us with a description of a trap and collecting method which they have found highly satisfactory. (See Fig. 3.) The trap consists of a 50-lb lard can at the bottom of which a large mass of bait (fermenting banana) is placed. Two sticks are laid across the open end; the lid is supported by these sticks and weighted with a stone. Flies are removed by suddenly lifting the lid and sweeping rapidly with a net back and forth above the opened trap can. The net is reworked so that the lower end is in the form of a tube, stiffened with cellulose dissolved in acetone. Flies are transferred directly to the food vial by means of a funnel fitted with a rubber stopper. Such a trap is collected from repeatedly; the bait is discarded and renewed before a new generation of flies emerges from eggs deposited on the bait mass. This trap is of particular value in studying populations over a long period; it has the distinct advantage of being so large that drying out of bait is not a problem. On the other hand, a large number of small traps make possible a wider sampling of material from a given region.

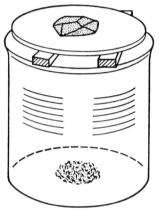

Fig. 3. Large Drosophila trap devised by J. T. Patterson.

Finally it is frequently possible to make collections of Drosophila by direct capture without recourse to setting traps. Sturtevant has collected many of the Drosophilinae by sweeping. This method yields certain species not often taken in baited traps. The capture of flies with shell vials, one specimen at a time, when they are found on bleeding trees, fungi, or other food provides small samples. Where flies are scarce and only a few coming to cup traps, they may be taken in this way from the cups. Cellophane sacks are also useful for this type of collecting. A few of these sacks may be carried conveniently and used when interesting specimens are seen. The open end of the sack is brought down over the fly; many specimens may be caught in one sack by shaking them into the end of the sack and twisting this to form a tiny compartment from which flies cannot escape. When collecting time is limited, bait may be spread out in a thin layer on moss or logs, and flies collected with vial or cellophane sack. This method allows for maximum attraction from the amount of bait used. For example, three large 2-qt-jar banana mash traps were set for 1 hour at mid-day on top of Clingmans Dome in the Great Smokies at an elevation of 6642 ft and a temperature of 18°C. These yielded 12 specimens of *Drosophila mahican*. In less time one collector, using the equivalent of the banana mash from one trap, spreading it on moss and, collecting with cellophane sack, secured 25 flies. Wind-fall fruits, pears, apples, peaches, etc., after beginning to rot often harbor Drosophila colonies. Flies can be collected in numbers by holding a vial or sack over openings to cavities in such decaying fruit. The bait and trap used will depend upon circumstances and should be adapted to the particular conditions and problems confronting the field worker.

TIME AND PLACE FOR TRAPPING

We shall now consider the problems of trapping which are most directly concerned with what Uvarov has referred to as "ecoclimate" and "microclimate."

Drosophila melanogaster in temperate climates is always an introduced species, as such generally overwinters indoors, and develops the largest populations in the autumn, when decaying fruits are at a maximum. Consequently from the standpoint of numbers the best collecting comes at this time of year. Owing to the short life cycle, and in spite of a lower fecundity than a number of other species, it shows less lag in developing peak populations when food stores are suddenly increased than any other species, with the possible exception of *simulans*. The investigator should understand that a population of this species under favorable circumstances may build up in a few weeks from a single pair of flies, and that many local populations may be expected to contain less genetic variability than slower-breeding forms. As the species feeds principally on decaying fruit, populations build up where this material is to be found. In temperate climates this species overwinters where there are stores of fruit, as in groceries and fruit cellars. It does not live as long in the adult stage as larger forms, and probably the winter populations are generally carried through as actively breeding colonies rather than as adults. When one collects this species, particularly in temperate climates, he should bear in mind the breeding structure of the populations: mostly micropopulations tending to homogeneity throughout the winter, spring, and early summer; then these micropopulations expanding and overlapping as decaying fruit becomes more abundant and food stores less localized; even with this panmixia little local and highly homogeneous populations rapidly developing in less than 2 weeks on small isolated food stores. Thus a rotting apple or pear lying in a grassy orchard by itself forms a food store of sufficient size for a population of 200 or 300 *melanogaster*, possibly the offspring of one female. With other species, such a tiny isolated food store would be much more likely to dry out before a population could come through on it. This point of the nature of *D. melanogaster* populations has been elaborated in connection with their breeding habits and life cycle because in some of the finest and most comprehensive work on the genetic variability of populations of *D. melanogaster* (Dubinin and collaborators, 1937) the investigators seem not to have given sufficient emphasis to the ecology of the species in the interpretation of their findings.

In the autumn *D. melanogaster* will almost certainly be collected if traps are hung in or near an inhabited house, grocery, or fruit store. They can generally be collected in great numbers in orchards, and in public parks, particularly around garbage pails. They may even be taken in the woods, but this is due to the pressure of populations from near-by orchards or dwellings, or from stray inoculations by picnic parties, etc. They are not found around fungus and do not apparently breed on leaf mold or humus. Most flies will be found in the traps when temperatures range from 20°C to 24°C, although some will be found at much higher or lower temperatures. Their feeding is not so sharply delimited by temperature fluctuations as that of many other species. They are an active fly and easily removed from traps by the "collector" method. In the winter, colonies of these flies may be found breeding in some fruit and vegetable cellars, but their distribution is spotty. It has been the writer's experience that more homes, restaurants, and fruit cellars harbor living specimens of *D. hydei* and *D. funebris* than of *D. melanogaster* toward the end of winter. Specimens may be taken in the spring around garbage pails outdoors. By the middle of summer almost all grocery and fruit stores will be colonized. Of course conditions differ in latitudes where the species overwinters outdoors. Here year-round collections may be made, and there is more likelihood of large populations at all seasons where food supplies are abundant.

The successful trapping of other species of Drosophila, like that of *melanogaster*, will be influenced by the breeding habits and the structure of populations. Thus in general species introduced into temperate climates, such as *repleta, hydei, funebris, immigrans, busckii,* and *simulans,* will tend to form larger populations in the autumn, as the breeding stock is cut to a minimum during the winter, when for the most part specimens outdoors are killed by the low temperature. In the spring and early summer only a few flies of these species will be collected. Later, and particularly in autumn, large populations may be found breeding on garbage heaps, in tomato patches, and around houses and fruit and grocery stores. *Drosophila immigrans* may spread out into woodland areas and even survive mild winters outdoors (Spencer, 1940b). To get these species traps should be set in or near homes, gardens, fruit stores, and garbage dumps, or in public parks. The distribution of these forms is likely to be much spottier than that of *melanogaster* because their longer life cycles give more lag between increasing food supply and peak populations. *Drosophila hydei,* in particular, may form very large populations on garbage dumps. In southern California citrus dumps harbor tremendous populations of this fly. Of all species this form is capable of thriving best under very hot conditions. It may be seen in grocery stores or about garbage pails in midsummer, when even *melanogaster* has retreated to cooler cover.

The name "Drosophila," lover of dew, is hardly a misnomer, for the species without exception seem to require considerable moisture in the environment. Some species, as *D. pseudoobscura,* appear to be exceptions. *Pseudoobscura* in the arid Southwest may be trapped during the summer in rather dry territory, but this is generally evergreen forest at high elevations. Even here it seems likely that these flies have found microenvironments where their surroundings are much more humid than nearby territory. Furthermore, populations of *pseudoobscura* invade the lower chaparral and desert regions during the rainy season, and in the mountains may be trapped in much larger numbers after a season of precipitation or in the spring after the snows have melted. Certainly in the Southwest *pseudoobscura* are taken in greater abundance in evergreen forests, particularly the upper Sonoran or piñon juniper zone and the transition zone where yellow pine abounds. However, the species invades the lowlands along the watercourses and spreads out into the deserts of the lower Sonoran during the rainy season.

In the eastern part of the United States many species inhabit woodland areas. The *affinis* group, *robusta,* and *melanica* are typical. These forms appear to be diffuse feeders and breeders, if this term may be applied to flies which apparently find in woods a uniform food supply which under favorable humidity allows for the development of well-distributed populations. Just as *D. pseudoobscura* may be trapped almost everywhere in the evergreen forests of the West, so the *affinis* group may be taken throughout the East in wooded areas. When summers are hot and dry, these woods species will best be taken by trapping in ravines, along small streams, or in deep woods with plenty of leaf mold and forest litter. In general one should trap where the humidity is high, to collect maximum numbers either of flies or of species. Like *pseudoobscura*, these eastern woodland species, in the spring when temperatures are mild and there is abundant precipitation, will invade areas, such as towns, where high temperature and dry air later in the summer preclude their continued existence. In Table 3 are given collecting records from the writer's garage in late June. Contrast these with much larger collections taken in September and note that *affinis*, *algonquin*, and

TABLE 3

Collections Taken June 21 and 22, and August 29 to September 3, from Banana Traps Exposed in a Town Garage, Wooster, Ohio

(Note seasonal variation in species represented, reflecting changes in population patterns)

Species	June 21 and 22	August 29 to September 3
affinis male	16	0
algonquin male	4	0
affinis group female	7	0
busckii male	4	0
busckii female	0	0
funebris male	26	25
funebris female	1	2
hydei male	0	1
hydei female	0	0
immigrans male	0	0
immigrans female	0	1
melanogaster male	82	912
melanogaster female	11	509
robusta male	4	0
robusta female	5	0
Total	160	1450

robusta, all typical woods species, were taken in June, but none of these were captured in September. It is evident that the population of these species in town suffers annihilation in summer, as that of *melanogaster* does in winter.

There are certain species to be found in woods around fungus. *Drosophila putrida, D. tripunctata,* and *D. transversa* are examples. They may readily be taken on banana yeast bait. Best collections of them and of the other forest forms will be made as the summer wears on toward autumn, providing there has been abundant and well-spaced rainfall. Otherwise, early summer collections, after expansion of populations in response to the favorable temperatures and moisture of spring, will be best for numbers of both individuals and species. Here again the length of the life cycle enters into the picture. Although peak populations of the *affinis* group may be reached in favorable years quite early in July, the much slower breeding *robusta* lags behind and reaches its maximum weeks later.

Where it is desired to collect maximum numbers of species from a given territory, as many different and varied habitats as possible should be trapped, such as marshes, deep woods, open woods along streams, patches of fungus, different elevations in mountainous country, and always towns for the introduced forms mentioned above and possibly others. From the easily observable populations of introduced species, breeding on garbage, fruit, etc., it seems safe to conclude that the numbers of individuals and the uniformity of their distribution in space and time will differ from species to species among forms which are native. One species survives through sheer weight of numbers, although individuals are short-lived and food supply is locally subject to great variation. Other species may never develop large populations, owing to the type of food and the length of the life cycle in relation to fluctuations in environmental factors. However, the second species may be so stabilized in relation to its environment that it has fully as good a chance of surviving throughout its total range as the first species and a much better chance of surviving locally.

Although it is difficult to get accurate data on the relation of numbers of flies coming to traps to the time these are exposed, experience over some years has led to the conclusion that traps left out for several days draw more flies than those left out for one day or part of a day. For those who are interested in data the following figures are presented. From 30 cup traps set in woods on Wednesday afternoon and collected from on Saturday morning, 568 flies were taken. From 29 such traps set on Friday afternoon and collected from on Saturday morning, 71 flies were taken. These 2 lots of traps were set at 10 different stations, and the Friday afternoon sets were made as nearly parallel to the Wednesday afternoon sets in regard to favorable habitat as possible. This single experiment may not be conclusive. Isolated cases could be cited to prove the opposite. The ecologist, however, is dealing with many variables. His conclusions are frequently drawn from long experience rather than isolated experiments which may give data leading to false conclusions.

Drosophila, particularly some species, show a diurnal rhythm of activity which is roughly correlated with temperature changes, although other factors such as wind, rain, and light enter into the picture. Where there is a sharp fluctuation between day and night temperatures, and providing the two extremes lie on either side of the optimum temperature for flies to feed, there will be an influx of flies to the traps as the temperature is approaching this optimum and an egress from the traps as the temperature is receding from this optimum. Let us consider an optimum temperature of 23°C. We are trapping in some locality in the East on an average July day. The traps have been set a few days previously. The habitat is open woodland with a generous infiltration of sunlight. Beginning at mid-day the temperature in the traps registers 31°C. There are no flies of the species in question in the traps. The temperature continues to rise, perhaps to 32°C by midafternoon. Around 4:00 P.M. the air begins to cool, at first very gradually. By 5:00 P.M. it has dropped to 28°C, by 5:30 to 26°C, and flies begin to appear in the traps, at first a stray specimen or two. By 6:30 P.M. they are coming in rapidly. By 7:00 P.M. the optimum temperature will be reached, and the trap will be alive with flies, feeding, courting, copulating, and laying eggs. By 8:00 P.M. it is getting dark, the temperature is now down to 20°C, and flies are leaving the traps. By the middle of the night the temperature may have dropped to 16°C or even lower, and very few flies will be left in the traps. The morning fluctuation is, of course, in the opposite direction and will generally be characterized by a sharper gradient. The sharper this gradient the more flies seem to concentrate in the traps at the optimum temperature. However, less time elapses at and near this temperature. In general then it may be said that the biggest catches can be made in the morning if it is possible to visit the traps at the right time. On the other hand, the slower evening gradient gives more time for getting around to traps.

A number of factors influence the temperature gradients. Thus a cloudy day will greatly lengthen the time at which flies may be in the traps, but this has a distinct tendency to lower the concentrate at any one time. The relative humidity is an important factor interacting with temperature in conditioning the behavior of flies. With a high humidity flies will be found in traps at a higher temperature than with low humidity. In semiarid regions, where humidity is low and morning and evening temperature gradients sharp, the migration of flies into the traps twice a day and their absence in the intervening period are somewhat more conspicuous phenomena than where the air is more moist. The humidity factor, as well as lowering of the temperature, probably has much to do with the fact that flies come to traps in large numbers immediately before or after a rain. Almost none, however, will be found in traps during rain. In high wind flies either do not or cannot venture to traps. Naturally the time of year will affect the temperature cycle. Thus in autumn flies may be collected later in the morning and earlier in the afternoon. In very hot weather they will be found in the traps only at night. Furthermore, the same species seems to become either permanently or temporarily acclimated to high or low temperatures in different parts of its geographical range. It is thus better to rely upon emptying traps at the time of sharpest temperature gradient than at any one temperature. *Drosophila affinis* was found in traps at 25°C and 26°C in the summer in Tennessee, when the same species would have been in cover at these temperatures in Ohio.

Since this section on the time and place of trapping was first written, Patterson (1943) and Dobzhansky and Epling (1944) have published extended accounts of trapping methods and records. Patterson and his coworkers used the large lard can traps already described. They took almost 700,000 specimens and 57 species of the Drosophilidae, including 47 species of Drosophila, in the state of Texas. They have also made very extensive collections in Mexico and many other states of the United States, particularly in the Southwest. Using large can traps, Patterson made monthly collections over a continuous period of 21 months at the Aldrich farm near Austin, Texas. During this period he took at this one point over 130,000 specimens of 39 species of Drosophilidae, including 31 species of Drosophila. He was thus able to study the seasonal fluctuations in population size in the more abundant species.

Dobzhansky and Epling concentrated on one species, *D. pseudo-obscura*. They collected intensively at two stations in the pine and pine-oak forests of the mountains of Southern California, using half-pint milk bottles or paper cups as traps. They set traps in a straight line at 20-meter intervals and also in large checkerboard plots with the traps 20 meters apart in some cases and 60 meters apart in others. After extensive study of the diurnal rhythm of activity in *D. pseudoobscura*, Dobzhansky and Epling concluded that light intensity determines the time at which flies come to the traps. They recorded temperature, humidity, and light intensity changes during the day and state: "The invariable rule to which no exceptions have so far been found is that no flies are active during the hours of darkness, from dusk to dawn."

At higher temperatures than any which they record it has been the author's experience that Drosophila were active only at night, when temperatures fell to a favorable minimum. In trapping at Mesquite Springs, Death Valley, flies began to come to traps sometime after nightfall, when the temperature was falling and had reached about 28°C. In Ohio during a long hot spell with daily temperatures above 32°C flies visited traps only after dark. At very high temperatures, therefore, flies may be active at night. On the other hand, under the optimum conditions of light there are temperature minima and maxima which inhibit activity. However, there is a wide range of temperatures under which flies may be active. Light, temperature, and humidity gradients probably all play a role in activity. At certain temperatures the light gradient appears to play the major role.

Most flies come to the traps from the immediate vicinity, although the range of effectiveness is probably increased as a trap remains set at a given point. Dobzhansky and Epling found that traps set 20 meters apart drew flies from overlapping territory, while those set at 40 meters did not. Apparently the normal range from which *D. pseudoobscura* is drawn to a banana trap is somewhere between a radius of 10 and 20 meters. In the eastern United States traps set in the open only a few yards from trees will bring in almost no flies, while traps hung under the trees are yielding flies. Harrison Stalker carried out an extensive series of observations on traps of different designs set in a large number of local habitats, including grass plots, open woods, thickets, deep woods, and ravines. He visited these traps at all hours of the day and night. Notes were taken on temperature, humidity, wind, and cloudiness. It is not feasible to present these data in detail. Many of the facts mentioned above were learned either directly from the data collected or later through further checks on hypotheses formed from a study of these data.

As to the distribution of species and trapping for particular species it may be said that one forms impressions of the kinds of habitats which will be likely to yield a particular species, and of those habitats which almost certainly will not. There is, however, a spottiness in the distribution of populations which leads to the conclusion that chance is a potent factor in the distribution of individuals in nature, and that in a given year there are many places well suited to the species but not occupied by it. Many factors of which we are ignorant probably help to determine species distribution. To conclude, the trapping of Drosophila gives a new insight into the problems of their ecology, and conversely a greater knowledge of the ecology and physiology of Drosophila would be a great aid in trapping.

METHODS OF SHIPPING

In shipping Drosophila it is advisable to use the most rapid means of transport. At the same time cultures should not be exposed to lethal high or low temperatures. Air mail is the most satisfactory method of transport over long distances, where this type of service is available. To reduce cost of mailing the writer has designed a light but strong container (Spencer, 1940a). Aluminum tubing, ½-in. outside diameter, 0.022-in. wall, is cut into 3½-in. lengths. (Aluminum tubing of this or other sizes may be purchased reasonably from the Aluminum Company of America, 2210 Harvard Avenue, Cleveland, Ohio.) The usual cornmeal-molasses or banana-agar culture medium is poured to a depth of ¾ in. into a beaker or other convenient container, and as many of the aluminum tubes as will fit into the container are set in the hot medium. If a container with a tight cover is used and a mold preventative (see culture media below) added to the medium, such a supply of tubes may be set away and kept for weeks in a refrigerator or taken in the icebox on a collecting trip.

In preparing flies for mailing, a tube is broken out of the agar medium, and a small cork is inserted in the end containing the medium. Specimens are shaken into the tube by use of a funnel, or etherized flies dropped into the tube. A cotton stopper is inserted. To make sure that flies do not stick to the aluminum wall the tube should be lined with a cylinder of paper. These tubes may be labeled with India ink. Two such tubes containing flies may be sent in an air-mail letter without further packing between points in the United States for the usual air-mail postage. Over 200 flies of one of the smaller species were sent from the California Institute of Technology laboratory in one of these tubes by ordinary mail service. The writer's wife collected flies in Rio de Janeiro, Brazil, and sent them through successfully by air mail, a distance of some 6000 miles.

To avoid difficulties in connection with interstate transport of insects, permits have been issued by the United States Department of Agriculture and may be secured from the Carnegie Institution of Washington, Department of Genetics, Cold Spring Harbor, New York. It is presumed that local restrictions in other countries may be covered in a similar manner by the pertinent departments.

CULTURE MEDIA

All species of Drosophila, like other Diptera, pass through four distinct stages in their life cycle: the egg, larva, pupa, and adult or imago. In two of these, the larva and adult, feeding takes place. Moreover, although there is no feeding in egg or pupal stages, certain environmental factors, as temperature, humidity, and aeration, are important to survival and proper development. Owing to the extremely rapid development of *Drosophila melanogaster* and to the relative ease with which this insect can be reared under a variety of conditions and media, even in pair matings of wild and many mutant types, relatively little emphasis has been placed on the problems of the culture of this fly. It is

only in recent years that a laboratory study of many other species
of the genus, among them some with much longer life cycles and
with feeding and breeding habits requiring special attention, has
served to emphasize the inherent problems in the rearing of
D. melanogaster. Furthermore, special techniques involving the
use of multiple stocks much weaker than wild flies and the securing and manipulation of accurately timed stages in development
—in short, the use of *D. melanogaster* in progressively refined
experiments in genetics, developmental embryology, and physiology—have served to focus attention on improved methods of
handling and rearing the several stages. It should be emphasized
that the development and use of refined and elaborate methods
are indicated only where they serve best to solve the problems
at hand. As a matter of fact, relatively satisfactory culture
methods have long been employed for *D. melanogaster;* otherwise
the advances that have come through the use of this animal as a
genetic tool would have been impossible.

Historically the first recorded use of this fly as an experimental
animal was by Castle, Carpenter, Clark, Mast, and Barrows
(1906). They cultured it in pair matings for 12 generations on
crushed banana and fermenting grapes and later for 47 generations on fermenting banana. These investigators made a study
of the effect of inbreeding on fecundity and fertility. The highest
recorded yield from a pair mating in their experiments was 397
flies, only a moderate record under present laboratory methods.
The first mutant to be discovered in any species of Drosophila
was reported by Delcourt (1909). Delcourt and Guyénot (1910)
pointed out that in order to study heredity in the genus Drosophila it seemed desirable to introduce a method of growing flies
which would give standard results in terms of culture medium
used. These authors discussed the chemical variability of foods
used as media: potato, raisins, bananas, apples, etc., and of the
microorganisms, yeasts, molds, and bacteria growing on them.
They found that flies grew well on dead yeast, and they were able
by repeated transfer to isolate lines of flies growing on dead,
sterile yeast which carried no microorganism and gave excellent
results in terms of fecundity and fertility. They made up an
artificial medium: 10 gr peptone, 18 gr glucose, 4 gr tributyrine,
1.25 gr sulphate of magnesia, 2.4 gr trisodium phosphate, 1.9 gr
potassium chloride, 2.3 cc acetic acid, and 1000 cc water. This

medium served well to rear flies when inoculated with yeast. They concluded that the same bacteriological methods used in the study of forms reproducing asexually could be applied to sexual types when they were reared on standard media. Guyénot (1913a, b, c, d, e) reported on an elaborate series of experiments designed to study the effects of live and dead yeast, with various types of sterile media, when fed to larvae and adult flies. He showed a remarkable insight into the problems of insect nutrition in general and of Drosophila nutrition in particular, and it is surprising that his work has received little notice by subsequent writers. He showed the importance of yeast as a food for both larvae and adults and emphasized the effect of larval nutrition on the early maturing of adults and the importance of a plentiful food supply in the adult stage for maximum reproductive activity. Guyénot secured excellent results by growing larvae on a suspension of dead yeast alone in water. Cotton was used as a base for the medium.

Bridges and Darby (1933) trace the development of Drosophila culture through the early use of fermented banana pulp, alcoholized banana pulp, and later banana agar and cornmeal-molasses agar. Bridges (1937) has given a clear and brief summary of culture media for Drosophila as used in several of the larger laboratories.

Where a general-purpose medium is being used in a laboratory for keeping stock cultures and for experiments where high yields and close approximations to Mendelian ratios among zygotes of differential viability are expected, this medium should combine several qualities. It should contain a high moisture content so that stock cultures may be carried over for a reasonable length of time without drying out. It should at the same time have a texture that makes possible shaking adult flies from culture vessels without danger of loosening the medium or having flies stick in it. It should have a high index of nutrition, forming a complete diet, and should be sufficient in quantity to bring all individuals through to their optimum size and vigor. It should retain enough nutrient value so that adult flies after emerging in a culture bottle may find adequate food for considerable periods of time in the residue left by the larvae. It should be relatively resistant to the growth of harmful molds and bacteria.

The three most important steps in the development of Drosophila media have been the addition of an agar base, forming a stiff food cake; the fortifying of the nutritive value by adding quantities of killed yeast, instead of relying entirely upon the growth of yeast on the medium as a source of food for larvae; and the addition of some mold preventative. The latter two steps are mutual improvements, as mold preventatives also cut down the growth of yeast, and the addition of yeast in quantity supplements the poor yeast growth.

An excellent general-purpose medium is prepared as follows: 25 gr of powdered agar are added to 4 liters of water and the mixture brought to a boil. Sixty cubic centimeters of brewer's yeast are stirred in, and the mixture is boiled for 15 minutes. This is an important step; unless the yeast is killed by boiling the medium will be unsatisfactory. Five hundred cubic centimeters of Karo syrup are added. Five hundred cubic centimeters of cornmeal are thoroughly mixed with 500 cc of water. This is stirred into the hot medium, and the mixture brought to a boil. Then 20 cc of dry Tegosept-M, the methyl ester of p-hydroxybenzoic acid, are added and thoroughly stirred in. The medium is then boiled for another 5 or 10 minutes and immediately poured into culture bottles. The recipe given above will furnish about 75 half-pint bottles with medium poured to a depth of 1 in. Tegosept-M may be purchased from the Goldschmidt Chemical Corporation, 153 Waverly Place, New York 14, New York; it is an excellent mold and bacteria preventative and non-toxic to flies.

In the author's laboratory for some years a medium consisting of cleansing tissue as a base and a suspension of baker's yeast in water (100 gr in 600 cc) has been used for experimental cultures. Eggs are laid by pre-fed females on cormeal-molasses-agar food chips. Fifteen cubic centimeters of the yeast suspension are poured into a 30 by 100 mm shell vial. One double sheet of cleansing tissue is folded and pushed into the bottom of the vial and drawn up along one side. The medium is soaked up by the paper, and the yeast held in suspension in its interstices. The food chip containing eggs and young larvae is then dropped in, and the vial stoppered with a cotton plug. This medium has the advantage of being porous, extremely concentrated, and not crumbling. Flies come through rapidly and are uniformly large and well nourished. They must be counted or transferred soon after emergence, however, as the residue from this medium does not furnish an adequate diet for adults. The method gives uniformly good results and makes possible the rearing of as many flies in a vial as may be grown in larger culture bottles by use of weaker food formulae.

Various ingredients have been suggested for media, particularly as substitutes for materials difficult to obtain locally. Thus Haskell (1940) recommends chondrus, a dried seaweed, as a substitute for agar. MacKnight and Roman (1940) suggest a food formula of 12 gr cornmeal, 8 gr sugar, 4 gr rolled oats, and 76 gr water with a little mold preventative added. Here the rolled oats add nutrient quality and take the place of agar as binding material. A number of synthetic media have been used but have proved less practical than media already mentioned. Thus the synthetic medium reported by Pearl and Penniman (1926) is considered by Bridges and Darby as less satisfactory than banana-agar. Lewis (1942) has described the following food formula, which has taken the place of the MacKnight formula used at California Institute of Technology: water to be put to boil, 2886 cc; water saved cold to moisten cornmeal, 1200 cc; cornmeal, uneven grind, 900 gr; molasses, 600 cc; rolled oats, not quick-cooking brand, to be stirred into food just before pouring, 90 gr; salt, 6 gr; Moldex in alcoholic solution containing 0.1 gr per cc, 42 cc. While no entirely satisfactory substitute for agar has been found, cornmeal and oatmeal in formulae such as the above serve the purpose fairly well.

Spassky (1943) has reported a satisfactory agarless medium in which Cream of Wheat is used as the binder. The formula includes water, 4500 cc; Cream of Wheat, 625 gr; molasses, 670 cc; salt, 7 gr; Moldex, 40 cc of a 10 per cent solution in 90 per cent alcohol. Three-fourths of the water, with salt, Moldex, and molasses added, is heated to a boil. Cream of Wheat is added to the rest of the water, and this stirred into the boiling water. The mixture is then boiled from 3 to 7 minutes and poured through a wide funnel into half-pint milk bottles. A fast worker can pour 200 bottles before the medium becomes too stiff. This medium will not shake out when cool and does not liquefy in old cultures as readily as most other agarless media.

The fact seems well established that fly larvae thrive best on a diet made up largely of yeast. It would seem reasonable that improvements in culture media will include better adjustment of yeast concentration to the nutrient needs of the species, additions of vitamins in proper concentrations (Tatum, 1939), and proper texture for tunneling of larvae. Some species of yeasts may prove

Fig. 4. Quarter-pint culture bottle and 4084 *Drosophila hydei* reared in such a bottle on a baker's yeast-raw oatmeal diet.

superior to others. Shimakura (1940) has recently given a preliminary account of the use of *Torula rosea* in Drosophila culture. This is a non-fermentative yeast which grows more vigorously than Saccharomyces without the formation of large quantities of carbon dioxide.

For experimental cultures it is feasible to grow many more and larger flies in a culture bottle without starving by using richer food. The author has reared 4084 normal size *D. hydei* in a quarter-pint bottle on a mixture of baker's yeast and raw oatmeal. (See Fig. 4.) By using concentrated food as many flies can be reared in a small vial as were formerly grown in a half-pint bottle. This, of course, conserves space and facilitates handling.

MOLDS, MITES, AND OTHER PESTS

It is not surprising that in the rearing of Drosophila difficulties arise from the infection of cultures by various types of organisms which may prove harmful. In earlier years molds were very troublesome, particularly in culturing some of the slow-breeding species. Often molds formed a mat over the culture medium before adult flies had laid eggs. In other cases where the number of larvae was small, flies emerging became covered with a coating of mold spores and soon died. Frobisher (1926) found a red torula or yeast growing symbiotically with a mold of the Penicillium group in some of his Drosophila cultures. Adult flies were killed by feeding on this yeast and mold. A study of the mechanism involved led to the conclusion that the mold growing among the yeast cells in the gut of the fly formed a dense mat which blocked elimination and also that the mold hyphae actually invaded the tissues of the fly.

Parker (1935), at the University of Texas laboratory, first reported experiments on the use of commercial mold preventatives in Drosophila culture medium. Moldex-A proved more efficient and less expensive than Nipagin-M or Nipagin-T. Moldex-A is the sodium salt of *para*-hydroxybenzoic acid. With the addition of small amounts of this substance, infections with molds ceased to be a problem. Today it is rare to see a culture bottle with heavy mold growth in a laboratory where a mold preventative is used. Furthermore, aging and transfer of adult flies to fresh food medium, now practiced in many laboratories, result in the development of young larvae before mold infections gain headway. Glass (1934) suggested adding 0.2 per cent formaldehyde to food medium in vials, and holding adult flies from mold-infected stocks in these for a few days to rid them of mold spores.

Sometimes bacterial infections occur in Drosophila cultures. These may cause a slimy exudate on the surface of the culture medium; flies may become mired in this exudate and in any case do not thrive under these conditions. Such bacterial infections are likely to spread through stocks. The best control seems to be frequent transfer of adult flies to fresh culture bottles, allowing them to feed well, and then addition of plenty of yeast to larval cultures. Such infections are less prevalent than formerly, owing to use of mold preventatives.

Mite infestations may prove to be more serious than either molds or bacteria. A rather large number of species of mites may infect Drosophila. Some of these attach themselves to the flies in the adult stage and may puncture the integument and suck tissue fluids. The author has found wild flies parasitized by several different species of mites. Although these may be harmful to the individual fly, they do not constitute a pest, as they fail to spread in laboratory cultures. Four species have been seen, all of which multiply in Drosophila cultures. Three species live in the culture medium but at no time attach themselves to the flies and are relatively harmless. However, large numbers of any of these may be detrimental, as they may use up fly food and may tend to dry out the surface of the medium. It is possible that they feed on Drosophila eggs. In any case a heavy infestation of any of these mites in a culture bottle generally coincides with a poor yield of flies. Of the three species the one least frequently seen is very small, almost transparent, and too rarely found in cultures to constitute a serious pest. A second species in the adult form is rounded, with a hard brownish or reddish dorsal surface, and runs rapidly. A third, and the commonest in laboratories in the United States, is a white mite in the adult form, body ovate and generously supplied with long white hairs, which extend out well beyond the posterior end. None of these three has a hypopus stage attaching to insects. However, eggs may accidentally be carried over into fresh cultures on the bodies of adult flies. These mites are certainly not symbionts in fly cultures, but under normal culture conditions do the Drosophila relatively little harm. Some workers believe that the adult mites feed on Drosophila eggs. In any case heavy infestations of these mites result in unhealthy culture conditions for Drosophila and in some types of experiments may prove disastrous. These mites spread from culture to culture by crawling; the young can crawl out of a bottle unless it is stoppered very tightly. Infestations of these forms may be cleared up by rapid turnover of stocks, isolation of clean stocks from sources of infection, sterilization of old culture bottles with heat, and caution in the use of etherizers or other apparatus which may have eggs or mites adhering to them. Two transfers of adults, a few days apart, will almost certainly rid a stock of these mites, even though eggs were carried over at the first transfer.

Wallace (1946) has reported that a 2–4 per cent solution of benzyl-benzoate in alcohol kills the long-haired or "white" mite in a few minutes when these are placed on paper toweling treated with this solution. Contact with this chemical in the same concentration does not harm *D. pseudoobscura*. Culture bottles and population cages may be sprayed with a 2 per cent solution with good results. Fly larvae will pupate on treated paper, and the adults emerge successfully. This chemical control is recommended but should be supplemented by careful avoidance of infestation and not relied on as the sole method of control.

By far the most serious pest is a form which among Drosophila workers has become known as the laboratory mite. There may be more than one species here, but if so they are very similar in morphology. This form has a hypopus stage which attaches itself to insects. Slides of the different stages of this mite were sent to the United States National Museum for identification. Dr. H. E. Ewing of the Bureau of Entomology and Plant Quarantine reported them to be (1) larvae and nymphs of Histiostoma sp., (2) Histiostoma sp. adults, and (3) migratory nymphs of some species of Tyroglyphidae, apparently the Histiostoma. The author has consulted several papers and monographs on the mites but finds figures or descriptions or both of the Tyroglyphidae to be rather unsatisfactory. The long-haired mite mentioned above resembles published figures of *Tyroglyphus longior*, but the latter is described as having a hypopus stage. A careful study of the life histories of the several mites associated with Drosophila, with figures of the various stages, would be useful. Figure 5 shows sketches of the dorsal aspect of the adult females of the Histiostoma and of the long-haired mite. These are sufficiently complete so that from them the two species can easily be distinguished. Histiostoma has a squattier body build and is easily separated from the other by the absence of long hairs.

The following observations made by the author (Spencer, 1937a) on the life history of the laboratory mite are of interest in connection with the problem of control. On September 16 one *Drosophila repleta*, carrying a single individual mite in the hypopus stage, was placed in a shell vial with banana-agar culture medium. By September 24 the mite, fully grown, was crawling in the culture medium. On September 25 a number of young mites were seen in the culture, thus indicating that the mite reproduces parthenogenetically (mating of the hypopus or earlier stage has never been observed and probably never occurs). By September 26 there were about 100 young mites in the culture vial. These moved rather rapidly over the surface of the culture medium or slowly if the legs became immersed in it. There was no tendency for these mites to wander far from the food surface and up the

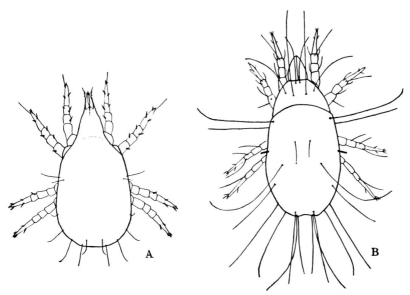

FIG. 5. Adult females of two species of mites. A: Dangerous laboratory mite. Histiostoma sp., the hypopus stage of which attaches itself to flies. B: Relatively harmless long-haired mite often found in Drosophila cultures and lacking hypopus stage.

sides of the vial. On September 27 they had grown to a size larger than the hypopus stage, and one pair was observed in copula. On September 28 several copulating pairs were seen, and a single specimen of the migratory (hypopus) stage had already developed. These observations indicate a life cycle in which parthenogenesis produces both males and females. Thus a culture or a laboratory might become infected from a single mite of the hypopus stage.

The hypopus of this mite is barely visible to the naked eye as a tiny brownish spot. These hypopi may develop in enormous numbers in old, infested cultures. They have three pairs of legs, which are used in crawling rapidly. Stalker has made the important observation that they may leap several inches, evidently using the legs in much the same manner as Collembola (springtails) use the furcula. This stage penetrates tiny crevices and readily crawls out of culture bottles unless very tightly stoppered. The hypopi are constantly on the move, and any object which touches them elicits a clinging reaction. When the hypopus comes in contact with an insect, it first clings with the legs and then attaches itself by a sucker on the ventral side. The hypopus may become attached at any point on the fly but more frequently on the legs, wings, or genitalia. One fly may carry hundreds of hypopi in badly infested cultures. After a period of a few days the hypopus generally detaches itself from the fly, crawls down into the food medium, and grows to the adult stage. Certainly a heavily infested fly may be weakened or rendered mechanically sterile by the presence of mites on the genitalia. It is true that in vigorous stocks of flies a few mites may do little harm. However, their presence in any laboratory constitutes a menace, particularly to valuable mutant stocks which may be below normal vigor and to species of flies which present difficulties in culture even under optimum conditions. Hughes (1946) has studied the life cycle of the laboratory mite, which he identifies as *Histiostoma genetica* or a closely related species. He finds that the hypopus stage may live for over 2 months in vials provided with water. This observation indicates that reinfestations might occur after a long period if the hypopus stage were to find a humid refuge in which to survive. Under conditions of low humidity the hypopi die much sooner. Any moist refuges present in the laboratory should be eliminated or treated with carbolic acid.

Any laboratory which raises Drosophila over a period of years is very likely to have experience with mites at one time or another. If all the workers in a laboratory will coöperate most mite infections may be forestalled, and the occasional one which occurs may be cleaned up quickly. The following directions are given for those who are interested in completely eliminating infestations of the laboratory mite. Others may find certain points of interest and worth adopting in their laboratory procedure.

Even when a laboratory has no mites, the following sources of potential infection must be considered:

a. Cultures received from other laboratories. Often only a few mites will be present in such cultures when first received and will not be detected. It should always be assumed that they are present.

b. Wild flies. If collections of wild flies are being made and brought into the laboratory, sooner or later flies with mites will be taken. Most wild flies are relatively free of mites, and this is particularly true of native species in a given region in contrast to introduced forms, which are likely to be breeding under artificial conditions favoring mite infestation. However, one mite brought in on a wild fly may result in a widespread infestation.

c. Other insects. The writer has on several occasions collected other insects, particularly a small hymenopterous form, *Eucoila drosophilae,* bearing the hypopus stage. As Eucoila is a parasitoid of Drosophila its association with the latter might transfer mite infestation. It is quite possible that these mites in the hypopus stage could be introduced into the laboratory on animals other than insects. Certainly laboratory workers carry them about and infect new cultures during periods of infestation. Patton and Evans (1929) report hypopi of one of the tyroglyphids commonly found on horses, and others have been known to infest man.

The first step, then, in dealing with mites is to reduce greatly the chance of an infestation by the proper handling of all wild collections and flies received from other laboratories. These should be examined preferably in a room not used in the regular course of rearing and manipulating stock and experimental cultures. In most laboratories such an "isolation ward" may be found. Then adult flies from any such cultures should be placed in fresh culture bottles and at the end of 10 days or a little less transferred to new culture media. All these cultures should be kept in the isolation ward and examined from time to time; if mites are found, the cultures should be destroyed as soon as breeding stock in the form of several adult flies is recovered. Repeated transfer of these adults to fresh media every few days and discarding of the old bottles as soon as larvae appear in the new ones should be practiced over a period of 3 or 4 weeks. This gets rid of the mites which were on the adult flies and does not allow for a new generation to come through to the migratory and hypopus stage. If this routine is practiced, and there are no other sources of infestation, the new stocks brought into the laboratory will be entirely free of mites in a month or less. There is no need to keep these stocks standing in Lysol, creosote, soap solution, or other preparation, as at one time suggested by the author (Spencer, 1936b), Demerec (1936), and others. The life cycle is definitely broken up by transfer of adult flies; no chance is given for fresh hypopi to develop, and if they did the solutions do not afford an effective barrier to their spread. After a period of 2 months these "foreign" stocks may be transferred safely to the regular stockroom.

If in spite of these precautions a mite infestation is discovered in the stockroom or among experimental cultures, the procedure should be to assume immediately that any culture in the stockroom may be infected. Then follows a rigid routine of rapid stock turnover, holding a good number of adult flies as breeders, and transferring them to fresh cultures under the time limit of the life cycle of the mite, discarding old cultures as soon as larvae appear in the new ones. If any bottle is found in which hypopi are collected on the surface, it should be quickly discarded, first saving adult flies from it as breeding stock. Shelves, trays, and tables should be wiped off with a weak solution of phenol or lysol, both of which are effective killing agents. A solution of 1:100 is sufficiently strong. Wherever bottles containing hypopi have been standing, a thorough application of the above solution will kill stray mites. The most important point in the procedure is to prevent any new culture from developing hypopi. This can, of course, be done by continued transfer of an adult breeding stock beyond the time that the last living hypopus remains attached to one of these breeders. Under the excellent nutritional conditions of these repeated transfers the adult breeders will thrive, lay many eggs, and can be counted on to live much longer than any of the original hypopi which they carried. Where a laboratory carries hundreds of stocks, the above procedure will seem laborious, but it must be carried through if there is to be any assurance of clearing up the infestation. The time is well spent, as a laboratory once free of mites can be kept free of them by proper isolation of any stocks brought in from elsewhere.

By far the best method, and in fact the only sure method of sterilizing bottles, is heat sterilization. Dipping infested bottles in various chemicals may kill many mites, but it will not kill them all and is far more expensive and troublesome than the use of heat. All discarded culture bottles are placed in metal trays

made for the purpose, and these are set in a dry sterilizer or auto-
clave and heated well beyond the critical time and temperature
necessary to kill mites and flies. This sterilization procedure
should always be followed, as it kills larvae and flies, molds, and
bacterial growths in discarded bottles, whether there are any
mites present or not.

Of course, it must be obvious that mites in the hypopus stage
may be transferred by the hands or clothes of the worker who has
handled a bottle containing them. Also brushes or other manipu-
lators, counting blocks, etherizers, and any other apparatus used
in handling infested flies will be a ready source of infestation.
The worker should take the position that, as long as a mite infestation exists in the laboratory, any move he or other individuals
may make is likely to result in spreading mites. A janitor dust-
ing off tables may carry infection from room to room. One care-
less worker who will not coöperate will constitute a constant
source of reinfestation, as his own cultures are almost certain to
carry mites. Such an individual will sooner or later carry infection to stockroom or food-preparation room. When one or more
individuals who are not interested in eliminating mite infestation
are working in a laboratory, then the rest will be fortunate indeed if their stocks are not constantly reinfected.

In brief, the most important items in control and elimination
of mites are heat sterilization of discarded culture bottles, isolation of all stocks brought into the laboratory from outside, rapid
turnover of all laboratory stocks as soon as a mite infestation is
discovered, and coöperation of all members of the laboratory
staff.

Although there are a number of predators on Drosophila in nature and various other forms which may compete for the food
supply, they are of minor importance in laboratory culture. Occasionally wild flies will carry nematode worms. These soil
nematodes may multiply in culture bottles, but transfer of adult
breeding stock soon eliminates them. In the summer, in laboratories where the humidity is high, large flies of the Muscidae,
species undetermined, may lay their eggs at the edge of bottle
caps. The larvae on hatching crawl down into the culture
medium. These large larvae may use up so much of the food
medium that Drosophila larvae present are starved out. This
seems to be a relatively rare type of infestation but has been

found troublesome for a few months in the summer. Obviously the control measure is to keep cultures in cabinets or under cover, where large flies do not have access to them, and if possible to screen such flies out of the laboratory. Among Drosophila there is one species, *D. busckii,* which will deposit eggs around culture bottle caps, with subsequent contamination of the culture by *busckii* larvae. Probably in a dry climate this sort of contamination never occurs, but it is a possibility in regions where the humidity is high.

CULTURE BOTTLES

In the pioneer work of Castle and colleagues (1906) glass battery jars 8 cm high by 6 cm wide, covered with glass plates, were used as culture vessels. Later glass tumblers of similar size were substituted. Bridges (1932) states:

> In the early work on the heredity of Drosophila the cultures were raised in quart fruit jars, in quart milk bottles, and in a very miscellaneous collection of museum jars and other laboratory glassware. . . . By 1913 pint and half-pint milk bottles had become standard, largely on account of their uniformity of mouths, their heavy, strong glass, and the ease with which they could be procured.

Until the introduction of Moldex and other mold preventatives, which also cut down yeast growth and fermentation, carbon dioxide often formed under the food cake and pushed it up against the cotton plug with disastrous results. This occurred most frequently in cultures held at temperatures of 26°C and higher, seldom when cultures were reared at 24°C or lower. This excess carbon dioxide formation is no longer a factor in Drosophila culture, owing to improved culture media. Special bottles were designed to prevent this dislodging of the food cake (see Bridges, 1932; Gottschewski, 1936; Shipman, 1935). These special designs are no longer needed, and except for particular problems the use of half- or quarter-pint milk bottles and shell vials of various sizes has been standard in most laboratories for some years.

However, a few observations on culture bottle size and shape may be of interest. For stock cultures a culture bottle should be big enough to hold a bulk of food sufficiently large to give a margin of safety against the loss of a stock through drying up or exhaustion of food, and small enough to conserve space in the stockroom. These requirements are admirably filled by half-pint milk bottles. These are also sturdy and not easily broken, and they give maximum visibility through clear glass walls.

For experimental cultures bottles and vials of different sizes will be found adapted to particular problems. With the use of yeast-enriched foods it is not true that better approximation of Mendelian ratios will be secured by rearing only a few flies per culture. The chief advantage in these heavily fortified media is that large numbers of well-nourished flies may be reared in small space. This is particularly important where careful temperature control is desired, and incubator space is at a premium. For example, using the yeast-paper technique described above, the writer secured in the F_2 of a cross of scarlet to wild *D. hydei*, in a 48-hour emergence period, 497 wild to 162 scarlet. In 2 more days 29 wild and 11 scarlet emerged. This made a total of 699 flies from a half-pint bottle with a deviation of 2 from the ideal ratio. Large numbers per bottle, close approximations to Mendelian ratios, and little spread in emergence are secured by the use of rich food, and the size or number of containers may be correspondingly cut down. Stalker (1940) has suggested 1-oz cream bottles as substitutes for shell vials. These may also be secured in 2-oz size. They may be purchased from Owens Illinois Glass Company at about 1¢ each. Stalker points out that they are easier to clean than vials and are less fragile; because of the wide mouth egg counts can be made in them; and they are much cheaper. They have the disadvantage of drying out more rapidly than vials, as the mouth is closer to the medium.

Certain species of Drosophila, particularly in pair matings, do better in small containers than in half-pint bottles. This is true of the *affinis* group and is probably accounted for by the fact that in the larger container adults and larvae do not keep down the growth of bacteria, and there tends to be a great excess of medium which will putrefy or dry before larvae start tunneling through it. However, for mass stock cultures half-pint bottles are recommended for all species.

Vials are stoppered with cotton wrapped into a tight plug and covered with a layer of cheesecloth to prevent sticking. Milk bottles or creamers may be closed with paper bottle caps or cotton plugs. Demerec in 1928 first used paper milk-bottle caps. Bridges (1932) suggested punching these with tiny needle holes for ample ventilation. In the writer's laboratory bottle caps are used. With a large-sized desk punch a hole about 7 mm in diameter is punched out. Into this is stuffed a small cotton plug. This hole serves for ventilation; and, by removing the cotton plug, yeast or flies may be added to the culture bottle without removing the paper cap. It is true that mites in the hypopus stage can crawl out of a bottle covered with a paper cap much more readily than out of one stoppered with a tight cotton plug. However, bottles containing hypopi should not be present in the laboratory.

ETHERIZERS AND OTHER APPARATUS FOR HANDLING

Delcourt (1911) describes a device by which he examined living flies. It consisted of an aspirator attached to a glass tube, which contained a flattened constriction. When flies crowded up into this constriction, they were examined under the binocular microscope for variant characters. In this way he reports having examined 13,000 *Drosophila confusa* and 8000 *D. ampelophila* (*melanogaster*). He also used a tapering prism tube, into the small end of which the flies were run for detailed examination. It is a far cry from these crude devices to the methods used today for examining living flies.

In the Columbia University laboratory, where the important early work on the genetics of Drosophila was done, flies were etherized in half-pint bottles before examination under the binocular. Bridges (1932), who was always alert to improvements in technique, has described an etherizer of his design, which represents the culmination of experimentation started in 1919 to construct an implement more efficient than the bottles then in use. Bridges' etherizer is shown in Fig. 6. A metal funnel (*A*) is mounted in a glass etherizing chamber (*B*) by means of a cork (*C*), these parts being cemented together with plastic wood. Asbestos fiber (*D*) fills a small chamber separated from the main chamber by the plaster-of-Paris partition (*E*), which is perforated with small holes. A cork (*F*) closes a short side tube into the fiber-filled chamber. Corks (*G*) and (*H*) close the intake funnel and the exit funnel, respectively. Before using, a few drops of ether are introduced into the fiber chamber, and the cork is replaced. The intake funnel is cut to fit inside the rim of culture bottles. The cork (*G*) is removed, and flies are shaken from the culture bottle into the funnel. When etherized, they are removed through the lower end of the etherizer. Cords tether the income and outgo corks to the etherizer to prevent loss. According to Bridges, this device "makes the intake of flies from culture bottles and their discharge to the sorting plate both safe and rapid, the etherization quick and uniform, and the amount of ether used and eventually liberated into the room very small." The visibility of flies is good, and the etherizer is easily kept clean.

Muller (1934) described a design consisting of a glass containing vessel (type to be chosen by the operator), into the mouth of which is fitted by means of a hollowed-out cork a metal funnel. At the lower end of this funnel is glued a large gelatin capsule, perforated with small needle-holes. In the bottom of the glass container is packed cotton, which must not touch the capsule. A pipette of ether is poured onto the cotton through the capsule, the inside of which is later wiped dry. Flies are poured in and

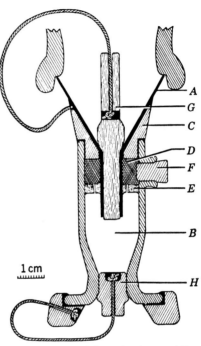

FIG. 6.   Drosophila etherizer. After Bridges (1932).

out through the same opening, eliminating manipulation of the stoppers. Flies are quickly etherized in the gelatin capsule by ether diffusing in from the glass chamber. Several modifications of the Bridges and Muller designs are used in various laboratories. The writer's laboratory uses (Spencer, 1936a) a simplification of the Muller design. This consists in a hollowed-out cork into the lower end of which is fitted, but not glued, a large gelatin capsule, the cork in turn fitting into a wide-mouthed specimen bottle of 60-cc capacity. In the bottom of this bottle is cotton for ether. An aluminum funnel is laid in place in the opening in the cork, and flies are poured or shaken in from the culture bottle. Funnel, cork, and capsule may all be easily detached, cleaned, and put back in place.

It should be stated that for some of the more rapidly etherized forms, as *D. putrida, D. americana,* and the *affinis* group, these rapid etherizers must be used with extreme caution, and many workers will find it preferable to etherize these species in a larger bottle with lower ether concentration, regulated by amount of ether placed on the cotton plug used to close the chamber. It might be well to suggest here that etherization takes place more slowly and that flies once etherized remain under longer at low than at high temperatures. Especially for species mentioned above and several others, it is a distinct advantage to work at temperatures below 24°C and almost impossible when the temperature is as high as 27°C. Etherization should be carried to a point where the flies are completely immobilized but not so far that the wings begin to turn up over the back and the legs to be extended and bunched together. In the latter condition the flies are overetherized, many of them will never revive, and even if none are to be saved they are extremely difficult to manipulate on the counting plate.

Etherized flies are examined on a counting plate which may vary in size and material, depending on the personal whims of the worker. A white opaque glass plate about 5 in. long by $3\frac{1}{2}$ in. wide is satisfactory. Where it is necessary to measure the approximate size of flies, Stalker uses a counting card consisting of a piece of heavy graph paper glued (with rubber cement) on heavy drawing board.

Flies are manipulated with a "pusher." Some workers use a camel's-hair brush, others a hard pencil, which can serve as a pusher and recording pencil. In the Amherst laboratory a pair of forceps is used with a small pencil attached to one blade by a small copper block soldered in place. A steel spear-point needle or strip of copper sheeting is sometimes used. After flies have been examined and recorded, those which are no longer to be used are emptied into a "discard" bottle, containing alcohol, light motor oil, or some other liquid of low surface tension. Muller (1936) suggests a broad dish containing light motor oil and covered with a coarse wire screen as a fly morgue requiring minimum trouble in discarding specimens. The oil is non-volatile and odorless.

In placing etherized flies in fresh culture bottles, care should be taken that they are not dumped directly onto the moist surface of the medium. A piece of paper toweling or cleansing tissue may be placed over part of the surface in such a way that flies can easily be deposited on it. Some workers use small paper cones for introducing etherized flies into culture bottles.

THERMAL CONTROL

While Drosophila may be reared over a rather wide temperature range, providing other culture conditions are satisfactory, the accurate control of temperature is frequently important and necessary. This is true of all studies dealing with growth processes and life histories, with the expression of many genetic factors, and with a host of problems in the closely related fields of developmental embryology, physiology, and genetics. Even in stock keeping it is an advantage to be able to manipulate the temperature in such a way as to lengthen the life cycle and cut to a minimum the work of stock maintenance.

For a few cultures the many designs of thermostatically controlled bacteriological incubators set at temperatures within the range of tolerance for Drosophila serve fairly satisfactorily and are used in many laboratories. However, these are quite expensive for the space available, and larger temperature-control cabinets have been devised for Drosophila work.

Bridges (1932) has described a four-shelf incubator for Drosophila work. This was designed to give uniformity of temperature and humidity and has been adopted by a number of laboratories. Figure 7 is a reproduction of the plan and elevation diagrams of Bridges' incubator. A complete description of the construction and operation of this incubator is given in the paper cited above; only a brief summary is presented here. The walls of the cabinet (A) are constructed of $\frac{3}{4}$-in. light pine or fir boarding, not more than 5 in. wide, tongued and grooved, and nailed to crosspieces (B) in such a way as to allow for expansion and contraction of the walls through the tongue-and-groove fittings on changes of temperature and humidity. Outside the pine wall are tacked two cardboard or Celotex walls, with air spaces between to increase insulation. In the chamber (C) are placed a battery of light bulbs (D) used as the heating unit. Light bulbs have little temperature lag. Humidity is controlled by evaporation of water from the large-area tray (E) which has dimensions of 25 in. $\times$ 22 in. and is raised above the bottom of this tempering

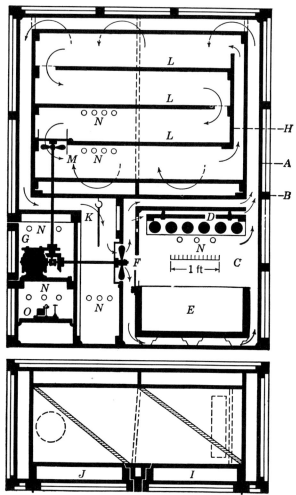

Fɪɢ. 7. Four-shelf Drosophila incubator. After Bridges (1932).

compartment. For refrigeration below room temperature 150 lb of ice may be placed in the tray which is $10\frac{1}{2}$ in. deep. By varying the area exposed through covers of different sizes the humidity of refrigeration can be controlled. Air from the tempering compartment is driven by a horizontal fan (F) attached to the $\frac{1}{10}$-hp motor (G) through a spiral course around the incubator chamber (H). The air passes up the right end, diagonally across the top, down the right half of the back, forward below the right half of the chamber, up the front through the hollow incubator door (I) on the right, diagonally across the top, down the back left rear wall, forward under the left half of the floor, up through the left door (J), diagonally across the top, down the left side, back through the thermostat compartment (K) to the fan. Within the incubator compartment the shelves (L) with screened connecting openings at alternate ends convert the space into a continuous air duct. Air is driven through this compartment by the vertical fan blade (M). The motor compartment is insulated from the incubating and tempering compartments. Holes bored in the back wall at (N) in the several compartments allow for renewal of air; the tongue-and-groove construction of side walls also aids in ventilation. The thermostat at (K) is connected with a relay at (O).

As may be seen from a study of this diagram, the air stream which heats the incubator chamber circulates mostly vertically and over all walls of this chamber, and the duct of air inside the chamber circulates for the most part horizontally. With this rapid double circulation air pockets and temperature layering are cut to a minimum. After using one of these incubators for a year the writer has only one criticism: when shelves are filled with culture bottles, at temperatures around 25°C, there is not sufficient air space to ventilate properly, and carbon dioxide tends to collect in the cabinet and cultures.

Bridges also designed a two-shelf incubator shown in Fig. 8. As he pointed out, a long two-shelf arrangement minimizes the vertical temperature differences due to the gradient in the room. This form is easily constructed and quite satisfactory where a small and inexpensive temperature control is required. The following is quoted from his description (Bridges, 1932):

The interior is divided horizontally by a shelf, with a supporting partition at each end. One of these partitions (see diagram) is complete below, except for a circular hole for the eight-inch fan blades. The other partition reaches only to within five inches of the floor, except at its sides, where two-inch pieces act as legs for the shelf. This end partition is set four inches from the end wall and forms an air passage to the upper shelf space. It reaches halfway to the roof, thus permitting the air to flow while keeping the bottles in position. In this end passage is hung the thermostat.

The fan is run continuously by a motor placed outside on a shelf at the end. This shelf is made large enough to hold the relay or other devices used in the heat control, and may be provided with a detachable cover which reduces the noise from the motor. The cover must be ventilated so that the motor does not overheat.

The heating lamps are in a special compartment next the fan in the lower shelf space. They are in series two-by-two, to operate at half

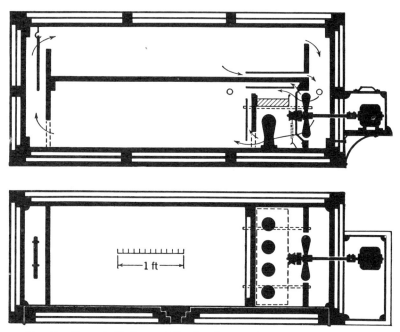

Fig. 8.   Two-shelf Drosophila incubator.   From Bridges (1932).

voltage. Two such gangs provide the total wattage required. The two lamps in series with each other should be of the same wattage. The total wattage should be adjusted as low as will just overcome, by continuous burning, the difference between the chosen inside temperature and the minimum outside temperature expected.

Since the heating elements are inside the incubator chamber, very special precautions must be taken to avoid higher temperatures next this compartment and in the shelf space immediately above it. This is accomplished by a succession of partitions with the interspaces swept by air streams. Directly above the lamps is interposed a metal tray for water. This rests on ⅜″ removable dowel pins that cross the heating chamber. The tray sets snugly against the partition away from the fan and extends toward the fan, leaving a passage for about one-third of the air to cross the water surface and pass over the partition, which rises only to the level of the top of the tray. Above the water tray is a horizontal partition of plywood that is hung by screws to the middle shelf or nailed to narrow strips parallel to the air stream. Finally the bottles on the shelf directly above do not rest upon the long main shelf itself but upon another plywood horizontal short partition raised upon narrow strips so that an air stream passes beneath it and into the return passage at the end. The top part of the end partition is mounted on this plywood to provide exit for this air. The partition separating the tempering chamber from the rest of the lower shelf is similarly reinforced by an accessory plywood partition so placed as to divert air through the space between it and the main partition.

The humidity can be controlled by covering over the water tray to an extent determined by trial.

The construction of the walls of this incubator from soft pine and $\frac{1}{4}$-in. builder's cardboard or Celotex is similar to that for the larger incubator.

Stalker (written communication) has devised and had installed a six-shelf incubator with carrying capacity of about 300 half-pint bottles, which includes certain advantageous features. The incubator cabinet is constructed of wood, and the outside dimensions are height 4 ft 7 in., width 2 ft 6 in., depth 1 ft 6 in. Shelves are of heavy metal screenwork, $\frac{7}{8}$-in. mesh, with steel edges $\frac{1}{2}$ in. high. These shelves are covered with heavy felt pads, except for a 4-in. gap at the end of each shelf for air to circulate. Gaps are at opposite ends of successive shelves, allowing for air circulation as in the Bridges-type incubator. Mounted above the cabinet is a $\frac{1}{20}$-hp motor driving a rotary blower. This blower is mounted in an air duct which leads from the top of the incubator cabinet down one side and in at the bottom. Near the top of this duct are two 100-watt bulbs which serve as heating units, and below them two long copper tanks connected by a series of flanges. These tanks are connected with a water faucet, so that a constant stream of tap water may flow through them and out. This heating and cooling system makes possible the maintenance of constant temperatures either above or below room temperatures and has a distinct advantage over cooling by ice, as in the large Bridges type.

A thermostat is placed within the cabinet and hooked up to a relay. An ether wafer, set in series with the lights and adjusted to turn off lights at 1°C above temperature of the thermostat set, provides safety in case of thermostat failure. An advantage of this design is that an adjustable opening in the wall of the blower duct allows for the sucking in of fresh air from outside the cabinet and its mixture with air drawn from the cabinet. There are several adjustable air outlets from the cabinet to the room as well as the main outlet to the blower duct. Shelves with their con-

Fig. 9. Interior view of large temperature cabinet especially constructed for rearing Drosophila.

tents may be lifted out, thus facilitating removal of culture bottles.

In a number of the larger laboratories constant-temperature rooms thermostatically controlled have been installed. The thermoregulation of these may be sufficiently accurate to allow for their use in temperature experiments. Generally, however, they have been used as stockrooms and as places for installing smaller and more accurate units, which may function more efficiently on

account of the slight temperature fluctuations in their surroundings. The Amherst laboratory, where much accurate temperature
work has been done, is equipped with a constant-temperature
room in which a large series of incubating and refrigerating cabinets very accurately controlled has been installed.

Through a grant from the Rockefeller Foundation a series of
four large temperature cabinets has been constructed at the labor-

Fig. 10. Detail of temperature cabinet (Fig. 9), showing movable open
shelves of angle iron, sliding wooden trays with wire-mesh bottoms, and
double thermostatic control.

atory of the College of Wooster. This series of units involves
several unique features. Figure 9 is a view of the interior of one
of these cabinets, and Fig. 10 shows some of the details of construction. The four cabinets are included in a single large unit
15 ft long, 45 in. deep, and 76 in. high. This unit is of heavy wood
construction and is set up in a cool basement room. The interior
of each cabinet (see Fig. 9) is 3 × 3 × 5 ft in height. The
cabinets are well insulated and lined with galvanized sheet metal.
The side walls and ceiling are covered with baffles. The light
bulbs which form the heating units are above the baffle in the ceiling. In the center of this baffle is a large aperture in which is
mounted a high-speed fan. This drives the air down through the

cabinet, and it circulates back up from the floor behind the side-wall baffles and over the battery of light bulbs.

Shelves are removable and consist of a skeleton framework of heavy angle-iron, with four tracks in which slide long trays for carrying culture vials or bottles. These shelves rest on four metal clips which can be set at any height. The trays are constructed of pine, with bottoms of ½-in. mesh metal screening (hardware cloth), and two hardwood strips nailed on to slide in the shelf tracks. These long trays hold 33 half-pint or 40 quarter-pint bottles. A tray full of cultures can be removed and set on the work table or quickly moved from one cabinet to another. The open construction allows free circulation of air down through all parts of the cabinet. Owing to the carrying capacity of the cabinets, the trays may be staggered on the shelves in such a way as to allow even better circulation than when shelves are completely filled. Each cabinet can carry between 600 and 700 half-pint cultures, making in all a total capacity of over 2500 half-pint culture bottles. The construction is such as to facilitate rapid and easy handling of large numbers of cultures. Each cabinet is regulated by a double thermostatic control. Two cabinets are supplied with heating units only and can therefore be run only at or above room temperature. A third cabinet is supplied with both heating and refrigerating units, and the fourth with refrigerating unit only. The refrigeration is by ¼-hp Reco-Mills condensing units for use with "Freon" and a set of copper and aluminum finned coils for cooling. The cabinets give a temperature range from 5°C to 50°C. Thermostats for each cabinet are in series with a relay. The writer recommends toluene-mercury thermostats rather than the bimetallic type shown in the figures.

These cabinets are sufficiently large so that several smaller units running at various temperatures could be placed within one of them. While there is some tendency for formation of air pockets at the corners, the size of the cabinet makes it unnecessary to use corners for experimental cultures. The long trays are particularly convenient in cutting to a minimum the labeling necessary for experimental cultures and in starting stocks at one temperature and then transferring them to another temperature with a minimum of manipulation.

The most accurate and satisfactory thermostats for careful temperature work seem to be of the type described by Bridges

(1932). These are made up of thin-walled glass grids filled with toluene, ether, or some other substance with high coefficient of expansion. The toluene grid is connected with a mercury trap and capillary tube containing a column of mercury, which on expansion of the toluene makes contact with a nickel-silver wire. These mercury-contact thermostats are generally run on a low voltage to prevent sparking and sticking at the mercury surface, and the thermostat circuit operates a relay which opens and closes the circuit containing the heating units. In general, thermal cabinets should be heated by the smallest units which will give the desired temperature and which also have little heat lag. For this purpose light bulbs serve admirably. Ideally the heating unit should run almost continuously, and with proper insulation and installation in a room with little temperature fluctuation this is quite possible.

COLLECTION OF EGGS

As an increasing number of workers are undertaking problems involving the collection and hatching of Drosophila eggs, a few notes on factors leading up to and inducing oviposition may prove of interest.

Prefeeding of Females

Starved flies will lay few or no eggs. It is important to furnish flies which are to be used in egg-laying experiments with an adequate supply of fresh food, particularly on the day or two days before the collection period. It is also well to use flies which have been matured for several days to 2 weeks, depending on the species. If flies are aged in vials, fresh food chips should be added or the old ones so cut as to furnish fresh surface, as the old chips soon dry out or form a film over the surface which cuts down food consumption.

Humidity

To elicit the ovipositing reaction the air in contact with the surface where the eggs are to be laid must have a high humidity, probably close to or at the saturation point. The scraping and roughening of the food surface, described by a number of workers, supply tiny humid valleys where the ovipositor meets an environment sufficiently moist to induce the reaction. However, eggs will be deposited in or on a smooth surface if the air in contact with it is saturated with moisture. Conversely no eggs are deposited in dry air. In a dry climate fresh medium may be left exposed to flies, which will feed on it but will not oviposit. If, however, the same dish of medium is covered so that the humidity rises, oviposition occurs.

Temperature

Ovipositing takes place through a wide range of temperatures, varying somewhat for different species of Drosophila. The range, however, is not as great as the upper and lower limits of temperature to which the fly is tolerant. Thus well-fed females may be kept at temperatures below 10°C for long periods without ovipositing. When the temperature is raised, the first eggs are small and abortive, indicating a resorption of material from these retained eggs.

Medium for Oviposition

When all other conditions are satisfied, i.e., females properly aged and well fed, temperature optimum, and humidity high, flies will ovipost readily on a great variety of substances from the most elaborately prepared medium to cellucotton or tissue paper soaked in distilled water. Yeasting of the medium is not necessary for oviposition. However, if flies are to continue to lay large numbers of eggs, they must have both sugar and yeast in their diet.

Various methods have been suggested for collecting Drosophila eggs. Small metal dishes (Beadle, 1936), paper spoons, milk-bottle caps, and paper ramekins (Williams, 1937) containing yeasted culture medium have been used for egg collecting. Addition of charcoal or lampblack to the medium increases visibility. The author (Spencer, 1937b) has described a small cage with underside covered by silk net, through which females lay their eggs on culture medium on which the cage is set. Although it is not difficult to secure a large initial lot of eggs from pre-fed females, it must be remembered that they must continue to have a balanced ration if they are to maintain heavy laying.

CAGES AND CONTAINERS FOR POPULATION STUDIES

In recent years there has been a growing interest in experimental studies on the genetics of populations. It seems advisable, therefore, to include a description of devices used in population

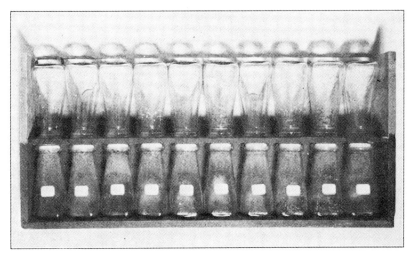

FIG. 11. Dr. Sheldon Reed's apparatus for population studies on Drosophila.

studies on Drosophila. Dobzhansky and colleagues (see Wright and Dobzhansky, 1946) have used a population cage, which is a modification of that used in the earlier studies of L'Héritier and Teissier (1933). Dobzhansky has kindly consented to our including here a description of the improved model in current use in his laboratory. The description given is taken from the paper cited above but includes certain improved features. The cage is a wooden box with outside dimensions 17½ in. × 11 in. × 6 in. The bottom has 16 circular openings 2¼ in. in diameter, evenly spaced in 4 rows, and closed by tightly fitting corks. On each cork rests a 2 × 1 in. Stender jar filled with the standard culture medium. One end of the box, 6 in. × 11 in., has a window, 8 in. × 2¾ in., covered with fine copper screen for ventilation. The opposite end has a metallic funnel opening inward and closed by a cork. The funnel is used for introducing or removing adult flies from the cage. A long glass pipette can be inserted through the funnel, and a yeast suspension in water added to cultures which appear dry. The top consists of a wooden rim $17\frac{1}{2}$ in. $\times$ 11 in. under which a large glass plate fits snugly. The rim is held in place by 6 set screws. When these are tightened, no flies can escape. When this plate becomes fouled with fly feces, the set screws can be loosened, and a clean plate inserted above the old one. When the top is covered with a black cloth and flies are shaken down, the old plate can be removed with little or no loss of flies from the cage.

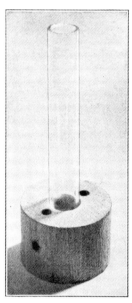

At set intervals the oldest dish is removed, and a new one containing fresh medium inserted. In this way a continuous population may be maintained, and from time to time samples of larvae, pupae, or adults may be removed from the population for study. The device has been used by Dobzhansky for studying changes in population structure over considerable time periods. By changing food dishes every other day a population of 3000–5000 adult *D. pseudoobscura* is maintained indefinitely at room temperature. The cage should be equally suitable for population studies on *D. melanogaster*

FIG. 12. Detail of stopper used in Reed's apparatus.

and other species. Mite infestations must be carefully guarded against.

Sheldon Reed has constructed a simple device for population studies, whereby the effect of fluctuating culture conditions may be investigated. He has kindly made available a description of the device and has furnished the photographs reproduced here. Figure 11 shows a series of these devices in use, and Fig. 12 illustrates the construction. Each culture bottle is supplied with a generous slant of the usual culture medium. The bottles lie horizontally in a wooden frame. The wooden stopper shown in Fig. 12 has a glass connecting tube open at both ends. Two T-shaped tunnels bored in the stopper admit air. In operation these are plugged with cotton. Of the set-up pictured in Fig. 11 Reed writes: "The old cultures in the black half of the box are two months old. The new cultures which are exposed to the light (when cover is on) are about a week old." When a replicated series such as that shown is used, the relative roles of selection and chance may be studied when certain gene frequencies have been introduced in the initial populations. The older cultures may be removed, and fresh culture bottles substituted at stated intervals. The device makes possible the study of a population fluctuating in size, a condition generally encountered in nature.

CONCLUDING STATEMENT

The author has tried to review the earlier methods of collecting and culturing Drosophila and to present current methods with possible improvements at some points. The special techniques used in transplantation work, the use of complex genetic stocks as balancers and in mutation studies, temperature shock, X-ray, radium and chemical treatments, cytological methods, and other special procedures have not been reviewed, as they seemed beyond the scope of this general report. The reader is referred to the literature in these fields for information on techniques. Demcrec and Kaufmann (1945) have prepared an excellent and well-illustrated guide for the culture of *D. melanogaster*, which includes instructions in cytological technique.

LITERATURE CITED

ARK, P. A., and H. E. THOMAS. 1936. Persistence of *Erwinia amylovora* in certain insects. *Phytopathology* **26**:375–381.

BEADLE, G. W. 1936. Collection of eggs. *Drosophila Information Service* 6:18.

BLISS, C. I., and B. M. BROADBENT. 1935. A comparison of criteria of susceptibility in the response of Drosophila to. hydrocyanic acid gas. I. Stupefaction time and mortality. *J. Econ. Entomol.* **28**:989–1001.

BOYCE, A. M. 1928. Studies on the resistance of certain insects to hydrocyanic acid. *J. Econ. Entomol.* **21**:715–720.

BRIDGES, C. B. 1932. Apparatus and methods for Drosophila culture. *Am. Naturalist* **66**:250–273.

BRIDGES, C. B. 1937. Revised data on culture media and mutant loci of *Drosophila melanogaster*. *Tabul. biol.* **14**:343–353.

BRIDGES, C. B., and H. H. DARBY. 1933. Culture media for Drosophila and the pH of media. *Am. Naturalist* **67**:437–472.

BROADBENT, B. M., and C. I. BLISS. 1936. Comparison of criteria of susceptibility in the response of Drosophila to hydrocyanic acid gas. II. Recovery time. *J. Econ. Entomol.* **29**:143–155.

CASTLE, W. E., F. W. CARPENTER, A. H. CLARK, S. O. MAST, and W. M. BARROWS. 1906. The effects of inbreeding, cross-breeding, and selection upon the fertility and variability of Drosophila. *Proc. Am. Acad. Arts Sci.* **41**:729–786.

DECOURSEY, R. M. 1925. A practical control for the pomace fly. *J. Econ. Entomol.* **18**:626–629.

DELCOURT, A. 1909. Sur l'apparition brusque et l'hérédité d'une variation chez *Drosophila confusa. Compt. rend. soc. biol. Paris* **66**:709–711.

DELCOURT, A. 1911. Sur un procédé permettant l'examen à un fort grossissement, à l'etat vivant, de mouches de petite taille, notamment de Drosophiles. *Compt. rend. soc. biol. Paris* **70**:97–98.

DELCOURT, A., and É. GUYÉNOT. 1910. De la possibilité d'étudier certains Diptères en milieu défini. *Compt. rend. acad. sci. Paris* **151**:255–257.

DEMEREC, M. 1936. Control of mites. *Drosophila Information Service* 6:69.

DEMEREC, M., and B. P. KAUFMANN. 1945. Drosophila Guide. 4th ed. *Carnegie Inst. Wash. Pub.*

DOBZHANSKY, TH. 1936. Collecting, transporting, and shipping wild species of Drosophila. *Drosophila Information Service* 6:28–29.

DOBZHANSKY, TH., and C. EPLING. 1944. Contributions to the genetics, taxonomy, and ecology of *Drosophila pseudoobscura* and its relatives. *Carnegie Inst. Wash. Pub.* 554:1–183.

DOVE, W. E. 1937. Myiasis of man. *J. Econ. Entomol.* **30**:29–39.

DUBININ, N. P., D. D. ROMASCHOFF, M. A. HEPTNER, and Z. A. DEMIDOVA. 1937. Aberrant polymorphism in *Drosophila fasciata* Meig. (-*melanogaster* Meig.). *Biol. Zhur.* **6**:311–354. (Russian and English text.)

FROBISHER, M. 1926. Observations on the relationship between a red torula and a mold pathogenic for *Drosophila melanogaster. Biol. Bull. Woods Hole* **51**:153–162.

GLASS, H. B. 1934. Control of molds and mites. *Drosophila Information Service* 2:61.

GOTTSCHEWSKI, G. 1936. Culture bottle and etherizer. *Drosophila Information Service* 6:50.

GUYÉNOT, É. 1913a. Études biologiques sur une mouche, *Drosophilia ampelophila* Löw. I. Possibilité de vie aseptique pour l'individu et la lignée. *Compt. rend. soc. biol. Paris* **74**:97–99.

GUYÉNOT, É. 1913b. Études biologiques sur une mouche, *Drosophilia ampelophila* Löw. II. Role des levures dans l'alimentation. *Compt. rend. soc. biol. Paris* **74**:178–180.

GUYÉNOT, É. 1913c. Études biologiques sur une mouche, *Drosophila ampelophila* Löw. III. Changement de milieu et adaptation. *Compt. rend. soc. biol. Paris* **74**:223–225.

GUYÉNOT, É. 1913d. Études biologiques sur une mouche, *Drosophila ampelophila* Löw. IV. Nutrition des larves et fécondité. *Compt. rend. soc. biol. Paris* **74**:270–272.

GUYÉNOT, É. 1913e. Études biologiques sur une mouche, *Drosophila ampelophila* Löw. V. Nutrition des adultes et fécondité. *Compt. rend. soc. biol. Paris* **74**:332–334.

HASKELL, G. M. L. 1940. The use of chondrus to replace agar in food preparation. *Drosophila Information Service* 13:79.

HUGHES, R. 1946. Notes on mites infesting Drosophila cultures. *Drosophila Information Service* 20:92–93.

KAMIZAWA, T. 1936. Studies on *Drosophila suzukii* Mats. *J. Plant. Prot.* **23**:66–70, 127–132, 183–191. (In Japanese)

KIKKAWA, H., and R. T. PENG. 1938. Drosophila species of Japan and adjacent localities. *Japan. J. Zoöl.* **7**:507–552.

LEBEDEFF, G. A. 1937. Methoden zur Züchtung von Drosophila. *Handb. biol. ArbMeth.* Abt. 9, Tl. 3:1115–1182.

LEWIS, E. B. 1942. Cornmeal-molasses-rolled oats food. *Drosophila Information Service* 16:71–72.

L'HÉRITIER, PH., and G. TEISSIER. 1933. Étude d'une population de Drosophiles en équilibre. *Compt. rend. acad. sci. Paris* **198**:770–772.

MACKNIGHT, R. H., and H. ROMAN. 1940. A culture medium without agar. *Drosophila Information Service* 13:80.

MULLER, H. J. 1934. Etherizing bottle. *Drosophila Information Service* 2:62.

MULLER, H. J. 1936. Fly morgue. *Drosophila Information Service* 6:58.

MULLER, H. J. 1939. *Bibliography on the Genetics of Drosophila.* Oliver and Boyd, Edinburgh.

PARKER, D. R. 1935. Moldex-A as a mold inhibitor. *Drosophila Information Service* 4:65.

PATTERSON, J. T. 1943. Studies in the genetics of Drosophila. III. The Drosophilidae of the southwest. *Univ. Tex. Pub.* 4313:7–216.

PATTERSON, N. A. 1935. Indices of toxicity for various poisons to *Drosophila ampelophila* Loew. *Rept. Entomol. Soc. Ont.* 65:78–80.

PATTON, W. S., and A. M. EVANS. 1929. Insects, ticks, mites and venomous animals of medical and veterinary importance. Part I. Medical. Ent. Dept. Liverpool School of Trop. Med., Liverpool.

PEARL, R., and W. B. D. PENNIMAN. 1926. Culture media for Drosophila. I. Changes in hydrogen-ion concentration of the medium. *Am. Naturalist* **60**:347–357.

SERGENT, E., and H. ROUGEBIEF. 1926. Moucherons (Drosophiles) et fermentations. I. Propagation des levures par les Drosophiles dans les vignobles. II. Disparition des moissisures sous l'action des Drosophiles. *Arch. inst. Pasteur Algérie* **4**:519–527.

SHIMAKURA, K. 1940. The use of non-fermentative yeast in Drosophila culture. *Drosophila Information Service* 13:74.

SHIPMAN, E. E. 1935. Bottle for Drosophila culture. *Drosophila Information Service* 3:54.

SPASSKY, B. 1943. Cream of wheat-molasses fly medium. *Drosophila Information Service* 17:67–68.

SPENCER, W. P. 1936a. Etherizing bottle. *Drosophila Information Service* 6:56.

SPENCER, W. P. 1936b. Mite and mold control. *Drosophila Information Service* 6:67–69.

SPENCER, W. P. 1937a. Life history and control of laboratory mites. *Drosophila Information Service* 7:92–93.

SPENCER, W. P. 1937b. A new technique for the study of Drosophila eggs and larvae. *Science* **85**:298.

SPENCER, W. P. 1940a. Shipping Drosophila. *Drosophila Information Service* 13:81–82.

SPENCER, W. P. 1940b. On the biology of *Drosophila immigrans* with special reference to the genetic structure of populations. *Ohio J. Sci.* **40**:345–361.

STALKER, H. D. 1940. Substitutes for shell vials. *Drosophilia Information Service* 13:82.

STURTEVANT, A. H. 1921. The North American species of Drosophila. *Carnegie Inst. Wash. Pub.* 301.

TATUM, E. L. 1939. Nutritional requirements of *Drosophila melanogaster*. *Proc. Natl. Acad. Sci. (Wash.)* **25**:490–497.

UVAROV, B. P. 1931. Insects and climate. *Trans. Entomol. Soc. London* **79**:1–247.

WALLACE, BRUCE. 1946. A new proposal for the control of mites. *Drosophila Information Service* 20:96–97.

WILLIAMS, C. R. 1937. Collecting eggs and larvae. *Drosophila Information Service* 7:99.

WRIGHT, SEWALL, and TH. DOBZHANSKY. 1946. Genetics of natural populations. XII. Experimental reproduction of some of the changes caused by natural selection in certain populations of *Drosophila pseudoobscura*. *Genetics* **31**:125–156.

Index of Names

591

Subject Index

PREPARED BY ALBERT MILLER

Page numbers in roman type refer to text; those in italic, to figures. L = in Literature Cited; n = footnote; t = table (timetable or data). References in general apply to *Drosophila melanogaster*. emb = embryo; lv = larva; pu = pupa; im = imago or adult.